D0952556

Science in America

Science in America

Historical Selections

Edited by
John C. Burnham
The Ohio State University

HOLT, RINEHART AND WINSTON, INC.
New York Chicago San Francisco Atlanta Dallas
Montreal Toronto London Sydney

Cover: *Boston Journal of Natural Science,* vol. 5, (1847)

Library of Congress Catalog Card Number: 71–151080
SBN: 03–085288–9
Printed in the United States of America
1 2 3 4 090 9 8 7 6 5 4 3 2 1

Preface

This book is intended to recapture the past of American science by bringing together a number of writings from and about scientific endeavor in American society from the seventeenth century to the early 1970s. In tracing the ideas of Americans of other times concerning their attempts to understand matter and life, such a volume unwittingly raises questions about science in the United States in the last third of the twentieth century as well as offering perspectives upon it. The intention is, nevertheless, not to illuminate our present dilemmas but to use words from former times to suggest in a concrete way how the study and understanding of nature has changed in three and a half centuries.

The constant in this volume is science: empirical and theoretical efforts to discover the contents and behavior of the world and the universe, arriving at answers that were couched in naturalistic terms and that could be verified by other, open-minded members of the scientific community. Ordinarily this endeavor, science, included both the physical and the biological sciences. Although it is possible in these selections to trace the major currents and developments in the various sciences, the substantive content is not technical, and an intelligent reader ought to be able to follow and understand all of the documents.

Often in the past, Americans were concerned not so much with new knowledge and ways of looking at nature as with how such knowledge could be utilized for the benefit of man. Applied science (even in the form of technology) and pure science are very difficult to separate in practice, and particularly within the American experience. This is reflected in the selections. The present history, however, is focused on science as newly discovered or newly systematized knowledge. Applied science is brought in only as it nourished or inhibited the enterprise and institutions of science. Thus, the histories of American industrial research, of American technology, and of American medicine are excluded insofar as possible. Each has a literature of its own. Each contributes as background, but the focus here is on the mainstream history of American science.

The topic is approached from the social and historical points of view. For the social point of view, the selections contain material showing how science affected American civilization and institutions, on the one hand, and, on the other hand, how the American environment affected the development of science as a body of knowledge and as a social institution. In order to suggest historical relations and developmental trends, the selections are arranged in a strictly chronological order.

In this volume the reader may search out as many or as few themes as he wishes; he may read the selections in any order that he desires; he may correlate his explorations as much or as little as he pleases with the standard literature in the field. The writings included are no more than a sample. Appropriate selections are not available to describe or discuss a number of important developments. In large parts of the book, for example, the discipline of geology is omitted for this very reason. Yet there are a number of problems and topics that can to a greater or lesser extent be traced in these documents:

(1) specific conditions in American society that have encouraged or discouraged scientific endeavor and interest; (2) the development of facilities and institutions for scientific work; (3) the changing nature of the scientist and his work; (4) the justification for science in a democratic society; (5) the place and role of the scientist in society; (6) pure versus applied science as argued in the American context; (7) mechanism and materalism and their critics; (8) the scientific attitude and its recurrent opponents. Hopefully each reader can find in any of the selections a great deal beyond one or more of these themes.

Both the introductions and the selections are designed to reflect the existing literature in the field of the history of American science. An attempt has been made not to duplicate, in so far as possible, such collections as Nathan Reingold, ed., *Science in Nineteenth-Century America. A Documentary History* (New York: Hill and Wang, 1964), George Daniels, ed., *Darwinism Comes to America* (Waltham, Mass.: Blaisdell Publishing Company, 1968), A. Hunter Dupree, ed., *Science and the Emergence of Modern America, 1865–1916* (Chicago: Rand McNally & Company, 1963), and James L. Penick, Jr., et al., eds., *The Politics of American Science, 1939 to the Present* (Chicago: Rand McNally & Company, 1965). By and large the tradition in historical writing about American science has until recently been biographical, as exemplified in Bernard Jaffe, *Men of Science in America. The Story of American Science Told through the Lives and Achievements of Twenty Outstanding Men from Earliest Colonial Times to the Present Day* (2d ed., New York: Simon and Schuster, 1958), and numerous other accounts of the lives of individual scientists. In a society in which science was until recent times almost exclusively the activity of independent investigators, the biographical mode has provided an appropriate approach to the subject. In the past few years historians viewing American science in other ways have produced a number of excellent topical monographs. Many of these are listed in the bibliographical notes of the selections. At present the best lists of relevant literature are to be found in the footnotes in a collection of essays, David D. Van Tassel and Michael G. Hall, eds., *Science and Society in the United States* (Homewood, Ill.: The Dorsey Press, 1966). That volume also contains a useful chronology.

The specialized field of study for which *Science in America* is designed, the history of American science, is explained and justified in the masterly essay of A. Hunter Dupree, "The History of American Science—A Field Finds Itself," *American Historical Review,* 71 (1966), 863–874. Scheduled for publication at about the same time as the present volume is the first comprehensive account of American science, by George H. Daniels, covering the history to about 1920.

A number of scholars have over the years very kindly suggested materials for inclusion or commented on drafts and projects. The editor is particularly deeply indebted to: June Z. Fullmer, of The Ohio State University; George H. Daniels, of Northwestern University; Samuel T. Suratt, formerly of The Smithsonian Institution and now with the Columbia Broadcasting System; Roger E. Bilstein, of Wisconsin State University, Whitewater; Vern Bullough, San Fernando Valley State College; and Michael Sokal, Worcester Polytechnic Institute.

Footnotes have been uniformly eliminated from all selections. In a few exceptional cases an annotation has been preserved from the original, but such footnotes are clearly labeled. All other annotations in this volume consist of explanations or definitions added by the editor.

In order to preserve the flavor of the selections, variations and errors in spelling and punctuation have for the most part been left intact except where it was felt that corrections (which have been indicated) might help in understanding.

J.C.B.

Contents

Science in America

Part 1

THE BEGINNINGS OF AMERICAN SCIENCE

One of the major factors leading to the rise of modern science was the discovery of the New World. After discovery and colonization, America continued to stimulate the growth of knowledge. Exotic American plants—beyond maize, tobacco, and potatoes—and animals, such as the fearsome and fascinating rattle snake, came to be well known among intellectuals in the Western world. A cultured European often owned a cabinet of curiosities, at best a miniature natural history museum, containing specimens collected from all over the globe. North American flora, fauna, and rocks were treasured and appreciated by such men, who often were more than mere collectors. They could be and were systematists and theoreticians—that is, scientists.

Along with curiosity about nature came the ability to ask questions concerning its operations and processes. The questions were not new, but the expectation of seventeenth and eighteenth century men that the answers to the questions should be entirely in naturalistic terms represented a significant change from the preceding age. Out of this naturalism rose modern science, and one of the characteristics of Enlightenment thinking was the attribution of natural phenomena to purely terrestrial and natural causes.

Since America's earliest settlers tended to be deeply, even fanatically, religious people, particularly in Puritan New England, one might well expect that naturalism and even scientific activity would have been suspect or proscribed among the pious first settlers and their descendants. Such was not at all the case. The Puritans believed that God speaks to men in two books—the Bible, His literal word, and the Book of Nature. The Puritans did see signs and portents in natural phenomena (just as medieval men found transcendent symbolism all about them), but they also welcomed a systematic study of God's phenomenal world. They were as capable of reading a lesson in natural laws that reflected God's general providence as they were of learning other lessons from some awful act of His special providence.

From the point of view of European science, the contribution of America to world knowledge in the Age of Reason consisted of various kinds of observations. The new areas provided an opportunity to examine novel plants and animals, giving context and greater meaning to those known already. In the New World, naturalists described new minerals and rock formations, verifying the universality of what could be seen of the impress of God's hand on the landscape of the Old World. From another spot on the globe came new perspectives and sightings of the heavenly bodies. Observations of the Great Comet of 1680 made with the first telescope in North America, for instance—once the property of John Winthrop, Jr., first governor of Connecticut, and later part of the equipment of Harvard College—were cited by Sir Isaac Newton in his great work, the *Principia.*

Yet the inhabitants of British North America were not merely passive observers and collectors for European science. Americans thought about the world around them, and they emulated their European correspondents in attempting to devise naturalistic explanations for what was going on. Even John Bartram, the botanical collector of Philadelphia, for example, tried to classify the plants that he brought back from his field trips. Some rare men, like two more of the Philadelphia community, James Logan and Benjamin Franklin, devised important and ingenious experiments to test their ideas. Others, more numerous than the classifiers and experimenters, were avid readers of European scientific intelligence and attempted their own syntheses of the ideas and conclusions published on the Continent and in England. The eighteenth century, especially, was a time when great theoretical and philosophical systems were put forth in every field of human knowledge and ignorance—science, medicine, cosmogony, economics, politics—and it is not surprising that Americans of a speculative turn of mind, like Hugh Williamson the physician and Cadwallader Colden the New York politician, should contribute theories, major and minor, about the physical world.

A scientist was, after all, called in those days a philosopher. He might conduct philosophical experiments (usually in physics and chemistry) and he was sure to savor, whenever possible, philosophical discussions. As in Europe, ingenious speculation and generalization was valued, often indiscriminately, alongside careful experimentation. But as time went on, Enlightenment men came to think that natural philosophy (the physical sciences) and natural history (biology and geology) would soon provide a total naturalistic explanation for all of the phenomena of the world and of the universe. Looking back upon the progress of science up to that age, men of the time could project into the future the probable progress not only of reason but of empirical investigation and testing. They understood that empiricism had validity only when the "philosophers" could gain its acceptance from their fellow scientists, those who were qualified to confirm experiments and observations. This collective scientific community to whom one could address one's findings gave functional actuality to the existence of the scientific endeavor.

By the end of the eighteenth century, a number of men of the Age of Reason—some scientists, but many others, too—in their private religious beliefs tended toward Deism (whatever public ceremonial stance they took). As Deists, these men believed that natural laws, on the model of those posited by Newton, govern the universe and that God is only the great Watchmaker who wound up the machinery in the beginning and regulated its movements through the mechanical action of natural law. Not surprisingly, many orthodox religious thinkers, such as David Daggett of New Haven, confused science with unreligious or antireligious views as well as other heterodox theories.

At any stage in the development of American science it would have been misleading to emphasize theoretical or pure science as opposed to the practical application of scientific knowledge. Indeed, from the seventeeth century on, technology and economic progress were almost universally confused with the pursuit of knowledge for its own sake, in the minds of citizens at large as frequently as in the minds of the practitioners of science. The confusion was also a reflection of the activities of those men. John Winthrop, Jr., was endlessly involved in small-scale mining and industrial ventures; Benjamin Franklin's prowess as an inventor is to this day better known than his "philosophical" inquiries and experiments. Geologists and mineralogists could never be far removed from mining. Chemists had ties to military technology, health, agriculture, and industry, even in those times. The work of botanists and zoologists affected medicine (which the entire adult population practiced to some extent) as well as farming and aspects of industry. There seemed to be no line drawn between physics and the mechanic arts. David Rittenhouse, America's foremost astronomer in the eighteenth century, also manufactured astronomical instruments, as well as clocks and similar devices for more general use. Mathematics and astronomy were part of the stock-in-trade of American navigators, and one of America's most substantial contributions to science was contained in the astronomical material in Nathaniel Bowditch's *The New American Practical Navigator* (1802).

In practice it is impossible to separate the natural curiosity of Americans who labored in the field of science from their practical turn of mind and inventive ingenuity. Even before 1800 any number of technological advances —mostly introduced from Europe—were increasing productivity and making life easier. Thomas Jefferson, for instance, combined in himself the philosophic and practical interest of a late eighteenth century gentleman of science. He also had a talent for gadget-making and a passion for agricultural and public improvement. It would be difficult to separate any one of these aspects of his life from any other. Scientific gentlemen often would not dirty their hands with mechanical work but could take a concrete interest in the improvement of activities of public concern such as animal breeding, crop rotation, iron processing, or the analysis of the chemical constituents of healthful mineral springs.

The first scientific writers tended to be of two kinds. Initially there were European travelers who came on business, or occasionally out of curiosity, and took time to learn about and describe American natural phenomena. The best known was the Swedish academician, Peter Kalm, who visited in 1748. Upon viewing Niagara Falls and the flora, fauna, and geology of the area, he wrote to a friend in Philadelphia: "The Hair will rise and stand upright on your Head when you see this! I cannot with words express how amazing this is!" Kalm's words convey the interest that Europeans took in North America in the first century of its accessibility. A second group of scientific reporters came from among the settlers themselves. Such were, for instance, the late seventeenth and early eighteenth century figures, John Winthrop, Jr., the governor, and Cotton Mather, the Boston clergyman. But both travelers and settlers were part of the European scientific community. Winthrop and Mather, for example, were Fellows of the Royal Society of London, as were a number of other Americans of the eighteenth century (still, until 1776 or so, English citizens). Within the general field of natural history, Europeans had by the middle of the eighteenth century established an international natural history network for the collection and trading of scientific specimens and information. Primarily through the influence of Peter Collinson, a London merchant and devoted amateur scientist, a number of American residents were thoroughly integrated into this international natural history circle, which included, for example, not only James Logan but Linnaeus himself. Because of these ties and the lines of communication of that day, Americans had far more contact with European scientific workers than with their fellow "philosophers" in other colonies.

The creation of an American scientific community as such, curiously enough, was in large part the outgrowth of politics. In the second half of the eighteenth century many Americans were aware that they were not just Englishmen living in an outpost of empire: They developed a sense not only of local pride but of their identification as Americans. Local pride was partly self-justification and partly a recognition that bustling urban centers such as Boston, New York, and Philadelphia were spawning their own intellectual circles. The beginnings of a larger American nationalism is harder to account for but was equally present. In writings about science, nationalism motivated a number of statements to the effect that science did, or would in the future, flourish better in American society than elsewhere. In terms of nationalistic politics, scientific endeavor was justified as an aid to making the American colonies economically and technically independent of Europe. This patriotic point of view reached full development during the period of the American Revolution.

From the middle of the eighteenth century on, a number of men suggested the founding of societies "for promoting useful knowledge among the British plantations in America" (the wording is from an early proposal by Franklin). These schemes culminated in the establishment of a number of local societies and finally, in 1768, the formation of the American Philosophical Society, for Promoting Useful Knowledge. Held at Philadelphia, the major impetus to its formation came from patriotic politicians who later figured conspicuously among the Founding Fathers of the Republic. In 1780, a group of patriotic New Englanders founded a similar but somewhat more general society, the American Academy of Arts and Sciences.

The establishment of the American Philosophical Society did not, of course, create an American scientific community; it did, however, reflect the fact that there was some basis for one. The *Virginia Gazette,* for example, printed a comment from Glasgow: "This is the first literary Establishment beyond the Atlantick Ocean, and gives a striking Proof of the Greatness and Prosperity of our Colonies, for Men seldom or never form themselves into Societies of that Kind where Ease and Affluence are not eminently enjoyed." The existence of the Society also did not in any way signify that American science was independent of European science. American science continued to be thoroughly colonial (regardless of political changes), primarily observational (rather than synthetic), and closely connected with "the useful arts." That there were a number of eminent American workers in science cannot change those facts.

Perhaps it was ominous that the first national scientific society should have been so closely tied to politics. In fact it, and most American scientific endeavors, languished after the Revolutionary period and in the closing years of the eighteenth century. Scientific luminaries such as Franklin passed from the scene, leaving few successors. The conservative and romantic period that began at about the time of the French Revolution, moreover, was not a favorable one for science. Because of its connection with rationalism, science in many eyes was associated with the Revolution in France, for the revolutionaries frequently invoked reason in justifying social upheaval. Progress in science (most conspicuously connected with French chemistry at the time) might be acceptable, but many Americans opposed the application of reason and progress to the realm of politics and society. French science, which had its partisans such as Felix Pascalis, along with Enlightenment thought in general met formidable opposition from those who already deplored the unorthodox religious tendencies of many scientists of that time.

The decline of science around 1800 raises questions. Why was it not substantial enough in American society to withstand better than it did the waning of revolutionary fervor and the lack of public enthusiasm associated with the era of reaction and romanticism? And more basically: Who were the scientists and where did the money come from to support their work?

Hardly any professional scientists existed before the nineteenth century. College teachers, for example, later such conspicuous members of the scientific community, were not important in that role in those days, although science was a regular part of the curriculum in every colonial college. Despite the fact that as early as 1766 there existed six professorships of natural philosophy, the preponderance of American scientists of the seventeenth and eighteenth centuries tended to be of two other types: gentlemen who had leisure time in which to study nature or perform "philosophical experiments"—or men whose vocations were closely associated with some aspect of science. The small number of men who had the leisure and money to pursue science did so as a hobby and a means of self-improvement. And, it is worth noting, one's philosophical activities and private cabinet of curiosities could count as a mark of gentility. It was usually men of wealth and leisure who supplied the funds to support botanical collection and astronomical observation. They helped support the handful of Americans such as Bartram and Rittenhouse who earned substantial parts of their incomes from their scientific activities. The affluent patrons of science enjoyed electrical, celestial, and chemical shows, and for them science was an enjoyable as well as an elevating and satisfying activity. Obviously when their interest dropped off, so did their support.

Of the Americans whose vocations seemed to generate an interest in at least some aspects of science, such as the sailors, the surveyors, and the farmers, by far the most important were the physicians. Not only were botanical and chemical preparations essential in medical practice, but for other not so obvious reasons, such as educational and personal predilection, physicians constituted the fundamental personnel involved in the natural history circle and all other scientific endeavor in America. It is significant that the only real scientific journal to survive the end of the eighteenth century without serious lapse was *The Medical Repository*. While primarily a medical journal, it included, as appropriate, important scientific publications, among them the last great controversy over the doctrine of phlogiston (defended by no less a man than the discoverer of oxygen, Joseph Priestley, who was living in religious-political exile near Philadelphia).

With few workers (only 21 full-time positions in 1802 in the whole country), little money, and practically no publications, science languished in America in the first years of the nineteenth century. As late as the second decade of the new century Philadelphia chemist Robert Hare was writing: "It really seems bad policy to publish any thing in this country upon science especially in the first instance. . . . It is rarely attended to in England & we are so low in capacity at home that few appreciate any thing which is done here unless it is sanctioned abroad." Americans, aware of their scientific activity in the age of Newton and their lack of it in the age of Napoleon, wondered what had happened to their science. They never answered the question to anyone's satisfaction. But before long the seeds sown in the colonial and Revolutionary years took root and flourished in the very different climate of nineteenth century American society.

1. A Puritan and Pioneer Reports to English Scientists. 1670

This letter is typical of reports of the pioneers in the New World to their scientific correspondents in the Old World. America was exciting to scientists because of new phenomena and exotic curiosities to be found there. The attitudes of the author of this letter, JOHN WINTHROP, JR. (1605/6–1676), exemplify the type of intellectual curiosity that led men to collect cabinets of natural curiosities and to compile observations and accounts of natural occurrences almost indiscriminately. It is notable that, in this letter describing wonders found in America, Winthrop, son of the Puritan leader of the Massachusetts Bay Colony, did not deviate from the purely natural level of explanation of what he saw.

Winthrop was born in Suffolk, England, and was graduated from Trinity College, Dublin. After trying his hand at the law and the Royal Navy, he spent the late 1620s wandering over Europe. During this period he met a number of leading scientists. In 1631 he came to America to join his father. In 1635 he served briefly as first governor of Connecticut, a post that he held again (except for one year) from 1657 until his death. While on colony business in England from 1661 to 1663, he associated with some of the founders of the great scientific society, the Royal Society of London, and in 1663 he was elected a Fellow. As initiator of a number of small industrial enterprises, including an iron mill, he has been called the father of American industrial chemistry. His medical knowledge was much in demand in the rough countryside, and he was a member of the international inner circle of alchemists of his day. He owned the first telescope in the colonies. In short, he was the leading scientific figure in seventeenth century America.

For further reading about Winthrop's activities see Winthrop Tilley, "The Literature of Natural and Physical Science in the American Colonies from the Beginnings to 1765" (Ph.D. dissertation, Brown University, 1933), 21, and Robert C. Black, III, *The Younger John Winthrop* (New York: Columbia University Press, 1966).

AN EXTRACT OF A LETTER, Written by John Winthrop Esq; Governour of Connecticut in New England, to the Publisher, concerning some Natural Curiosities of those parts, especially a very strange and very curiously contrived Fish, sent for the Repository of the R. Society.*

*The Royal Society of London, *Philosophical Transactions,* 5 (1670), 1151–1153. Italics omitted.

I Know not, whether I may recommend some of the productions of this Wilderness as rarities or novelties, but they are such as the place affords. There are, amongst the rest, 2. or 3. small Oaks, which though so slender and low (as you may see, if they come safe) have yet Acorns and cups upon them, so that it may be truly said, that there is a Country, where Hoggs are so tall, that they eat acorns upon the standing growing Oakes. This is every year visible in many parts here, there being of this sort of dwarf-Oak whole Forrests in the Inland Country; too many for the Husband-man, who finds that sort of land most difficult to break up at first with his plough, in regard that the whole surface is fill'd with spreading strong roots of this sort of Oak. Neither must it be thought, that they are small shoots, which in time would grow big trees; for, where these grow, there are no great Oaks, or very few amongst them. But whether it be a novelty to see such kind of dwarf-trees bearing acorns, I know not: It was to me, having not seen the like (as far as I remember) in England or France, or other parts. Mean time I have observed, that in some Plains, full of these shrubs, there have been no acorns on most of them; but whether in other years they were not fruitful, I know not. Some years, we know, even the great Oaks bear no fruit, which are very full at other times; but this year throughout the whole Country there is plenty of acorns; and I should be glad to be informed, whether this year they have been also abounding in England, or other parts of Europe; and if so, or not so, possibly something not altogether inconsiderable may be thence inferr'd. Besides, if such dwarfish Oakes, as these, should be found in other parts of the World, it were not amiss, me thinks, to inquire, Whether it be not some Mineral ground, where these grow; and if so, what sorts of Minerals those places afford.

There are also sent you some pieces of the Bark of a Tree, which grows in Nova-Scotia, and (as I hear) in the more Easterly parts of N. England. Upon this bark there are little knobs, within which there is a liquid matter like Turpentine (which will run out, the knob being cut open) of a very sanative[1] nature, as I am informed by those, who affirm, that they have often tryed it.

In the same Box are Pods of a Vegetable, we call Silk-grass, which are full of a kind of most fine down-like Cotton-wool, many such flocks in one and the same pod ending in a flat Seed. 'Tis used to stuff up Pillows and Cushions; being tryed to spin, it proves not strong enough. The Seeds 'tis like may grow with you, if set in some Garden; whereby the whole Plant may be seen.

You'l find also a Branch of the Tree, call'd the Cotton-tree, bearing a kind of Down which also is not fit to spin. The Trees grow high and big. At the bottom of some of the Leaves, next to the stalk of them, is a knob, which is hollow, and a certain fly, some-what like a pis-mire-fly, is bred therein.

More-over, there are some of the Matrices, in which those Shels are bred, of which the Indians make the while Wampanpeage,[2] one sort of their mony:

[1] Healthful.

[2] That is, wampumpeag or wampum.

They grow on the bottom of Seabays, and the shels are like Periwinkles, but greater. Whilst they are very smal, and first growing, many of them are within one of the concave receptacles of these Matrices, which are very tough, and strong, so contrived, that they are separate from one another, yet so, that each of them is fastened to a kind of skin, subtended all along to all these cases or baggs.

There is, besides, in a large round Box, a strange kind of Fish, which was taken by a Fisherman, when he was fishing for Codfish in that Sea, which is without Massachuset Bay in N. England. It was living, when it was taken, which was done, I think, by an hook. The name of it I know not, nor can I write more particularly of it, because I could not yet speak with the Fisherman, who brought it from Sea. I have not seen the like. The Mouth is in the middle; and they say, that all the Arms, you see round about, were in motion, when it was first taken.

We omit the other particulars here, that we may reflect a little upon this elaborate piece of Nature, the Fish, which, since it is yet nameless, we may call Piscis Echino-stellaris Visciformis; its Body (as was noted by M. Hook) resembling an Echinus or Egg-fish, the main Branches, a Star, and the dividing of the branches, the Plant Missel-toe. This Fish spreads itself from a Pentagonal Root, which incompasseth the Mouth (being in the middle at *a*)[3] into 5 main Limbs or branches, each of which, Just at the issuing out from the Body sub-divides itself into two (as at 1.) and each of those 10 branches do again (at 2.) divide into two parts, making 20. lesser branches: Each of which again, (at 3.) divide into 2. smaller branches, making in all 40. These again (at 4.) into 80; and those (at 5.) into 160; and they (at 6.) into 320; they (at 7.) into 640; at 8, into 1280; at 9, into 2560; at 10, into 5120; at 11, into 10240; at 12, into 20480; at 13, into 40960; at 14, into 81920; beyond which, the farther expanding of the Fish could not be certainly trac'd, though possibly each of those 81920 smalsprouts or threds, in which the branches of this Fish seem'd to terminate, might, if it could have been examined when living, have been found to subdivide yet farther. The Branches between the Joynts were not equally of a length, though for the most part pretty near: But those branches, which were on that side of the Joynt, on which the preceding Joynt was placed, were always about a 4th or 5th part longer than those on the other side. Every [one] of these branchings seemed to have, from the very mouth to the smallest twiggs or threads, in which it ended, a double chain or rank of pores, as appears by the Figure. The Body of the Fish was on the other side; and seemed to have been protuberant, much like an Echinus (Egg-Fish or Button-Fish) and, like that, divided into 5. ribbs or ridges, and each of these seemed to be kept out by two small bony ribbs.

In the Figure is represented fully and at length but one of the main branches, whence 'tis easy to imagine the rest, cut off at the 4th sub-dividing

[3] A diagram accompanied this letter.

Branch, which was done to avoyd Confusion as well as too much labour and expence of time in the Engraving.

The Figure, well viewed and considered will be more instructive, than a larger Discourse thereon; though other particularities might have been mention'd if the dryness, perplexedness and brittleness of the Fish had not hindred it.

2. Gentlemen Carry Out Exotic Experimental Zoology in a New World Community. 1720-1723

This has been called a report of "the first experimental scientific study carried out in this country." It is interesting because while experimental in nature, including a control experiment, the form of presentation is flavored by natural history observation reports rather than laboratory description. CAPTAIN HALL's attempt to give a full and objective account might be compared with reports of experimenters of a much later time who also used small animals for scientific experiments.

The identity of Captain Hall is not ascertainable from any standard source. He was obviously a correspondent of Sir Hans Sloane (1660–1753), a distinguished physician and one of the leaders of Enlightenment science. Sloane succeeded Isaac Newton as president of the Royal Society in 1727. Many naturalists, amateur and otherwise, addressed their observations to Sloane in the hopes that he would accord recognition to their efforts, as he did in Captain Hall's case.

For further information, see Jacob Rosenbloom, in *Annals of Medical History,* 4 (1922), 396; *Dictionary of National Biography,* XVIII, 379–380.

AN ACCOUNT of some Experiments on the Effects of the Poison of the Rattle-Snake.

By Captain Hall. Communicated by Sir Hans Sloane*

In *South Carolina* on the 10*th* of *May, Anno* 1720, having got a fine healthful Rattle-Snake about four feet long, I perswaded three or four Gentle-

*The Royal Society of London, *Philosophical Transactions,* 35 (1727), 309–315.

men, and one Mr. *Kidwell* a Surgeon, to assist me in making some Experiments on the Effects of its Poison.

We got three Curr-Dogs, the biggest not larger than a common Harrier, and the least about the bigness of the largest siz'd Lap–Dog, all of them smooth haired.

The Snake being ty'd and pinn'd down to a Grassplat, we took the largest of them, which was a white one, and having tied a Chord round his Neck, so that it should not strangle him, another Person held one end while I held the other; the length was not more than four Yards each way from the Dog.

Immediately on our bringing the Dog over the Snake, the Snake raised himself near two feet, and bit the Dog as he was jumping; the Dog yelpt, by which I perceived he was bitten; and upon it I pull'd him to me, as fast as I could, and perceived his Eyes fixt, his Tongue between his Teeth, which were closed, his Lips so drawn up as to leave his Teeth and Gums bare: In short, he was quite Dead in a quarter of a Minute; but one Person (beside my self) was of Opinion it was in half that time: The first was the Opinion of the By-standers, who were five or six; but I believe, none of them so much used to measure time as the Gentleman and I were, from our constant making use of the half Minute, and quarter Minute Glass at Sea. We could not see where the Dog was bitten, nor any Blood: Upon which we ordered some hot Water to scald the Hair off; when we could find but one Puncture, which look'd of a bluish Green a little round it; it was just between his fore Leg, and his Breast; where (when the Legs are distended) the Hair is much thinner than in some other places.

Half an Hour after the first Bite we took a second Dog, which was somewhat less, of a Liver-Colour, and in like manner brought him over the Snake, which in a very little time bit his Ear, so that we all saw it; he yelpt very much, and soon shew'd the signs of being very sick, holding that Ear that was bit uppermost. He reel'd and stagger'd about for some time; then he fell down, and struggl'd as if Convulsed, and for two or three times got up, each time wagging his Tail, tho' slowly, and attempting to follow a Negro-Boy, who used to make much of him. We put him into a Closet, and order'd the Boy to look after him.

About an Hour after the second was bitten, we took the third Dog in like manner: The Snake bit him on the right side of the Belly, about two Inches behind the long Ribs; for we saw he had drawn Blood there. The Dog at first, I mean for about a Minute, seem'd not to be hurt; so we let him go, being one we could get again when we pleased. For that Day we put up the Snake, imagining his Poison was very near, if not quite expended.

In a little time after which was just two Hours after the second Dog was bit, the Boy told us he was dead.

About an Hour after, I perswaded Mr. *Kidwell* to open him, and I was in no small haste to examine the Heart, where I perswaded my self, I should discover something extraordinary; but could not perceive any remarkable

Difference between that and many others I had seen, where there was no Poison in the Case. Mr. *Kidwell* laid open the Skull, and was of Opinion, that the Brain was more red and swoln than any he had ever seen; and he told me a little while after, that the Blood turn'd very black.

For that Day we heard no more of the third Dog which was bitten; but the next morning the Woman who own'd him came to me, complaining of my Cruelty for killing her Dog. She did not know when he died, but said she saw him at seven that Evening, which was about 3 hours after he was bit; and that he was so sick he could scarce wagg his Tail. None of these Dogs were swoln before they died.

On the *Saturday* following, which was the 14*th,* we got two Dogs both as big as common Bull-Dogs. The first Dog, which he bit on the Inside of his left Thigh, died in half a minute exactly, in the Opinion of two Gentlemen, who kept their Watches in their hands all the while: There were two very small Punctures in his Thigh, which lookt livid, tho' no Blood was drawn. This Dog did not swell for four hours after he was dead. I saw him, and order'd him to be bury'd.

The second Dog was bit about an Hour after the first, on the out side of his Thigh, where we perceived the Blood at two places: He soon sicken'd, and died in four Minutes.

We thought his Poison was not spent; so we got a Cat (for we could get no more Dogs) which he bit about an hour after, though I can't say where. The Cat was very sick, and we put her up in a Closet: By some means the Cat was let out in less than an Hour and a half after she was bitten. The next Morning early she was found dead in the Garden, and much swoln; so that nobody cared to examine or search where she was bit.

About a quarter of an Hour after he had bitten the Cat, he bit a Hen twice: The Hen seem'd very sick and drooping, and could not, or did not fly up to her usual place of Roost among the rest that Night; but the next Day she seem'd very well, and continued so till Evening, when I order'd her to be Killed, and her Feathers scalded off: There were 2 Punctures in her Thigh, and a Scratch on her Breast over the Craw, all which lookt livid.

About a Week after, having got a large Bull-Frog, we brought that over him as usual: He bit it with much force; so that he seem'd to fasten for a small space. The Frog died in two minutes or thereabouts. In less than a quarter of an Hour he bit a Chicken, which was hatcht the *Feburary* [*sic*] before, that died in 3 minutes; I can't say where it was bit, and I was at a loss to try any further Experiments for a long time, for want of proper Subjects. Dogs and Cats were not to be had; for the good Women, whose Dogs had been killed, exclaimed so much, that I durst not meddle with one afterwards.

About the Middle of *June* I took him out according to Custom, and having got a common black Snake, not of the of Viper-kind, about two and a half or near three Feet long, in good Health, just taken; I put them both together, and irritated them both, that they bit each other, and I perceived the black Snake had drawn Blood of the Rattle-Snake before I took them asunder.

In less than 8 Minutes the black Snake was dead, and I could not perceive the Rattle-Snake at all the worse or sick.

On the last Day of *June*, being *Friday*, I took him out to try, whether if he bit himself, it would not prove mortal to him. I hang'd him so, that he was not above half his Length on the Ground; and with two Needles at the End of a Stick, one to prick, the other to scratch, irritated him so much that he soon bit himself, after having attempted to bite the Stick many times. I then let him down, and he was quite dead in 8 minutes or thereabouts, but am sure it did not exceed 12 Minutes.

A Gentleman perswaded me to cut the Snake in five Pieces which he gave to a Hog, the Head-part first, in Sight of many of us. The Hog eat up all the Snake, and 10 or 12 Days afterwards I saw the same Hog alive and in Health.

This was no more than I had seen before; but doubted there might be a Mistake, as that they had taken some other Snake for a Rattle-Snake: For being at the House of *Charles Hart,* Esq; they shew'd me a Snake, which a Negro told me he had kill'd just before; it was in three pieces, the Head of it bruised into the Ground. While I was looking on, a Sow came and eat it up very greedily, tho' the Negro-man endeavour'd to hinder her, being afraid it would kill her; for she had Pigs following her.

I never heard she was sick for it, tho' I inquired; and about ten Days after I saw her in very good Health. I have heard fifty Relations of the same kind, and am told that those Hogs which feed in the Marshes will run after the common sort of Water-Snakes, which are not poisonous, and will feed on them greedily: And I assure you, in *Maryland* last *August* was two years I saw a Hog eat up the Head of a Rattle-Snake just cut off, and while it was gasping very dreadfully; and I was told, it was a common thing, and it would do them no harm.

On the 10*th* of *June* 1723, Mr. *Thomas Cooper,* a Gentleman who practises Physick at *Charles Town,* and who was late of *Wadham-College Oxon,* a very Ingenious Man, sent to me to let me know, he had got a fine Rattle-Snake which had been taken not above 4 Days, was about three feet and a half long, and that he design'd to try whether he could save some of the Dogs after the Snake should bite them. He provided a large quantity of *Venice-Treacle* or *Mithridate,* I can't positively say whether, which he divided into two Potions, each about two Ounces; to one of them he put a large quantity of *Diaphoretic Antimony.*

The first Dog which the Snake bit on the Inside of the Thigh, died so soon (*viz.*) in about half a minute, that we could not get the Potion, which was that *without* Antimony, down his Throat soon enough to expect it could have Effect.

Above an Hour after, the second Dog was bitten by him, and had two

Punctures or Holes in the fleshy part of the Inside of his fore left Leg, which did bleed more than any I had seen before: We immediately got down his Throat that Preparation *with Antimony.* He soon grew very sick and strove to vomit; but I think brought up very little, if any; he froth'd at the Mouth and bit at the Grass, which he champed, as if he were mad; and indeed we were all afraid of him. We therefore put him into a Room and there kept him till next Morning, where I saw him as I thought recover'd: We throw'd him some Meat, which he eat, so we let him out and he went home. About a Month after that, the Dog's Hair came off, and his Master killed him, being so ugly to look at; he told me, he lookt like a Leprous Person, (that was his very Expression;) I never heard this Dog swell'd.

The third Dog which he bit was a Shaggy Spaniel, about an hour and a quarter after the second. He was bitten on the foremost part of his right Shoulder, as we perceived by the Blood. The Dog seem'd to bite at the place himself, and was very sick for that whole Evening (*viz.*) about 2 or 3 hours; but without any means or application he recover'd, and I never heard he was sick afterwards.

3. A Merchant Produces First-Rate Science in a Philadelphia Garden. 1735

JAMES LOGAN (1674–1751) conducted his experiments showing the sexual nature of plant reproduction before he knew that such a view was common among well-informed scientists of his day. His work was, however, very elegant and won the praise of Linnaeus, among others.

Logan was a wealthy Quaker merchant of Philadelphia who corresponded with prominent scientists of his day and contributed greatly to the culture of colonial Philadelphia in many ways.

Peter Collinson (1694–1768) was a London merchant and science enthusiast. He was a friend of Sir Hans Sloane and others in the inner circle of the Royal Society, and he played a vital role in early American science by encouraging his New World correspondents in their scientific inquiries (see also the next selection).

For further material on Logan see Frederick B. Tolles, "Philadelphia's First Scientist, James Logan," *Isis,* 47 (1956), 20–30; Frederick B. Tolles, *James Logan and the Culture of Provincial America* (Boston: Little, Brown, 1957), especially 199–202.

SOME EXPERIMENTS concerning the Impregnation of the Seeds of Plants,

by James Logan, Esq; Communicated in a Letter from him to Mr. Peter Collinson*

Philadelphia, Nov. 20, 1735.

SIR,

As the Notion of a Male Seed, or the *Farina Fœcundans* in Vegetables is now very common, I shall not trouble you with any Observations concerning it, but such as may have some Tendency to what I have to mention—— And, first, I find from *Miller*'s Dictionary, that M. *Geoffroy,* a Name I think of Repute amongst *Naturalists,* from the Experiments he made on *Mayze,* was of Opinion, that Seeds may grow up to their full Size, and appear perfect to the Eye, without being impregnated by the *Farina,* which possibly, for ought I know, may in some Cases be true; for there is no End of Varieties in Nature:——But in the Subject he has mention'd I have Reason to believe it's otherwise, and that he applied not all the Care that was requisite in the Management.

When I first met with the Notion of this Male Seed, it was in the Winter Time, when I could do no more than think of it; but in the Spring I resolved to make some Experiments on the *Mayze,* or *Indian Corn.* In each Corner of my Garden, which is forty Foot in Breadth, and near eighty in Length, I planted a Hill of that Corn, and watching the Plants when they grew up to a proper Height, and were pushing out both the Tassels above, and Ears below; from one of those Hills, I cut off the whole Tassels, on others I carefully open'd the Ends of the Ears, and from some of them I cut or pinch'd off all the silken filaments; from others I took about half, from others one fourth and three fourths, &c. with some Variety, noting the Heads, and the Quantity taken from each: Other Heads again I tied up at their Ends, just before the Silk was putting out, with fine Muslin, but the fuzziest or most Nappy I could find, to prevent the Passage of the *Farina;* but that would obstruct neither Sun, Air or Rain. I fastened it also so very loosely, as not to give the least Check to Vegetation.

The Consequence of all which was this, that of the five or six Ears on the first Hill, from which I had taken all the Tassels, from whence proceeds the *Farina,* there was only one that had so much as a single Grain in it, and that in about four hundred and eighty Cells, had but about twenty or twenty-one Grains, the Heads, or Ears, as they stood on the Plant, look'd as well to the Eye as any other; they were of their proper Length, the Cores of their full Size, but to the Touch, for want of the Grain, they felt light and yielding. On the Core, when divested of the Leaves that cover it, the Beds of Seed were in their Ranges, with only a dry Skin on each.

*The Royal Society of London, *Philosophical Transactions,* 39 (1736), 192–195.

In the Ears of the other Hills, from which I had taken all the Silk, and in those that I had cover'd with Muslin, there was not so much as one mature grown Grain, nor other than as I have mentioned in the first: But in all the others, in which I had left Part, and taken Part of the Silk, there was in each the exact Proportion of full Grains, according to the Quantity or Number of the Filaments I had left on them. And for the few Grains I found on one Head in the first Hill, I immediately accounted thus: That Head, or Ear, was very large, and stood prominent from the Plant, pointing with its Silk Westward directly towards the next Hill of *Indian* Corn; and the *Farina,* I know, when very ripe, on shaking the Stalk, will fly off in the finest Dust, somewhat like Smoak. I therefore, with good Reason, judged that a Westerly Wind had wafted some few of these Particles from the other Hill, which had light on the Stiles of this Ear, in a Situation perfectly well fitted to receive them, which none of the other Ears, on the same Hill, had. And indeed I admire that there were not more of the same Ear than I found impregnated in the same manner.

As I was very exact in this Experiment, and curious enough in my Observations, and this, as I have related it, is truly Fact, I think it may reasonably be allowed, that notwithstanding what M. *Geoffroy* may have deliver'd of his Trials on the same Plant, I am positive, by my Experiment on those Heads, That the Silk was taken quite away, and those that were cover'd with Muslin, none of the Grains will grow up to their Size, when prevented of receiving the *Farina* to impregnate them, but appear, when the Ears of Corn are disclosed, with all the Beds of the Seeds, or Grains, in their Ranges, with only a dry Skin on each, about the same Size as when the little tender Ears appear fill'd with milky Juice before it puts out its Silk. But the few Grains that were grown on the single Ear, were as full and as fair as any I had seen, the Places of all the rest had only dry empty Pellicles, as I have described them; and I much question whether the same does not hold generally in the whole Course of Vegetation, though, agreeable to what I first hinted, it may not be safe to pronounce absolutely upon it, without a great Variety of Experiments on different Subjects. But I believe there are few Plants that will afford so fine an Opportunity of observing on them as the *Mayze,* or our *Indian Corn;* because its Stiles may be taken off or left on the Ear, in any Proportion, and the Grains be afterwards number'd in the Manner I have mentioned.

4. A Great Mind Turns Uplifting Entertainment into Theoretical Innovation. 1747-1748

The following excerpts from the classic work of BENJAMIN FRANKLIN (1706–1790) on electricity suggest a great deal about science in pre-Revolutionary America. Like Logan's, these experiments were the work of an amateur and they were written to the same scientific inquirer—Peter Collinson. They show the public and recreational aspects of eighteenth century science, including that science which gave rise to major theoretical advances.

Franklin is too well known to need a biographical introduction. These experiments took place shortly before he retired from the printing business and devoted himself to philosophical (that is, scientific) studies and public affairs. As soon as he read deeply in the electrical theory of the day his theoretical ingenuity seemed to disappear.

A full account of Franklin's experiments can be found in I. Bernard Cohen's *Benjamin Franklin's Experiments* (Cambridge, Mass.: Harvard University Press, 1941).

EXPERIMENTS AND OBSERVATIONS on Electricity, Made at Philadelphia in America*

Benjamin Franklin

EXTRACT OF LETTER I.

From Benj. Franklin, Esq; at Philadelphia,
to Peter Collinson, Esq; F.R.S. London.

Philadelphia, March 28, 1747.

SIR,

Your kind present of an electric tube, with directions for using it, has put several of us[1] on making electrical experiments, in which we have observed some particular phænomena that we look upon to be new. I shall therefore communicate them to you in my next, though possibly they may not be new

* Benjamin Franklin, *Experiments and Observations on Electricity, Made at Philadelphia in America* (5th ed., London: F. Newbery, 1774), pp. 1–15, 21, 24–29, 37–38. Slightly abridged. Some italics omitted.

[1] i.e., of the Library-Company, an institution of the Author's, founded 1730. To which company the present was made. [*Footnote in original.*]

to you, as among the numbers daily employed in those experiments on your side the water, 'tis probable some one or other has hit on the same observations. For my own part, I never was before engaged in any study that so totally engrossed my attention and my time as this has lately done; for what with making experiments when I can be alone, and repeating them to my Friends and Acquaintance, who, from the novelty of the thing, come continually in crouds to see them, I have, during some months past, had little leisure for any thing else.

I am, &c.

B. FRANKLIN

LETTER II.
From Benj. Franklin, Esq; in Philadelphia,
to Peter Collinson, Esq; F.R.S. London.

July 11, 1747.

SIR,

In my last I informed you that, in pursuing our electrical enquiries, we had observed some particular phænomena, which we looked upon to be new, and of which I promised to give you some account, though I apprehended they might not possibly be new to you, as so many hands are daily employed in electrical experiments on your side the water, some or other of which would probably hit on the same observations.

The first is the wonderful effect of pointed bodies, both in *drawing off* and *throwing off* the electrical fire. For example,

Place an iron shot of three or four inches diameter on the mouth of a clean dry glass bottle. By a fine silken thread from the cieling [*sic*], right over the mouth of the bottle, suspend a small cork-ball, about the bigness of a marble; the thread of such a length, as that the cork-ball may rest against the side of the shot. Electrify the shot, and the ball will be repelled to the distance of four or five inches, more or less, according to the quantity of Electricity.—When in this state, if you present to the shot the point of a long, slender, sharp bodkin,[2] at six or eight inches distance, the repellency is instantly destroyed, and the cork flies to the shot. A blunt body must be brought within an inch, and draw a spark to produce the same effect. To prove that the electrical fire is *drawn off* by the point, if you take the blade of the bodkin out of the wooden handle, and fix it in a stick of sealing-wax, and then present it at the distance aforesaid, or if you bring it very near, no such effect follows; but sliding one finger along the wax till you touch the blade, and the ball flies to the shot immediately.—If you present the point in the dark, you will see, sometimes at a foot distance and more, a light gather upon it, like that of a fire-fly, or glow-worm; the less sharp the point,

[2] A large needle or leather punch.

the nearer you must bring it to observe the light; and at whatever distance you see the light, you may draw off the electrical fire, and destroy the repellency.—If a cork-ball so suspended be repelled by the tube, and a point be presented quick to it, though at a considerable distance, 'tis surprizing to see how suddenly it flies back to the tube. Points of wood will do near as well as those of iron, provided the wood is not dry; for perfectly dry wood will no more conduct electricity than sealing-wax.

To shew that points will *throw off*[3] as well as *draw off* the electrical fire; lay a long sharp needle upon the shot, and you cannot electrise the shot so as to make it repel the cork-ball.—Or fix a needle to the end of a suspended gun-barrel, or iron-rod, so as to point beyond it like a little bayonet;[4] and while it remains there, the gun-barrel, or rod, cannot by applying the tube to the other end be electrised so as to give a spark, the fire continually running out silently at the point. In the dark you may see it make the same appearance as it does in the case before-mentioned.

The repellency between the cork-ball and the shot is likewise destroyed. 1. By sifting fine sand on it; this does it gradually. 2. By breathing on it. 3. By making a smoke about it from burning wood. 4. By candle-light, even though the candle is at a foot distance: these do it suddenly.—The light of a bright coal from a wood fire; and the light of a red-hot iron do it likewise; but not at so great a distance. Smoke from dry rosin dropt on hot iron, does not destroy the repellency; but is attracted by both shot and cork-ball, forming proportionable atmospheres round them, making them look beautifully, somewhat like some of the figures in *Burnet*'s or *Whiston*'s Theory of the Earth.

N.B. This experiment should be made in a closet, where the air is very still, or it will be apt to fail.

The light of the sun thrown strongly on both cork and shot by a looking-glass for a long time together, does not impair the repellency in the least. This difference between fire-light and sun-light is another thing that seems new and extraordinary to us.[5]

We had for some time been of opinion, that the electrical fire was not created by friction, but collected, being really an element diffused among, and attracted by other matter, particularly by water and metals. We had even discovered and demonstrated its afflux to the electrical sphere, as well as its efflux, by means of little light windmill wheels made of stiff paper vanes, fixed obliquely, and turning freely on fine wire axes. Also by little wheels of the same matter, but formed like water-wheels. Of the disposition and appli-

[3] This power of points to throw off the electrical fire, was first communicated to me by my ingenious friend Mr Thomas Hopkinson, since deceased. . . . [*Footnote in original.*]

[4] This was Mr Hopkinson's Experiment, made with an expectation of drawing a more sharp and powerful spark from the point, as from a kind of focus, and he was surprized to find little or none. [*Footnote in original.*]

[4] This was Mr Hopkinson's Experiment, made with an expectation of drawing a more sharp the particles separated from the candle, being first attracted and then repelled, carrying off the electric matter with them; and from the rarefying the air, between the glowing coal or red-hot iron, and the electrised shot, through which rarified air the electric fluid could more readily pass. [*Footnote in original, added as a revision from the first edition.*]

cation of which wheels, and the various phænomena resulting, I could, if I had time, fill you a sheet.[6] The impossibility of electrising one's self (though standing on wax) by rubbing the tube, and drawing the fire from it; and the manner of doing it, by passing the tube near a person or thing standing on the floor, &c. had also occurred to us some months before Mr. Watson's ingenious *Sequel*[7] came to hand, and these were some of the new things I intended to have communicated to you.—But now I need only mention some particulars not hinted in that piece, with our reasoning thereupon: though perhaps the latter might well enough be spared.

1. A person standing on wax, and rubbing the tube, and another person on wax drawing the fire, they will both of them (provided they do not stand so as to touch one another) appear to be electrised, to a person standing on the floor; that is, he will perceive a spark on approaching each of them with his knuckle.

2. But if the persons on wax touch one another during the exciting of the tube, neither of them will appear to be electrised.

3. If they touch one another after exciting the tube, and drawing the fire as aforesaid, there will be a stronger spark between them than was between either of them and the person on the floor.

4. After such strong spark, neither of them discover any electricity.

These appearances we attempt to account for thus: We suppose, as aforesaid, that electrical fire is a common element, of which every one of the three persons abovementioned has his equal share, before any operation is begun with the tube. *A,* who stands on wax and rubs the tube, collects the electrical fire from himself into the glass; and his communication with the common stock being cut off by the wax, his body is not again immediately supply'd. *B,* (who stands on wax likewise) passing his knuckle along near the tube, receives the fire which was collected by the glass from *A;* and his communication with the common stock being likewise cut off, he retains the additional quantity received.—To *C,* standing on the floor, both appear to be electrised; for he having only the middle quantity of electrical fire, receives a spark upon approaching *B,* who has an over quantity; but gives one to *A,* who has an under quantity. If *A* and *B* approach to touch each other, the spark is stronger, because the difference between them is greater: After such touch there is no spark between either of them and *C,* because the electrical fire in all is reduced to the original equality. If they touch while electrising, the equality is never destroy'd, the fire only circulating. Hence have arisen some new terms among us: we say *B* (and bodies like circumstanced) is electrised

[6] These experiments with the wheels, were made and communicated to me by my worthy and ingenious friend Mr Philip Syng; but we afterwards discovered that the motion of those wheels was not owing to any afflux or efflux of the electric fluid, but to various circumstances of attraction and repulsion. 1750 [*Footnote in original.*]

[7] Sir William Watson was one of the foremost English writers on electricity in the eighteenth century. The paper referred to was read to the Royal Society in 1746 and appeared in the *Philosophical Transactions* for 1746–1747.

positively; A, negatively. Or rather, *B* is electrised *plus; A, minus*. And we daily in our experiments electrise bodies *plus* or *minus,* as we think proper.—To electrise *plus* or *minus,* no more needs to be known than this, that the parts of the tube or sphere that are rubbed, do, in the instant of the friction, attract the electrical fire, and therefore take it from the thing rubbing: the same parts immediately, as the friction upon them ceases, are disposed to give the fire they have received, to any body that has less. Thus you may circulate it, as Mr. Watson has shewn; you may also accumulate or subtract it, upon, or from any body, as you connect that body with the rubber or with the receiver, the communication with the common stock being cut off. We think that ingenious gentleman was deceived when he imagined (in his *Sequel*) that the electrical fire came down the wire from the cieling to the gun barrel, thence to the sphere, and so electrised the machine and the man turning the wheel, &c. We suppose it was *driven off,* and not brought on through that wire; and that the machine and man, &c. were electrised *minus; i.e.* had less electrical fire in them than things in common.

As the vessel is just upon sailing, I cannot give you so large an account of American Electricity as I intended: I shall only mention a few particulars more.—We find granulated lead better to fill the phial[8] with, than water, being easily warmed, and keeping warm and dry in damp air.—We fire spirits with the wire of the phial.—We light candles, just blown out, by drawing a spark among the smoke between the wire and snuffers.—We represent lightning, by passing the wire in the dark, over a china plate that has gilt flowers, or applying it to gilt frames of looking-glasses, &c.—We electrise a person twenty or more times running, with a touch of the finger on the wire, thus: He stands on wax. Give him the electrised bottle in his hand. Touch the wire with your finger, and then touch his hand or face; there are sparks every time.[9]—We encrease the force of the electrical kiss vastly, thus: Let *A* and *B* stand on wax; or *A* on wax, and *B* on the floor; give one of them the electrised phial in hand; let the other take hold of the wire; there will be a small spark; but when their lips approach, they will be struck and shock'd. The same if another gentleman and lady, *C* and *D,* standing also on wax, and joining hands with *A* and *B,* salute or shake hands. We suspend by fine silk thread a counterfeit spider, made of a small piece of burnt cork, with legs of linnen thread, and a grain or two of lead stuck in him, to give him more weight. Upon the table, over which he hangs, we stick a wire upright, as high as the phial and wire, four or five inches from the spider: then we animate him, by setting the electrified phial at the same distance on the other side of him; he will immediately fly to the wire of the phial, bend his legs in touching it; then spring off, and fly to the wire in the table: thence again to the

[8] This refers to the Leyden jar or condenser in a very primitive form.

[9] By taking a spark from the wire, the electricity within the bottle is diminished; the outside of the bottle then draws some from the person holding it, and leaves him in the negative state. Then when his hand or face is touch'd, an equal quantity is restored to him from the person touching. [*Footnote in original.*]

wire of the phial, playing with his legs against both, in a very entertaining manner, appearing perfectly alive to persons unacquainted. He will continue this motion an hour or more in dry weather.—We electrify, upon wax in the dark, a book that has a double line of gold round upon the covers, and then apply a knuckle to the gilding; the fire appears every where upon the gold like a flash of lightning: not upon the leather, nor, if you touch the leather instead of the gold. We rub our tubes with buckskin, and observe always to keep the same side to the tube, and never to sully the tube by handling; thus they work readily and easily, without the least fatigue, especially if kept in tight pasteboard cases, lined with flannel, and fitting close to the tube. This I mention, because the *European* papers on electricity frequently speak of rubbing the tube as a fatiguing exercise. Our spheres are fixed on iron axes, which pass through them. At one end of the axis there is a small handle, with which you turn the sphere like a common grindstone. This we find very commodious, as the machine takes up but little room, is portable, and may be enclosed in a tight box, when not in use. 'Tis true, the sphere does not turn so swift as when the great wheel is used: but swiftness we think of little importance, since a few turns will charge the phial, &c. sufficiently.[10]

I am, &c.

B. FRANKLIN

LETTER III.

From Benj. Franklin, Esq; at Philadelphia,
to Peter Collinson, Esq; F.R.S. London.

Sept. 1, 1747.

SIR,

The necessary trouble of copying long letters, which, perhaps, when they come to your hands may contain nothing new, or worth your reading, (so quick is the progress made with you in Electricity) half discourages me from writing any more on that subject. Yet I cannot forbear adding a few observations on M. *Muschenbroek*'s wonderful bottle.[11]

1. The non-electric[12] contain'd in the bottle differs when electrised from a non-electric electrised out of the bottle, in this: that the electrical fire of the latter is accumulated *on its surface,* and forms an electrical atmosphere round it of considerable extent; but the electrical fire is crowded *into the substance* of the former, the glass confining it.[13]

[10] This simple easily-made machine was a contrivance of Mr. Syng's. [*Footnote in original.*]
[11] The Leyden jar.
[12] Conductor.
[13] . . . The fire in the bottle was found by subsequent experiments not to be contained in the non-electric, but *in the glass.* 1748 [*Footnote in original.*]

2. At the same time that the wire and top of the bottle, &c. is electrised *positively* or *plus,* the bottom of the bottle is electrised *negatively* or *minus,* in exact proportion: *i.e.* whatever quantity of electrical fire is thrown in at the top, an equal quantity goes out of the bottom.[14] To understand this, suppose the common quantity of electricity in each part of the bottle, before the operation begins, is equal to 20; and at every stroke of the tube, suppose a quantity equal to 1 is thrown in; then, after the first stroke, the quantity contain'd in the wire and upper part of the bottle will be 21, in the bottom 19. After the second, the upper part will have 22, the lower 18, and so on, till, after 20 strokes the upper part will have a quantity of electrical fire equal to 40, the lower part none: and then the operation ends: for no more can be thrown into the upper part, when no more can be driven out of the lower part. If you attempt to throw more in, it is spued back through the wire, or flies out in loud cracks through the sides of the bottle.

3. The equilibrium cannot be restored in the bottle by *inward* communication or contact of the parts; but it must be done by a communication formed *without* the bottle between the top and bottom, by some non-electric,[15] touching or approaching both at the same time; in which case it is restored with a violence and quickness inexpressible; or, touching each alternately, in which case the equilibrium is restored by degrees.

4. As no more electrical fire can be thrown into the top of the bottle, when all is driven out of the bottom, so in a bottle not yet electrised, none can be thrown into the top, when none *can* get out at the bottom; which happens either when the bottom is too thick, or when the bottle is placed on an electric *per se.*[16] Again, when the bottle is electrised, but little of the electrical fire can be *drawn out* from the top, by touching the wire, unless an equal quantity can at the same time *get in* at the bottom. Thus, place an electrised bottle on clean glass or dry wax, and you will not, by touching the wire, get out the fire from the top. Place it on a non-electric, and touch the wire, you will get it out in a short time; but soonest when you form a direct communication as above.

So wonderfully are these two states of Electricity, the *plus* and *minus,* combined and balanced in this miraculous bottle! situated and related to each other in manner that I can by no means comprehend! If it were possible that a bottle should in one part contain a quantity of air strongly comprest, and in another part a perfect vacuum, we know the equilibrium would be instantly restored *within.* But here we have a bottle containing at the same time a *plenum* of electrical fire, and a *vacuum* of the same fire; and yet the equilibrium cannot be restored between them but by a communication *without!*

[14] What is said here, and after, of the *top* and *bottom* of the bottle, is true of the *inside* and *outside* surfaces, and should have been so expressed. [*Footnote in original.*]

[15] Conductor.

[16] Nonconductor.

though the *plenum* presses violently to expand, and the hungry vacuum seems to attract as violently in order to be filled.

5. The shock to the nerves (or convulsion rather) is occasioned by the sudden passing of the fire through the body in its way from the top to the bottom of the bottle. The fire takes the shortest[17] course, as Mr. *Watson* justly observes: But it does not appear from experiment that in order for a person to be shocked, a communication with the floor is necessary: for he that holds the bottle with one hand, and touches the wire with the other, will be shock'd as much, though his shoes be dry, or even standing on wax, as otherwise. And on the touch of the wire, (or of the gun-barrel, which is the same thing) the fire does not proceed from the touching finger to the wire, as is supposed, but from the wire to the finger, and passes through the body to the other hand, and so into the bottom of the bottle. . . .

LETTER IV.
From Benj. Franklin, Esq; in Philadelphia,
to Peter Collinson, Esq; F.R.S. London. . . .

1748.

SIR. . . .

8. When we use the terms of *charging* and *discharging* the phial, it is in compliance with custom, and for want of others more suitable. Since we are of opinion that there is really no more electrical fire in the phial after what is called its *charging,* than before, nor less after its *discharging;* excepting only the small spark that might be given to, and taken from the non-electric[18] matter, if separated from the bottle, which spark may not be equal to a five hundredth part of what is called the explosion.

For if, on the explosion, the electrical fire came out of the bottle by one part, and did not enter in again by another, then, if a man, standing on wax, and holding the bottle, in one hand, takes the spark by touching the wire hook with the other, the bottle being thereby *discharged,* the man would be *charged;* or whatever fire was lost by one, would be found in the other, since there was no way for its escape: But the contrary is true.

9. Besides, the phial will not suffer what is called a *charging,* unless as much fire can go out of it one way, as is thrown in by another. A phial cannot be charged standing on wax or glass, or hanging on the prime conductor, unless a communication be formed between its coating and the floor.

10. But suspend two or more phials on the prime conductor, one hanging on the tail of the other; and a wire from the last to the floor, an equal number of turns of the wheel shall charge them all equally, and every one as

[17] Other circumstances being equal. [*Footnote in original.*]
[18] Conductor.

much as one alone would have been. What is driven out at the tail of the first, serving to charge the second; what is driven out of the second charging the third; and so on. By this means a great number of bottles might be charged with the same labour, and equally high, with one alone, were it not that every bottle receives new fire, and loses its old with some reluctance, or rather gives some small resistance to the charging, which in a number of bottles becomes more equal to the charging power, and so repels the fire back again on the globe, sooner in proportion than a single bottle would do.

11. When a bottle is charged in the common way, its *inside* and *outside* surfaces stand ready, the one to give fire by the hook, the other to receive it by the coating; the one is full and ready to throw out, the other empty and extremely hungry; yet as the first will not *give out,* unless the other can at the same instant *receive in;* so neither will the latter receive in, unless the first can at the same instant give out. When both can be done at once, it is done with inconceivable quickness and violence.

12. So a strait spring (though the comparison does not agree in every particular) when forcibly bent, must, to restore itself, contract that side which in the bending was extended, and extend that which was contracted; if either of these two operations be hindered, the other cannot be done. But the spring is not said to be *charg'd* with elasticity when bent, and discharged when unbent; its quantity of elasticity is always the same.

13. Glass, in like manner, has, within its substance, always the same quantity of electrical fire, and that a very great quantity in proportion to the mass of glass, as shall be shewn hereafter.

14. This quantity, proportioned to the glass, it strongly and obstinately retains, and will have neither more nor less though it will suffer a change to be made in its parts and situation; *i.e.* we may take away part of it from one of the sides, provided we throw an equal quantity into the other.

15. Yet when the situation of the electrical fire is thus altered in the glass; when some has been taken from one side, and some added to the other, it will not be at rest or in its natural state, till it is restored to its original equality.—And this restitution cannot be made through the substance of the glass, but must be done by a non-electric communication formed without, from surface to surface.

16. Thus, the whole force of the bottle, and power of giving a shock, is in the GLASS ITSELF; the non-electrics[19] in contact with the two surfaces, serving only to *give* and *receive* to and from the several parts of the glass; that is, to give on one side, and take away from the other.

17. This was discovered here in the following manner: Purposing to analyse the electrified bottle, in order to find wherein its strength lay, we placed it on glass, and drew out the cork and wire which for that purpose had been loosely put in. Then taking the bottle in one hand, and bringing

[19] Conductors.

a finger of the other near its mouth, a strong spark came from the water, and the shock was as violent as if the wire had remained in it, which shewed that the force did not lie in the wire. Then to find if it resided in the water, being crouded into and condensed in it, as confin'd by the glass, which had been our former opinion, we electrified the bottle again, and placing it on glass, drew out the wire and cork as before; then taking up the bottle, we decanted all its water into an empty bottle, which likewise stood on glass; and taking up that other bottle, we expected, if the force resided in the water, to find a shock from it; but there was none. We judged then that it must either be lost in decanting, or remain in the first bottle. The latter we found to be true; for that bottle on trial gave the shock, though filled up as it stood with fresh unelectrified water from a tea-pot.—To find, then, whether glass had this property merely as glass, or whether the form contributed any thing to it; we took a pane of sash-glass, and laying it on the hand, placed a plate of lead on its upper surface; then electrified that plate, and bringing a finger to it, there was a spark and shock. We then took two plates of lead of equal dimensions, but less than the glass by two inches every way, and electrified the glass between them, by electrifying the uppermost lead; then separated the glass from the lead, in doing which, what little fire might be in the lead was taken out, and the glass being touched in the electrified parts with a finger, afforded only very small pricking sparks, but a great number of them might be taken from different places. Then dextrously placing it again between the leaden plates, and compleating a circle between the two surfaces, a violent shock ensued.—Which demonstrated the power to reside in glass as glass, and that the non-electrics[20] in contact served only, like the armature of a loadstone, to unite the force of the several parts, and bring them at once to any point desired: it being the property of a non-electric,[21] that the whole body instantly receives or gives what electrical fire is given to or taken from any one of its parts. . . .

19. I perceive by the ingenious Mr. *Watson*'s last book, lately received, that Dr. *Bevis* had used, before we had, panes of glass to give a shock; though, till that book came to hand, I thought to have communicated it to you as a novelty. The excuse for mentioning it here is, that we tried the experiment differently, drew different consequences from it (for Mr. *Watson* still seems to think the fire *accumulated on the non-electric*[22] that is in contact with the glass, p. 72) and, as far as we hitherto know, have carried it farther. . . .

Chagrined a little that we have been hitherto able to produce nothing in this way of use to mankind; and the hot weather coming on, when electrical experiments are not so agreeable, it is proposed to put an end to them for this season, somewhat humorously, in a party of pleasure, on the banks of

[20] Conductors.
[21] Conductor.
[22] Conductor.

Skuylkil.[23] Spirits, at the same time, are to be fired by a spark sent from side to side through the river, without any other conductor than the water; an experiment which we some time since performed, to the amazement of many.[24] A turkey is to be killed for our dinner by the *electrical shock,* and roasted by *electrical jack,* before a fire kindled by the *electrical bottle:* when the healths of all the famous electricians in England, Holland, France, and Germany are to be drank in *electrified bumpers,*[25] under the discharge of guns from the *electrical battery.*

April 29,
1749.

5. A Well-Read Physician's Facts and Fancies as Respectable Science. 1770

This essay was presented to the American Philosophical Society by a member who served on a commission to study the transit of Venus. Williamson's work is a good example of the speculative nature of much of the science of the day and also shows the real difficulties of scientists of the late eighteenth century who tried to deal with ideas of heat. Williamson's amateur theorizing on this occasion presents a strong contrast to the careful experimentation and deduction of some of his contemporaries who, like Franklin, were also members of the American Philosophical Society.

[23] The Schuylkill River.

[24] As the possibility of this experiment has not been easily conceived, I shall here describe it. —Two iron rods, about three feet long, were planted just within the margin of the river, on the opposite sides. A thick piece of wire, with a small round knob at its end, was fixed on the top of one of the rods, bending downwards, so as to deliver commodiously the spark upon the surface of the spirit. A small wire fastened by one end to the handle of the spoon, containing the spirit, was carried a-cross the river, and supported in the air by the rope commonly used to hold by, in drawing the ferry-boats over. The other end of this wire was tied round the coating of the bottle; which being charged, the spark was delivered from the hook to the top of the rod standing in the water on that side. At the same instant the rod on the other side delivered a spark into the spoon, and fired the spirit. The electric fire returning to the coating of the bottle through the handle of the spoon and the supported wire connected with them.

That the electric fire thus actually passes through the water, has since been satisfactorily demonstrated to many by an experiment of Mr. Kinnersley's, performed in a trough of water about ten feet long. The hand being placed under water in the direction of the spark (which always takes the strait or shortest course, if sufficient, and other circumstances are equal) is struck and penetrated by it as it passes. [*Footnote in original.*]

[25] An *electrified bumper* is a small thin glass tumbler, nearly filled with wine, and electrified as the bottle. This when brought to the lips gives a shock, if the party be close shaved, and does not breath on the liquor. [*Footnote in original.*]

HUGH WILLIAMSON (1735–1814) was a native of Pennsylvania who studied theology in America and then medicine in Europe. He was a prominent citizen of Pennsylvania and afterwards represented North Carolina in the Continental Congress.

Background and biography are to be found in Brooke Hindle, *The Pursuit of Science in Revolutionary America, 1735–1789* (Chapel Hill: University of North Carolina Press, 1956), especially 172; S. A. Mitchell, "Astronomy During the Early Years of the American Philosophical Society," *Science,* 95 (1942), 489–495; *Dictionary of American Biography,* XVIII, 298–299; David Hosack, *Biographical Memoir of Hugh Williamson, M.D. Ll.D.* (New York: C. S. Van Winkle, 1820).

AN ESSAY on the Use of COMETS, and an Account of their Luminous Appearance; together with some Conjectures concerning the Origin of HEAT.*

Hugh Williamson

Read before the Society, Nov. 16th, 1770.

A COMET is a solid dark body revolving round the Sun in stated periods, receiving light and heat from the Sun. Comets revolve as other planets do in an ellipsis, one part of which is much farther from the Sun than another; some of them are very eccentric; that which appeared Anno 1680 was twelve thousand millions of miles from the Sun in aphelio, it was not half a million in perihelio.[1] The period of the comet which appeared Anno 1758 is 75 years. That of 1661 is 120 years. And that of 1680 is 575 years. Though Comets doubtless move in an ellipsis, yet from the extreme length of their path, the small part that falls under our observation, the difficulty in determining the Comet's absolute distance or velocity, &c. we have obtained no certainty concerning the period of any Comet except the three I have mentioned, nor shall we ever determine their periods in all probability, except by a series of observations on the return of each particular Comet, which may require several thousands of years.

Comets receive their light and heat from the Sun, for they appear to have no light of their own, and are thence invisible, except on their near approach to the Sun. In the year 1723, an Astronomer had the fortune to discover a Comet by means of his telescope before it was bright enough to become visible by the naked eye. The great Comet which appeared Anno 1743 seemed no larger than a star of the fourth magnitude when first discovered; as it came down towards the Sun it acquired a tail, and increased gradually in size and lustre till it obtained that amazing form with which it terrified half the world. As this Comet departed from the Sun, its tail decreased, it lost its brightness, till in a short time it became invisible; this has also been

Transactions of the American Philosophical Society, 1 (1769–1771), Appendix, 27–36.

[1] Aphelio and perihelio signify most remote from the sun and the closest approach to the sun.

the fate of every other Comet; hence it is plain that their light, like that of other planets, is borrowed from the Sun.

Having just mentioned those general properties in which Comets evidently agree with other planets, I shall now try to account for that luminous train which attends them on their approach to the Sun, from which they are generally denominated Blazing Stars, and are supposed to differ essentially from every other planet or star. If I should be singular in any part of my opinion on this subject, I presume I shall be indulged, since it is matter of mere hypothesis.

Comets are not Blazing Stars, they do not burn at all, nor is there any remarkable heat in that tail which has so often terrified the nations, and been thought to portend dissolution to the world itself. The Comet of 1743 had acquired a tail some thousands of miles long about two months before he passed the Sun, while he was yet three hundred millions of miles from the Sun. Surely this could not be a flame of fire kindled by the Sun, else Comets take fire in a place where every drop of water on this globe would instantly freeze. There is no greater reason to think that Comets burn by their own heat, since their tail, whatever it be, as well as their light, evidently depends on the Sun, as we have already explained.

Philosophers[2] have differed greatly in their attempts to account for the tail of a Comet. One imagines that Comets are surrounded on all sides by a lucid fiery vapour, or atmosphere, which on account of the Sun's superior light, is only visible in the dark, whence we see no part of it but that which is in the shadow of the Comet on the side opposite to the Sun. According to him their atmosphere extends in all directions seventy or eighty millions of miles, for some Comets have appeared with a tail of that length, so that from the near approach of Comets to the earth we must frequently have been enveloped in that same lucid atmosphere.

From the extreme vicissitudes which Comets seem to endure, at one time penetrated with intolerable cold, at another time blazing with destructive heat, some have irreverently conjectured that they were designated as a place of future residence for the unhappy transgressors in this state, and thus vainly suppose that fifty or an hundred worlds were created for the sake of punishing the inhabitants of this little globe. It is sufficient to have mentioned such conjectures.

The great Sir Isaac Newton was of opinion, that Comets were designated, among other purposes, to nourish and refresh this earth and all the neighbouring planets. He imagined that by vegetation and putrefaction, a great deal of radical moisture is consumed or changed into earth; that the tail of a Comet is a thick vapour exhaled from the Comet by the heat of the Sun, which vapour is scattered through the planetary regions, and part of it being received within our atmosphere, occasionally supplies our loss of moisture.

[2] "Scientists" would be the modern term.

Whatever properties have been ascribed to heat, it seems very clear that evaporation cannot be performed unless by means of an atmosphere whereby the fluid is attracted, suspended and carried off. Therefore if we suppose that the earth and all the Planets are supplied with radical moisture from the Comets, we must also suppose, that the solar system is universally filled with an atmosphere sufficient for attracting and suspending fluids, which hypothesis would certainly destroy our present system of Astronomy. Besides this we may observe, that from the most accurate chymical analysis, there seems great reason to believe, that all the apparent changes in matter depend on combination and solution alone. That water may be combined with earth and again separated from it; but, that since the Creation, this Globe has not sustained the absolute loss of one ounce of water, or gained one ounce of earth. Therefore we do not require any nourishment from the vapour of Comets.

I see no reason to doubt that Comets were created like this world, to be the residence of intelligent beings; some of them no doubt which travel to immense distances through the Heavens, may be inhabited by an order of beings, greatly superior to this short-lived race of mortals, and much better fitted for comprehending and admiring the works of their divine original, which they behold in greater perfection. One of the primary ideas we form of the Supreme Being is, that he is the source of life, intelligence and happiness, and delights to communicate them; the earth we tread, the water we drink, and the very air in which we breathe, swarm with living creatures, all fitted to their several habitations. Are we to suppose that this little globe is the only animated part of the Creation, while the Comets, many of which are larger worlds, and run a nobler course, are an idle chaos, formed for the sole purpose of being frozen and burnt in turns. We cannot admit the thought; that Comets are doubtless inhabited. The great vicissitudes of climate, is the only plausible objection that has been made to this opinion. The Comet of 1680 came within one hundred thousand miles of the Sun, but the Sun's whole diameter is more than seven hundred thousand miles. The Comet's heat was then supposed to have been two thousand times hotter than red hot iron; but the same Comet was about twelve thousand millions of miles from the Sun, at his greatest distance, when it is supposed, that he perceived ten thousand times less heat than we usually enjoy. Hence it is supposed, that such a Planet could never afford a comfortable residence for rational creatures.

But here philosophers have taken for granted that the heat of every body is inversely as the square of its distance from the Sun, a proposition which I greatly suspect; for I apprehend that it is contrary to experiment.

Were heat a certain body proceeding immediately from the Sun, the quantity of heat in any space would doubtless be inversely as the square of its distance from the Sun. But I see no reason to believe that Heat comes from the Sun, while there is much reason to think that it does not. We perceive that Light comes from the Sun. We also perceive that Heat is produced in the bodies on which the rays of light fall, hence we are apt to confound Light and Heat together, though it be demonstrable that Light is not Heat and that Heat is not Light. So contracted is our knowledge of the primary constituent parts of bodies, that we cannot readily determine why any particular cause should not excite Heat with equal facility in all bodies. But we are taught by experience that different quantities are produced by the same cause, according to the medium on which it operates. It also appears that the particular aptitude of any body to be heated is nearly as the elasticity of that body, or the cohesion of its parts. Whatever produces a tremulous motion in the particles of any body, excites Heat in that body, and vice versa whatever excites Heat produces a tremulous motion in the particles of the body. Does Heat therefore consist in nothing else than the rapid vibrations of the minute particles of any body? or is there an elementary principle of fire diffused through all bodies, which is only excited or brought into action by any cause which produces a tremulous motion in the particles of those bodies? The latter seems most probable, though in solving the present hypothesis there is no difference whether Heat depends on the simple vibration of the particles of matter, or whether it depends on the fire which was only brought into action by the vibration of those particles, provided it should appear that the Heat in every body is uniformly as the vibratory motion of the particles of that body.—This I apprehend is the case, and shall beg leave to mention such evidence as seems to render the matter at least very probable.

Philosophers have enumerated five methods by which Heat is generated, viz. 1, by attrition, 2, chymical mixture, 3, fermentation, 4, inflammation, and 5, by the Sun. In all these cases it appears that the Heat depends on a vibratory motion which by one means or another is excited in the particles of the body.

1. Heat is produced by attrition, or by the striking or rubbing of one body against another. In this case there can be no doubt that the Heat depends on the vibratory motion of the particles, hence bodies are soonest heated where the friction is considerable, provided the bodies have also a proper degree of electricity. For the motion once communicated to the particles of an elastic body, are retained a considerable time, and increased by every succeeding stroke of the cause which put them into motion. The quantity of Heat produced in any body by friction, depends greatly on the body being fit to preserve the motion once communicated. Thus a saw fixed in a handvice so that it may long retain its tremulous motion, will soon be heated, whilst the file with which it is rubbed is not soon heated, being held in the soft unelastic hand, whereby the vibratory motion of its particles are immediately destroyed. The facility with which some bodies are heated before others, and with which the same body may be heated in one position rather than in another, abundantly prove that the quantity of Heat produced in any body by friction will not be as the motion communicated, but as the strokes com-

municated, together with the number of vibrations retained and communicated in consequence of each stroke.

2. The Heat which is produced by chymical mixture has been the subject of much speculation.—There are sundry bodies which joined together produce considerable Heat, as water with oil of vitriol;[3] others produce cold, as salt of nitre[4] with water. Why should one union produce heat the other cold? It appears in general that all mixtures properly so called, produce heat, all solutions produce cold. But in every mixture the bodies undergo a certain change in their qualities, whereas bodies undergo no change by solution. This may point out to us the true origin of heat in one case, and cold in the other. When two bodies have an attraction to one another, and the pores of the one body are so constituted as that the minute particles or atoms of the other body may penetrate into them, a general dissolution of the constituent parts of the body must ensue, the minute particles being rent asunder by the attractive force of the parts; such dissolution of the constituent parts of a body necessarily alters the qualities of that body. We may easily perceive that in the rapid union of such bodies by which the minutest particles are rent asunder, the vibratory motion of those parts must be greatly increased. Hence the generation of heat by mixtures. Hence too the heat in such mixtures, seems to be in proportion to the number of particles, which in any body of a determined bulk rush into union with and destroy the texture of one another.

In solutions or cooling combinations no change is produced in the qualities of the bodies. Thus by a solution of nitre in water cold is produced, and the salt may be deposited from the water, or the water be evaporated, and neither of the bodies undergo the least change. In this case it appears, that there is no dissolution of the constituent parts of either body, by the attractive force of the other, or by the construction of their parts; but that the globules of one body adhere superficially to those of the other, and the particles of the fluid are simply charged with those of the solid, by which means the vibrator motion of the particles is diminished, whence cold is necessarily produced.

It has been observed that spirit of nitre[5] mixed with water produces heat, while the same spirit mixed with snow produces the most intense cold. This may be probably urged as an objection to the above theory of heating and cooling combinations, under the apprehension that snow being nothing else than frozen water, should on these principles produce the same effects, on combination with any third body. But it must be observed, that one is a mixture, the other a solution. Water joined with spirit of nitre produces a mixture, the bodies undergo a change of qualities, and heat is generated. Pour the spirits of nitre into snow and nothing will follow, at least nothing has followed but a solution of the snow in the spirit. For these experiments have

[3] Sulfuric acid.

[4] Potassium nitrate.

[5] A volatile substance made from nitric acid and alcohol.

always been made when the temperature of the spirits was much below the freezing point of water, so that the snow could not be melted by such combination. Hence there being no intimate union of the parts, nor any thing else than a proper solution, cold was generated as in all similar cases.

3. Heat produced by fermentation or putrefaction, may be accounted for in the same manner as that produced by chymical mixture, there being no doubt that new mixtures are constantly forming in every putrescent or fermenting body.

4. Heat which is produced by inflammation seems also to depend on the chymical mixture of bodies. In all bodies which blaze there is found an acid and mephytic air,[6] which seem to abound in those bodies in proportion to their different degrees of inflammability. The separation of these two bodies constitutes a flame; this we observe can only be effected by means of a third body, *viz.* common air. The union of the acid with the water that is suspended in the air, and the union of the mephytic with the common air, produces two heating mixtures. Hence Heat is excited by flame.

5. Heat is produced by the Sun: Does that Heat proceed immediately from the sun, as is generally supposed, or is it mechanically excited by the action of the rays of light? The latter is most probable. We have seen a variety of methods by which Heat is produced. They appear in different forms, but they all terminate in the same thing; they are different methods of exciting a tremulous motion in the particles of the body. By some of them the most intense Heat is produced, and yet in no case is there any actual addition of fire. When Heat is excited by the Sun, there is also a tremulous motion excited in the particles of the body, they are expanded, &c. The phenomena resemble those of Heat excited by other means, whence it seems unphilosophic[7] to suppose that there should be an accession of fire[8] in this case more than in the others. I therefore suppose that *all the heat which is caused by the Sun, depends on a tremulous motion excited by the rays of light, in the particles of the body which is heated.* Hence it will follow that *the heat of any body will not be according to its distance from the Sun, but according to the fitness of that body, to retain and propagate the several vibrations which are communicated to its particles by the rays of light.* Hence it is that the air which is very elastic, when well compressed by the weight of the incumbent atmosphere, will receive a great degree of Heat near the surface of the earth, while the light thin air whose particles are removed to a considerable distance, as on the top of a high mountain, is always in a freezing state within the torrid zone.

Let us see how this theory of the generation of Heat may be subservient to the inhabitants of the Cometary worlds.

[6] Literally, noxious air; usually, carbon dioxide (which in solution in water makes carbonic acid).

[7] That is, unscientific.

[8] That is, the inherent "principle of heat."

It is evident that Comets are surrounded with an atmosphere very different from that of our globe; the heigth of our atmosphere is hardly supposed to exceed 60 or 70 miles, while that of a Comet is frequently 8 or 10,000 miles. Why should they have such a weight of atmosphere more than us? This is doubtless subservient to some very extraordinary purpose. We may also suppose with great probability, that the atmosphere of a Comet differs greatly from ours. The particles may be smaller, more subtile, elastic, and much more easily heated, whence the Sun's rays may be enabled to warm such an atmosphere compressed together by the weight of eight or ten thousand miles, at a distance from the Sun, in which we should perceive the most intense cold. This will explain the manner in which the inhabitants of a Comet may be sufficiently warm at their greatest distance from the Sun; but if they were proportionably heated on their nearest approach to the Sun, their summer heats would be intolerable; but this must certainly be the case if their atmosphere were in a permanent state, and continued in all seasons of equal density and weight. We are certain however from observation, that this is not the case; for as the Comet approaches the Sun, we can easily perceive its atmosphere greatly rarify'd, and thence rendered less fit for generating or retaining heat. But this is not the principal relief which Cometarians receive from the summer's Heat. The atmosphere of a Comet seems to undergo a change which is peculiar to itself. It is removed by the rays of light, and thrown off to a considerable distance behind the planet. It is demonstrable that the rays of light pass with amazing velocity, they travel above thirteen millions of miles in a minute; such amazing velocity multiplied into their weight, however small they be, must give them a considera[b]le momentum or impelling force, which must be great in regions near the Sun; by this force they repel the extremely subtile and light particles of air, and drive them off to such a distance behind the Comet that their weight is hardly perceived on its surface. The atmosphere being thus repelled by the Sun's rays, and thrown as it were into a shelter behind the planet, will be there extended longitudinally in the form of a shadow, being very rare towards the top. Every particle near the surface of this immense stream of air must be illuminated by the refraction and reflection of the Sun's rays, whence they will exhibit the faint appearance of a blaze. Thus we are apt to imagine that a Comet is intensely hot, and that a prodigious flame proceeds from it, while we see nothing else than its enlightened atmosphere.

As the inhabitants of Comets are not pressed by day, when they come near the sun, with a thousandth part of the atmosphere which usually surrounds them, and which is doubtless the mediate and principal cause of their perceiving Heat, we may easily see how they may be tolerably cool at noon day, on their nearest approach to the Sun.

If we might form any conjecture concerning the life of a Cometarian, from the annual periods of the world which he inhabits, we should apprehend that he far exceeds the years of an antedeluvian [*sic*]. Or might we attempt to measure the continuance of this globe, from the length of time which will be necessary to bring the astronomy of Comets, as well as every other science to that perfection at which they must doubtless arrive, we should infer that a small portion of that time is yet elapsed. On which ever of these subjects the mind is suffered to stray for a few minutes, it will find sufficient subject of pleasing speculation.

6. A Political Conservative Discredits Enlightenment Science. 1799

This oration presents a good example of the way in which enthusiasm for science can be connected with a certain type of political attitude. Daggett himself makes a classic statement of antiscientific sentiment and raises the perennial question of the extent to which science and social innovation go together.

DAVID DAGGETT (1764–1851) was a native of Massachusetts who attended Yale and afterward practiced law in New Haven. He was elected a member of the legislature, eventually served as Kent Professor of Law at Yale, and for many years was a distinguished judge in the courts of the state of Connecticut. As this oration shows, Daggett in 1799 was a Federalist opposed to the allegedly pro-French party of Jefferson. He was one of those horrified by the violent destruction of social institutions during the French Revolution.

More information can be found in Merle Curti, *The Growth of American Thought* (New York: Harper & Brothers, 1943), 205–206; John C. Greene, "American Science Comes of Age, 1780–1820," *Journal of American History,* 55 (1968), 22–41; Marjorie Nicolson and Nora M. Mohler, "The Scientific Background of Swift's *Voyage to Laputa,*" *Annals of Science,* 2 (1937), 299–334; and *Dictionary of American Biography,* v, 26–27.

SUN-BEAMS MAY BE EXTRACTED FROM CUCUMBERS, But the Process Is Tedious. An Oration, Pronounced on the Fourth of July, 1799. At the Request of the Citizens of New-Haven.*
David Daggett

* David Daggett, *Sun-Beams May Be Extracted From Cucumbers, But the Process Is Tedious. An Oration, Pronounced on the Fourth of July, 1799. At the Request of the Citizens of New-Haven* (New Haven: Thomas Greene and Son, 1799), pp. 5–19, 25–28. Italics omitted.

History informs us, that at Lagado, in Laputa, there was a grand academy established, in which there was a display of much curious learning.[1]

"One artist, of a very philosophic taste, was racking his invention to make a pin-cushion out of a piece of marble.

"Another had formed an ingenious project to prevent the growth of wool upon two young lambs, by a composition of gums, minerals and vegetables, applied inwardly, and thus he hoped in a reasonable time to propagate the breed of naked sheep throughout the Kingdom.

"A third had contrived a plan to entirely abolish words; and this was urged as a great advantage in point of health as well as brevity. For it is plain that every word we speak is an injury to our lungs, by corrosion, and consequently contributes to the shortening of our lives. An expedient was therefore offered, that since words were only names for things, it would be more convenient for all men to carry about them such things as were necessary to express the particular businesses on which they were to discourse;" and the Historian adds, "that he had often beheld two of those sages almost sinking under the weight of their packs, who, when they met in the streets, would lay down their loads, open their sacks, and hold conversation for an hour together; then put up their implements, help each other to resume their burdens, and take their leave.

"A fourth appeared with sooty hands and face, his hair and beard long, ragged and singed in several places. His clothes, shirt and skin were all of the same colour. He had been eight years upon a project for extracting sunbeams out of Cucumbers, which were to be put into vials, hermetically sealed, and let out to warm the air in raw inclement summers. He said he did not doubt but that, in eight years more, he should be able to supply the Governor's gardens with sunshine at a reasonable rate."

These Theorists were very patient, industrious and laborious in their pursuits—had a high reputation for their singular proficiency, and were regarded as prodigies in science. The common laborers and mechanics were esteemed a different race of beings, and were despised for their stupid and old fashioned manner of acquiring property and character. If the enquiry had been made whether any of these projects had succeeded, it would have been readily answered that they had not; but that they were reasonable—their principles just—and of course, that they must ultimately produce the objects in view. Hitherto no piece of marble had been made into a pin-cushion, and few, very few sun-beams had been extracted from Cucumbers; but what then? Are not all great, and noble, and valuable things, accomplished with immense exertion, and with an expense of much time? If a farther enquiry had been made what would be the great excellence of a marble pin-cushion,

[1] The account which follows is paraphrased from Jonathan Swift's *Gulliver's Travels,* in which Laputa was pictured as a flying island peopled by philosophers. Lagado was a city ruled over by the king of Laputa.

or the superior advantage of a breed of naked sheep, the answer would have been, it is philosophical to ask such questions.

In more modern times we have witnessed projects not unlike those of the learned of Laputa, above mentioned. A machine, called an Automaton, was, not long since, constructed. This was designed to transport from place to place, by land, any load without the aid of horses, oxen, or any other animal. The master was to sit at helm, and guide it up hill and down, and over every kind of road. This machine was completed, and proved demonstrably capable of performing the duties assigned to it, and the only difficulty which attended it, and which hath hitherto prevented its universal use, was, that it would not go.——Here, if any ignorant fellow had been so uncivil, he might have doubted why, if wood and iron were designed to go alone and carry a load, the whole herd of oxen, horses and camels were created.

A few years ago the Learned insisted that it was grovelling to travel either by land or water, but that the truly philosophical mode was to go by air. Hence, in all parts of the world, speculatists were mounted in balloons, with the whole apparatus of living and dying, and were flying through the Heavens, to the utter astonishment and mortification of those poor illiterate wretches who were doomed to tug and sweat on the earth. To be sure this method of travelling was somewhat precarious.——A flaw of wind, regardless of the principles of this machine, might destroy it, or, by the giving way of one philosophical pin, peg, or rope, it might be let into the sea, or dashed against a rock, and thus its precious contents miserably perish. But doubtless reason will, in time, provide sufficient checks against all these casualties. Here again some "busy body in other men's matters" might ask, if it was intended that men should fly through the air, why were they not made with feathers and wings, and especially why are there so many who are justly called Heavy moulded men?

Another class of the literati of our age, scorning to travel either on the sea, or on the land, or in the air, have constructed a submarine boat or diving machine, by which they were constantly groping among shark, storgeon and sea-horses. To say nothing of the hazard which these gentlemen encounter of running on rocks or shoals, or of being left in the lurch, on the bottom of the sea, by a leak, may we not wonder that they were not made with fins and scales, and may they not esteem themselves very fortunate that they have hitherto escaped being cut up to be made into oil?

These are a few among many modern inventions. All the principles of these various machines are capable of defence, and the inventors are all great, and learned, and ingenious men. Yet, strange as it may seem, the stupid, foolish, plodding people of this and other countries, still keep their oxen and their horses—their carriages are still made as they were an hundred years ago, and our coasters will still go to New-York on the surface of the sound, instead of sinking to the bottom or rising into the clouds—and they still prefer a fair wind and tide to the greatest profusion of steam, produced in the most scientific manner.

This species of enterprise, and this spirit of learning, has entered deeply into the business of agriculture. Discoveries have been made which have rendered sowing and reaping unnecessary. The plow, harrow, spade, hoe, sickle and scythe, have undergone a thorough change, on mathematical principles, and the speculative husbandman has yearly expected to see the fields covered with grass, and the hills and vallies with corn and wheat, without the clownish exercise of labour. With Varlow on Husbandary,[2] in his hands, and a complete collection of philosophical farming utensils, he has forgotten that by the "sweat of his brow he was to eat his bread," and is hourly expecting to "reap where he hath not sown, and gather where he hath not strawed". ——Still here and there an old fashioned fellow, and New-England abounds with them, "will rise early and set up late, and eat the bread of industry; will sow his seed in the morning and in the evening withhold not his hand," and is secretly flattering himself that this is the surest road to peace and plenty.

Hypocrates, Galen, and Sydenham, have been successively and conjointly attacked by the Physicians of the present refined age, and the medical learning of ancient times, or even of the last century, pronounced quackery and nonsense. A few years since, if a man were attacked with a most violent disease, he was directed to stimulate. Stimulants, powerful stimulants, were all the fashion, and, instead of Apothecaries shops and Lancets, the nurse was directed to the brandy pipe and the gin-case. Thus the Brownonian[3] system had superceded all others, and it was proved demonstrably, that the reason why the children of men were subject to death was, that they did not sufficiently fortify against its attacks, with beef steaks and wine. These principles had slain but a few when they were universally exploded, and men, going into the opposite extreme, were literally bled to death; and thus, lest the system should be overcharged, all its props were cautiously but entirely removed.

At length reason, unerring reason, appeared, and patients, writhed with agonies by the most subduing maladies, were solemnly directed to the Points.[4] Yes to the Points, as the great antidote against disease, and the certain restorer of health; and thus it was found, to the everlasting contempt of all the learned of the faculty of ancient and modern days, that the materia medica was useless, for that being pluss electrified, in one part of the body,

[2] Charles Varlo (or Varley) in 1770 published a three-volume work on farming as he had observed it in his travels over Europe. He advocated some agricultural reforms and invented several agricultural machines.

[3] In the late eighteenth century the great ancient authorities in medicine, such as Hippocrates and Galen, as well as their later counterparts such as Sydenham, were superseded by several theoretical systems of medicine, including that of John Brown (1735–1788), who contended that life is maintained by continuous stimulation.

[4] Points and tractors were instruments by means of which "animal magnetism," discovered by F. A. Mesmer, was utilized in the cure of patients. Later it appeared that Mesmer was dealing with hypnotic phenomena, and he was discredited by scientists in his own day.

and minus in the other, was the true radix of every disease, and that the sovereign remedy was, to restore an equilibrium by an external application of brass and steel.

Yet there are many so bigotted to the customs and practices of their ancestors, that they insist on the foolish habit of temperance, industry and exercise, and express some doubts respecting the entire efficiency of tractors.

A more extensive field for the operation of these principles, has been opened, in the new theories of the education of children. It has been lately discovered that the maxim, "Train up a child in the way he should go, and when he is old he will not depart from it," is an erroneous translation, and should read thus—"Let a child walk in his own way, and when he is old, he will be perfect." Volumes have been written and much time and labor expended, to shew that all reproof, restraint and correction, tend directly to extinguish the fire of genius, to cripple the faculties and enslave the understanding. Especially we are told, (and the system of education now adopted in the great gallic nursery of arts, is entirely on this plan) that the prejudices of education, and an inclination to imitate the example of parents and other ancestors, is the great bane of the peace, dignity and glory of young men, and that reason will conduct them, if not fettered with habits, to the perfection of human nature. Obedience to parents is expressly reprobated, and all the tyranny and despotism in the world ascribed to parental authority. This sentiment is explicitly avowed by Mr. Volney,[5] who is the friend and associate of many distinguished men in the United States, and who has, in this opinion, shewed that Paul was a fool or knave when he said, "Children obey your Parents in the Lord, for this is right."

If any person, groping in darkness, should object to these sentiments and enquire, how it is possible that children should become thus excellent if left entirely to themselves, when the experience of ages has been, that with great and continued exertions, no such facts have existed, it may be replied, the projector of Laputa had not been able in eight years to extract sun-beams from Cucumbers, but he was certain it would be done in eight years more.

We all recollect when these principles began to impress our Colleges—when it was seriously contended that the study of mathematics and natural philosophy was ruinous to the health, genius and character of a young gentleman—That music, and painting, and dancing, and fencing, and speaking French, were the only accomplishments worth possessing; and that Latin and Greek were fitted only for stupid divines, or black letter-lawyers. An indispensable part of this philosophical, and polite, and genteel, and pretty education was, to travel into foreign countries, and there reside long enough to forget all the early habits of life—to forget all domestic connexions—to forget the school-house where he was first taught his New-England primer—to

[5] A French Enlightenment revolutionary and intellectual, who while in the United States in the 1790s was denounced as a dangerous foreign agent.

forget the old fashioned meeting-house where he was first led to worship God, and especially to forget his native country, and to remember only, but remember always and effectually, that he was a polished cosmopolite, or citizen of the world.

The system of morals which has been reared by the care, anxiety, and wisdom of age, has, in its turn, been assailed by these Theorists. The language of modern reformers to those who venerate ancient habits, ancient manners, ancient systems of morals and education, is, "O fools, when will ye be wise." To first shake, and then destroy the faith of every man on these interesting subjects, has been attempted by many distinguished men, with an industry, labor and perseverance, which deserved a better cause, and has been for many years a prime object of pursuit in that nation[6] which has been the great hot-bed of premature and monstrous productions. To particularize on this subject would be impossible, but I cannot forbear to hint at a few of those doctrines now strenuously supported.

That men should love their children precisely according to their worth, and that if a neighbor's child be more deserving, it should be preferred.

That men are to regard the general good in all their conduct, and of course to break promises, contracts and engagements, or perform them, as will conduce to this object.

That to refuse to lend a sum of money, when possible, and when the applicant is in need of it, is an act equally criminal with theft or robbery, to the same amount.

If a difficulty should here be [stated], that men may judge erroneously as to the defect of a neighbor's child——the demands of the public as to the fulfilment of a promise, or the necessity for the loan in the case mentioned, the answer is ready, reason, mighty reason, will be an infallible guide. A plain old fashioned man will say, this is indeed a beautiful system, but there appears one difficulty attending it, that is, it is made for a race of beings entirely different from men. Again, says he—Why for six thousand years the love of parents to children, has been considered, as the only tie by which families have been connected; and families have been considered as the strongest band and most powerful cement of society—destroy then this affection, and what better than miserable vagabonds, will be the inhabitants of the earth?——This part of the project really strikes me, he adds, like the attempt to propagate the breed of naked sheep. Then again, it is quite doubtful, whether parents, of ordinary nerves, can, at once, divest themselves of natural affection.——Indeed, there is a strong analogy between this part of the scheme, and making a pin-cushion out of a piece of marble.——But to the cosmopolite, who belongs no where, is connected with nobody, and who has been from his youth, progressing to perfection, these sentiments are just, and the exercise of them quite feasible.

[6] That is, France.

But these modern theories have appeared, in their native beauty, and shone with the most resplendent lustre in the science of politics. We are seriously told that men are to be governed only by reason. Instruct men and there will be an end of punishment. It is true, since the world began, not a family, a state or a nation, has been, on these principles, protected; but this is because reason has not been properly exercised. The period now approaches when reason unfolds itself—one more hot-bed will mature it, and then behold the glorious harvest!

But it may be stupidly asked what shall be done in the mean time? men are now somewhat imperfect.——Theft, burglary, robbery and murder, are now and then committed, and it will be some years before the perfection of human nature will shield us from these evils. This interregnum will be somewhat calamitous.——And also, is it certain that the commission of crimes has a tendency to refine and perfect the perpetrator? These questions never should be asked at the close of the eighteenth Century.——They are manifestly too uncivil.

Again, say modern theories, men are all equal, and of course no restraints are imposed by society—no distinctions can exist, except to gratify the pride of the ambitious, the cruelty of the despotic. Hence it is the plain duty of every individual, to hasten the reign of liberty and equality. It is not a novel opinion, that men are by nature possessed of equal rights, and that "God hath made of one blood all nations of men to dwell on the face of the earth," but 'tis somewhat doubtful whether every man should be permitted to do as he pleases.——Such liberty, it may be said, is unsafe with men who are not perfect.——A cosmopolite, to be sure, will not abuse it, because he loves all mankind in an equal degree: but the expediency of the general principle may be questioned—any opinion of great and learned men in any wise, to the contrary notwithstanding.

If, however, by liberty and equality is intended the power of acting with as much freedom as is consistent with the public safety—and that each man has the same right to the protection of law as another, there is no controversy; but these terms, as now explained, advocated and adopted, mean the power of acting without any other restraint than reason, and the levelling all distinctions by right or wrong, and thus understood, they are of rather too suspicious a character for men, of ordinary talents, to admit.

But these principles extend still farther—their grasp is wider. They aim at the actual destruction of every government on earth.

Kings are the first object of their attack—then a nobility—then commons.

To prepare the way for the accomplishment of these objects, all former systems of thinking and acting, must be annihilated, and the reign of reason firmly established.

But it will be enquired, where have these novel theories appeared? I answer—They have dawned upon New-England—they have glowed in the southern states—they have burnt in France. We have seen a few projectors in Boats, Balloons and Automatons—A few philosophical farmers—A few attempts to propagate the breed of naked sheep—and we have at least one Philosopher in the United States,[7] who has taken an accurate mensuration of the Mammoth's bones—made surprising discoveries in the doctrine of vibrating pendulums, and astonished the world with the precise gauge and dimensions of all the aboriginals of America.

But in France, for many years, these speculations in agriculture, the mechanic arts, education, morals and government, have been adopted and pursued. It is there declared and established, by law, that ancient habits, customs and manners, modes of thinking, reasoning and acting, ought to be ridiculed, despised and rejected, for that a totally new order of things has taken place. All those rules of action which civilized nations have deemed necessary to their peace and happiness, have been declared useless or arbitrary, unnecessary or unjust. The most distinguished treatises on the laws of nations—treatises which have been considered as containing rules admirably adapted to the situation of different countries, and therefore of high authority, have not only been disregarded, but publicly contemned as musty, worm-eaten productions. Even that accomplished Cosmopolite, Mr. Genet,[8] who came the messenger of peace and science to this guilty and deluded people, and who treated us precisely according to those assumed characters, opened his budget with an explicit renunciation of the principles of Puffendorf, Vattel,[9] and other writers of that description, and declared that his nation would be governed by none of their obsolete maxims.

Indeed, this learned nation, have yielded implicitly to the sentiments of Mr. Volney, Mr. Paine, and Mr. Godwin,[10] in all questions of morals and policy; and in all matters of religion there is associated with them that learned and pious divine, the Bishop of Autun,[11] who had the Cosmopolitism to boast he had preached twenty years, under an oath, without believing a word which he uttered.

To aid the establishment of these projects, the credulity of the present age has become truly astonishing. There appears to be a new machinery for the mind, by which its capacity at believing certain things is perfect. It is believed that Socrates, and Plato, and Seneca—Bacon, Newton and Locke, and all who lived and died prior to the commencement of the French Revolution, were either fools or slaves. That in no country but France is there science or virtue. That the body of the people in England are now groaning under the most oppressive bondage and tyranny. That this was precisely the

[7] That is, Thomas Jefferson.

[8] Edmond Charles Genêt, Citizen Genet as he was known, was appointed French chargé d'affaires to the United States in 1793. In that capacity he tried to involve the United States in the war of Revolutionary France against England and her allies. He also acted as enthusiastic advocate of the ideals of the French Revolution and Enlightenment.

[9] Legal conservatives and traditionalists.

[10] All three held unorthodox opinions regarding religion.

[11] That is, the famous French statesman and wily diplomatic negotiator, Talleyrand.

case in Holland, Italy and Switzerland, till France introduced them to their present happy condition. It is believed by all the Cosmopolites in Europe, and by many in America—by all genuine Jacobins, by many Democrats, by the greater part of the readers of the Aurora, the Argus and the Bee,[12] and by an innumerable multitude who don't read at all, that the citizens of these States, and particularly of New-England, are miserable, benighted, enslaved and wretched dupes; and that the President and his adherents, are in a firm league to injure and destroy them. That our members of Congress, and the Heads of departments, are bribed with British gold, and are exerting all their faculties to forge chains for their posterity. That all, in any way, connected with the government, are constantly plundering the Treasury—amassing wealth—becoming independent—and thus establishing an abominable, cruel, wicked, despotic and devilish aristocracy, which is to continually enlarge its grasp, till it shall embrace all the valuable interests of America, and leave the people "destitute, afflicted, tormented." And, finally, it is believed by many, that John Adams has entered into co-partnership with John Q. Adams, his son, now minister at Berlin, for the express purpose of importing Monarchy, by wholesale, into this country: And to increase and perpetuate the stock of the house, that the son is to marry one of the daughters of the King of England.

If you enquire respecting the truth of these things, they cite Gallatin, Nicholas and Lyon[13]—They quote from the Aurora, the Argus and the Bee; and who can doubt these sources of information . . . within a year past, respecting Connecticut, this City, and our College?

But it may be asked, where is your proof that the sentiments and theories which you have been describing, in fact, have an existence? Where is your proof, Sir, that the modern Literati are attempting to extract sun-beams from Cucumbers—to travel without exertion—to reap without sowing—to educate children to perfection—to introduce a new order of things as it respects morals and politics, social and civil duties, and to establish this strange species of credulity? I reply—those who have not yet become Cosmopolites, need no proof. They have seen, and heard, and read these wild vagaries, and are therefore satisfied of their existence. As to the others, I have only to remark, that this same new machinery of the mind, by which certain things are believed, necessarily, and by the plain axiom, that action, and reaction, are equal, produces absolute incredulity, as to certain other things, and of course, no testimony will have any effect. Thus genuine Jacobins do not believe a word published in the Spectator, the Connecticut Journal, the Connecticul Courant, or the Centinel.[14] They do not believe that France has any intention to destroy the government of this country—They do not believe

[12] Jeffersonian Republican newspapers.

[13] Supporters of Jefferson.

[14] Federalist newspapers.

that our Ministers at Paris were treated with any neglect, or contempt.[15]——Indeed, some doubt whether Mr. Pinckney ever was in France. They do not believe that Italy, or Holland, or Germany, has ever been pillaged by the armies of the Republic,[16] or that the path of those armies has been marked with any scenes of calamity and distress. In short, they do not believe but that the Directory,[17] with their associates, are a benevolent society, established in that regenerated country, for the great purpose of propagating religion and good government through the world; and that their armies are their missionaries to effect these glorious objects.

And now, my Fellow-Citizens, let me ask, what effects have been produced by these theoretic, speculative, and delusive principles? France has made an experiment with them. Under pretence of making men perfect—of establishing perfect Liberty—perfect Equality—and an entirely new order of things, she has become one great Bedlam, in which some of the inhabitants are falling into the water, some into the fire, some biting and gnashing themselves with their teeth, and others, beholding these acts, are chanting "Rights of Man! Ca-Ira!"[18]

With the pleasant, but deceptive sounds, of Liberty and Rights of Man, on their tongues, they have made an open and violent war upon all the valuable interests of society.

Their own country, Italy, Belgium, Batavia and Switzerland, making together, the fairest portion of Europe, have been despoiled by the arms of these reformers, and they are now plundering the wretched Arabs.

No place has been too sacred for them to defile—no right too dear for them to invade—no property too valuable for them to destroy.

They have robbed and plundered, because they could rob and plunder.

They have conquered, not to bless their subjects, but to aggrandize the Republic, and gratify a lust of domination.

There is not a man, woman or child, whom they have attempted to render wiser, better or happier. There is not a family, a neighborhood, a village or a country, from which there now ascends, to God, one act of sincere praise, for the establishment of this new order of things, among them: but to weep and bewail their condition, is the ceaseless employment of millions.

When their conduct, from any circumstance, in their opinion, needed justification, they have resorted to that unmeaning defence, "imperious necessity." . . .

I have made these observations, my Fellow-Citizens, that we may, on this anniversary of our National existence, a day which I hope may be kept sacred

[15] This is an allusion to the famous XYZ affair.

[16] That is, France.

[17] The ruling agency in France.

[18] "Ca ira" was one of the great songs of the French Revolution.

to that solemn employment, contemplate the labours, the exertions, and the characters of those venerable men, who founded, and have, hitherto, protected this nation. I wish them to be seen, and compared with the speculating theorists, and mushroom politicians of this age of reason.

It is now less than two hundred years, since the first settlement of white people was effected, in these United States; less than one hundred and eighty, since the first settlement was made, in New-England, and less than one hundred and seventy, since the first settlement was made, in Connecticut. The place where we are now assembled, was then, a wild waste. Instead of cultivated fields, dens and caves. Instead of a flourishing city, huts and wigwams. Instead of polite, benevolent, and learned citizens, a horde of savages. Instead of a seat of science, full of young men qualifying to adorn and bless their country, here was only taught the art of tormenting ingeniously, and here were only heard the groans of the dying.

What is here said of New-Haven, may, with little variation, be said of all New-England, and of many other parts of the United States.

We have now, upwards of four millions of inhabitants, cultivating a fertile country, and engaged in a commerce, with 876,000 tons of shipping, and second only, to that of Great Britain.

How has this mighty change been effected?——Was it by magic? By supernatural aid? or was it by ingenious theories in morals, economics and government? My Fellow-Citizens, it was accomplished by the industry, the labour, the perseverance, the sufferings and virtues, of those men, from whom we glory in being descended.

These venerable men spent no time in extracting sun-beams from cucumbers—in writing letters to Mazzei,[19] or perplexing the world with the jargon of the perfectability of human nature.

They and their illustrious descendants pursued directly, and by those means which always will succeed, for they always have succeeded, those which common sense dictate, the erection and support of good government and good morals. To effect these great objects, they stood like monuments, with their wives, their children, and their lives in their hands.——They fought—they bled—they died.——At this expence of ease, happiness and life, they made establishments for posterity—they protected them against savages—they cemented them with their blood—they delivered them to us as a sacred deposit, and if we suffer them to be destroyed by the tinselled refinements of this age, we shall deserve the reproaches, with which, impartial justice will cover such a pusillanimous race.

Look particularly at the various complaints, remonstrances and petitions made by these States, on various occasions, from the first settlement of this

[19] This refers to a letter Jefferson wrote in which he spoke regarding political conditions; his words were incautious and were taken out of context to show him up as a critic of American institutions.

country, to the 4th of July 1776, and compare them with the state papers, of the great Republic.[20] In the one, you will see the plain, pointed language of injured innocence, demanding redress—in the other, the sly, wily, ambigious, camelion dialect of Jesuits, curiously wrought up to mean every thing and nothing, by a set of mountebank politicians, headed by a perjured Bishop of Autun.

At this day there exist two parties in these United States. At the head of one are Washington, Adams, and Ellsworth.——The object of this party is to protect and defend the government from that destruction, with which, they believe it threatened, by its enemies. To preserve and transmit to posterity those establishments, which they believe important to the happiness of society.

At the head of the other, is the gentleman who drank toasts at Fredericksburgh in May 1798, in direct contempt of our government, who wrote the letter to Mazzei, with Gallatin, and Nicholas, and Lyon, and to grace the company they shine, with the borrowed lustre of Talleyrand, that dissembler to God and Man. The object of this party is to destroy ancient systems—ancient habits—ancient customs—to introduce a new liberty, new equality, new rights of man, new modes of education, and a new order of things.

Let them meet and make a full, fair, and perfect exposition of their principles—their objects, and the means by which they are to be accomplished—And let there be present at this display, the departed spirits of Davenport, Hooker, Winthrop, Wolcott, Hopkins, Haynes, and Heaton, and let there also appear a Lawrence, a Warren, a Mercer, and a Wooster,[21] and to which of these parties would they give their blessing?——For which of these causes, if it were possible to bleed and die again in the cause of America, would the beloved WARREN AGAIN BLEED AND DIE?

7. A French Émigré the Legacy of Enlightenment Enthusiasm and Achievement. 1801

FELIX PASCALIS (1750–1833) here presents a survey of science at the turn of the nineteenth century as seen from Philadelphia. In general, he focuses on France—especially French science—and a view of the rela-

[20] That is, France.

[21] These were all New England fathers and heroes.

tionship between science and liberty that contrasts with that of Daggett. In specifics he surveys current problems: how physics had dealt satisfactorily with gravity but not with heat; how the chemical revolution had come, leaving organic chemistry as a major field for exploration; how physiology was in a state of fundamental change.

This versatile man was also known as Felix Pascalis-Ouvrière. A physician educated at Montpelier, he practiced in Santo Domingo until the slave insurrection of 1793 drove him to Philadelphia. He lived and practiced there for seventeen years and then went to New York, where he lived out the rest of his days. In Philadelphia he was active in the short-lived Chemical Society of Philadelphia before which, as vice-president, he delivered this oration.

See *Dictionary of American Biography,* XIII, 286–287; Edgar Fahs Smith, *Chemistry in America, Chapters from the History of Science in the United States* (New York: D. Appleton and Company, 1914); Windham Miles, "Early American Chemical Societies," *Chymia,* 3 (1950), 95–113; William S. Middleton, "Felix Pascalis-Ouvrière and the Yellow Fever Epidemic of 1797," *Bulletin of the History of Medicine,* 38 (1964), 497–515.

ANNUAL ORATION, Delivered Before the Chemical Society of Philadelphia, January 31st. 1801.*

Felix Pascalis

Gentlemen of the Chemical Society,

I come, this day, to fulfil the honourable task of addressing your society, on subjects relative to Chemistry.

Permit me, at first, to congratulate you on the return of another year to be added to your commendable exertions for the improvement of a science which constitutes true philosophy, and imparts so many advantages to enlightened and polite nations.

Of these none have remained unknown to you, for as soon as this institution was formed, by the talents and diligence of its members, it stood adequate to the experimental researches pursued by the celebrated schools of Europe.—Less censurable than those venerable seminaries of learning, which have spent so many years in false theories, in idle and useless systems, you have had the satisfaction of participating with them, in their discoveries, as early as your members could controvert, with the learned of all the world, any of the subjects or causes of the revolution that that science has experienced, within these few years. By whatever consideration personal praise could be waved [*sic*], in your assembled Body, no doubt, it remains deserved by those who early promoted the sedulous cultivation of the most useful science. I subscribed to this laudable intention when I had the honour of being called

* Felix Pascalis, *Annual Oration, Delivered Before the Chemical Society of Philadelphia, January 31st. 1801* (Philadelphia: John Bioren, 1802), pp. 3–31, 37–48. Slightly abridged.

as a member, in your institution. But I scarcely can confide in myself, this day, when by your appointment I am to display some of the admirable laws and numerous advantages of Chemistry. Was I adequate to the task, I would publicly declare that I became so among you, and after many years of our scientifical intercourse.

Necessity was the parent of our science in the most distant ages that historical records can trace; from avarice and cupidity it afterwards received some slow and obscure improvements, but accurate analysis only has lately brought it to perfection. The ancient history of Chemistry offers such a lamentable view of ignorance, superstition and empiricism, that its pages seem no more useful but to prove how laborious, slow and uncertain is the advancement of human understanding, unless it is aided by the correct results of observation and by an unprejudiced love of truth. Happy is our age, in which at last, we are acquainted with the elementary laws of existing bodies! Those laws which extend to all material objects, visible or invisible [,] known or still concealed from our observation; those laws, the limits of which, we do not not know, because we cannot trace where the limits of nature are to be marked; those laws form and constitute the science of Chemistry;—indeed by its principles it is connected with all the branches of natural philosophy, and by its comparative results, it dictates the rules of arts and the processes of manufactories. Under this twofold view, Chemistry [,] embracing all wants and comforts of mankind, is now to be contemplated as the most important and interesting subject, altogether to do honour to your pursuits in its study, and to encourage many more votaries to the acquisition of its numerous advantages.

You know, Gentlemen, how numerous are the branches of universal Philosophy! they form like a beautiful tree, which often has been drawn by the hand of genius; under its shade, all sages vied to find shelter or repose, and of its fruits they all wished to partake. When all these branches have been severally examined and studied, they appear so well connected and so much depending on the same laws, that they compose but one science, the mysteries of which cannot be disclosed, unless we enter into the laboratory of the chemist, and there we explore its processes. From these, all sciences in the physical order, must receive their tenets and elementary doctrines. Would the Astronomer refuse to witness our invariable results of attraction and gravity, because we cannot explain their operations, as he does for the motions of the rolling sun, and planets, by the *direct ratio of masses,* and by the *inverse ratio of the square of distances*: we must then propose some difficulties arising from the known characters of Caloric.[1] That this element is not ponderous, and has no gravity to any given centre of density, that no law but that of a projectile power, can be attributed to it, is beyond doubt. How

[1] Caloric was at this time believed to be the substance that constituted heat. Caloric was thought to be a chemical element but without certain properties of other elements; it could, for example, be added to, or taken away from, other substances without adding to their weight.

is it then, that all the bodies of the planetary system, the greatest number of which are solar or ignited, could equally obey the same law of the direct ratio of masses, and of the inverse ratio of the square of distances? Newton had calculated that the comet of the 8th of December, 1680, when in its perihelium[2] had received a heat 2000 times greater than that of red hot iron. What matter, we ask, can be conceived to exist, as a centre of gravity, at a still more and infinitely higher degree of heat? The known laws of nature leave no room to any conjecture, except to that of an elementary and homoggeneous fire, which cannot exert but an immense projectile power, that of course, it excludes entirely any share of attraction, by the direct ratio of masses, and the inverse ratio of the square ratio of distances. A conclusion more than probable is therefore to be drawn, that to chemistry is perhaps reserved to disclose a different order of the primary laws of the universe.

In this, as well as in other investigations, the Philosopher must be aided by the results of experimental chemistry. Any proportion for instance, of the attributes of matter—is it well understood, unless the laws of Caloric are correctly defined? Caloric! astonishing principle of destruction and life! To describe well its activity and operations, it would be necessary to advert to mountains which it undermines, to the frightful craters it opens on their most elevated regions, and to the immense torrents of lava with which it inundates afterwards cities and empires. We might thence follow its burning streams into the abyss from whence it again breaks open its barriers, lifting mountains, raising islands, or if at liberty, uniting to water, boiling up in the shape of whirlwinds or clouds which darken the firmament. This indestructible agent then divides itself in various scintillating meteors, or by its sudden combination with air, forms the lightning, and by tremendous electrical detonations, spreads terror and devastation among mankind, threatens nature with the perturbation of all established order. However, you may master Caloric, Gentlemen, you may attract it from the regions above, under your lens and concentrate it in your crucibles, where to your command it will melt or volatilize the hardest metals, reduce rocks to the elasticity of clay, and clay itself to the adamantine hardness: where it will be disposed by your processes, into fixed and opaque substances, to arrange their pores for an easy passage of the light!—With you it will create as it were aerial and invisible bodies, among all the known substances you can enumerate and those that can be suspended in a gazeous state. After such comparative results, Philosophers may explain elementary laws concerning the *attributes of matter;* they are able to explain likewise the revolutions of seasons, and the phoenomena of the atmosphere. With the capillary tubes of observation they may measure spaces, density of air, and gravity of fluids. How often scolastic philosophy blundered on those simple subjects, when chemistry had not yet controled all other sciences! The first who measured the parabola of a

[2] At its closest approach to the sun.

bullet launched into the air, by the thundering explosion of a cannon, attributed the almost incandescent heat of the metallic globe to the detonating mixture, because to chemistry only it belonged to demonstrate the power of friction in disengaging heat from its latent recesses. Other erroneous doctrines on that noble agent of nature which so often has been mistaken for a modification or produce of matter, have been exploded!—The supreme laws of the existence, and diffusion of Caloric in nature, are its necessary tendency to equilibrium with external temperature, and its power of overcoming the cohesion of particles, to satisfy its affinity with them.—But let us not pass over that other phenomenon of affinity, unnoticed. Without the knowledge of its laws, Philosophy would be mute at the view of the stupendous works of nature, or she would be obliged to conceive as many deities as there are prodigies in the creation.

This [affinity] is the great power which arranges, unites, and hardens homogeneous particles of matter, and exerts itself likewise with due proportion, upon heterogeneous substances. Under the heavy foundation of mountains, in the deepest subterraneous cavities of the earth, in the bottom of the seas, no where it finds obstacles sufficient to oppose its operation, except Caloric accumulated could suspend their effects, to a certain degree. Yet distance can truly impair or weaken the power of affinity: aproximation of course will augment it. In itself, it is composed of many tendencies, but these remain incommensurable. Their insulated effect is nul, and as soon as they are simultaneous, they become effectual. In fine affinity, which is attraction, adhesion, cohes[i]on, and aggregation, by the immediate result of elective power, is the primary cause of dissolution and decomposition. Now let the Philosopher[3] listen to the laws of affinity, and the minutest circumstances will be explained, from the spheric form of a bubble of air, or of a drop of water to the various degrees of elevation of fluid, in capillary tubes, or to their pressure, by the same, both effected in inverse ratio of the squares of diameters. To these we may add the phenomenon of blaze in ignited bodies, the mechanism of breathing and renewing animal heat, by the decomposition of common air, but more especially, the effects of the refraction of Light. The Philosopher may, no doubt, well explain how the density of the rays of light is like the square of the distance from its focus; how it is transmitted to the retina; in what angle it is reflected, and at last refracted;—to these elementary laws of Optics, Catoptricks and Dioptricks,[4] the chemist adds a question ready to answer it; What is light? Light, which adorns the whole creation by an infinite variety of colours, because, each colour is light presented in a different angle of refraction or refrangibility; light is the element of all the worlds, of innumerable solar systems and comets; Light and Caloric are as often united, as vital air and Azot[5] are aggregated to form the

[3] That is, natural philosopher or physical scientist.

[4] Branches of optics dealing with reflected and refracted light.

[5] That is, oxygen (vital air) and nitrogen (azot).

atmospheric orbit of the earth; yet, caloric may be present without light, and viceversa: light ranks the first among the elements of nature by its tenuity and elasticity, for more rapidly than any other, it can traverse the immensity of spaces. It cannot be said to be like a modification of a diffused subtile matter, because we trace substances, with which it has an elective effinity; it stimulates animated bodies, and to vegetables, it is as a last component part necessary to develope their succulent juices, their splendid flowers, their robust fibres, and their luxuriant colours; No paradoxical doctrine will ever be able to erase from the books of Philosophy such truths inscribed in it, by the hand of the chemist.

Behold! another field is open; but to the most ingenuous philosopher it will appear a dreary solitude, an immense desert, until Chemistry discovers in it, all the elements of the productions of nature. Air, that invisible fluid, but so sensibly preceived by our organs, has long been the object of innumerable investigations, yet in our days only, they have been successful. The learned among the ancients, filled up their atmosphere with their Genii, to explain contrary effects of that invisible orbit; and among moderns, it has long been a desideratum to account for an element which appeared necessary to fertilize the fields, and to nourish life, while it was dreaded as the destroyer of agricultural labour, or the vehicle of contagion and death among mankind. The attributes of its particles were still more incomprehensible, when it was considered that with a rapidity to be computed only by the imagination it could transmit light and sound, although they would be agitated in a thousand contrary directions. The pressure which it exercises around the globe, to the center of which it gravitates, was, at first, discovered by Toricelly, who traced to what height a column of fluid could be equal by its weight to the superior pressing column. More of its comparative gravity with other fluids was still better defined, when Monge declared, that if the whole atmosphere could disappear, all the liquids on earth, would suddenly rise, be converted into vapour, and form another atmosphere; but all these progressive views scarcely unveiled one corner of a more extended view of the laws of nature. The chemist attempted the analysis and sinthesis of atmospheric air, and it was performed; moreover, he found out, and demonstrated, that its component parts could be equally concrete and fixed in organic bodies, or combined with other various solid substances; it has then been proclaimed that the Oxygen and Azot constituting common air, were nothing but another modification of matter itself, and a continuation of the chain of existing bodies. It was no more problematic afterwards, to ascertain how this fluid administered life to us . . . or how it effected the most principal changes and modifications in the animated or [inanimated] creation. The pneumatic observer could soon disengage the component parts of atmospheric air, from common materials, and imitate a new creation. Let honour be given to the memory of the great Newton; he was the first who predicted the prodigies of our laboratories: *Si se tangerent* said he, *particulae aëris—aër evaderet in marmor,"*[6] but without the science of Chemistry, that oracle could never be explained, and philosophy could not advocate as a true proposition, that the atmosphere is the last reservoir where the elements of all bodies are ultimately received, and from where again they are substracted to create bodies and to support organic life: there is demonstrated another axiom long ago proclaimed by Lucretius, *"nothing is annihilated in nature:"* As soon as she has effected a dissolution, she exhibits in her very bosom the consoling power of a new creation!

But from the elevated regions of Heaven, and from all the primary elements and phenomena of the world, the mysteries of which have been disclosed by Chemistry, let us more minutely fix our attention on the surface of our globe. A more admirable view of the analysis and combinations of all existing bodies; their formation, growth, alteration and destruction, their treasures and deleterious qualities, every thing will be unveiled by our processes. More evidently we will then enumerate the infinite improvements which are applied to arts and manufactures by mineral, animal, and vegetable Chemistry.

You are fully acquainted, gentlemen, with the difficulties and confusion which attended the science of mineralogy, when laborious researches and observations were transmitted through external and erroneous characters. Some such as Gellert and Wallerius could not distinguish any thing else but vitrescible matters, or argillaceous or apyres, or alkalines and calcarious:[7] others would prefer the general division of earths, sands, and stones. Some simply divided minerals into earths, salts, combustible and metallic substances. As late as the year 1784, the famous Daubenton, took another erroneous classification, consisting of negative characters, of insoluble, incombustible, metallic, then transparent, and crystallized, smooth, &c. Now we may observe, that these descriptive methods, could not add the least improvement to science, and that they contributed to many erroneous assertions. It is not, for instance, by the presence of a metal that the ponderous spar is to be judged, since its basis is a primitive earth, and if this is a Tungstat of lime, or if the Wolfram[8] itself (which is thought by many, another kind of primitive earth) is nothing but an ore of Tungstein, we are to be determined in these various opinions by accurate analysis only, and in no ways by external characters. Were these to fix our mineralogical definitions, by what given habitudes, would we be justified to class the beautiful fluoric spars of Derbyshire, or fluat of lime, among insoluble neutral salts? Error would be still more egregious and unavoidable, between the Sulphures of Molybdena and the Carbures of iron, which are so absolutely resembling each other. It has been found, that a Borate of lime may spark and scintillate as well as quart-

[6] If the particles of air should touch one another, then the air would escape into marble.
[7] That is, glass producing minerals, clayey minerals, and limes.
[8] That is, tungstate of iron and manganese, often found in tin mines.

zose substances; that metallic carbonates must effervesce with acids, in short, there is not one descriptive method, but it will in many points, lead us to contradictory facts. So sensibly was Daubenton convinced of their deficiencies, that he expressed his wishes for a method founded on analysis of constituent principles. Fourcroy begun it in 1780, but the Scyagraphy of Bergman, still better demonstrated the superiority of the method of constituent principles.[9] This doctrine was embraced by Monge who retained external character for varieties only; by Kirwan of England, Werner, Debern, Cromstedt and Chaptal; thus the science of Mineralogy was forever added to the dominion of Chemistry.

With this conquest, we will not however consider it as the chief point in Chemisty, to form theories on the primitive state of the earth, to explore the ruins of extinguished volcanos, nor the most antique works of nature, from the grotto of Fingal to that of Antiparos. We will not indulge to the idle curiousity of enumerating the causes of the granitic ridges of mountains and of the calcareous secondary ones; we may dispense with contemplating whether the ocean has once been an atmospheric orbit, during the original conflagration of the Earth, or how, after its condensation it insensibly retired from the polary regions, and elevated plains of Tartary, to its [P]acific and Atlantic bed, leaving every where immense masses of its animal and vegetable productions. However useful to science, these mineralogical essays and deductions would be, with more real advantage, the chemist will be contented to analyze mineral substances, and to procure those which satisfy our wants, constitute our wealth, augment our comforts, which enriches our arts, and manufactures. Among these, Nitre may justly occupy the first rank. Any thing relative to the production of that precious saline matter is interesting as much to save expensive importations, as to perfect manufacturing processes.—Let me remind you of that awful period of the French revolution, when the fate of 30 millions of people, divided, confused and famished, ruined and surrounded by sea and land, seemed to depend on a sudden formation of saltpetre, that only means of defence, which was exhausted. Various ingenious ways were devised by the learned,—saltpetre was formed,—and the tyrants phalanx being dispersed, Liberty was triumphantly obtained, thanks to analytic science.—To all the known processes which may procure Nitre either from artificial beds, or lixiviation[10] of certain substances, we now may add that of the decomposition of common metallic oxyds, and of ammoniac; but the most surprising, if further experiments could evince its efficacy, would be that of procuring the formation of nitric . . . acid by an only tenfold *compressibility* of the aggregate Oxygen and Azot composing atmospheric air. Various attempts with ingenious apparatus promise great success, and if Citizen Guyton is not disappointed in his expectation, with gratitude, Chem-

[9] This is a reference to a 1782 book in which Bergman advocated the study of minerals in terms of chemical composition as opposed to, for example, crystalline structure.

[10] Leaching, usually of ashes, to obtain lye.

istry will receive from him the power of accelerating the operations of Nature.

It was during that lamentable period when the friends of Liberty were with importunity, compelling nature to supply them with some new means of defence against their foes; that the discovery of Berthollet, on the tremendous fulmination of Oxygenated muriat of Potash[11] were again resumed.—In this age of reason and benevolence, let us never boast of increasing the power of destruction, since the prevailing Philant[h]ropy exerts itself against any system of warfare. We only remark therefore, that from various habitudes of the *Hyper*-oxygenated muriat of Potash, the most surprising effects are to be reckoned among the laws of nature. If you take about 20 Centigrammes, or 3 grs. of that saline matter, with one third of pulverized Sulphur, you may by a slight trituration, produce several detached detonations: but wrap up the mixture with paper, put it between an anvil and a hammer, strike a blow and the detonation will be equal to that of a cannon, surrounding you with a purple blaze, and white smoke.—With several other mixtures of that salt, shock or percussion will equally operate with the most tremendous power. This mechanical effect is therefore equal to that of caloric, or of fire communicated from one body to another. This singular phenomenon had been witnessed, although in a very small degree in the gun powder, but it had not been attended, in this point of violence, that by compression, the particles of Oxygen could unite to those of inflammable bodies, and form an abundant gazeous fluid to which a great quantity of caloric, gives such electricity as to strike on the air with an unparralleled violence.

Under the unremitting exertions of chemists, Mineralogical enquiries have developed as many curious facts as there were improvements to be added to all known processes, for obtaining muriatic acid, for cheaply crystallizing the sea salt, for extracting the Soda, for preparing the muriat of ammoniac, for purifying all combustible minerals that can be converted to our use, for preparing metals, and converting them to the most useful purposes, for adding new ones to the treasures we were already possessed of; but one of the discoveries of the latter kind, is too remarkable not to be fully detailed here.

Long before the Phlogiston of Stahl[12] had been exploded by the new analysis, and in spite of all the beautiful experiments of Bergman and Priestley, nothing very certain was known on the fabrication of various sorts of steel. Not longer, than twelve years ago, it was discovered at last that carbone only, in due proportion, would constitute various sorts of fused iron, and of steel: But the English who had been long in possession of a more perfect kind of cast steel, could not in the least be disturbed by the repeated re-

[11] Potassium chlorate.

[12] A widely accepted theory for a long time, according to which combustion can be explained by the presence of the substance phlogiston in all flammable materials, combustion consisting of phlogiston leaving the burning material. Chemical theory following the discovery of oxygen finally superseded Stahl's theory.

searches of Vandermonde, Monge and Berthollet. The latter had ever declared that he was at a loss to account for the character of that precious metal. But a mystery which had been perhaps fortuitously discovered, which avarice and cupidity had so long concealed, was at last disclosed by the ingenious Citizen Clouet,[13] and made public with a liberality, which, altogether honours science and the National character, the whole secret was I believe, found out by analogy; as different quantities of fused iron and of steel were known to be the result of various proportions of Carbone and of few vitrescible substances of the original ore. It was therefore concluded that a more intimate union of that element, and of the purest clay, with the best iron, effected spontaneously at a due degree of heat, could form the famous English cast steel. Liberal private means, and even national expenditures have been largely and sedulously applied to numerous experiments, at the wind furnace of Macquer, to the 150th degree of Wedgwood's Pyrometer. The results, Gentlemen, have magnificently equalled the most admired steel of Huntsman and Marshall. The new theory of our chemistry has besides acquired another illustration of the elective affinities of iron, for carbone, since the Carbonate of lime, or marble in dust has been offered to it. The fixed air[14] has been decomposed, and the results have been equally successful. So much for the wonderful secret of the cast steel of Sheffield, Yorkshire.

Before I terminate this article of Mineral chemistry, let me remind you, gentlemen, of the great advantages obtained by other nations, from the class of alumines[15] only. Among the compounds in which that primitive earth is predominating, they have found all the materials from the common tile and brick to the most elegant works of Porcelain. Not only various kinds of clays have rendered infinite services for the commodities of life, but they may be turned into the best instruments of arts and utensils of manufactures of glass. Some clays have been found likewise to be good manures, and others are an excellent substitute for soap in the Fuller's art.

That America possesses such treasures, it is needless to prove, and that they have not yet been applied to the use of the community, it is also an object of regret. Do, gentlemen, let a liberal patriotism animate your scientifical pursuits, among you who are to see the glorious days of the trans-Atlantic Republic. Encourage and repeat miner[a]logical experiments on all kinds of Alumine, the first who will successfully procure manufactured works of the kind and tolerably good earthen wares, will deserve well of his country and be rewarded by the gifts of fortune.

By offering you, a few more observations on Vegetable chemistry, I hope I will further illustrate the dominion of that science over all the branches of

[13] Clouet heated wrought iron with diamond powder and thus discovered the secret of making steel, namely, controlling the carbon content.

[14] Carbon dioxide.

[15] At this time the common occurrence of aluminum compounds was known, but Pascalis in general here was referring to the clays from which alum rather than aluminum was derived.

natural Philosophy. Indeed, not only the most interesting materials upon which we operate must be obtained from vegetables through the processes of combination, but the chemist moreover is called upon to explain the phenomena of vegetation.

Vegetable [chemical] analysis has offered us three elements only, in vegetation; Carbone, Hydrogen, and Oxygen. The number and proportion of these principles will afterwards be sufficient to answer any question on different other results of vegetation.

Observation demonstrates that water and Carbone are the real nutritive principles of vegetation, that is to say; Hydrogen and Oxygen are afforded to them by the decomposition of water, while Carbone is procured from the decomposition of animal and vegetable matter. All this is confirmed by the analysis of the fibrous part of plants which are a mere aggregate of Carbone. But in what manner is Carbone carried into all the parts of a plant or of a tree? by what means can it circulate in them? by what solvent is it rendered the precious food of vegetables? The solution of all these problems, gentlemen, is provided by correct and incontrovertible experiments.

Pure charcoal such as it is left in our hearths, or mixed with any kind of pure and dry earth, could not certainly be spread on the ground, and depended on as the best manure. But on the other hand, do not we know that dead vegetables (from which much carbone can otherwise be procured by combustion) when relaxed or softened by maceration and putrefaction, are indeed the best materials for an excellent manure! How shall we account then, for the striking difference of pure carbone being ineffectual for vegetation, and of compound[s] of carbone becoming so evidently necessary to it? why, in the latter case, that element is truly held in solution, by oily, extractive, alkaline and resinous vehicles. Now water, which has the faculty to dilute those natural combinations, becomes itself the solvent which carries carbone through all the system of vegetation, and by which nutrition and digestion are accomplished.

To proceed to the last stage of formation and growth of vegetables, let us mention, that air, and perhaps azot, caloric, acids, motion, and even rest, may suffice to precipitate the carbone. These other eventual agents therefore, which every where, are found active, are those that effect the concretion and growth of any fibrous matter—these are wholly explaining the principles and mysteries of life in vegetation, because they support and animate its organs, distribute the nutritive matter, modify the action of perturbating causes, preside over all the operations of that living laboratory of nature, just as the Chemist directs the operations of his own, and changes the results by altering the form or number of his reagents.

These general principles being established by analysis and sinthesis, who can deny that Chemistry teaches the natural philosopher, how in the system of Nature, primary substances can be arranged so as to form organized bodies, ... and from thence, obtain such attributes with which we were formerly unacquainted, active attributes! which we never dare assigne to matter! But if nutrition, growth, and decomposition of plants, are explained by the laws of chemistry, how much better any precepts respecting the choice of their soil, climate and temperature, of their management and multiplication, will be derived from the same science? not only agricultural rules are connected with Chemistry, but it dictates likewise a kind of vegetable Medicine, which has, its institute in Agriculture, of Hygiene, of Clinics and of Theurapeutics.

Vegetables as objects of analysis are to be considered as containing substances and compounds necessary to our pleasures and comforts, to arts and manufactures. Such are mucilages, oils, rosins, fecula, gluten, Sugar, various acids, colouring matter, wood, extractive substance, the Aroma, and the Oxygenous gas which all vegetables emit by their excretory organs; of these subjects I can now notice but few.

It was very difficult formerly to define which of the component parts of vegetables, would afford a wholesome and nutritive food. The mucilage which exists in them, variously elaborated with acids gums or sugar, has been at last pointed out. Analysis has even traced that food in the abundant and pulverulent fecula[16] which constitutes the greatest quantity of substance in our grains, roots and seeds, the same principle admirably combined or deposited in different organs of plants is the very means of their propagation, growth, and fructification. In whatever plant it is more or less abundant, in what parts it can be more or less concentrated, it cannot escape the analysing investigation of the chemist who might extract a wholesome food from rejected vegetables and insipid roots. . . .

Passing through a great number of discoveries and useful results, for which we are indebted to vegetable chemistry, I come to the last part of our division, that of animal Chemistry.

With that definition, you anticipate me in the series of late discoveries and extensive improvements which in all its branches, Medicine has received from Chemistry. That science which directs all its pursuits to the preservation of health and life, to the relief and cure of human diseases, had been reduced during many ages, (and by the most unaccountable fatality) to absurd systems, or had been composed in its institutes, of few good precepts, and so few accurate observations that it was inadequate to the wants, or little entitled to the confidence of mankind. About the beginning of the last century, however, the institutes of Medicine appeared to be connected with as many of the physical laws, as can be applied to Physiology. Yet various contradictory systems continued to be advocated, without much retarding the progress of that science. But alas! in a period not much remote from our days, it has been again confused with metaphysical entities that have no connection at all with the laws of nature, and which as attributes of animated matter are unintel-

[16] That is, starch.

ligible, erroneous and absurd. At last, in imitation of chemistry, the spirit of analysis has prevailed in all the branches of natural philosophy, and consequently the friends of the healing art, who wished independently and usefully to pursue the career of their labour, have renounced all logical systems, and composed their institu[t]es of medecine of such facts and aphorisms of the ancient and modern, that experience had rendered incontrovertible with all the results that physical laws and analytic animal chemistry, could consistently offer to their medical investigations.

But here, Gentlemen, let me bear evidence against the unrestrained spirit of novelty and the unphilosopical theories which of late have been mistaken for physiological and Chemico-medical improvements. . . . Indeed those Philosophers of Germany who pretended once to explain the case of a child born with a golden tooth, without having previously ascertained the evidence of the fact, were no less ridiculous than new theories deserve to be thought, when they pretend to explain animal irritability[17] by Oxygenation, and to cure diseases, with certain gases,[18] without having in the least established how much and in what respect the attributes of animated matter, could be assimilated to any of the laws of elective affinity, nor to the habitudes of the elementary principles in nature. No, no, the Loco-motive power of life, or rather, animal irritability has nothing in its effects that can be connected to those of the substances that we can torture by chemical processes. The simple contractibility of a muscular fibre, or the active secretion effected in the cavity of a viscus, will ever baffle our enquiries, and all the vast comparative systems, which we would derive the most ingeniously, from the laws, habitudes and combinations of the existing inanimated bodies. But if theories on organized and animated matter, or physiological causes of life and health, and on the origin of diseases, are to form a book of aphorisms perfectly distinct and different from the elements of Chemistry, there are however, certain points of contact, between the functions of our organs, and all the laws of nature, in which, much good has been done; important discoveries have been made and many more are to be expected, in medical science, from Analysis and animal Chemistry. To it, for instance, we are indebted for the knowledge of that kind of combustion of atmospheric air, which is effected in the lungs, of the Oxygenation of the blood, of the origin of animal heat, and of interesting conjectures, supported by experimental observations concerning the causes and characters of malignant fevers, and of their inflammatory or anomalous symptoms. To these great acquisitions we must add the Analysis of animal solids and fluids, the component parts of which have been admirably enumerated, and likewise some mysteries of their growth, distribution and final dissolution. Of course, new Pathological views have been

[17] Animal irritability refers to the theory current then that living animal tissue had the property of responding to irritation (typically, a muscle contracted), which was the fundamental element of being alive.

[18] Most amusingly, nitrous oxide (laughing gas).

judiciously offered, respecting certain visible or invisible agents, which cause perturbation of animal life; among them, deleterious gases, by their operation on animal irritability, stimulant, sedative or poisonous, have really disclosed a long series of our diseases. Animal Acids chiefly, and other primary combinations in the blood, in the bile, in the bones, in earthly concretions and others, do form, Gentlemen, the most precious collection of facts and observations, that ever medical science could be improved with, for the relief and cure of a great many diseases. Let me mention one only, to prove the useful applications of chemistry to the science of medicine.—The great difference of characters of the Phosphoric acid, obtained from human bones, and of that produced by the deflagration of Phosphorus, lately induced Vaucquelin and Fourcroy, minutely to investigate the habitudes of that Animal substance. . . . Phosphoric acid has . . . the power of holding in solution the Phosphate of Lime, and strange to say, in that state is but partly attacked by mineral acids, while it entirely yields to the power of vegetable acids. Therefore, Phosphoric acid, must be powerful enough to soften, decay and distort the bones; and to its superabundant presence only, such disorders in those solids must be ascribed.—Moreover, if by any cause whatever, the usual secretion of Phosphoric acid effected through urine, is interrupted, the consequences will necessarily be injurious to the very support of our frame, by attenuating the Phosphate of lime of the bones, by causing it to deviate in our fluids, or in membranous parts, where it will accumulate, or obstruct and torture the most delicate organs. Now, such is, in this instance, the help we have received from Chemistry, that we may oppose to the greatest ravages in animal oeconomy, if we know which is the superabundant principle that must be counteracted: and thus, in an infinite number of other cases, Chemistry teaches the Physician, on what necessary combinations animal functions are depending, by what assemblage of substance, Providence has marked the order of succeeding periods of Life.—I had almost said, even of Death; because, when that other modification of existence takes place.—Chemistry still more triumphantly can operate on the remnant of a Divine work, disunite all its aggregate compounds, and trace their particles to their original elements!

This is not all; it is interesting to consider a moment with what simplicity and regularity, Chemistry has in general, composed our precepts of Theurapeuticks and methods of Pharmacy. How long and how often the exhibition of remedies has been confused by empiricisms, embarrassed by ignorance and endangered by avarice! The learned of all ages ever lamented the almost inseparable evils of that branch of the healing art, but their regrets or their cares were ever inadequate to the accurate knowledge of the virtual properties of remedies, and of their proper classification. Behold! now Chemistry has swept off all the dregs of quackery and ignorance; it has detected the imposition of useless compounds, the fallacy of celebrated nostrums, and has exposed the danger, or the inutility of wondrous specifics which had been handed down by a credulous or fanatic care, as the most powerful agents in the cure of diseases. The Revolution in Science, as effectual as that in the political order, has equally silenced every kind of assuming authority, and of fanatic delusion. The universality of Analysis among all the productions and bodies in nature, has traced, all laws, all virtual properties, and almost all possible combinations. Away, therefore, with the vender of nostrums, who cannot rank among philosophers and chemists; away with the physician whose incapacity or equivocal qualifications could have been formerly usurped under the garb of Science, and of literary titles. These, and all propagators of evil and errors, cannot stand the test of science, because like truth, science has impressed its features on the physiognomy of its votaries. Science is possessed of its own language, and that of chemistry, which has substituted its nomenclature to the absurd and unmeaning definitions of the old, is the last trait, gentlemen, which distinguishes its adepts from the vulgar or unskilled, as much as it guards its discoveries against any unfounded innovation, and exalts its supremacy, among philosophical sciences, by the language of truth, which belongs to it alone.

Here I conclude, Gentlemen, an imperfect survey of the improvements procured by Chemistry, to Philosophical Sciences, Arts and Manufactures. I may, very justly say, that society has received more real advantages from it, in a few years, than during all the preceding ages of ignorance, or of imperfect knowledge of the Laws of Nature. I could not advert to every interesting view of that Science, within the short space of time I have fixed, and perhaps, fatigued your attention.[19] With regret thus, I have almost omitted to describe its flourishing cultivation in the Universities and Colleges of this great Republic. The splendid talents of several of their Professors, have still more promoted the sedulous emulation and the distinguished abilities of many students, in their numerous classes. But in the name of your institution, I must notice that in the Professor of Chemistry of this university, our worthy President,[20] that science has not only gained a strenuous vindicator of its doctrines, but also a liberal inquirer after truth, an elegant, and successful experimenter.—You remember what a great man he has had to contend with; Priestley to whom our science is so much indebted, and whose opinions and experiments are to be consulted, in any of the elementary processes of our Laboratories; Priestley, who commands respect in his Chemical controversies, because, as long, as some mysteries in nature will perplex the Philosopher, he is entitled to the same degree of evidence, which he has exhibited in his doctrines; Priestley, the persecuted friend of Liberty, of Religion, and the model of all social and private virtues. Honored is our

[19] In the Philadelphia Laboratories new experiments have been lately instituted, relative to the tremendous effects of the fulminating mercury. The galvanic influence also, that astonishing phenomenon of PERPETUAL MOTION, has been minutely investigated by Prof. Woodhouse, both through the METALLIC PILE and the CHAIN OF CUPS OF VOLTA. [*Footnote in original.*]

[20] James Woolhouse.

Society with such Members; congratulated is the Republic, with such Citizens, and happy is the rising generation with such philanthropic examples, which have already opened to you the Golden Æra of SCIENCE and LIBERTY.

Part 2

THE NINETEENTH CENTURY

For science, the nineteenth was the grand century. Reinforced by industrialism and the wealth produced by industry, science developed from a philosophical diversion into a specialized occupation, from probing into nature into a way of life itself. This was the period of great discoveries—the principles of thermodynamics, the periodic table of the elements, Faraday's discoveries leading to the dynamo, radioactivity, the physicochemical basis of physiology—and of great theories—geological uniformitarianism, atomic theory, and, above all, the theory of evolution. Only the latter involved American science in a major way and, in general, scientific thinking in the United States remained derivative from that of the great European scientific workers. American scientific endeavors—outside of applied science—seemed in most fields to be amateur, second-rate, and starved. Physicist Joseph Henry really was no match for Michael Faraday nor biologist Henry Fairfield Osborn for Hugo de Vries. Yet behind these facts, and in the exceptions, lies a significant story, for by 1900 American science was flourishing, decidedly, and the basic institutions within which it would operate were well developed.

Two kinds of specialization formed American science in the nineteenth century. On the one hand, scientists became full-time professionals and specialists within the realm of science (geologists, biologists, and eventually botanists and paleontologists, for example); on the other hand, scientific thinking became an explicitly independent venture, and "science" a distinct category of human endeavor.

This latter specialization had the effect of removing, for many decades, philosophical questions from the domain of science and, in so far as possible, scientific questions from the domain of philosophy and religion. This separation in turn had two effects. First, science could pass into the hands of the religiously orthodox without suffering damage, and this is exactly what happened in America. Where in the early nineteenth century there were attacks on science, the object of attack was eighteenth century Deism and rationalism, not learning about nature and the laws of nature. In the end, the revival of orthodox religion that occurred in the romantic period early in the century turned out not to be inimical to the growth of science but irrelevant to it, at least until mid-century. Mechanism and even materialism were permitted to flourish—as long as orthodox religion remained unchallenged, or not challenged explicitly. Scientists could freely show that man himself was composed of ordinary chemical substances and functioned mechanically according to the laws of nature, as long as no one denied that each man had a soul. The sentiments of the mechanist Timothy Walker, for instance, were understood to be entirely orthodox.

The orthodox were careful to make sure that science stayed in the hands of their own. Thus President Dwight of Yale, instead of recruiting a foreigner for a professorship of chemistry and natural history, chose one of his own young graduates, Benjamin Silliman. Silliman's religious views were irreproachably orthodox, but he had no training in science. Such training, it was assumed, could easily be picked up; character could not. And, indeed, science was continually considered to glorify God in that it revealed in ever more detail the wonders of His ways. Mineralogy, asserted Parker Cleaveland in 1818, was "by no means destitute of the impress of the Deity['s] . . . wisdom, power, and benevolence. . . . " Protestant church journals were generally friendly to scientific endeavor, confident that it was in good hands and being turned to good purposes.

There was a somewhat different point of view (represented, for example, friendly to scientific endeavor, confident that it was in good hands and being by physiologist Charles Caldwell) held by scientists who were not particularly orthodox but who believed that nature is not reducible to purely mechanical and particularly not to purely chemical processes. These workers did not necessarily believe in a soul, but they did believe that both the nature of life and the processes of living organisms were in some way mysterious and inexplicable in terms other than life. These men were, however, also scientists and therefore as far removed from nineteenth century philosophy and religion as their orthodox but mechanistically minded colleagues. By the 1840s the new term, "scientist," was largely replacing the older word, "philosopher," in recognition of the distinction between those dealing with proximate (immediate and natural) and those dealing with final (spiritual or metaphysical) causes.

In the middle of the century, and specifically with the appearance of Charles Darwin's work in 1859, the truce between religion and philosophy, as opposed to science, broke down. The extent of the change is illustrated by one of the great intellectual documents of the last part of the century, a book entitled, *History of the Conflict between Religion and Science,* by John Draper, an American physician who was also a notable pioneer in the science of spectroscopy. Why should a peace of half a century's standing break into the open warfare characterized by Draper's title?

In fact, science and speculation had not signed a treaty of peace but had simply divided the field. The religionists made the error of moving over into a no-man's-land (not recognized at the time as such), namely, the question of the origins of living beings. Also, clerical writers and especially orthodox popularizers of science had made perhaps too much use of the wonders of nature to try to impress their readers with the mind and hand of God. When science, then, in the guise of the words of Darwin, was able to suggest a new theory of origins and the means by which nature operated to give form to life, the pious orthodox were suddenly challenged in an area that they felt that they had occupied for a long time. Moreover, the scientists, who had previously emphasized their reliance on "the facts," now showed an increased willingness to introduce theory, and even inference, as proper parts of the scientific way of looking at the world. Professional philosophers, who already often shared the preoccupation of nineteenth century thinkers with origins, change, and development, were not severely affected; religionists, however, either had to change their claims of knowledge or come into overt conflict with science. For many religiously orthodox scientists, such as the botanist Asa Gray, it was possible to adapt religion to the new science (so, as he suggested, faith might survive); for others, special creationists like the great naturalist Louis Agassiz, adaptation was impossible.

Within the scientific community, resistance to Darwin died with the passing of Agassiz and the rest of his generation. Outside the community, and particularly among some religionists, opposition persisted, far out of proportion to the actual contents of ideas about evolution and natural selection. The furor can be ascribed to the fact that Darwin's work provided only the occasion for narrowly orthodox people to express awareness that the modern world was overwhelming Western society and particularly American society. For a time, the evolutionary theory appeared to be the most vulnerable representative of global intellectual changes that the conservatives were trying to avoid or deny.

The younger generation of scientists and most American intellectuals accepted evolutionary theory with gusto. While the intellectuals applied ideas of survival of the fittest to subjects perhaps inappropriate, American field workers were digging up the "missing links" that they believed sustained Darwin's contentions, and American biologists in general were engaged in research and discussion aimed at perfecting and extending evolutionary theory.

By the last decades of the century the enthusiasm for science was such that it appeared there was no field in which scientific endeavor and method could not supply answers to all of the questions anyone might want to ask. To some thinkers, such as the naturalist John W. Powell, the promise of science could be fulfilled by an endless accumulation of facts, the process by which ignorance and superstition is pushed back and science, in time, solves all problems. Earlier in the century, scientists had been preoccupied with aggregating facts but tended to lack the fanatical faith in the adequacy of science that later flourished. To many thinkers of the last part of the century, such as the great physician and literary figure, Oliver Wendell Holmes, the materialism and mechanism on which research was inexorably based called for a new philosophical adaptation, known eventually in America as pragmatism. Like the positivism of Powell, Holmes's pragmatism represented not an accommodation of science to philosophy but a way of doing without traditional religion or philosophy.

In truth, advocates of science as an exclusive way of life had impressive evidence by the late nineteenth century to support their views. Scientific discovery (almost all of it, of course, European but followed avidly by Americans) was touching the secrets and, therefore, ultimately, control of even life and death. The reduction of life to its chemical and physical elements appeared to be a feasible goal for late nineteenth century research. Man could "prepare organic bodies," as chemist Ira Remsen put it. He could, as Holmes realized, show thinking to be literally a mechanical process. And the germ theory of disease, so reluctantly accepted by many physicians, brought with it more hope for the utter conquest of nature by man than any other theory or discovery of the day.

What impressed men of that century was not the activity of scientists in the field and laboratory but the material changes that they saw around them —telegraphy, steamships, railways, mountain moving, and the endless harnessing of power by machinery. For scientists, as well as the general American, science was primarily applied science, technology, and invention. Collection, observation, theory, and experiment were secondary or derivative activities justified in that each one contributes to making life easier and work grander. This was true not only for editor-physicist John Trowbridge in 1869 but for most writers on science for decades before and after him.

Many American scientists scorned applied science and glorified pure science (Remsen once refused to claim patent rights on a discovery because it would remove his work from the realm of "pure" science). William Watts Folwell in 1883 described such men as comprising "a lay priesthood, devoted to the search for truth for its own sake and its own value." Yet even purists were willing to use any argument at hand to appeal to other Americans to support science. Earlier in the century the moral effects of instruction in science were often cited: "A true naturalist cannot be a bad man," wrote geologist Edward Hitchcock in 1845. Cultural patriotism was continually being invoked, and Americans felt their inferiority in science keenly. But above all, throughout the century scientists—including the strongest advocates of pure research—justified their work on the basis of the material progress that grew out of it. Sometimes such reasoning represented mere tactics; at other times it was unwitting surrender to the anti-intellectual currents of a society that emphasized quick returns and economic exploitation. Some proponents of science (like chemist F. W. Clarke) blamed the educational system for not nourishing science. Others (like the famous mathematical astronomer, Simon Newcomb) blamed American society in general when they described the chronic lack of adequate support for American research, and particularly research in "pure" science.

There were two major potential sources of funds for science in the nineteenth century which require special comment: colleges and governments. Although science courses appeared increasingly in the curricula of colleges, in the first half of the century the inclusion of such instruction was often only nominal. At Dartmouth, for instance, science was apt to be offered during the winter term when most students were out teaching school. At Harvard, chemistry disappeared from the curriculum for many years when the professor, John Webster, was hanged for murder in 1849. Even where instruction flourished, the scientists employed in it were unable to introduce laboratory instruction or, often, to carry out research themselves. Science was considered to be a gentlemanly and humane study, not a preparation for a life in research. As late as the 1870s the president of Williams College informed his chemist, Ira Remsen, that "this is a college and not a technical school. . . . The object aimed at is culture, not practical knowledge."

Enrollment in American colleges was actually declining when, largely through the influence of proponents of science, substantial changes ensued. The first symptom of change was the rise of independent science curricula alongside regular bachelor of arts courses, rather akin to vocational night schools. At Harvard and Yale special science schools were actually established within the university before the Civil War. Elsewhere special technical schools proliferated, such as Rensselaer Polytechnic Institute and Massachusetts Institute of Technology. The federal government, beginning in 1862, helped subsidize a system of state universities devoted, at least in part, to the agricultural and mechanical arts, which were usually understood to embrace applied, if not pure, science. Finally, with the rise of the so-called elective system, science was able to enter into the regular curriculum of the colleges, which in turn were becoming universities after the German model. Under the elective system, a student no longer followed a prescribed general course of study but could choose from a number of specialized routes leading to a degree. Instructors, also, obviously could be more specialized. By the end of the century American scientists often were still overworked teachers, but many enjoyed subsidies of their research as a part of their university appointments. After 1894, for example, the Ryerson Laboratory of Physics of the University of Chicago supported important research in a way that would have fulfilled the dreams of earlier college scientists. Chicago was the exception, however, and much was left to be done in the country as a whole before even the new, well endowed universities and graduate schools could meet the demands of their faculties for more money for ever more sophisticated scientific research.

Above and beyond the higher education system and the agricultural experiment stations that grew up with the agricultural colleges, the federal and state governments often supported science in diverse and substantial ways throughout the nineteenth century. Both Newcomb and Clarke worked for the federal government, for example. Government science, however, tended with some notable exceptions to be applied science, whether applied to war, navigation, or agriculture. In one major effort in government subsidy of applied science, the geological surveys, a substantial percentage of the money was utilized, contrary to the intent of the legislatures involved, not for discovering minerals but for pure research. The extent to which such funds were available to contribute to knowledge can be suggested by the fact that by 1850 every state and some territories had initiated a geological survey. Yet this diversion of government funds, presaging the mid-twentieth century "bootlegging" of military development appropriations, was atypical. More symptomatic was the formal attempt of the founders of the National Academy of Sciences in the 1860s to obtain government money for their research. This effort ended in resounding failure, and by and large neither governments nor universities satisfied the hunger of scientists for financial support.

The complaints and the reality concerning support for science reflected two facts. First, by tradition American science was an arena for amateur endeavor. Almost anyone could walk in the woods and with patience discover a new species of something. Part-time physicists like Draper and his son could play with photography and telescopy and make international reputations. It was appropriate that science should not appear to need support beyond what any man spent on a hobby. Local natural history societies flourished everywhere, even near the frontier, to foster and express the scientific interests of well educated and even self-educated Americans. The amateurs resented the professionalization and discipline that, for instance, as one man complained in 1842, "takes botany from the multitude, & confines it to the learned." The second fact is that in spite of niggardly support there were full-time professional scientists in America, and they had every desire to nourish science in general and the profession in particular. As early as 1838 Joseph Henry, the physicist, was writing to his friend, geophysicist Alexander Dallas Bache, "that the real working men in the way of science in this country should make common cause and endeavor by every proper means unitedly to raise our scientific character, to make science more respected at home, to increase the facilities of scientific investigators and the inducements to scientific labours."

Professional science was for all practical purposes born in the period near the end of the Napoleonic wars. Scientific pioneers felt as if they were beginning anew in establishing science in America. Benjamin Silliman, for example, wrote in 1810: "Chemistry is here almost a new pursuit, and hence it is not uncommon to find even intelligent men manifesting an entire ignorance of its nature and utility." In 1818 Silliman was able to launch a major scientific journal, *The American Journal of Science and Arts,* which lasted the entire century. By the 1840s, an American scientific community had emerged, numbering between three and six hundred with an elite of half a hundred or so of the most productive workers. Three-quarters or more of the elite were professors of science of one kind or another. In contrast to the amateur status of scientists of another day, only a handful of the leading figures were practicing physicians or otherwise not full-time scientists. By 1870 there were well over 400 Americans listed in Poggendorff's prestigious international directory of scientists, and this number continued to expand steadily. By 1906, another, purely national, directory, *American Men of Science,* listed 4,000 names. (The American Association for the Advancement of Science, which included nonprofessionals to some extent, grew in membership from 759 in 1851 to 5,058 in 1906; the population, by comparison, increased in those years from 24 million to 85 million.)

Even before the Civil War, the professionals tended to be specialists, that is, identified not only as scientists but as either natural scientists or physical scientists. At the first meeting of the American Association for the Advancement of Science in 1848, the Association resolved to have two sections, one for the physical sciences and one for the life sciences. By the turn of the twentieth century, virtually every major specialty was well enough established to be represented by a national society: The American Chemical Society (founded in 1876), the American Physical Society (1899), the Geological Society of America (1888–1889), the American Society of Naturalists (1883), the American Society of Zoologists (1903), the Botanical Society of America (1894). Subspecialists such as entomologists and astrophysicists had also established their identity through professional organizations and specialized journals.

In order to find out how to live the life of science, Americans had turned to Europe. There the specialties of science flourished not only in the institutional framework of research and publication but in the universities, and especially the German graduate schools. Thousands of American scientists went to learn their physics, chemistry, physiology, microbiology, or any other specialty in Old World universities. Overseas, eager students learned, as chemist Evan Pugh recalled, "a contempt for that superficial smattering of everything without even an idea of what thoroughness is in anything which is too characteristic of our American system of education." They returned for the most part fired with enthusiasm for research, publication, and specialization and carried their enthusiasm into the construction of American graduate education. The better American scientist of the late nineteenth century tended to be a professional, a specialist, and a holder of a

Ph.D. degree, from either a German university or an American institution set up on the German model.

Such Americans were not, however, satisfied with the achievements of American science. They lionized Hermann von Helmholtz and even the iconoclastic Thomas Henry Huxley but they could boast of no residents of the United States except, perhaps, Louis Agassiz, who were giants in world science. Nor, despite the efforts of those trained abroad, was American science the same as European science. Some of the differences—and the reasons for the differences—are instructive.

Although throughout the century many Americans were responsible for theoretical writings, except for the unique and atypical physical chemist, J. Willard Gibbs of Yale, none was significant for producing such work. Americans, when they did gain recognition from the world scientific community, did so because of observation, collection, and instrumentation. The idea took hold that Americans might excel in applied science and technology but that creativity in pure science was beyond them. Particularly in the later part of the century both Americans and Europeans expressed this stereotyped image of American scientific activity, and with some justification. Everyone agreed that when scientists proposed problems not immediately practical their funding (particularly from government and industry) tended to decline. The European tradition was that great science was done by men who, like Darwin, never had to work for a living or who were supported by governments in magnificent research institutes. In the United States there were no institutes, and personal fortunes rarely brought scientists to flower (although Gibbs and his colleague at Yale, paleontologist Othniel C. Marsh, were conspicuous exceptions to this rule, and earlier Nathaniel Bowditch had spent a large part of his private financial resources publishing his edited translation of Laplace's *Mécanique céleste*). As physicist W. A. Anthony of Cornell noted in 1887: "In this country men devoted to science purely for the sake of science are and must be few in number. Few *can* devote their lives to work that promises no return except the satisfaction of adding to the sum of human knowledge."

But the uniqueness of American science was not just that of money. A later observer could point out that when there was money available, as, for example, in the new university laboratories, Americans still tended to excel not in theory but in experimentation. A. A. Michelson, the physicist, for instance, was a genius with instruments, not with theories. Likewise Henry A. Rowland of Johns Hopkins made more of an impression on Europeans with his method and tools than with his ideas. Such work was of course good science, but by the 1890s not only physicists but all of the scientists, like their overseas colleagues, were becoming so advanced in their laboratory techniques and in the problems to which they had to address themselves that interpretation, synthesis, and speculation were eclipsing the positivistic approach to nature predominant in the United States. Not until a long time after the end of the century was the notion dissipated that national character unfitted Americans for theoretical science.

In the nineteenth century the American areas of excellence were relatively easy to identify. The physical sciences were contained to a surprising extent within the geophysical tradition: astronomy, astrophysics, geology, and a physics concerned with matters such as terrestrial magnetism. Natural history also flourished, as it had earlier, because of the abundance of flora, fauna, and geological material available; because Americans became deeply involved in evolutionary theory; and because U.S. trade and exploration were not limited to North America but carried merchants and missionaries to all parts of the world. Finally, that curious quirk of American character, interest in the soundness and health of the body, led to intense interest in life processes. Where there was economic support for scientific endeavor, American scientific accomplishments were if not epochal at least notable and sometimes distinguished. This limited glory was the hope and the despair of American scientists who desperately wanted to be like the Europeans but at the same time to exist in their own culture.

8. The President of a Scientific Institution Surveys the Study of Natural History. 1826

JAMES ELLSWORTH DeKAY, (1792–1851) here surveys the state of natural history in America in 1826. He made the address only a few years after natural science had begun to establish itself in a professional, institutionalized way in the United States but recognizes explicitly the process of institutionalization and professionalization. That DeKay begins with mineralogy and geology suggests that achievements of American scientists were primarily in the realm of observation, in which geologists can excel. His lengthy history of botanical observations (so long and unenlightening that it is omitted here) likewise shows American accomplishments in collecting. DeKay's analysis of the shortcomings of domestic zoological work and the large place that he has to give to foreign workers is most suggestive. But he shows that Americans, along with other preevolutionary scientists, were thinking about the larger implications of their work.

DeKay was a physician, educated in Connecticut and at Edinburgh. He practiced in New York but is better remembered as a naturalist and one of the pillars of the Lyceum of Natural History, to which group he directed this address.

For more information concerning DeKay and his evaluation see George H. Daniels, *American Science in the Age of Jackson* (New York: Columbia University Press, 1968), and *Dictionary of American Biography*, v, 203–204.

ANNIVERSARY ADDRESS on the Progress of the Natural Sciences in the United States: Delivered before the Lyceum of Natural History, of New-York

James E. DeKay

ADDRESS

The progress made by our countrymen in those departments of knowledge which are more immediately connected with the wants of society, has been the theme of frequent discussion. That such progress has actually been made, is too obvious to be denied; and those who have been unwillingly constrained to admit its truth, have assigned other causes than are to be found

* James E. DeKay, *Anniversary Address on the Progress of the Natural Sciences in the United States: Delivered before the Lyceum of Natural History of New-York* (New York: G & G. Carwell, 1826), pp. 5–12, 14–19, 21–24, 29–30, 37–55, 65–70.

in the active, enterprising spirit of our citizens, happily co-operating with the genius of our free political institutions. It has accordingly been urged that the strong stimulus of necessity, and a thirst for personal wealth and aggrandizement, have led to these results. Let us then inquire whether those sciences which are considered rather as ornate than useful, and which are evidently unconnected with personal advantages, let us examine whether these have not also received a proportionate share of attention from our countrymen.

On the present occasion it is proposed to give a brief outline of the progress and present state of the Natural Sciences in the United States. Such occasional exhibitions are something more than mere appeals to national vanity. They are consonant with the usages of other nations, and if faithfully executed, are not always flattering to national pride. They become useful records of the labours of our cotemporaries, may indicate sources of information which might be overlooked by the inquirer, and often serve as an excitement to greater exertion with the rising generation.

Previous to the epoch of the late war with England,[1] although a few works of merit had appeared at distant intervals, yet the Natural Sciences were but partially cultivated. The few individuals who had turned their attention to such pursuits were too widely scattered over this extensive country, to allow of that familiar interchange of opinions which necessarily elicits further inquiries and discoveries. At a still earlier period we may refer to our colonial situation, the embarrassments arising from our exposed and peculiar position, and the example of the mother country, as among the most prominent causes which impeded the cultivation of Natural History in the United States.

Since, however, the period to which we have alluded, and the general peace which subsequently ensued, a spirit of inquiry has been awakened. The forest, and the mountain, and the morass have been explored. The various forms and products of the animal, vegetable, and mineral kingdoms have been carefully, and, in many instances, successfully investigated. A proper feeling of nationality has been widely diffused among our naturalists; a feeling which has impelled them to study and examine for themselves, instead of blindly using the eyes of foreign naturalists, or bowing implicity to the decisions of a foreign bar of criticism. This, if restrained within due bounds, if it is not perverted into a narrow and bigotted sentiment, that has not unfrequently been mistaken for national feeling, must be attended with beneficial consequences. To those who feel disposed to undervalue the useful and meritorious labours of foreigners, it may be suggested whether some deference is not due to the judgments of those learned individuals who have spent long and laborious lives, often in the investigation of a single group of phenomena, in the illustration of a single class, order, or genus of natural objects.

[1] The War of 1812.

In the following pages we propose a sketch, which must necessarily be brief, of the progress made in Mineralogy, Geology, Botany, and Zoology, and shall then conclude with a notice of the travels performed by individuals, or under the auspices of the government, with a view of enlarging the boundaries of Natural Science.

MINERALOGY

The first attempts to arrange minerals by a certain supposed resemblance in their properties, was evidently unsatisfactory and insufficient. The external characters and crystalline structure shed further light on this subject; but it was not until the composition of minerals by chemical analysis was attended to, that Mineralogy was fairly entitled to take the rank of a science. No part of Natural History, perhaps, presents greater charms to the youthful student. The attempt to classify and arrange, in a connected series, the different stones and earths, is often the first essay of the young inquirer into nature; and, as in a new country, it was natural to anticipate the discovery of new forms and combinations of minerals, the attention of our naturalists was early directed to this science.

The experiments of Mr. Cloud on the fusion of metals, the treatise of Mr. Cooper on the blue earth of New-Jersey, and those of Mr. Cleaveland, as contained in the *American Phil. Trans.* are among the early American essays in this department. In recurring, however, to the early history of Mineralogy in the United States, we cannot pass over in silence the labors of Dr. Archibald Bruce of this city. Early imbued with a taste for this pleasing study, he improved the opportunity afforded by an intercourse with the learned in Europe, and by a careful investigation of the mineral products of his own country. Equally zealous in the acquisition of knowledge and liberal in imparting it to others, he commenced in this city the first journal of a purely scientific nature ever established in North America. Ill health, and the duties of an arduous profession, prevented its continuance beyond a single volume, but no one who is desirous of studying with advantage the Mineralogy or Geology of this country, will fail to refer to the *American Mineralogical Journal*. . . .

An interesting essay under the title of *Outlines of the Mineralogy and Geology of Boston,* by Mr. Dana, appeared in 1818. This contains a useful catalogue of the minerals found in Boston and its vicinity, arranged in systematic order. A list of the rocks after the arrangement of Werner, and a map illustrating their geographical distribution accompanies the work.

Mineralogy has now become a popular branch of science, lectures are delivered in every considerable town in the Union, and extensive cabinets of minerals are to be met with in every direction. Increased attention has been paid to the composition of minerals, and several of our countrymen, with a view of perfecting themselves in this delicate branch have enrolled themselves in the School of Mines at Paris. Many new mineral species have been firmly established, and others which were doubtful have been re-examined, and restored to their proper places in the system. The analyses of our president Dr. Torrey, of Seybert, Keating, Bowen, Vanuxem, and others, have thrown much light upon the Mineralogy of our country, and the pages of the *American Journal of Science and the Arts* bear frequent and honorable testimony to the industry and talents of our Mineralogists. . . .

GEOLOGY

This does not consist, as many have imagined, in mere ingenious speculations concerning the origin of the globe, or in idle conjectures about the changes it has subsequently undergone. It is, or should be, the result of actual examination into the arrangement and structure of the various materials composing our world, or in deductions drawn from such examination. As a consequence of the general diffusion of mineralogical knowledge, much attention has been directed to the investigation of the rocks and mountain masses of our country.

Previous to the year 1812, the published notices on our Geology were few and unimportant, and little more was known in a geological point of view, of the United States, than that they extended along the coast of the Atlantic, and were bounded on the north by a chain of mighty lakes. Mr. Maclure, a gentleman peculiarly qualified for such a task by a familiar acquaintance with the most interesting formations of Europe, commenced a personal examination of this country, which he traversed in various directions. The result of his labors was given to the public in 1817, under the title of *Observations on the Geology of the United States.* This is a bold outline, sketched by a masterly hand, and replete with the most valuable and interesting information.

The Geologists of Europe, who had been chiefly occupied with their theories and speculations, and from the phenomena of a mole hill had not unfrequently deduced principles on which depended the formation of a world, were struck with the simple, yet grand features, presented by this geological map of America. The different sects of Geognosts[2] at first only perceived in this outline, a further confirmation of their peculiar views, but it soon became apparent that the facts there detailed were not strictly in accordance with the views of either of the two great parties which at that time divided the Geological world.[3]

Not the least instructive parts of this essay are his observations upon the fertility of the soil in different parts of the Union, as connected with the

[2] Those concerned with the major configurations and features of the crust of the earth.

[3] That is, the vulcanists, who accounted for the formation of the features of the earth by volcanic activity, and the neptunists, who invoked as causation the action of water.

nature of the subjacent rocks, or as dependant upon their decomposition. As an evidence of the accuracy of his deductions on this subject, it may be mentioned that his conjectures respecting the nature of the soil in the country west of the Mississippi, founded on the imperfect data at that time in his possession, have been verified by travellers who have subsequently explored these regions.

Much of this outline has been filled up by the labors of succeeding geologists in different sections of the Union, and the materials for this purpose are daily accumulating. Among the most important of these we may particularize the various geological notices of Dr. Mitchill in the *Medical Repository,* and more at large in the appendix of an edition of *Cuvier's Preliminary Essay,* published in this city in 1818. From the presence of marine organic relics in the soil and rocks adjacent to the great lakes, the learned professor conjectures that the ocean once filled the basins of the latter, and covered the surface of the former. He has exercised much ingenuity in tracing the barriers of this imaginary inland sea, and has indicated the principal spots where breaches are supposed to have been made, which drained the extensive country now included in the states of New-York, Pennsylvania, Ohio, and part of Virginia. The alluvial described by Maclure, is extended to the east end of Long Island, and this appendix is remarkable as containing the first attempt at a systematic arrangement of the organic relics of the United States.

An *Index to the Geology of the United States,* by Amos Eaton, appeared in 1818. This was prepared as a text book for the pupils of the author, and is accompanied by a geological section extending from the Kaatskill mountains to Boston, or through five degress of longitude. It is valuable as containing the first attempt at a general arrangement of the geological strata in North America. Three parallel sections, at the distance of fifty miles on each side, were carefully examined, in order to confirm his geological profile. It is no small praise of this work, and at the same time exhibits a gratifying proof of the interest taken in this science, when we state that a second edition was required and published in 1820, and formally recommended by the Troy Lyceum of Natural History, as an authentic record of geological facts. The section of strata is again given with many corrections and additions, and is extended to the Susquehanna, comprising nearly five degrees of longitude. A useful grammar of geology serves as an introduction to the work. . . .

[In 1820] . . . appeared the valuable treatise of Mr. Hayden of Baltimore, entitled, *Geological Essays, or an inquiry into some of the Geological Phenomena in various parts of America.* As Geology had, in a manner, grown out of Mineralogy, it was to be expected that the order and arrangement of rocks and mountain masses should have almost exclusively occupied our attention. Hence the nature and extent of that portion of the globe, consisting of loose sand and gravel, or what is technically called alluvion, has been much neglected. Hitherto these had been considered as occasioned solely by deposits from the ocean and rivers, and much ingenuity has been exercised to explain many appearances which could scarcely have depended on such comparatively trifling causes. The attention of Mr. Hayden was long since directed to the vastness of this formation in the United States, and as early as 1817, he commenced a personal examination of this immense deposit, which extends over twenty degrees of longitude. In his endeavours to account for its origin, he has satisfactorily separated it from the alluvial, distinguished it by the epithet *ternary,* and is entitled to the merit of having anticipated the important distinction recently established by the European geologists, in regard to this information. In connection with this subject, Mr. Hayden has examined, with great attention, the phenomena presented at the embouchures of our principal rivers, and has satisfactorily accounted for the peculiar appearance along the banks and shores. A vast collection of facts is accumulated to prove that all our great alluvial district has been formed by a sudden and violent deluge, accompanied by powerful and irregular currents, which buried numerous vegetables and animals, whose remains are now frequently disinterred. This flood, contrary to the speculations of Mr. Hill, in his *New Theory of the Earth,* published at Baltimore in 1823, is supposed by Mr. Hayden to have been caused by the melting of the ice at the poles, in consequence of a change in the axis of the globe. The general direction of this great overwhelming current, is strongly indicated to have been from the northeast to the southwest, and the evidences in favor of this hypothesis, are numerous and plausible. The disintegration of rocks, a favorite source of supply with all writers on alluvial districts, is very clearly shown by Mr. Hayden to be more limited than is generally supposed. . . .

BOTANY

This department of Natural Science has been prosecuted with much success. It would not be too much to assert that it has attracted more attention than its collateral branches. This may be in some measure owing to the greater perfection displayed in the systematic arrangement, and physiological history of plants, to the superior attractions displayed by the varied and beautiful forms of the vegetable kingdom, or to the facility with which extensive collections may be arranged and preserved.

In the notices which we propose to give of the progress of this science among us, we shall necessarily be brief, as the same ground has been occupied by the writer of an able article on the history of American Botany in the thirteenth volume of the *North American Review,* and more recently by Professor Hooker, of Glasgow, in the *Edinburgh Journal of Science* for 1825, conducted by Dr. Brewster. Nothing has contributed so much to extend our acquaintance with the plants of this country, as the publication of local Floras, which have been already numerous, and new ones are now of frequent occurrence. . . .

ZOOLOGY

Has Zoology, or the history of animals, been cultivated in the United States with the same success as botany, or geology and mineralogy? The answer must be in the negative. Its progress has been impeded by the operation of the same causes which have affected the other branches; but in addition to these, it has had to contend against the unjust views which have been taken of its relative importance, and to a want of concert in nomenclature and systematic arrangement between the laborers in the different subdivisions of this science. The fact, that until the last few years it has been neglected in England, affords another mortifying, but sufficiently obvious reason, why it has received so little attention here. A formal exposition of the utility of this science is hardly requisite before a society, expressly established for its cultivation. Thus much, however, we may be permitted to say, that few departments of knowledge demand more varied acquirements, more accurate habits of investigation, or a more familiar acquaintance with the labors of the learned in every part of the world.

Ten years ago, our animals were little known or carelessly described, and perhaps a stronger proof of the ignorance or indifference prevailing on this subject cannot be adduced, than the fact that the *Cervus virginianus,* or common deer of this country, was not satisfactorily known or identified, until within a very recent period. Shall it be added, that even for this we are indebted to foreigners? The formidable grizzly bear, the terror of all western travellers, is not to the present day sufficiently determined to be a different animal from the *U. arctos,* or northern bear of Europe.

Pursuant to the proposed plan of discourse, we should proceed regularly through every separate genus of this class, and exhibit the labors of our naturalists in each. This would, however be inconsistent with the brevity requisite on this occasion. It will be sufficient to indicate, in a general manner, the efforts made to illustrate the several great subdivisions of the animal kingdom.

Mammalia. Few works professedly on the animals of this class, have appeared in this country. To this division is naturally to be referred all investigations connected with the natural history of man. The history of the American race has received considerable elucidation from the labors of MacCulloch, Heckewelder, and the naturalists attached to the two expeditions under Major Long.[4] It was natural to expect that the physical history of man would have attracted the attention of the first visitors to this country, but their limited views are evinced by a perusal of their respective voyages. Even in those which have been projected for the purpose of extending our acquain-

[4] In 1817 Major Stephen H. Long commanded an army exploratory expedition to investigate the upper Mississippi region. In 1819–1820 he commanded the War Department exploration expedition sent to the Rocky Mountains. In 1823 he led still another expedition, this one to explore the northern boundary of the United States west of the Great Lakes.

tance with the different varieties of the human race, we may observe that the inhabitants of newly discovered countries are described, more with reference to their dress and manners, than to their organization as evinced by their external characters. Indeed, it must be admitted, that Zoologists themselves have too often overlooked the history of man, as if he was not a link in the great chain of animated nature.[5] A common opinion is prevalent that little more is to be discovered respecting the former inhabitants of this country; and that every thing beyond their manners and unmeaning ceremonies must be left to idle conjecture. This, however, is far from being the case. The question of their origin and descent, whether they are to be considered either directly or indirectly of Tatar origin, or, as some have maintained, are an original people, is capable of further illustration by a careful study of their peculiar physical characters. As much light will be derived from this source, as from the examination of their curious forms of language, which have but a very remote analogy with the dialects of any other nation on the face of the earth.

Of the quadrupeds distributed over the whole surface of the globe, about six hundred species are known and described. South America possesses one hundred and eighty, and North America about one hundred and thirty species. Doubtless this number will be much increased by the future exertions of our naturalists; the smaller quadrupeds will furnish, of course, the largest proportion, and the examination of our fossil relics, will probably add to the number.

During the past year we have been favored with the first attempt at a systematic arrangement of our mammalia, under the title of *Fauna Americana,* or a description of the mammiferous animals inhabiting North America. One hundred and forty-seven species are described in this work, of which fourteen are new, and of these four are now extinct. The author, Dr. Harlan, has exercised much ingenuity and industry in assembling together the scattered notices and descriptions of his cotemporaries, arranging them in systematic order, and adding his own discoveries, which are numerous and valuable. Previous to the appearance of this volume, little attention had been paid by our naturalists to the only sure and firm basis on which the genera of Mammalia can be constructed. Various attempts had, indeed, been made, but hitherto without rigorously examining the dental formula. We shall take occasion again to recur to this subject.

As if the vast territory of the United States was too limited for their exertions, several of our naturalists have examined and given faithful descrip-

[5] This refers to the belief that God had created an orderly world with rational series of plants and animals, the categories of which reflected the orderly progress of God's thinking—the great chain of being. There were, at this time, only a few glimmerings of the idea that the logical arrangement of plants and animals might be placed in a progressive time sequence. Most people, like DeKay, believed that all parts of nature came into being at once and, following ancient philosophical ideas, that rocks are at the bottom of the stable hierarchy of nature, and plants, animals, and man higher up in the hierarchy of existence, which is crowned by the angels and God himself.

tions of the animals of other regions. The Isodon of Cuba, by Mr. Say, and the very curious Chlamyphorus from the interior of Chili,[6] by Dr. Harlan, deserve particular attention. We look impatiently for the promised work of Dr. Godman and his able coadjutors; which is to furnish us with a complete *History of all the hitherto known quadrupeds of North America.*

Birds. These have early claimed the attention, not only of foreigners, but of our own naturalists. Indeed, in all ages and countries, their beautiful forms often decorated with the gayest plumage, their attractive habits and artless song, have never failed to render them objects of universal interest. A few inconsiderable and imperfect lists have been made, and of these we may indicate such as are contained in Jefferson's Notes on Virginia, and in the works of Bartram, Belknap, and Williams. The work of Vieillot, splendid and useful as it really is, [loses] much of its value from the circumstance of its being incomplete, and from the unnecessary changes introduced in the names of long established species. When will naturalists learn to shun the barren honors of a synonyme?

The great work of Wilson may be considered as having created a new era in American Ornithology. In this we have descriptions of two hundred and seventy-eight species, of which fifty-six are described as new. Perhaps no work contributed in a more eminent degree to create a taste for Natural History in this country, than the publication of these splendid volumes. The peculiar disadvantages under which Wilson labored in the progress of his work would have dampened and disheartened any spirit but his. His ardent enthusiasm for his favorite pursuits, and his noble disdain of the most appalling obstacles, are finely exhibited in his reply to a friend who endeavoured to dissuade him from the publication, "I shall at least leave a beacon to show where I perished."

Wilson was not, strictly speaking, a systematic naturalist; but he evinced great acuteness in the determination of species. He was an enthusiastic observer of the manners and habits of the feathered tribe; which he describes in the most vivid and appropriate language, and his almost living figures in this particular have never been surpassed. His friend Mr. Ord, has recently published a new edition of the three last volumes, in which the errors of nomenclature are corrected, and those improvements introduced which were rendered necessary by the advances made in this science, since the first publication. The ninth volume, entitled a *Supplement to Wilson's American Ornithology,* contains an enlarged biography of Wilson, which gives much additional interest to the work.

The American ornithologist will feel grateful to the Prince of Musignano for the very thorough manner in which he has corrected errors and settled synonymes in his *Observations on the nomenclature of Wilson,* as contained

[6] A small marsupial and an armadillo.

in the Journal of the Academy of Natural Sciences. Other important additions may be found in this valuable Journal. The expedition to the Rocky Mountains, under the command of Major Long, has furnished us with twelve new species carefully described by Mr. Say. The plan of that excellent work unfortunately did not admit of figures to illustrate the department of Natural History. This deficiency is, however, now splendidly supplied by the *American Ornithology* of Charles Lucian Bonaparte, whose zeal and profound acquirements, as evinced in this volume, have added new titles of distinction to those already acquired by his illustrious family.

This work, which may be considered as a continuation of Wilson, will, when completed, leave but scanty gleanings to the future inquirer. Twenty-two species are figured and described in this volume, and should the two succeeding volumes contain as many, the whole number of species may be roughly estimated at three hundred and sixty. Our Annals contain notices of Mr. Clinton, relative to a beautiful species of swallow which has very lately appeared in the United States; but which, from present appearances will probably ere long be common over the whole country. To the same work we are indebted for the description, by Mr. Cooper, of a singular and interesting species of Fringilla,[7] which had hitherto escaped the researches of our naturalists. The synopsis of all the North American birds, now publishing in the Annals by C. L. Bonaparte, will furnish an excellent manual to the American ornithologist.

Reptiles. Until within the two last years, the examination of our reptiles has been generally neglected. The confused and contradictory statements in the systematic works, and the difficulty of observing their habits in their native haunts, have combined to deter the inquirer. The article entitled *Description of several species of North American Amphibia,* by Professor Green, in the Journal of the Academy, contains the first attempt by an American naturalist, to describe and arrange some of our reptiles in a systematic order. They were at that time in a state of chaotic confusion, and Mr. Green has conferred no trifling obligation upon our herpetologists by his original and judicious observations. We hope he may be induced to resume his labors.

The essays of Messrs. Lesueur and Say have illustrated, in an eminent degree, the history of the genus Testudo,[8] which now includes fifteen North American species. The Saurian[9] division of reptiles, comparatively few of which are found in this country, has been much enlarged by the addition of several species, of the genera Agama, Lacerta, Scincus, and Ameiva. Among the Ophidia[10] our researches have been extremely limited. The expedition before alluded to has furnished us with excellent descriptions of seven, and the Journal of the Academy with three additional new species. The Batra-

[7] Finch.

[8] Tortoises.

[9] Lizards and their relatives.

[10] Snakes and their relatives.

chia[11] have only very recently been investigated, and although their study is attended with peculiar difficulties, as they frequently change their color and markings when alive, and still more after death; yet our naturalists have not been discouraged by these unpropitious circumstances. We are chiefly indebted to our associate, Captain Le Conte, for the additions made to this department. His long and familiar acquaintance with species, has given additional interest and value to his observations on the genera Rana and Hyla. The same subject has been further investigated by Dr. Harlan, who has contributed *Descriptions of several new species of Batrachian animals, with observations on the larvæ of Frogs,* which may be found in the tenth volume of the American Journal of Science.

The doubtful reptiles, of which five species are already known to exist in the United States, having been attentively observed by Dr. Harlan, and his able papers on that subject in our Annals, will be constantly referred to by the American herpetologist. In the same Journal will be found remarks by Captain Le Conte on the genus Siren,[12] and the description and figure of a striking species. Our former president, Dr. Mitchill, has described in the American Journal of Science, that singular reptile which he has designated as the Proteus of the Lakes. Another allied animal, the Amphiuma[13] of Dr. Garden, has been known for fifty years, but by some oversight had not been arranged in the systems, and, indeed, was entirely forgotten until it was again brought forward and described under the name of Crysodonta. Additional observations on this highly interesting animal are to be found in the Transactions of the respective societies of Philadelphia and New–York.

Fishes. From the nature of the medium in which these animals exist, and the consequent difficulty of studying their peculiar habits, and likewise from the limited observations which have hitherto been made, to determine such differences as may depend on age or sex, this branch of Natural Science perhaps presents less of interest to the philosophical inquirer. The artificial and highly complex systems of Bloch, Schneider, and Lacepede, and even the more philosophical arrangement proposed by Cuvier,[14] have all failed to a certain extent in securing the requisite support of naturalists. Hence the confusion in this department is such, that it is often difficult to determine with much accuracy the species designated by our predecessors. Under all these disadvantages, they have been very industriously studied by Mr. Lesueur of New Harmony,[15] and his descriptions, with his excellent figures, have furnished many materials for a future American Ichthyology. His *Memoir*

[11] Amphibians.

[12] Tailed amphibians.

[13] A congo eel, actually, as it turned out, a salamander.

[14] All four were European naturalists and systematists.

[15] One of a number of very able naturalists attracted to the utopian community of the English industrialist and social thinker, Robert Owen, established at New Harmony in Indiana. The mutual interaction and support of these men, including Say and Troost, at New Harmony was an important event in early American science.

on the Chondropterygious Fishes of America, in the Transactions of the American Philosophical Society, is one of the best zoological monographs with which we are acquainted, and like the work of Broussonet, will be found a useful model for the ichthyologist. Nor have separate treatises been wanting in this department. Mr. Rafinesque in his *Ichthyologia Ohioensis,* and Dr. Mitchill in his *Report in part on the Fishes of New–York,* have each endeavoured to illustrate the Ichthyology of these respective regions. The latter work, much enlarged and improved, has appeared in the *Transactions of the Literary and Philosophical Society of New–York.* Our *Annals* contain a few papers on this subject; the Stylephorus, and the Cephaloptera Vampirus, by Dr. Mitchill, and the description of an interesting species from the pen of Mr. Clinton. The pages of the *American Monthly Magazine* contain further descriptions of species, more particularly from the neighbourhood of this city.

Among the *invertebrated* animals, or those not furnished with a bony spine separable into many parts, much has been done, but when we consider the vast field yet remaining to be explored, we must admit that there will still be left enough to employ our naturalists for many years. Our Crustacea have been thoroughly investigated by Mr. Say in the *Journal of the Academy of Natural Sciences.* We are not aware that any other American has devoted the least attention to this very singular group of animals. The extensive labors of Mr. Say leave us the less to regret on this account. Some idea may be conceived of the vast extent of the department of Entomology when we are assured that our naturalists are already acquainted with about nine thousand North American species. Mr. Say and Captain Le Conte, in the Transactions of the societies of New–York and Philadelphia, have furnished materials towards a system of American Entomology. The want of a good entomological manual is still felt by our young naturalists; this is indeed partially supplied by the *American Entomology* of Mr. Say, of which the second volume has recently appeared, but its expensive form puts it beyond the reach of most private individuals. To the same able and indefatigable zoologist, and to our colleague Mr. Barnes, we are indebted for ample illustrations of our marine and fresh water shells. The Monograph of the extensive and obscure family of the Uniodeæ in the American Journal of Science, by Mr. Barnes, has equally facilitated the inquiries of the student, and elevated our scientific reputation. The marine mollusca have been, with the exception of the observations of Lesueur, almost entirely neglected by our naturalists. We know of no department in which more interesting discoveries are to be made, or which would secure a more honorable distinction to the young naturalist than the investigation of these curiously organized beings.

Having thus passed in rapid review the whole kingdom of living nature, we should consider our remarks as incomplete unless we adverted to those extinct animals whose study has received a new impulse from the sublime genius of Cuvier. The great abundance of the relics of beings which have now perished from the face of the earth, early attracted the attention of the first explorers of this country. The most remarkable for its size is the Mastodon, which has been found as far north as the 50th degree of north latitude, and from the shores of the Atlantic to the great lakes. Hitherto but one species has been found.

The remains of another, but very different quadruped, an account of which was published by Mr. Jefferson, were described and figured in the fourth volume of the Transactions of the American Philosophical Society. The Megalonyx, as it has been happily named, respecting whose habits we can only form vague conjectures, was furnished with tremendous organs of defence and attack. Very recently the bones of the Megatherium or gigantic Sloth of South America, has been discovered in the neighbourhood of Savannah, and the industry and talents of our associate, Mr. Cooper, has enabled him to determine in a positive manner, their identity with the South American species. For our acquaintance with other extinct mammalia, we are indebted to Dr. Wistar, for two new species of the genus Bos; and to Dr. Harlan for one of the Elephant, Tapir, Deer, and a gigantic Manatus or Sea Cow from the eastern shore of Maryland.

We are not aware that any Ornitholites or remains of birds, have yet been discovered in our country. Specimens have been presented to the Lyceum, from the sandstone of Nyack in the vicinity of this city, which, although much comminuted, were evidently the bones of birds. The circumstances under which they were found, and their appearance, led to a belief that they had been fortuitously deposited in open fissures of the rock. The Reptiles, from some cause to us inexplicable, have rarely been found in a fossil state. The greater part of those discovered within a few years belong to the class of marine reptiles, and have but a remote analogy with any of the present living species. Among these we may distinguish the Saurocephalus of Dr. Harlan, and the Monitor, described by Dr. Mitchill, from the tertiary formation of New Jersey.

Many parts of our country present extensive deposits of fossil fishes. A remarkable locality of this kind is to be found almost in the neighbourhood of this city. Let us hope that some of our naturalists will soon favour the public with their observations on this extensive deposit. All the specimens which we have examined, whether from Westfield or this latter deposit, are very closely allied to the genus Lepisosteus of Lacepede.

No region of the globe presents a greater number or variety of the remains of invertebrated animals, than our own. So numerous and varied are these relics, that the bare enumeration of those only which are already known, would occupy more space than the limits assigned to this discourse would admit. The investigation of these organic relics will amply repay the curious inquirer.

Having thus in a summary manner terminated this sketch of the progress made in the Natural Sciences by our countrymen during the last few years, it remains to be seen how far the government has extended its patronage towards these objects. . . .

CONCLUSION

In taking this rapid, and, I am sensible, imperfect review of the labors of our countrymen, several thoughts are naturally suggested. It will be perceived that without any greater incitement to exertion than what is derived from the laudable curiosity which prompts us to investigate the operations of nature, and with no other reward than the satisfaction derived from the investigations themselves, our naturalists have been industriously employed. In other countries, from the unequal distribution of property, it not unfrequently happens that large fortunes in the hands of private individuals are munificently expended for the encouragement of the Natural Sciences, and the fostering hand of government is liberally extended towards these objects. Splendid establishments are founded and amply endowed, affording gratuitous instruction in the most minute branches, exhibiting brilliant prospects to the zealous student, and securing to the ripe scholar a secure and honorable retreat in his old age. In our own country, notwithstanding the peculiar constitution of society, which affords little leisure to the mere scholar, and the meagre recompense which awaits the student of science, yet Natural History has not failed to attract much attention from numerous votaries. Indeed a stronger evidence cannot be given of the interest which is taken in this study than the fact that numerous institutions for the cultivation of Natural History have been for many years in active operation in the United States, and every year adds several new associations to the list.

In Mineralogy more accuracy has been introduced, and analyses have been much improved, many new forms and combinations have been brought to light, and species hastily introduced have been speedily restored to their proper places. The loose and confused attempts at analysis, of which some examples might be adduced, ten years ago, would not be tolerated at the present day. In Geology ample materials are daily accumulating for a complete history of our different formations, and the efforts of our geologists are unwearied in adding to the stock of our postive knowledge on this subject. Unbiassed by the theories of European naturalists, they attach themselves exclusively to the study of the nature, arrangement, and connection of the different strata without attempting to seek for proofs of their identity with similar formations in Europe. The Botany of our country has been carefully studied, and although the attempts hitherto made to introduce the natural orders in preference to the Linnean arrangement, have failed of success, yet there is every reason to anticipate that ere long our botanists will generally adopt this only philosophical mode of studying the vegetable kingdom. In Zoology, from the peculiarity of our situation, naturalists have been more occupied in discovering and describing new species, than in investigating the natural affinities and relations of beings, the chief end of all zoological studies. Indeed it is but natural to expect that more attention should be given to the examination of new species, than to a rigid criticism of genera. A knowledge of the former is doubtless of great importance, but the latter will enable us to detect the delicate affinities by which the different classes of organized beings are approximated, if not brought into absolute contact. We are aware that the idea of a chain of beings has been ridiculed as a philosophical reverie, but the more this question is examined with the light afforded by modern observation, the firmer will this opinion be established. Already we hear the terms, "natural series" "annectant groups" "regular series," and other expressions which mark the first glimmerings of light on this hitherto obscure subject. And when we reflect that these affinities have been for the most part drawn from external and obvious characters, that we have yet much to learn from the internal anatomy, that new species are continually discovered which connect hitherto separate genera, and finally, that every day brings with it the discovery of some extinct animal, whose structure varies more or less from those of any living being, we are insensibly led to admit that the idea of a chain of beings is neither visionary nor unphilosophical.

As naturalists we have much reason to be satisfied with our peculiar position. Placed on a comparatively virgin soil, with new forms and objects constantly presented to our view, suggesting new trains of thought, and giving rise to new associations, we are more highly favored than the naturalists of older countries. As pioneers in the Natural History of the United States, reputation and after-fame, those powerful incentives to active and honorable exertion, is more immediately within our reach than it will be to the numerous naturalists who shall but tread in our footsteps. With such incentives before us, let us apply ourselves diligently to the work,—

—*dum loquimur, fugerit invida aetas.*[16]

In another point of view our situation offers some striking advantages. Removed as we are, from the scenes of those rivalries and contentions, which unfortunately too often intrude even upon the peaceful domains of science, where unworthy national prejudices are sometimes associated with private jealousies, we are enabled to examine controverted points with coolness and impartiality. The remoteness of our situation supplies the place of time, and we may be supposed to decide between the conflicting opinions of European naturalists, with the same justice and impartiality as if we were removed from them by intervening centuries.

[16] While we speak, time ebbs away.

9. A Religiously Orthodox Lawyer Defends Science and Material Progress. 1831

As science became better established in nineteenth century America both institutionally and intellectually, the philosophical and moral consequences of a scientific attitude became the object of discussion. In general, Americans were friendly to science and to scientific explanations of all phenomena. It is significant that when TIMOTHY WALKER (1802–1856) wanted an opponent of a mechanistic view, he could not find an American adversary but had to use an Englishman (Thomas Carlyle, as it happened to turn out). Just how thoroughgoing a mechanist Walker was, the reader will have to judge for himself.

Timothy Walker was born in Massachusetts of Puritan stock. He graduated first in his class at Harvard in 1826 and then attended law school there. After 1830 he practiced in Cincinnati and became a major figure in the legal institutions of the area. He was noted for his closely reasoned legal arguments.

See George H. Daniels, *American Science in the Age of Jackson* (New York: Columbia University Press, 1968); *Dictionary of American Biography*, XIX, 363; Leo Marx, *The Machine in the Garden, Technology and the Pastoral Ideal in America* (New York: Oxford University Press, 1964), 170–190, says that Walker represented the "rational and, in the main, politically conservative, theological liberal opinion of the New England Establishment."

DEFENCE OF Mechanical Philosophy. Signs of the Times.—Article VII. in the Ninety-eighth number of the Edinburgh Review, for June, 1829.*

Timothy Walker

The article[1] which we have just named raises the grave and solemn question, whether mankind are advancing or not, in moral and intellectual attainments? The writer expresses his opinion, with sufficient distinctness, in the following words; 'In whatever respects the pure moral nature, in true dignity of soul and character, we are, *perhaps,* inferior to most civilized

* *North American Review,* 33 (1831), 122–136. In the original the author was mistakenly identified as "F. Walker." The author was Timothy Walker.

[1] The author of the article in the *Edinburgh Review* was, as noted in the introduction, Thomas Carlyle, (1795–1881). Carlyle was noted as a critic of many of the modern tendencies witnessed by the generations of the first half of the nineteenth century.

ages.' If this be true, it is a truth of deep and melancholy import. But is it true? Well may we pause, and ponder the matter carefully. What are the petty controversies which agitate sects, parties, or nations, compared with one which concerns the destinies of the whole human race? When we essay to cast the world's horoscope, and interpret auguries for universal man, it becomes us to approach the task with diffidence. And we do approach it with unfeigned diffidence. We despair of being able to rise to the height of the theme, on which we are to speak. Yet we feel that good may come even from the attempt.

Are we, then, in fact, degenerating? Has the hand been moved backward on the dial-plate of Time? Has the human race, comet-like, after centuries of advancement, swept suddenly round its perihelion of intelligence, and commenced its retrogradation? The author of the article before us, as we have seen, expresses though with a *perhaps,* his belief of the affirmative. Throughout the whole article, with the exception of the last paragraph or two, of which the complexion is somewhat more encouraging, he draws most cheerless conclusions from the course which human affairs are taking. If the writer do [*sic*] not, as he humanely assures us in the end, ultimately despair of the destinies of our ill-starred race, he does, nevertheless, perceive baleful influences hanging over us. Noxious ingredients are working in the caldron. He has detected the 'midnight hag' that threw them in, and her name is Mechanism. A more malevolent spirit, in his estimation, does not come from the hateful abodes. The fated inhabitants of this planet are now under her pernicious sway, and she is most industriously plotting against their weal. To countervail her malignant efforts, the author invokes a spirit of character most unlike the first. Her real name, as we shall see, is Mysticism, though this is not pronounced in the incantations.

Now we cannot help thinking, that this brilliant writer has conjured up phantoms for the sake of laying them again. At all events, we can see nothing but phantoms in what he opposes. In plain words, we deny the evil tendencies of Mechanism, and we doubt the good influences of his Mysticism. We cannot perceive that Mechanism, as such, has yet been the occasion of any injury to man. Some liberties, it is true, have been taken with Nature by this same presumptuous intermeddler. Where she denied us rivers, Mechanism has supplied them. Where she left our planet uncomfortably rough, Mechanism has applied the roller. Where her mountains have been found in the way, Mechanism has boldly levelled or cut through them. Even the ocean, by which she thought to have parted her quarrelsome children, Mechanism has encouraged them to step across. As if her earth were not good enough for wheels, Mechanism travels it upon iron pathways. Her ores, which she locked up in her secret vaults, Mechanism has dared to rifle and distribute. Still further encroachments are threatened. The terms uphill and downhill are to become obsolete. The horse is to be unharnessed, because he is too slow; and the ox is to be unyoked, because he is too weak. Machines are to perform all the drudgery of man, while he is to look on in self-complacent ease.

But where is the harm and danger of this? Why is every lover of the human race called on to plant himself in the path, and oppose these giant strides of Mechanism? Does this writer fear, that Nature will be dethroned, and Art set up in her place? Not exactly this. But he fears, if we rightly apprehend his meaning, that mind will become subjected to the laws of matter; that physical science will be built up on the ruins of our spiritual nature; that in our rage for machinery, we shall ourselves become machines. This we take to be the import of the following unusually plain passages; 'Not the external and physical alone is now managed by machinery, but the internal and spiritual also.'—'Philosophy, Science, Art, Literature, all depend upon Machinery.'—'Men are grown mechanical in head and in heart, as well as in hand.'—'Their whole efforts, attachments, opinions, turn on Mechanism, and are of a mechanical character.' These are pretty broad and sweeping assertions, and we might quote many equally positive, and of the same style and meaning. In fact, the whole article is a series of repetitions of this leading idea, under various shapes; and this idea we propose to examine and controvert.

And, on the face of the matter, is it likely that mechanical ingenuity is suicidal in its efforts? Is it probable that the achievements of mind are fettering and enthralling mind? Must the proud creator of Mechanism stoop to its laws? By covering our earth with unnumbered comforts, accommodations, and delights, are we, in the words of this writer, descending from our 'true dignity of soul and character?' Setting existing facts aside, and reasoning in the abstract, what is the fair conclusion? To our view, directly the contrary. We maintain, that the more work we can compel inert matter to do for us, the better it will be for our minds, because the more time shall we have to attend to them. So long as our souls are doomed to inhabit bodies, these bodies, however gross and unworthy they may be deemed, must be taken care of. Men have animal wants, which must and will be gratified at all events; and their demands upon time are imperious and peremptory. A certain portion of labor, then, must be performed, expressly for the support of our bodies. But at the same time, as we have a higher and nobler nature, which must also be cared for, the necessary labor spent upon our bodies should be as much abridged as possible, in order to give us leisure for the concerns of this nature. The smaller the number of human beings, and the less the time it requires, to supply the physical wants of the whole, the larger will be the number, and the more the time left free for nobler things. Accordingly, in the absolute perfection of machinery, were that attainable, we might realize the absolute perfection of mind. In other words, if machines could be so improved and multiplied, that all our corporeal necessities could be entirely gratified, without the intervention of human labor, there would be nothing to hinder all mankind from becoming philosophers, poets, and votaries of art. The whole time and thought of the whole human race could be given to inward culture, to spiritual advancement. But let us not be understood as intimating a belief, that such a state of things will ever exist. This we do not believe, nor it is necessary to our argument. It is enough, if there be an approach thereto. And this we do believe is constantly making. Every sober view of the past confirms us in this belief.

In the first ages of the world, when Mechanism was not yet known, and human hands were the only instruments, the mind scarcely exhibited even the feeblest manifestations of its power. And the reason is obvious. As physical wants could only be supplied by the slow and tedious processes of hand-work, every one's attention was thereby completely absorbed. By degrees, however, the first rudiments of Mechanism made their appearance, and effected some simple abbreviations. A portion of leisure was the necessary result. One could now supply the wants of two, or each could supply his own in half the time previously required. And now it was, that mind began to develop its energies, and assert its empire over all other things. Leisure gave rise to thought, reflection, investigation; and these, in turn, produced new inventions and facilities. Mechanism grew by exercise. Machines became more numerous and more complete. The result was a still greater abridgment of labor. One could now do the work of ten, or each could do his own, in one tenth of the time before required. It is needless to follow the deduction farther. Every one knows that now, in many of the departments of labor, one can perform, by the help of machinery, the work of hundreds; or, supposing no division, each could perform his own in a hundredth part of the time before required. The consequence is, that there has never been a period, when so large a number of minds, in proportion to the whole, were left free to pursue the cultivation of the intellect. This is altogether the result of Mechanism, forcing inert matter to toil for men. And had it been reached gradually, commencing at the Creation, and continuing until now, the blessing would have been without alloy. But unhappily the progress has not been gradual. Of late, Mechanism has advanced *per saltum,*[2] and the world has felt a temporary inconvenience from large numbers being thrown suddenly out of employment, while unprepared to embark in any thing else. But this evil must be from its nature temporary, while the advantage resulting from a release of so large a proportion of mankind from the thraldom of physical labor, will be as lasting as the mind. And hence it is, that we look with unmixed delight at the triumphant march of Mechanism. So far from enslaving, it has emancipated the mind, in the most glorious sense. From a ministering servant to matter, mind has become

[2] By leaps and bounds.

the powerful lord of matter. Having put myriads of wheels in motion by laws of its own discovering, it rests, like the Omnipotent Mind, of which it is the image, from its work of creation, and pronounces it good.

When we attempt to convey an idea of the infinite attributes of the Supreme Being, we point to the stupendous machinery of the universe. From the ineffable harmony and regularity which pervade the whole vast system, we deduce the infinite power and intelligence of the Creating Mind. Now we can perceive no reason, why a similar course should not be pursued, if we would form correct conceptions of the dignity and glory of man. Look at the changes he has effected on the earth; so great, that could the first men revisit their mortal abodes, they could scarcely recognize the planet once inhabited. Fitted up as it now is, with all the splendid furniture of civilization, it no more resembles the bleak, naked, incommodious earth, upon which our race commenced their improvements, than the magnificent palace resembles the low, mud-walled cottage. From the effect, turn your attention to the cause. Examine the endless varieties of machinery which man has created. Mark how all the complicated movements co-operate, in beautiful concert, to produce the desired result. Before we conclude that man's dignity is depreciated in the contrivance and use of this machinery, let us remember, that a precisely analogous course of reasoning must conduct us to the conclusion, that the act of Creation subtracted from the glory of the Creator; that the Infinite Mind, as it brooded from eternity over chaos, was more transcendently glorious, than when it returned from its six days' work, to contemplate a majestic world. We accordingly believe there is nothing irreverent in the assertion, that the finite mind in no respect approximates so nearly to a resemblance of the Infinite Mind, as in the subjugation of matter, through the aid of Mechanism, to fixed and beneficial laws—to laws ordained by God, but discovered and applied by man.

If the views now presented be correct, it follows that the mechanical enterprise, with which our age is so alive, far from being unfavorable to our spiritual growth, is the one thing needful to furnish the freedom and leisure necessary for intellectual exercises,—to establish mankind in the *otium cum dignitate,*[3] in a higher sense than even Cicero conceived it. But we may be referred, by way of refutation, to some of the renowned nations of antiquity, for which Mechanism effected little or nothing, but which, nevertheless, 'in true dignity of soul and character,' would be pronounced by the writer, whose views we are examining, superior to any of the present day. Greece may be selected as the most prominent illustration. To Greece, then, let us look. But let it be borne in mind, that we are speaking of society in the mass, and that our doctrine is, that men must be released from the bondage of perpetual bodily toil, before they can make great spiritual attainments. And now the question is, how came Greece to achieve her high intellectual

[3] Ease with dignity.

supremacy, when all her work was performed by hand? The answer, so far as it respects this discussion, is ready. The Greeks themselves did not toil. Every reader of their history knows, that labor, physical labor, was stigmatized as a disgrace. Their wants were supplied by levying tribute upon all other nations, and keeping slaves to perform their drudgery at home. Hence their leisure. Force did for them, what machinery does for us. But what was the condition of the surrounding world? It is explained in a word. All other men had to labor for them; and as these derived no helps from Mechanism, manual labor consumed their whole lives. And hence their spiritual acquisitions have left no trace in history.

Now if we are willing to recur to that barbarous principle, that one nation may purchase itself leisure, as the Greeks did, by aggressions upon the rest, and if all other nations can be persuaded to submit to the experiment, we may no doubt behold a people, spurning all mechanical improvements, and yet attaining to a surprising 'dignity of soul and character.' But so long as it continues to be settled by compact among the nations, that each shall produce the means of subsistence within itself, or else an equivalent to exchange with others; and so long as the balance of power continues to be so adjusted, as to prevent any one from living upon the rest through the force of superior numbers; we see not how we can avoid the conclusion, that that nation will make the greatest intellectual progress, in which the greatest number of labor-saving machines has been devised. It may not produce a Newton, Milton, or Shakspeare, but it will have a mass of thought, reflection, study, and contemplation perpetually at work all over its surface, and producing all the fruits of mental activity.

But this writer has not confined his warfare to the world as a whole. He has divided mankind into classes, and attacked them in detail. We shall try to follow him through his campaign. One remark, however, upon the name which he has given to the age. 'It is not an Heroical, Devotional, Philosophical, or Moral Age, but, above all others, the Mechanical Age. It is the Age of Machinery, in every outward and *inward* sense of that word.' It may puzzle our readers as much as it does ourselves, to understand what is meant by the 'inward sense of machinery.' We are still more perplexed to understand how the following charge, which seems intended as unusually severe, can be construed by thinking men into any thing else than substantial eulogy. 'With its whole, undivided might, it [this age] forwards, teaches, and practices, *the great art of adapting means to ends*. Nothing is now done directly, or by hand; all is by rule, and calculated contrivance. For the simplest operation, some helps and accompaniments, some cunning, abbreviating process is in readiness.' Now take away the lurking sneer with which this is said, and we see not how it would be possible to crowd more praise into a smaller compass. It is no small part of wisdom, to possess 'the capacity of adapting means to ends.' What would the writer have us do? Pursue ends without regard to means?

But to the specific charges. And first, the world is full of literary, scientific, and religious associations. It is one of the mechanical features of the age, that large numbers of men are in the habit of combining together to effect those objects, which no individual could accomplish alone. Now we have been accustomed to consider this prevailing tendency, as one of the greatest modern improvements. In no propensity do we discover *a more prudent adaptation of means to ends.* We employ the mechanic lever to lift weights, which our unassisted strength could not lift. Why not employ the *social lever* in the same way? We are aware that some great and good men have expressed apprehensions, that the individual is in danger of being lost in the mass. But, for aught we can perceive, the individual is as free as ever, and his influence is even greater. Let him unite with those whose opinions agree with his, and he adds another unit to the sum. Let him stand out alone, and he must be transcendently gifted, if he do not lose his unit of influence. And as to his freedom, there is no reason why he should part with that, when he joins himself to a society. He may act with it so long as he approves its course. When he disapproves it he may attempt a change; and if he cannot prevail, he may separate, and, at worst, he will stand in the position in which he was placed before he joined the society.

The writer next indulges in pleasantry at the expense of Physical Science and its votaries. 'No Newton, by silent meditation, now discovers the system of the world from the falling of an apple; but some quite other than Newton stands in his Museum, his Scientific Institution, and, behind whole batteries of retorts, digesters, and galvanic piles, imperatively "interrogates nature"; who, however, shows no haste to answer.' If this means any thing, it is to cast ridicule upon the universal practice of demonstrating and illustrating scientific truths by actual experiments. And in what school has the writer been brought up, if he really need to be reminded, that nature *does* answer, and hastily too, when thus interrogated? Again and again did she make haste to answer to Franklin, Priestley, Black, Lavoisier, Davy, and a host of other *imperative interrogators.* Where was this writer, when she was questioned as to the cause of lightning, the composition of water, the nature of heat, the mode of guarding against the fire-damp of the mine, and a hundred other equally momentous secrets?

The Mathematics are next subjected to our author's criticism. 'Its calculus, differential and integral, is little else than a more cunningly constructed arithmetical mill, where the factors being put in, are, as it were, ground into the true product under cover, and without other effort on our part, than steady turning of the hand. We have more mathematics, certainly, than ever; but less *mathesis.*[4] Archimedes and Plato could not have read the *Mécanique Céleste;*[5] but neither would the whole French Institute see aught

[4] That is, learning, especially knowledge of mathematics (with the connotation of philosophical knowledge).

[5] The great work of Laplace, in which he extended Newton's system of mechanical interrelationships between heavenly bodies to show the eternal stability of the solar system.

in that saying, "God geometrizes!" but a sentimental 'rodomontade.' Now we are much in the same predicament with regard to this passage, as the French Institute with regard to that saying. We can see naught in it but 'rodomontade.' We learn from it, that Newton, Leibnitz, and Laplace, were nothing more than mill-wrights, and that their work was very easy. Indeed, the author had just before asserted, that to excel in the higher departments of mathematics, required no great natural gifts. Did we entertain our author's opinion of the facility with which a man, by setting himself patiently to work, could produce a treatise like the *Principia*[6] and the *Mécanique Céleste,* we would certainly give up writing for the reviews; and we almost wonder that our author adheres to it, instead of placing his name by the side of those of Newton and Laplace. As to the remark about *mathesis,* it is true that Plato had the honor of *saying* 'God geometrizes!' but to *prove* it, was reserved for the mechanicians above mentioned.

The next thrust is made at Metaphysics. And here we are informed that nobody has gone to work right. The whole world are now, and always have been, totally in the wrong. Even Locke, the great master, was at fault in the outset. But to avoid mis-statement, let the reviewer speak for himself. 'The whole doctrine of Locke[7] is mechanical, in its aim and origin, in its method and results. It is a *mere* discussion concerning the origin of our consciousness, or ideas, or whatever else they may be called; a genetic history of what we see *in* the mind. But the *grand secrets* of Necessity and Free-will, of the mind's vital or non-vital dependence on matter, of our *mysterious relations* to Time and Space, to God, to the Universe, are not in the faintest degree touched on.' So because Locke confined his inquiries to what can be known, instead of meddling with 'grand secrets,' and 'mysterious relations,' he is a mere mechanic. Commend us to such mechanics. Give us Locke's Mechanism, and we will envy no man's Mysticism. Give us to know the 'origin of our ideas,' to comprehend the phenomena 'which we see in the mind,' and we will leave the question of the mind's essence to transcendental speculators. So of Necessity and Free-will; mechanical as the age is, we have heard of no machinery which can be brought to bear upon their explanation. And as to 'the mind's vital or non-vital dependence upon matter,' we are compelled to plead ignorance of what it means. We are bound, however, to suppose it has a deep meaning, since Locke did not get at the bottom of it. And should the writer give some of his leisure moments to the investigation, we hope the world may have the benefit of his researches. He may next find it profitable to undertake with Entities, Quiddities, Essences, and Sensible Forms, those stubborn secrets which did so puzzle some of the schoolmen.[8] After brushing away these mists, there will still remain a rich field for discovery, in 'our mysterious relations to

[6] That is, Newton's *Principia.*

[7] John Locke advocated a theory of knowledge in which everything that a person knows is built up from nothing by the process of association of sensory impressions.

[8] That is, medieval scholastics.

Time and Space.' And these relations being fully ascertained, the way will be cleared for a discussion of the celebrated question,—Whether spirits can pass from one point of space to another, without passing through the intervening space.

From Metaphysics, the reviewer passes to Politics. 'But the trail of the serpent is over them all.' Mechanism has twisted his coils fast even about the slippery politician. 'No where,' complains the writer, 'is the deep, almost exclusive faith we have in Mechanism more visible, than in the politics of the time.' If this had been written within the last month or two, we should request to be informed, by what rare combination of mechanical powers the recent political changes have been effected in Europe.[9] Truly, there is something in these vast movements, which rather looks as if mind were the mover. Mr. Canning predicted that the next war in Europe would be a war of opinions. That war is not yet commenced. But, to use the language of our author, the *revolutionary machine* is working with a tremendous *momentum,* such as the world never before witnessed. But that we may not mistake the writer on this subject, we quote his words. 'It is no longer the moral, religious, spiritual condition of the people that is our concern, but their physical, practical, economical condition, as regulated by public laws. Thus is the *body politic,* more than ever, worshipped and tended; but the *soul politic,* less than ever.' We are almost tempted to believe that this was intended particularly for the United States, where 'public laws' have left the 'religious and spiritual concerns' of men, exclusively to themselves and their God. But then the writer must, on his own principles, allow us due credit for managing to get along without that expensive and complicated piece of machinery, a *Church Establishment.* Time was, it must be confessed, when the 'soul politic' was more cared for by government, than it is any where at present. But this sort of care has always been found to require a great deal of machinery, and among the rest, the rack, the fagot, and the axe. The writer, therefore, in his anxiety for the *soul politic,* seems to be placed in a dilemma. He has such an antipathy to machinery in general,—and to that above mentioned, we will do him the justice to suppose, in particular,—that he would probably reject the only means, by which governments have hitherto been able to 'tend' the souls of their subjects.

Having proceeded thus far in his assault upon notions, which must be allowed to possess a very general currency, the writer proposes a new illustration, or, perhaps we should say, theory of our nature, which he supports with great vivacity and learning; and which, that we may do him no injustice, we state in his own words. 'To speak a little pedantically, there is a science of Dynamics in man's fortunes and nature, as well as of Mechanics.' This would hardly carry perfect conviction to the mind, without the following lucid explanation. 'There is a science which treats of, and *practically* (?)

[9] The revolutions of 1830.

addresses the primary, unmodified forces and energies of man, the *mysterious* springs of Love, and Fear, and Wonder, of Enthusiasm, Poetry, Religion, all of which have a truly *vital and infinite character,* as well as a science which practically addresses the finite, modified developments of these, when they take the shape of immediate "motives," as hope of reward or fear of punishment.' Having thus stated his theory, our author illustrates it by several examples, of which we shall notice one or two. Among others, the French Revolution—not the recent glorious one—is drawn in. 'The French Revolution had something higher in it than cheap bread and a *habeas corpus* act. Here too, was an Idea; a Dynamic, not a Mechanic force. It was a struggle, though a blind, and at last an insane one, for the infinite, divine nature of Right, of Freedom, of Country.' We do not exactly understand what is meant by 'a struggle for the infinite, divine nature of Country.' If by saying, that 'cheap bread and a *habeas corpus* act' were not the motives of the Revolution, the author mean[s] that neither the wants of the populace, oppressed by misgovernment, nor the political theories of the Philosophers occasioned this explosion, we can only say that he denies it to have been produced by the action of those causes, to one or both of which it is universally ascribed.

Another example, scarcely less unfortunate, is the Christian Religion. We desire always to approach this subject with the most profound reverence. And when we are told by the reviewer, that 'the Christian Religion, under every theory of it, in the believing or the *unbelieving* mind, is the crowning glory, or rather the life and soul of our whole modern culture,'—we most cordially concur in what we can understand of the panegyric. Of religion in the believing mind, too much cannot be said. How religion in the *unbelieving* mind can be a crowning glory, we are at loss to conceive. But our chief concern is with the assertion, that the Christian Religion has been promulgated 'Dynamically, and not Mechanically.' This is in direct conflict with our historical information. It seems to us, that there has been, unfortunately, altogether too much machinery employed in the propagation of Christianity. In the beginning, we know that it was not so. Then the simple but powerful 'preaching of the word' was sufficient. But afterwards, external helps were employed in such a degree, as to suggest to the sceptical historian, Gibbon, the idea of accounting for the establishment of Christianity, exclusively by human means.

We can notice but a few more complaints of this dissatisfied writer, and those briefly. He is very much scandalized, that 'we should have our little theories on all human and divine things,' and particularly that 'even Poetry is no longer without a scientific exposition.' But wherefore should it be? Does the poet merely rave? Is his mind lawless in its wanderings? Or does it, according to Dr. Channing, obey higher laws than it transgresses? If so, we can perceive no harm or absurdity in a 'scientific exposition.' Such an one has been given by the eloquent author just mentioned, in his remarks on the poetry of Milton; and a passage of more transcendent beauty is not to be found in our mother tongue.

Another cause of complaint with our author is, that 'our first question with regard to any object is not, *What is it?* but *How is it?*' This is equivalent to saying, that it is the fashion of the present age not to analyze; and a suggestion wider of the fact could not easily be made. Every one knows that Chemistry, which more than any other science is the offspring and growth of this age, is one perpetual reiteration of the question, *What is it?* The author would find it difficult to name a substance to which Chemistry has not put this question, and received a satisfactory answer. But he would have us go still further, and waste our investigating energies upon fruitless inquiries into the essence of matter and mind. We know that there has been a strong propensity among men to press their discoveries to this verge; and that even Newton was so far beguiled by his 'wish to know,' as to speculate upon the nature of the cause of gravitation, after he had ascertained its laws. But we had supposed, that mankind were now generally agreed as to the inutility of thus invading the Deity's inscrutable mysteries. If the clear teachings of Lord Bacon on the subject[10] were not sufficient, one would think that the practical *reductio ad absurdum* exhibited in the incompatible, yet equally plausible hypotheses of Berkeley and Hobbes, would be sufficient. One resolves every thing into matter, the other resolves every thing into mind; while the only satisfactory proposition, which both conclusively demonstrate is, that the resolution of such questions is beyond the capability of the human mind. Reason affords no clue to guide those who plunge into the labyrinths of mystic speculation.

On the whole, we have no wish to disguise the feeling of strong dissatisfaction, excited in us, by the article under consideration. We consider its tendency injurious, and its reasoning unsound. That it has some eloquent passages must be admitted, but when we hear distinguished philosophers spoken of as 'logic-mills,'—the religion of the age as 'a working for wages,'—our Bible societies as 'supported by fomenting of vanities, by puffing, intrigue, and chicane,'—and all descriptions of men 'from the cartwright up to the code-maker,' as mere 'mechanists;' when we further hear 'the grand secrets of necessity and free-will,'—'our mysterious relations to time and space,'—and 'the deep, infinite harmonies of nature and man's soul,'—brought repeatedly forward under the most varied forms of statement, as the legitimate objects of philosophical inquiry, and the most illustrious of the living and the dead, men whom we never think of but as benefactors of our race, made the objects of satire and ridicule, because they have preferred the *terra firma* of mechanical philosophy to the unstable quagmire of mystic conjecture;—we find it difficult not to regard the Essay rather as an effort of paradoxical ingenuity,—the sporting of an adventurous imagination with settled opinions,—than as a serious inquiry after truth.

[10] Bacon was believed to have advocated the study of facts, as opposed to "mere" speculation.

Indeed the writer himself seems to think, towards the end, that he has gone too far; and deems it prudent, in contradiction, as it seems to us, to the assertion first quoted, as well as to the whole tenor of the article, to insert the following *saving clause:*—'It seems a well ascertained fact, that in all times, reckoning even from those of the Heraclides and Pelasgi,[11] *the happiness and greatness of mankind at large, have been continually progressive.'* This is one of the few assertions in the article, in which we altogether agree with the author. We do entertain an unfaltering belief in the permanent and continued improvement of the human race, and we consider no small portion of it, whether in relation to the body or the mind, as the result of mechanical invention. It is true, that the progress has not always been regular and constant. In happy times it has been so rapid, as to fill the benevolent with inexpressible joy. But anon, clouds have gathered over the delightful prospect,—evil influences, but not mechanical, have operated,—evil times have succeeded,—and human nature has undergone a disastrous eclipse. But it has been only an eclipse, not an extinguishment of light. And frequent as these alternations have been, mankind are found to have been constant gainers. The flood has always been greater than the ebb. Each great billow of time has left men further onward than its predecessor. This could be proved, if necessary, by a thousand references. Darkness has indeed given a name to some ages, but light on the whole has immensely preponderated; and it is this conviction which nerves the heart and invigorates the arm of philanthropy. They who feel this divine impulse, know that the labors of kindred spirits in past ages have not been in vain. They see Atlantis, Utopia, and the Isles of the Blest, nearer than those who first descried them. These imaginary abodes of pure and happy beings, which have been conceived by the most ardent lovers of their kind, we delight to contemplate; for we regard them as types and shadows of a higher and better condition of human nature, towards which we are surely though slowly tending.

But let us not be misunderstood. The condition we speak of, is not one of perfection. This we neither believe in, nor hope for. Supposing it possible in the nature of things, it would be any thing but desirable. For with nothing left to achieve nor gain, existence would become empty and vapid. But if, with this explanation, our views should pass for visionary, we cannot help it. We cannot go back to the origin of mankind and trace them down to the present time, without believing it to be a part of the providence of God, that his creatures should be perpetually advancing. The first men must have been profoundly ignorant, except so far as the Supreme Being communicated with them directly. But with them commenced a series of inventions and discoveries, which have been going on, up to the present moment. Every day has beheld some addition to the general stock of information. When the exigency of the times has required a new truth to be revealed, it has been revealed. Men gifted beyond the ordinary lot, have been raised up for the

[11] The ancestors of the ancient Greeks.

purpose; witness Cadmus, Socrates, and the other sages of Greece, Cicero and the other sages of Rome, Columbus, Galileo, Bacon, Newton, and the other giant spirits of modern times. We cannot regard it as an abuse of language to call such men inspired, that is, preeminently endowed beyond all their contemporaries, and moved by the invisible agency of God, to enlighten the world on subjects, which had never till they spoke, occupied the minds of men. In other words, we believe that the appearance of such men, at the exact times when all things were ready for the disclosures they were to make, was not the result of accident, but the work of an overruling Providence. And if such has been the beneficent operation of Providence upon the minds of men in all past times,—if whenever a revelation was needed, He has communicated it, and in the exact measure in which it was needed,—how can we, without irreverence, adopt any other conclusion, than that He, who changeth not, will still continue, through all future time, to make known through gifted men, as fast as the world is prepared to receive them, new truths from His exhaustless store?

10. A Distinguished Physiological Theorist Attacks New Applications of Chemistry. 1843

Justus von Liebig (1803–1873) was a great innovator in chemistry. Likewise he was an enthusiastic, effective, and dogmatic advocate of the importance of chemistry in general and organic chemistry in particular. When his book, *Animal Chemistry* (American edition, 1842), was published, great numbers of scientists the world over were exposed to his scorn, for Liebig asserted that life processes could be shown to be chemical in nature, and the physiological theories of that day were, therefore, unnecessary. He stirred up opposition from those who felt that he represented bad science rather than scientific progress.

In this selection from the work of a representative American opponent, three major issues can be discerned: the nature of the materials of organic chemistry; whether chemical processes could explain life processes better than established physiological systems; and if there is a difference—the presence of the "vital principle"—between living and dead matter (Liebig never did really deny this himself). Some of the concern here is religious. And some of it is directed primarily to the preservation of well-established physiological theory.

CHARLES CALDWELL (1772–1853) was born in North Carolina and orphaned while still a boy. After teaching school for some time he studied medicine at the University of Pennsylvania. He practiced medicine in Philadelphia and was active in medical circles there. For a time he edited a literary journal. In 1819 he helped reorganize Transylvania University in Lexington, Kentucky, and he remained there until 1837, when he went to another medical school, in Louisville.

See Frederic L. Holmes, "Introduction," in Justus Liebig, *Animal Chemistry, or Organic Chemistry in Its Application to Physiology and Pathology,* trans. by William Gregory and John W. Webster (New York: Johnson Reprint Corporation, 1964 reprint), vii-cxvi; Timothy O. Lipman, "Vitalism and Reductionism in Liebig's Physiological Thought," *Isis,* 58 (1967), 167–185; Emmet Field Horine, *Biographical Sketch and Guide to the Writings of Charles Caldwell, M.D. (1773–1853)* (Brooks, Kentucky: High Acres Press, 1960); Herbert S. Klickstein, "Charles Caldwell and the Controversy in America over Liebig's 'Animal Chemistry,'" *Chymia,* 4 (1953), 129–157; George H. Daniels, *American Science in the Age of Jackson* (New York: Columbia University Press, 1968), chap. VII.

PHYSIOLOGY VINDICATED, In a Critique on Liebig's Animal Chemistry*

Charles Caldwell

Having experienced, during a lifetime, the toils and annoyances of controverting unsound and pernicious doctrines, it can hardly be thought unexpected, much less surprising, should I sicken at the prospect of a new collision. Nor is this the only reason that restrained me. The interruptions and avocations, arising out of my connection with a young school of medicine, whose claims on the labor and services of its officers were numerous and pressing, have left me but little leisure, either to collect the materials necessary in such a contest, or, when collected, to arrange and employ them with an efficiency corresponding to the important interest which truth and science have at stake on the issue.

Under these circumstances I have repeatedly recommended the undertaking to others, and urged them to engage in it, with all the arguments and incentives to action I could devise and bring to bear on them. But in vain. No effectual effort was made by them to protect truth from the deluge of error that was steadily advancing.

Nor had I yet taken any action myself, preparatory to the conflict, in which I was destined to be ultimately involved. Though fully aware of the encroachments of mischief, I had made no attempt, except by occasional

* Charles Caldwell, *Physiology Vindicated, In a Critique of Liebig's Animal Chemistry* (Jeffersonville, Ia. [*sic*]: *A. S. Tilden,* 1843), pp. vi-xvi, 1–7. Slightly abridged.

animadversions in my public lectures, even to impede its progress—much less to encounter and suppress it. Still looking and hoping for aid in the enterprise, from the exercise of other pens, I was constrained to withhold the tribute of my own.

During this state of inaction and expectancy on my part, a new foe to the *philosophy of life*[1] appeared in resolute and vigorous operation, in the person of Professor Liebig, much more formidable than any of his predecessors. Without any inordinate extravagance of hyperbole, well might he, in his relation to the science of living organized matter, be likened to Attila, of whom, under the sobriquet of the "Curse of God," it was figuratively declared by him, that beneath the hoofs of his war-horse the very verdure of the earth where he trod was blighted. So certain, sudden, and numerous were his conquests, and so complete his triumphs, that the *"veni, vidi, vici"* of Cæsar seemed fitly enough to characterize the sweep of his achievements. He had but to speak, and all were enchanted—but to proclaim his dogmas, and they were swallowed as pearl-drops from the fountain of truth.

By those who are strangers to the circumstances of the case, it will be very naturally supposed that this immense flood-tide of popularity and influence, accompanied by such unprecedented success in the work of proselytism, must have been owing to the uncommon merit, scientific or literary, or both, of Professor Liebig's volume on Animal Chemistry. No such thing! His success was attributable almost entirely to two other very different causes—the high reputation, whether solid or not, which he had already acquired by his work on Agriculture—and to the bold and confident tone and manner, in which he made known his statements, calculations, and opinions.

It requires but a very moderate acquaintance with mankind to know, that the mass of readers are much more influenced by the strength of resolution, firmness, and self-confidence of an author, which every one can *feel,* than by the strength of his reasoning, the value of his matter, or the excellence of his style, of which but few can *judge.* And in such resolution, firmness and confidence, Professor Liebig can hardly be surpassed. Of the matter, argument and style of his book, I have spoken in the [present] essay. And I venture to say, without hesitation, that it is the impressive manifestation, in Animal Chemistry, of the former qualities, much more than of the latter, that has so fascinated the *multitude* of readers, and commanded their approval. It is thus that the chieftain of the most confident address, the boldest aspect, and the most gallant bearing, is most certainly and cheerfully followed by his troops, whether he lead them to defeat or victory—to glory or the grave.

Nor is the worst yet told. The Professor's own voice was not raised alone in the diffusion of his doctrines. Hosts of propagandists soon issued forth in Europe and America, and, waxing warm in the new faith, became apparently

[1] That is, vitalism, the belief that no vital phenomenon, such as digestion, can be explained on purely physical and chemical principles, but that every action is conditioned by an unknown force, higher in its nature and distinct in kind from all other forces.

more ardent in their zeal, and more authoritative in their pretended apostleship to instruct and reform, than their idolized leader. Even numerous periodicals, some of them of high and long-established reputation, gave their sanction to the notion which metamorphoses man, with all his attributes, corporeal and mental, into a *chemical product!* For, notwithstanding the declaration, by Professor Liebig, of his belief in the superintendence of vital laws over the formation and general economy of living organized beings, every exposition he gives of the functions of those beings, is *strictly chemical—and he himself has pronounced it so.* And this I say is as true in relation to the brain, the special organ, by which alone the mind manifests, or, in our present state of being, *can* manifest itself, as it is of any other portion of the body. It is therefore I have said that the Professor virtually refers to chemistry, as the source of man's mental, as well as of his corporeal, attributes. In truth, *in the details of his hypothesis,* he makes the entire man as explicitly and positively a *chemical machine,* as is the stove or the steam-engine, to which he compares him, and by which he even attempts to illustrate some of his most important functions. Though I shall not assert that the hypothesis of our author, if completely carried out, would involve *materialism,* in its grossest condition; yet, were I actually to do so, I should find but little difficulty in making good my position.

The interests of the *philosophy of medicine,* which I am deputed to teach, being thus endangered, by the celebrated German Professor and his followers, it did not comport with my sense of either moral or official duty, to remain longer inactive. Insufficient, therefore, as my preparations were, I no further hesitated, but resolved, that, however limited my tribute might be, it should be emptied, without further delay, into the coffer of truth. And of this resolution, two or three public lectures, delivered last winter, were the earliest product. And my second contribution to the same end, is this essay, in which are embodied the entire substance of those lectures, in a style, manner, and arrangement somewhat altered, and accompanied by much additional matter.

Though the entire drift of the essay may seem to be levelled at the writings of Professor Liebig alone; yet such is not the fact. Though I was first moved to engage in it, by the Professor, it is notwithstanding pointed alike against the whole school of chemical physiologists. For, in whatever manner, or to whatever extent, they may be varied in detail, all the fundamental principles of that school are the same. Its settled creed, which every member of it professes, and to which he strictly adheres, is, that real chemical influences are in some way concerned, in the original institution, and the subsequent performance, of the functions of living organized beings—that if those influences do not themselves perform the entire work, they are *essential* to its performance.

Of such a creed it may be safely asserted, and proved if requisite, (as clearly appears, if I mistake not, in my [present] essay) that is it condemnable in its nature and all its relations—that it is groundless in itself—degrading in its tendency—and pernicious in its effects. It is groundless, because it is at war with truth and the order of things devised and established by the

Creator—degrading, because its influence is to lower the nature and standing of our race—and pernicious, because it saps the foundation of that sentiment of dignified and salutary pride, which all men should cherish, as one of the strong-holds of honor and virtue.

Convince any man, however high his political and social rank, and his influence among his fellows, that he is nothing but an aggregate of oxygen, nitrogen, carbon, and hydrogen, and a few other lifeless ingredients, put together, fashioned, and held together, by the same affinities, and governed, in the performance of his functions, by the same laws that preside *in* and *over* masses of dead matter—convince *any* man of this, and you necessarily diminish his self-respect, and render him comparatively indifferent to his actions, and regardless of his destiny. Convince mankind at large of this, and you brutify them. Degraded in their own estimation, and approximated in their belief to masses of brute matter, their feelings and conduct will conform to their view of their humbled condition. For, that a sense of high and honorable descent and condition influences morals and actions, as well as manners and bearing, is a maxim as true as any other that belongs to the history and philosophy of man.

Not only then is chemical physiology physically groundless; it is morally pernicious. Most assuredly it would be so, could the world at large be induced to believe in Professor Liebig's extraordinary modification of it. But in the cool sobriety of his common-sense moments, we deem it impossible for him to believe in it himself. He does so only in his system-building—I should rather perhaps call them his *castle-building* moments. Nor do his followers believe in it, except *as* his followers, who, without examination, receive *as oracular* all that he utters. Those gentlemen no more believe that their systems are warmed, as their parlors are warmed and lit, by the *real chemical* action of oxygen on carbon and hydrogen—that their chyme[2] is formed in their stomachs on the same principles, and by the same sort of process, with those of "putrefaction" and "fermentation"—and that bile is secreted from the hepatic blood by chemical agency—they no more soberly and deliberately believe these notions, than they believe in the fansasy of *perpetual motion,* or in the practicability of forming or discovering the *philosopher's stone.*[3] In truth the two latter sorts of hallucination are less gross and mischievous than the former one. On them no practice involving danger or destruction of health and life can be founded. But of the other the reverse is unquestionably true. Chemical physiology and pathology, and chemical practice, the result of them, have slain their millions. From them have arisen the unspeakable mischiefs and miseries of humoralism.[4]

[2] The mass of food in the stomach mixed up with the digestive juices as it passes from the stomach.

[3] A substance known from alchemy which if discovered would be able to transmute or change the inferior metals into gold.

[4] That is, medical practice based upon the belief that life, health, sickness, disease, and personality are all dependent upon the quantities and qualities of the body juices such as bile, blood, and phlegm.

I do not positively *assert* that oxygen, carbon, nitrogen, and hydrogen have no concern in the production of vital temperature, or in the formation of bile or urine; because I do not positively *know* that they have none. And my assertion never transcends my knowledge. But I do assert, that those gases do not act on each other in the living system of man, and other forms of vital organized matter, precisely as they do in the laboratory of the chemist. Nor can the reason of this assertion be concealed or resisted. It is that the gases do not, in their mutual action, in the two places, produce the same results. In the laboratory they produce water, carbonic acid gas, nitric or nitrous gas, ammoniacal gas, and a few other aeriform substances. And in the living systems of animals and vegetables they give rise to thousands of different modifications of matter, under the names, in the former, of chyle,[5] blood, flesh, brain, bone, tendon, cartilage, skin, hair, hoof, and secreted fluids—and, in the latter, of sap-juice, wood, bark, leaves, flowers, fruit, and also of innumerable kinds of secreted substances.

Such, I say, are the incalculable differences between the products, from the same elementary matters, in the dead laboratory of the chemist, and in the living one, composed of the systems of animals and vegetables. And yet, according to the notion of Liebig and his disciples, chemistry is the immediately productive source of the whole!!

My appeal is to the deliberate judgment and discrimination of the enlightened and reflecting portion of my readers, and of the world, whether such a fantasy be not tantamount to *actual monomania!* and whether it ought not to be so regarded! As far as my own view of the matter may avail, I positively consider it a form of mental derangement as genuine in character and nearly as extraordinary in degree, as that of the bedlamite, who believes his head to be a globe of glass, and his legs rolls of butter—and who implores his visitants not to touch the former, lest they break it, and turns the latter from the fire, lest the heat should melt them! Nor is it less coarse and debasing in its nature! A figure made up out of butter and glass, is as elevated and refined, as one consisting of charcoal and gas, drawn together and held together by chemical attraction.

For the *practical* overthrow of the hybridous doctrine respecting life and living productions, maintained by chemists, and, at the same time, to render it an object of ridicule, the following expedient was, many years ago, devised and executed.

A chemical man-maker, having publicly boasted of the soundness of his science, and the paramount value and standing of his art, was challenged to a debate on the subject, by a vitalist, who was resolved to test the resources of his antagonist by the sternest of ordeals, and to bring the matter at issue to a decision, from which there would be no appeal. The challenge was accepted, the parties met, and the assemblage of spectators and listeners was large and highly intelligent. The vitalist entered the lists, provided with every

[5] A white or milky substance that represents the nourishment drawn out of the chyme in the process of digestion and conveyed by the circulatory system to the various parts of the body.

thing necessary to enable him to conduct the trial in a straight-forward, practical, and demonstrative manner.

The chemicalist (may I coin a word for the occasion) who was himself an expert practical chemist, rose at the call of the presiding officer, adjusted his costume, assumed a magisterial air and attitude, and opened the exercise with what all present considered a fluent, some an ingenious and perhaps an able, and himself an irresistable [*sic*] and conclusive harangue. When he had finished and resumed his seat, the vitalist, with a curl of his lip half expressive of scorn and half of sarcasm and mischief, thus addressed the chairman of the meeting.

Do I understand the gentleman, who has just concluded his discourse, to contend, that every function and every functional product of living organized matter are the results of *chemical* agency? Such, Sir, is my doctrine, replied the chemicalist, with a tone of confidence, and an air and look in no ordinary degree consequential and sapient. And then came on the pith and point of the contest.

The vitalist, lifting a small covered dish, removed the lid, and, presenting it to his opponent, said; here, Sir, is a small quantity of well-prepared beef, potatoes and gravy—Will you oblige me by taking it into your laboratory, and changing it into genuine chyme? With an air of embarrassment, the chemicalist replied—really, Sir, I cannot, otherwise than by eating it. That I can do myself, returned the challenger, and a *pig* can do the same, as well as either of us.

The vitalist, sensible of the advantage he had gained by this manœuvre, from the effect produced by it on the spectators, proceeded in his scheme—Here, said he, is some chyme taken from the stomach of a dog, and in this vessel is another quantity procured from the stomach of a pig; will you convert both, or either of them, into chyle for me? C. Pray, excuse me, Sir—I cannot.

V. Here, then, is some chyle, extracted from a calf; and here is some more taken from a full grown cow—Do me the favor to change them into blood. C. Sir, it is out of my power to comply with your request.

V. That is very singular, considering the sentiments you uttered in your address, and the express avowal you subsequently made. I present you again with four several vessels, each containing a portion of blood. This is the blood of a half-grown hog; will you convert it for me into pork—this of a cow, from which I will thank you to prepare me a bit of beef—this from the vena portæ[6] of an ox, from which I wish you to secrete me some bile—and this from the renal artery of a dog, from which I ask the favor of you to prepare me some urine.

I am surprised, answered the discomfitted chemicalist, with a look of mortified pride, and in a tone of embittered pettishness, that you so un-

[6] The large vein which conveys the blood from the intestines to the liver.

reasonably persist in asking *me* to do what you know that no chemist in the world can do.

Indeed! said the vitalist, with apparent surprise—You told us just now, that the processes I have proposed to you are purely chemical. Why then cannot you and other chemists perform them?

C. Because, Sir, it is impossible. Nothing but the nature of a hog can turn hog's blood into pork, and nothing but the nature of a cow, the blood of a cow into beef. And you know that the liver alone can make bile out of vena-portæ blood, and that the kidneys alone can make urine out of renal blood—I thought so myself, was the laconic reply, which the chemicalist received.

V. Here then is some goose-blood. Pray, Sir, oblige me, by making an egg out it. This was too much for the discomfitted chemicalist to bear. He therefore, answered with great bitterness—Sir, do you mean to insult me! By no means, replied the vitalist; my only object is, to give you a fair opportunity to prove the soundness of the doctrines you hold, and to manifest the extent of your own chemical skill and dexterity.

Notwithstanding this open disclaimer, on the part of the chemicalist, accompanied by such mortified pride and blighted hope, the pertinacious vitalist, resolved on the irretrievable overthrow of his *victim* (antagonist now no longer) was about to propose to him other vital metamorphoses, for the trial of his chemical efficiency, when an irrepressible explosion of laughter from the audience on account of what they supposed to be the meaning of the *goose-egg* proposition, put an end to the contest, and suggested to the crest-fallen chemicalist the expediency of retiring from the place of his own defeat, the downfall of his doctrine, and the withering of the bay-wreath his fancy had woven—a measure which he unceremoniously carried into effect.

And could Professor Liebig, backed by the boldest and ablest of his followers and advocates, be induced to engage in a similar encounter, he would experience a similarly humiliating defeat. Under this head I shall only add, that were chemical physiologists to acknowledge publicly, and in a spirit of good faith, that Animal or Organic chemistry differs essentially from the chemistry of the laboratory, in its principles and laws, modes of action and forms of products, and that it is therefore fundamentally a different branch of science—could they be induced to comply with this requisition of truth and duty, vitalists would cease to be their opponents, and would cordially harmonize with them, as joint laborers in a common cause—the promotion of science, and the benefit of man. But as long as they shall continue, in defiance of all testimony to the contrary, to proclaim the identity with each other of the processes of chemistry and the organic functions of living matter, and thus palm on the unobserving and unthinking portion of the public their groundless and delusive notions, their fate ought to be, and I cannot doubt will be, to experience at all times determined and resistless opposition, and ultimately to sustain a hopeless discomfiture—the unavoidable doom of error and its effects. . . .

It has been suggested to me that, in consequence of the unceremonious and perhaps severe tone and manner, in which I have expressed myself, in some parts of my essay, I may be deemed wanting, in friendliness and respect toward chemistry and its cultivators. Though a charge to that effect would in no degree disquiet me, I shall notwithstanding so far notice it, as to say, that, should it be made, it will be groundless and unjust. Between chemists and myself the friendliest terms have always subsisted. Toward chemistry itself, in its true character, and when operating within its proper sphere, my disposition is as amicable and respectful as that of its most enthusiastic votaries. It is only to its misapplications, perversions, and other abuses of it, that I am inimical. And my reason for being so is, that, when thus distorted and abused, it is itself an enemy of truth, an abundant source of false doctrines, and a perpetrator of vast and diversified mischief. . . .

Notwithstanding *the extravagance* of the hosannas, with which "Animal Chemistry" has been ushered to the public, and is still idolized by a large class of readers, we neither pronounce them altogether undeserved, nor are inclined, in a spirit of disparagement, to pluck even a leaflet from the overloaded bay-wreath they are designed to bestow.

Not to receive the production (faulty as, in many respects, we think and shall endeavor to prove it,) with a welcome both courteous and cordial, and speak of it in terms of respectful consideration, would be not only too unfashionable, but in some measure also, *hypercritical* and *unjust* in us. While, therefore, a becoming degree of observance and regard toward the decision of others, who differ from us in opinion, withholds us from the *former* act of delinquency, the determination of our own judgment and the mandate of our conscience deter us from the *latter*. With perfect sincerity then, as relates to certain topics, and to a given extent, we concur in the verdict the work has received.

We cannot however, as already intimated, so far defer to fashion in opinion, as to echo back every trumpet-note of acclamation lavished on our author by the enthusiasm of readers. Far from it. On the contrary, from the correctness of many of them we positively dissent. Those of them we thus refer to (as will hereafter appear,) are not merely extravagant in degree; most of them are also groundless in fact. So true is this that they remind us of the well known apothegm of the satirist, as pungent as it is sound, that, *"Praise undeserved, is satire in disguise."*

We are compelled to believe that the exuberance of applause bestowed on Professor Liebig, as a philosopher and a writer, will prove injurious to him. It is much easier to procure praise by error and accident, than to retain it by truth and merit. As relates to the Professor's abilities and qualifications, expectation will be rendered so unreasonably high, that he will, in some future work, almost inevitably fail to meet and fulfil it. If so, then will the rapidity of his fall equal, probably, that of his rise.

These sentiments we deem especially true, as respects the *prophetic and sweeping allegation* of Dr. Gregory, the translator and annotator of the work, of the Editor of the London Quarterly, and of the American Editor, Professor Webster, that Animal Chemistry is destined to mark the "commencement of a new era in Physiology." To produce this effect, (supposing the era to be an improved one,) we fearlessly pronounce it as utterly unqualified, as it is to revolutionize theology, mathematics, or political economy.

That the work may usher in a new and improved era in the analysis of both the organic and inorganic *products* of physiology, as well as of the substances out of which those products are formed, we neither affirm nor deny. Our reason for thus forbearing to decide, is, that we profess not to be such a competent judge in the matter, as to be justifiable in assuming the office of an umpire. And we deem it much less discreditable to confess our ignorance and inability, than to make *an empty pretension* to knowledge and competency. Yet are we strongly apprehensive that many persons will promptly decide for themselves the question before us, and obstinately persist in their decision, who, to say the least, are as moderately versed in physiology, as we are in chemistry.

Nor, in denying to Professor Liebig the honor of having introduced a new era in the science of physiology, do we either speak or design to speak of him, in terms of disparagement. We only treat him as a man who, though he leads his contemporaries in some things, does not lead them in every thing. We acknowledge that he *may have,* and probably *has,* surpassed both his predecessors and contemporaries in the analysis of the products of *living* organic substances, and also in the analysis of those substances themselves, when in a *lifeless condition.* But farther, in his career of improvement, we feel persuaded that he has not yet advanced. And we are strongly inclined to think, that, with the glory of an achievement so distinguished as that which we are willing to concede to him, both he and his admiring friends and followers have reason to be satisfied. To be, at this enlightened and rapidly improving period of the world, an acknowledged leader, in any important branch of science, is glory sufficient for a man of his age—or indeed of *any age.*

That we may treat our subject with the greater clearness and precision, it is necessary for us to observe, that, according to the definition and understanding of technical terms, at the present time, the names *chemistry* and *physiology* are given to branches of science whose processes are not only widely and essentially different in their principles, modes of action, and uniform products, but, in many respects, the actual opposites of each other.

Taken in the full extent of its *original* and *true* signification, physiology is an *exposition of the knowledge of nature.* As such, therefore, chemistry is so far identified with it, as to be included in it as one of its departments. According, however, to the present acceptation of the term, the signification of physiology is much more restricted. It means an exposition of the state, action, and products of living organized matter, whether vegetable or animal, when natural and healthy in its condition and functions—and nothing more.

True; we now also employ the phrase, *morbid physiology,* as synonymous with pathology. To the designation, therefore, of the state and functions of living organized matter, in a *deranged* and *unnatural* condition, the expression is likewise applied. But to no form or condition of dead matter, whether organic or inorganic, is it ever appropriated, at the present period. The branches of science, which alone treat of that description of matter, are all included under the name, *Physics.*

Not only different from physiology then, but, in most, if not all, respects, as already mentioned, the very *opposite of it,* is chemistry. To *dead matter alone* is it applicable in its principles, action, and laws. And to that its applicability is universal. Of the whole mineral kingdom it holds the absolute and undisputed control. And no sooner does animal and vegetable matter become perfectly lifeless, than it invades that also, takes full possession of it, and acts on it in conformity to its own laws, without restraint, and almost without limit. And, as if still discontented with the immensity of its possession, and its undisputed sway, it is ardently struggling, through the agency and ambition of its "high-reaching" Professors, to usurp possession of every object and element of the sublunary creation, whether living or dead.—Such is the palpable aim of this aspiring branch of physics, which, far within the memory of individuals now living, was confined, we believe, exclusively to the restricted limits of the mineral kingdom.

Before proceeding further in our inquiry, it may not be amiss, but may perhaps aid us in shaping our course, and attaining our end, to inquire first, into the true meaning of the phrase—*improvement of physiology*—and to ascertain *wherein such an amendment of its condition consists.* And having, as best we may, accomplished this, to compare, in the next place, the result of our inquiry with what Professor Liebig has (not merely *written about or reported,* but) *actually done.* In pursuance of this purpose, we may here remark, what no one will deny, that improvement in science, of whatever description, can be effected only by one of the two following achievements—*the discovery of new facts—or a new and more pertinent, practical and useful construction and application of such facts as are already known—or by both united.* And, in the present case, the facts, their construction, application, and uses, must be in strict conformity to *vital* laws. For, as relates to the economy of living organized beings, our author himself concurs with all educated and enlightened physiologists of the day, in the general doctrine, that every function and phenomenon pertaining to it, whether healthy or diseased, is under the control of a *vis vit, or vital force.*

To be entitled then to the *name* and *character* of an *improver* of physiology, much more to that of the *introducer* of a new era in that science, Professor Liebig must have discovered either a new function or functions in living organized matter, a new use of some function or functions already known, or a new, improved, and more profitable exposition of the manner in which one or more faculties or functions are performed. He must have made some new and valuable disclosure, as to the action and economy of some of those organs, which are subject to the government of *vital laws.* But, that he has performed neither of these services, it shall be our business to evince. And, if we mistake not, we shall further make it appear, that his *special expositions* are, perhaps without an exception, in direct contradiction of his *general confession* or acknowledged theory—we mean, of his declaration of his belief in the supremacy of *vital laws,* in the economy of living organized matter.

In further illustration of this topic, and as additional evidence of the soundness of our views in relation to it, we shall here observe, that it neither is, nor can be, by *experiments* in the laboratory of the chemist, that improvements in the science of physiology are to be made. It is by attentive and well directed observations, and well devised and skilfully performed experiments on the *living laboratory* of the animal economy. By this means alone can physiological improvements be made. By this alone can new organs and new fuctions be detected.

When Gall and Spurzheim discovered, in the brain, sundry phrenological organs and their functions, they improved physiology. So did Sir Charles Bell, when he discovered the functions of the sensitive, motory, and respiratory nerves.[7] So did Hall and Muller, when they discovered the *Excito-motory* nerves and their functions.[8] And Dr. Buchanan has done the same, in his discovery of the mode of acting on various parts of the body, by the mesmerical excitement of given and corresponding portions of the brain. The latter gentleman has, with due significancy and propriety, added to the nomenclature of physiology the term Neurology, as expressive of the contributions he has made to that science. Such contributions not only justify the formation of a new technical name—they call for it.

These several discoveries, taken *in mass,* have been truly productive of a "new era" in physiology. But the same cannot be correctly said of all that chemistry has done (yet it has done much,) from the time of Dr. Black of Edinburgh,[9] to that of Dr. Liebig of Giessen. And the same is equally true of whatever it shall, or can, do in all future time. It will operate greatly to the benefit of man, but not by improving the science of physiology.

Were any one to ask us what we mean by the phrase *vital laws*—or a *vital force?* we should be unable definitely to answer the question; because of vitality—its effects alone excepted—nothing is known. We *might* reply however, that, by the expression, we intend to designate a source or power

[7] Bell's discovery (along with Magendie) that motor and sensory nerves comprise separate systems was just beginning to be appreciated at this time.

[8] These eminent anatomists (although Muller is better known as a physiologist) made substantial contributions to knowledge about reflex action.

[9] Joseph Black (1728–1799), a Scottish chemist who rediscovered carbon dioxide.

of action, differing, in its principles, manner, and products, from every other power. We might also pronounce it of a higher order than any other earthly power inherent in matter. Of these truths, most of which may be regarded as so many fundamental principles, or rather maxims, a brief application is now to be made.

That Professor Liebig may have discovered many new facts in *chemistry,* or applied old ones, under improved constructions, to new purposes, more appropriate and useful, than those to which they had been previously applied, we do not, as already stated, either affirm or deny. But we defy the whole of his followers and expositors, to point out a single instance, in which he has effected either of these forms of improvement, actually or avowedly, under the influence and control of a *vital principle.* That he has *not* done this, but has derived from chemistry alone his special and express exposition of every phenomenon, which, from the beginning to the end of his book, he has *attempted* to expound, it shall be our business to demonstrate. To return therefore from this digression (as if it had not been made) to the tenor of our inquiry.

If it be true, as heretofore stated, that physiology is but an exposition of the functions of living organized matter, while chemistry, treats only of dead matter, its properties, laws, and changes (and no competent judge will question the position,) how is it possible that any advancement of the latter branch of science, however original, extensive and valuable, can constitute the "commencement of a new era" in the condition of the former? After the slightest consideration of the matter, no one can possibly, we think, be the advocate of a notion so utterly groundless, not to say preposterous. As well might it be contended that improvements in geography, navigation, or natural history are calculated to effect like improvments in the art of painting, or to produce a new and brighter era in the sciences of mathematics and moral philosophy. If chemistry and physiology be so intimately allied *to,*—we might say identified *with*—each other, that improvements in the former can improve also the latter, of necessity the *converse* must be equally true. Improvements in physiology must, in like manner, contribute to the improvement of chemistry—a position which no chemist, we presume, is willing to admit; and to which we are confident no enlightened physiologist will subscribe.

11. A Local Scientific Society Receives the World's First Description of the Gorilla. 1847

The first description of the gorilla appeared in an American periodical, the *Boston Journal of Natural History,* in December, 1847. The article itself contains many clues as to why this journal was an appropriate place for such a publication. There is also much indirect evidence of the status of natural science—and what kind of natural science—in Boston and its connection with other interests of Americans of that day. The interest in the relationship of the apes to man—which in later years became an overwhelming one—already is evident in this pre-Darwinian work.

THOMAS S. SAVAGE (1804–1880) was a native of Connecticut and attended Yale and Yale Medical School. Before beginning practice he attended a seminary and became also an Episcopal clergyman. As a missionary physician he traveled extensively, especially in Africa. He was also known as a naturalist, not only through this article but also through his work on the chimpanzee. JEFFRIES WYMAN (1814–1874) was a native of Massachusetts and a graduate of Harvard College. He took up the practice of medicine and then taught anatomy at Hampden-Sidney College and, after 1847, as Hersey Professor at Harvard. Much of the later part of his life he devoted to his position as curator of the museum of archeology and ethnology at Harvard.

For interesting discussions see John C. Greene, *The Death of Adam, Evolution and Its Impact on Western Thought* (New York: The New American Library, 1961), especially chap. 6; *Dictionary of American Biography,* XV, 391–392; XIX, 583–584; Robert M. and Ada W. Yerkes, *The Great Apes, A Study of Anthropoid Life* (New Haven: Yale University Press, 1929), chap. 4.

NOTICE OF THE EXTERNAL CHARACTERS and Habits of Troglodytes Gorilla, A New Species of Orang from the Gaboon River

Thomas S. Savage

Osteology of the Same

Jeffries Wyman

* *Boston Journal of Natural History,* 5 (1847), 417–427, 429–430, 432–433, 435, 436–441. Slightly abridged.

Four species of anthropoid Simiae, commonly known as Orangs,[1] have been described by naturalists; of these, three are found in the eastern world either on the continent of Asia, or the islands of Borneo, Sumatra and Java, and a fourth on the western coast of Africa. In the East there exists

I. *Simia satyrus,* Lin. Pithecus satyrus, Geoff.; S. Abellii, Fisch.; L'Ourang outang, Cuvier; Red ourang. This is the species most frequently exhibited in an immature condition, in America and Europe, and is obtained in Cochin China, Malabar, and Borneo.

II. *S. Wurmbii,* Fisch. Pongo Wurmbii, Kuhl. Grand Ourang-outang, Geoff. Dusky orang; from Borneo.

III. *S. Morio,* Owen. Of this species the cranium has been described by Prof. Owen, but its external characters are not yet well known to naturalists.

In Africa is found

IV. *Troglodytes niger,* Geoff. Chimpanzée, Black Orang, Engé-eco, Jocko. This doubtless is the *Barris* of Pyrard de Laval, the *Smitten* of Bosman, *Quimpésé* of De la Brosse, and the *Quojas moras* of Tulpius.[2]

The existence of a second species in Africa, does not appear to have been recognized by naturalists, nor in fact has there been hitherto adduced any evidence on which its existence might be predicated, except the vague statements of the different voyagers and travellers. But these, resting principally on information derived from the natives, and not on the personal observation of the narrator, are in general so mingled with absurd and marvellous accounts, that they have been deservedly regarded as unworthy of credence.

In two of the published narratives, however, the existence of a second species has been distinctly referred to. Andrew Battell, an English sailor, while a prisoner of the Portuguese, in Angola, speaks of the "two monsters," the "Pongo and Engeco."[3] This last, or as it is called by the natives of the

[1] The term Orang, more commonly but incorrectly written Ourang, is strictly applicable to the Eastern species only. Orang is a Malay word, which means a *reasonable being,* and is also given to man and the elephant. Outan means *wild* or *of the woods;* Orang-outan, *wild man.* . . . [*Footnote in original.*]

[2] Later on these apes were all reduced to two types, the orang and the chimpanzee.

[3] The greatest of these two monsters is called Pongo, in their language, and the lesser, is called Engeco. This Pongo is in all proportion like a man, but that he is more like a giant in stature than a man; for he is very tall, and hath a man's face, hollow eyed, with long haire upon his brows. His bodie is full of haire but not very thicke, and it is of a dunnish colour. He differeth not from man but in his legs, for they have no calfe. Hee goeth alwaies upon his legs, and carrieth his hands clapsed on the nape of his necke, when he goeth upon the ground. . . . They cannot speake, and have no understanding, more than a beast. . . . They goe many together and kill many negroes that travaile in the woods. . . . These Pongoes are never taken alive, because they are so strong ten men cannot hold one of them; but they take many of their young ones with poisoned arrows. . . . The strange adventures of Andrew Battell. Purchas, His Pilgrims, Part II. p. 984. London, 1625. [*Footnote in original.*]

Gaboon, the Enché-eco, is the common name for the Chimpanzée, and it is highly probable, though Battell has given no characters by which it might be recognized, that the Pongo was the animal which forms the subject of our communication.

The *"Ingena,"* referred to by Bowdich,[4] in his mission to Ashantee, is probably the Engé-ena of the natives of the Gaboon, though his statement, that the animal was "five feet high and four across the shoulders," detracts from the credibility of his narrative. Whatever doubt may have heretofore existed, the following notices of habits, and external characters, and descriptions of the crania and some of the bones, will serve most satisfactorily to confirm the statements of Battell and Bowdich, with regard to the existence of a second African Orang, and to demonstrate that it is as specifically distinct from the *Troglodytes niger,* as from the Orangs of Borneo and Sumatra. The specific name, *gorilla,* has been adopted, a term used by Hanno, in describing the "wild men" found on the coast of Africa, probably one of the species of the Orang.[5]

I. NOTICE OF THE EXTERNAL CHARACTERS AND HABITS OF THE ENGÉ-ENA.

While on the voyage home from Cape Palmas, I was unexpectedly detained in the Gaboon river, latitude 15′N., and the month of April (1847) was spent at the house of the Rev. J. L. Wilson, Senior Missionary of the American Board of Commissioners for Foreign Missions to West Africa. Soon after my arrival Mr. Wilson showed me a skull, represented by the natives to be that of a monkey-like animal, remarkable for its size, ferocity and habits. From the contour of the skull, and the information derived from several intelligent natives, I was induced to believe that it belonged to a new species of Orang. I expressed this opinion to Mr. Wilson, with a desire for further investigation, and if possible, to decide the point by the inspection of a specimen alive or dead. He entered with interest into the matter, and promised his hearty coöperation. Having been a resident at that point for several years, well acquainted with the chiefs and people, possessing in an eminent degree their regard, and speaking freely their language, he was

[4] "The favorite and most extraordinary subject of our conversations on Natural History was an Ingena compared with Orang-outan, but much exceeding it in size, being generally five feet high and four across the shoulders. Its paw was said to be more disproportionate than its breadth, and one blow of it to be fatal. . . ." Mission to Ashantee, by T. Edward Bowdich, Esq., p. 440. 4to. London, 1819. [*Footnote in original.*]

[5] We arrived at a bay called the Southern Horn; at the bottom of which lay an island like the former, having a lake, and in this lake another island, full of savage people . . . whose bodies were hairy, and whom our interpreters called Gorillae. . . . Though we pursued the men, we could not seize any of them; but all fled from us, escaping over the precipices, and defending themselves with stones. Three women were taken, however; but they attacked their conductors with their teeth and hands, and could not be prevailed upon to accompany us. Having killed them, we flayed them and brought their skins with us to Carthage. The Voyage of Hanno. . . . London, 1797, p. 13. [*Footnote in original.*]

enabled to secure to me advantages of signal importance to my investigations. I did not succeed however, in obtaining the animal, but several crania of the two sexes, and of different ages, with other important parts of the skeleton were received. These have been forwarded for examination, and I now proceed to give the results of my investigations on its *External Characters and Habits.*

It should be borne in mind that my account is based upon the statements of the aborigines of that region. In this connection it may be proper for me also to remark, that having been a missionary, resident for several years, studying from habitual intercourse, the African mind and character, I felt myself prepared to discriminate and decide upon the probability of their statements. Besides, being familiar with the history and habits of its interesting congener, (*Troglodytes niger, Geoff.*) I was able to separate their accounts of the two animals, which, having the same locality and a similarity of habit are confounded in the minds of the mass, especially as but few, such as traders to the interior and huntsmen, have ever seen the animal in question. In this last fact stated we find an explanation of the confusion, inaccuracy and exaggeration which characterize the occasional references that have been made in books to both animals, the sources of such information being transient visitors and voyagers. If it be admissible to base and sustain a proposition on human testimony, then to my mind the existence of this new species of orang, aside from the evidence of its skeleton, is established, and the account that I now submit of its habits, authentic. It is on such grounds, and with such convictions I venture to place my statements on record, leaving them to the future for confirmation or correction.

The tribe from which our knowledge of the animal is derived and whose territory forms its habitat, is the *Mpongwe,* occupying both banks of the river Gaboon from its mouth to some fifty or sixty miles upward. The face of the country as you proceed inward is undulating and hilly, well watered with streams and rivers, and abounding with indigenous fruits. The river is visited for purposes of trade in ivory, ebony and dye-woods by vessels from different parts of America and Europe. In view of this fact it may seem surprising that the animal should be unknown to science, and without its proper place in systems of Zoology. But this is accounted for by the fact that its immediate habitat is back some distance from the coast, and its habits and ferocity such that it is not often encountered. The natives stand greatly in fear of it, and never attempt its capture except in self-defence.

If the word 'Pongo' be of African origin it is probably a corruption of *Mpongwe,* the name of the tribe on the banks of the Gaboon, and hence, applied to the region they inhabit. Their local name for the Chimpanzée is *Enché-eko,* as near as it can be anglicised, from which the common term "Jocko" probably comes. The Mpongwe appellation for its new congener is *Engé-ena,* prolonging the sound of the first vowel and slightly sounding the second.

The habitat of the *Engé-ena* is the interior of Lower Guinea, while that of the Enché-eko or Chimpanzée is nearer the seaboard.

Its height is above five feet, it is disproportionally broad across the shoulders, thickly covered with coarse black hair, which is said to be similar in its arrangement to that of the *Enché-eko.* With age it becomes gray, which fact has given rise to the report that both animals are seen of different colors.

Head. The prominent features of the head are, the great width and elongation of the face, the depth of the malar region, the branches of the lower jaw being very deep and extending far backward, and the comparative smallness of the cranial portion; the eyes are very large, and said to be like those of the Enché-eko, a bright hazel; nose broad and flat, slightly elevated towards the root; the muzzle broad and prominent, lips and chin with scattered gray hairs, the under lip highly mobile, and capable of great elongation when the animal is enraged, then hanging over the chin; skin of the face and ears naked, and of a dark brown approaching black.

The most remarkable feature of the head is a high ridge or crest of hair in the course of the sagittal suture, which meets posteriorly with a transverse ridge of the same, but less prominent, running round from the back of one ear to the other. The animal has the power of moving the scalp freely forward and back, and when enraged, is said to contract it strongly over the brow, thus bringing down the hairy ridge, and pointing the hair forward so as to present an indescribably ferocious aspect.

Neck short, thick and hairy; chest and shoulders very broad, said to be fully double the size of the Enché-eko's; arms very long, reaching some way below the knee, the forearm much the shortest; hands very large, the thumbs much larger than the fingers.

Abdomen very broad and prominent, the hair thinner than on the back, legs bowed like the Chimpanzée's, but the muscles larger or better developed.

No tail nor callosities; a small tuft of hair at the extremity of the os coccygis; the genitalia similar in both sexes to the same parts in the Chimpanzée except their larger size in the male.

Their gait is shuffling, the motion of the body, which is never upright as in man, but bent forward, is somewhat rolling, or from side to side. The arms being longer than those of the Chimpanzée it does not stoop as much in walking; like that animal it makes progression by thrusting its arms forward, resting the hands on the ground and then giving the body a half jumping, half swinging motion between them. In this act it is said not to flex the fingers as does the Chimpanzée, resting on the knuckles, but to extend them, thus making a fulcrum of the hand. When it assumes the walking posture to which it is said to be much inclined, it balances its huge body by flexing the arms upward. They live in bands, but are not so numerous as the Chimpanzée's; the females generally exceed the other sex in number. My informants all agree in the assertion that but one adult male is seen in a band; that when the young males grow up, a contest takes place for mastery, and the strongest by killing and driving out the others, establishes himself as the head of the community. The silly stories about their carrying off women from the native towns, and vanquishing the elephants, related by voyagers and widely copied into books, are unhesitatingly denied. They have been averred of the *Chimpanzée,* but this is still more preposterous. They probably had their origin in the marvellous accounts given by the natives, of the *Engé-ena,* to credulous traders.

Their dwellings, if they may be so called, are similar to those of the Chimpanzée, consisting simply of a few sticks and leafy branches supported by the crotches and limbs of trees; they afford no shelter, and are occupied only at night.[6]

They are exceedingly ferocious, and always offensive in their habits, never running from man as does the Chimpanzée. They are objects of terror to the natives, and are never encountered by them except on the defensive. The few that have been captured were killed by elephant hunters and native traders as they came suddenly upon them while passing through the forests.

Is is said that when the male is first seen he gives a terrific yell that resounds far and wide through the forest, something like *kh—ah! kh—ah!* prolonged and shrill. His enormous jaws are widely opened at each expiration, his under lip hangs over the chin, and the hairy ridge and scalp is contracted upon the brow, presenting an aspect of indescribable ferocity. The females and young at the first cry quickly disappear; he then approaches the enemy in great fury, pouring out his horrid cries in quick succession. The hunter awaits his approach with his gun extended; if his aim is not sure he permits the animal to grasp the barrel, and as he carries it to his mouth (which is his habit) he fires; should the gun fail to go off, the barrel (that of an ordinary musket, which is thin) is crushed between his teeth, and the encounter soon proves fatal to the hunter.

The killing of an *Engé-ena* is considered an act of great skill and courage, and brings to the victor signal honor. A slave to an Mpongwe man, from an interior tribe, killed the male and female whose bones are the origin of this article. On one occasion he had succeeded in killing an elephant, and returning home met a male Engé-ena, and being a good marksman he soon brought him to the ground. He had not proceeded far before the female was observed, which he also killed. This act, unheard of before, was considered almost superhuman. The man's freedom was immediately granted to him, and his name proclaimed abroad as the prince of hunters.

It is said that this animal exhibits a degree of intelligence inferior to that of the Chimpanzée; this might be expected from its wider departure from the organization of the human subject. I could not ascertain that more than

[6] The natives ridicule this habit of the Engé-ena. They call him a fool, to make a house without a roof, in a country where they have so much rain. They say he has not so much sense as a certain bird, which makes a large nest with a tight roof. . . . [*Footnote in original.*]

one or two at most of the young had ever been captured. One was taken and kept for a year by a native, and then sold to a Frenchman, but it died on the passage home. Whether the skeleton was preserved is not known. No information respecting its habits in a state of domestication could be had upon which reliance might be placed.

In the wild state their habits are in general like those of the *Troglodytes niger,* building their nests loosely in trees, living on similar fruits, and changing their places of resort from the force of circumstances. . . .

Here, as at all other points on the coast the orangs are believed by the natives to be human beings, members of their own race, degenerated. Some few, who have put on a degree of civilization above the mass, will not acknowledge their belief in this affinity; such profess to view them as embodied spirits, the belief in transmigration of souls being prevalent. They say that the Enché-eko or Chimpanzée has the spirit of a *Coastman,* being less fierce and more intelligent, and the Engé-ena that of a *Bushman.* The majority however, fully believe them to be men, and seem to be unaffected by our arguments in proof of the contrary. This is especially true of the tribes in the immediate vicinity of the locality. They believe them to be literally "wild men of the woods."

They are generally eaten, and their flesh, with that of the Chimpanzée, and monkeys at large, occupies a prominent place in their "bill of fare."

II. DESCRIPTION OF THE CRANIA AND SOME OF THE BONES OF THE ENGÉ-ENA.

The collection of crania and bones brought from Africa by Dr. Savage, and which served as the basis of the following descriptions, consists of four skulls, two males and two females, one of each in a perfect condition, and all of them adult; a male and female pelvis, the long bones of the upper and lower extremities, and a few vertebrae and ribs.

The crania of the males are much larger than those of the females, and exceed in their longest diameter the skull of a well characterized Negro by two and a quarter inches, and by nearly one inch the diameter across the zygomatic arches.[7]

The sutures were entirely obliterated in one, and nearly so in the other, a condition similar to that of the cranium of the adult Simia satyrus, and of the older crania of the Troglodytes niger, in both of which all sutures sooner or later disappeared.

When viewed laterally, the incisive alveoli[8] in both the Orang and Chimpanzée form a strong projection below the nasal orifice, but are most conspicuous in the former, giving the lower part of the face that remarkable

[7] Bones extending from before the ear to the cheekbone.

[8] Cavities in the jawbone in which the teeth are fixed.

degree of prominence which is so characteristic. In the Engé-ena the outline of the face is straight from below the superciliary ridges to the edge of the incisive alveoli, and when the head is so placed that the edge of the lower jaw is horizontal, the facial line makes with it an angle of about 45°. The facial angle according to the usual method of measuring it (the superciliary ridges excluded) is about 30°. According to Mr. Owen that of S. satyrus is 30°, and that of Troglodytes niger 35°.

The most remarkable peculiarity, one which strikes the observer at sight, is the great development of the interparietal and occipital crests,[9] as well as of the superciliary ridges, all of which give the head great angularity of outline, causing it somewhat to resemble that of S. satyrus, and to contrast with that of T. niger, on which there are no crests, and the superciliary ridges of which, though well developed, are much more curvilinear and smooth. . . .

If the length of the bony roof of the mouth compared with its breadth is an index of inferiority, the Engé-ena certainly occupies a lower position in the animal scale than the Chimpanzée; in the latter the breadth is to the length as 1.5 inches to 2.8 inches, and in the former as 1.5 inches to 3.9 inches. The incisive canals, which in the Chimpanzée open into the mouth by two distinct orifices, in the Engé-ena are but imperfectly separated from each other at their termination; a single foramen on each side exists midway between the incisive foramen and the edges of the alveoli laterally, which is represented in the Chimpanzée by two or more smaller foramina on each side. . . .

The lower jaw presents a degree of massiveness and strength which the anatomist would be led to anticipate from an examination of the great surface for the attachment of the temporal muscles, and the great size of the zygomatic arches. In its general conformation it resembles that of the Simia satyrus, but surpasses it in size, although the projection of the face in the last is the greatest. Its ascending portion is nearly vertical, and contrary to that of the Chimpanzée and Orang; the coronoid process is more elevated than the condyloid.[10]

The dental formula, as in the Orangs and the Catarrhine Quadrumana,[11] is generally the same as that of man, viz. incisors $\frac{2-2}{2-2}$, canines $\frac{1-1}{1-1}$, premolars $\frac{2-2}{2-2}$, molars $\frac{3-3}{3-3}$, total, 32. . . .

Trunk. Of the vertebral column only a few bones were preserved; two adjoining cervical vertebræ probably the fifth and sixth, were remarkable for the great length of the spinous processes, the longest of which, measured from the inside of the spinal canal (posterior face) had a length of 2.4 inches, the longest process in the neck of the Chimpanzée was only 1.1 inches

[9] Bone intersections at the top and back of the skull.

[10] Bony prominences connecting parts of the larger skull bones together at the top and back of the skull.

[11] American monkeys, characterized by nostrils twisted or curved at the end of the snout.

The last dorsal, and the first two lumbar vertebræ were also preserved, all of which present a peculiarity obviously in relation to the natural semi-erect attitude of the animal. In man, the vertical diameter of the body is greater on the anterior than on the posterior face, and it is from this that results the anterior convexity of the lumbar region; in the Engé-ena the anterior diameter is shorter than the posterior, so that there would exist a concavity instead of a convexity anteriorly, thus rendering it almost impracticable to balance the trunk on the base of the sacrum, or to maintain with ease an erect attitude unaided by the arms.

All the long bones of the upper extremity are remarkable for their great size and strength. By reference to the table of admeasurements it will be seen that the scapula is two inches longer, and nearly two inches broader than that of an ordinary man. Instead of having the narrow and elongated form of the Chimpanzée, the bone is more nearly equilateral. . . .

If, in the upper extremity, the Engé-ena approaches nearer to man than his congeners in the relative lengths of the ulna and humerus, he recedes much farther in those of the humerus and femur. In the Chimpanzée the humerus and femur are almost exactly of the same length, but in the Engé-ena the humerus exceeds the femur by three inches, a disproportion very nearly the same as that which exists in the corresponding parts of the *S. satyrus*. . . .

III. GENERAL REMARKS.

From the preceding descriptions there can be no reason to doubt that the Engé-ena is specifically distinct from the Enché-eco or Chimpanzée, the only member of the sub-genus Troglodytes hitherto recognized by naturalists. From the Enché-eco it is readily distinguished,

1. By its greater size;
2. By the size and form of the superciliary ridges;
3. By the existence of the large occipital and interparietal crests in the males, and by rudiments of the same in the females;
4. By the great strength and arched form of the zygomatic arches;
5. By the form of the anterior and posterior nasal orifices;
6. By the structure of the infra-orbitar canal;
7. By the existence of an emargination on the posterior edge of the hard palate;
8. The incisive alveoli do not project beyond the line of the rest of the face as in the Chimpanzée and Orang;
9. The distance between the nasal orifice and the edge of the incisive alveoli is less than in the Chimpanzée;
10. The ossa nasi are more narrow and compressed superiorly;
11. The scapula is more nearly equilateral;

12. The ulna is shorter in proportion to the humerus;
13. The ossa ilia are much broader, more concave on the anterior face, and the anterior spines project farther forward.

The Engé-ena in the strength of the zygomatic arches, in the existence of the occipital and interparietal crests, and in the strength and size of the lower jaw, approximates the Orangs, but is readily distinguished from all those yet described,

1. By its large superciliary ridges;
2. By the straight outline of the face;
3. By the existence of a fifth tubercle on the last molar of the lower jaw;
4. By the existence of a round ligament in the hip joint;
5. By the more anthropoid conformation of the pelvis;
6. In having the cerebral cavity more depressed behind the face;
7. In having the ulna shorter than the humerus.

It should be borne in mind that the above distinctions are not based upon observations made upon a single specimen, but upon the examination of four adult crania of the Engé-ena, two males and two females, and upon six adult crania of the Enché-eco.[12] In no one of these last has there been found any approach to an interparietal crest, nor have they in any instance deviated materially in their dimensions from those given. . . . The temporal ridges are generally separated from each other by a space varying from half an inch to one or two inches, according to age, but in none of them is to be seen even a rudiment of the interparietal ridge.

The skull of the Engé-ena recedes much farther from the human type than that of the Enché-eco, in its greater development of the cranial crests and ridges, in the greater elongation of the upper jaw downwards and forwards, in the length of the bony palate, no less than in the much more brutal and ferocious expression of the face; in this last respect it even surpasses the Orangs of Borneo and Sumatra.

In the conformation of the pelvis, as indicated by the broad and concave iliac bones, the projection forwards of the anterior spines, it is, on the other hand, the most anthropoid of all the Simiadæ. The central portion of the ilia acquires a certain degree of transparency in both sexes, and the same has been noticed, though to a less extent, in the pelvis of the Enché-eco. This observation is interesting in connection with the results obtained by Vrolick with reference to the marks of degradation in the Negro, an index of which he finds in the absence of this character.[13] Certainly, a much more satisfactory

[12] Of the adult crania of the Chimpanzée, there are two in the Cabinet of the Boston Society of Natural History, one in that of Dr. J. C. Warren, two in that of the Academy of Natural Sciences in Philadelphia, and one in my collection. [*Footnote in original.*]

[13] "The pelvis of the male Negro, in the strength and density of its substance, and of the bones which compose it, resembles the pelvis of the wild beast, while on the contrary, the pelvis of the female of the same race combines lightness of substance and delicacy of form and structure. Yet the pelvis of the female Negro, though of light and delicate form, when compared to

index of degradation is to be derived from the general shape of the bones, and their approximation in form to those of the semi-erect animals with which they have been compared; and the exact measure would be the amount of deviation from the Caucasian type.

From an examination of the narratives of travellers, and the works of some naturalists, it will be found that no one peculiarity is more strongly insisted upon than the ability and disposition of the Orangs to assume and maintain the erect attitude. An attentive examination however, of their organization, as compared with that of man, gives the most conclusive evidence on the other hand, that they are not constituted like him for the erect position, a conclusion abundantly supported by the observation of living specimens exhibited from time to time in America and Europe. Evidence is yet to be adduced that any Quadrumanous animal whatever, assumes or maintains the erect attitude in its ordinary and natural movements. The gait of an Orang walking on its legs alone, is one of great instability, the animal showing by its bent position like that of an infirm old man, by the attitude of the arms, the constant effort to balance itself, and its disposition to get assistance from surrounding objects, that such a mode of progression is one of extreme difficulty.

The conditions for walking erect, as manifested in the human skeleton, are as follows:

1. The head must be balanced, or very nearly so, on the atlas;
2. The curves, and the general direction of the vertebral column must be such that the centre of gravity of the trunk shall be over the plane of the pelvis, passing through the heads of the thigh bones;
3. The lower extremities on which the pelvis rests should have the axes of the thigh and leg in one and the same vertical plane;
4. The feet should be directed at right angles to the axes of the legs, the sole resting on the ground.

These conditions are found to co-exist in man, and in man alone; they are not found in any of the anthropoid animals hitherto described, nor is there any nearer approximation to them in the species now under consideration, unless it be the existence of a pelvis a little more perfect in its conformation than in the congeners. The natural attitude of the Orangs on the ground is *semi-erect,* aided and supported by one or both of their long arms; the Chimpanzées and Orangs resting on their callous knuckles, and the Engé-enas on the palms of their hands. If they at any time support themselves on their legs alone, their heads droop, the trunk is bent forwards, the thighs

that of the male, is destitute of that transparent portion in which, in the pelvis of the European female the tables of the bone are so closely united. . . . Delicate however as is the form of the pelvis in the female, it is difficult to separate from it the idea of degradation in type, and an approach towards the form of the lower animals. . . ." Pritchard, Researches into the Phys. Hist. of Mankind, 4th edit. vol. i. p. 324. [*Footnote in original.*]

are flexed, and their feet inverted, all which necessarily results from the mechanical arrangements of their skeleton. The *foramen magnum* and (consequently the occipital condyles), instead of being situated in the middle of the base of the skull as in man, is situated in the middle of the posterior third, from which results the greater preponderance of the head forwards.[14]

The vertebral column is concave throughout nearly the whole of its anterior face, and in the lumbar region especially, deviates from the form of that of the human body, in which it is strongly convex. This results from the anterior portions of the bodies of the lumbar vertebræ in the Orangs having the vertical diameter of the anterior face shorter than that of the posterior, so that when they are piled one above another, the superior ones incline forwards, and will necessarily cause the whole superimposed trunk to preponderate in that direction, consequently throwing the centre of gravity forward in a proportional degree.

The bent position of the body necessarily involves a greater or less flexion of the legs, in order that a portion of its lower part should be thrown behind the centre of gravity, to compensate in a measure for the upper portion including the head, which is thrown in front of it.

Lastly, the feet are always inverted, in consequence of the mode of their articulation with the leg. Living habitually in trees, and the natural locomotion being that of climbing or swinging from limb to limb by the aid of long and powerful arms, their feet are so constructed as to enable them to apply the soles against the sides of the trunks and branches, consequently requiring them to be in planes, either really in, or approximating to a vertical direction. When on the ground therefore they are from necessity obliged to walk on the outer edge of the foot, and this with the other peculiarities of their organization, gives them an unstable gait, contrasting with that of man, who, habitually walking erect on a horizontal surface, has the soles of the feet necessarily in a horizontal plane.

The organization of the anthropoid Quadrumana justifies the naturalist in placing them at the head of the brute creation, and placing them in a position in which they, of all the animal series, shall be nearest to man. Any anatomist, however, who will take the trouble to compare the skeletons of the Negro and Orang, cannot fail to be struck at sight with the wide gap which separates them. The difference between the cranium, the pelvis, and the conformation of the upper extremities in the Negro and Caucasian, sinks into comparative insignificance when compared with the vast difference which

[14] The foremen magnum is the large opening in the occipital bone at the back of the skull where the medula oblongata passes to become the spinal chord. An original footnote continues the discussion: The position of this foramen in the Orangs is correctly regarded as an evidence of degradation. Sœmmering has imagined that an approach to it existed in the crania of Negroes, and his statement has been frequently repeated by subsequent anatomists. The more recent observations of Owen and Pritchard however, have a tendency to show that this foramen does not occupy a place in the Negroes materially different from that of the Caucasians; so that the difference between the Negroes and Orangs with regard to this peculiarity is vastly greater than between any two of the human races. . . .

exists between the conformation of the same parts in the Negro and the Orang. Yet it cannot be denied, however wide the separation, that the Negro and Orang do afford the points where man and the brute, when the totality of their organization is considered, most nearly approach each other.

12. A Botanist Introduces Darwin's Theory of the Origin of Species. 1860

When Charles Darwin's *Origin of the Species* appeared in 1859, his friend, ASA GRAY (1810–1888), undertook to explain and defend Darwin's theory of evolution by means of natural selection. The only general scientific periodical in the United States, *The American Journal of Science,* carried what was essentially a debate on Darwinism, with Gray the advocate and Louis Agassiz (see next selection) the adversary. Gray's review is still considered the best summary of Darwin's work. Here he indicates clearly the basic ideas of Darwin and the evidence that contributed to his thinking. Gray also takes up and answers both religious and scientific objections to Darwin's theory. Tone is as important as logic in evaluating Gray's judgments.

Gray was born in upper New York. He received an M.D. at Fairfield Medical School but instead of practicing set about becoming a botanist. He supported himself for some years by teaching school. In 1842 he was called to Harvard, where he founded the department of botany. Although he chiefly contributed to North American botany, he was a world figure in botany, as his friendship with Darwin indicates.

See A. Hunter Dupree, *Asa Gray* (Cambridge, Mass.: Harvard University Press, 1959); and Edward J. Pfeifer, "The Reception of Darwinism in the United States, 1859–1880"(Ph.D. dissertation, Brown University, 1959).

REVIEW OF DARWIN'S THEORY on the Origin of Species by means of Natural Selection.*

Asa Gray

* *The American Journal of Science and Arts,* 79 (1860), 153–184. Slightly abridged.

This book[1] is already exciting much attention. Two American editions are announced, through which it will become familiar to many of our readers, before these pages are issued. An abstract of the argument,—for "the whole volume is one long argument," as the author states,—is unnecessary in such a case; and it would be difficult to give by detached extracts. For the volume itself is an abstract, a prodromus of a detailed work upon which the author has been laboring for twenty years, and which "will take two or three more years to complete." It is exceedingly compact; and although useful summaries are appended to the several chapters, and a general recapitulation contains the essence of the whole, yet much of the aroma escapes in the treble distillation, or is so concentrated that the flavor is lost to the general, or even to the scientific reader. The volume itself,—the proof spirit—is just condensed enough for its purpose. It will be far more widely read, and perhaps will make deeper impression than the elaborate work might have done, with its full details of the facts upon which the author's sweeping conclusions have been grounded. At least it is a more readable book: but all the facts that can be mustered in favor of the theory are still likely to be needed.

Who, upon a single perusal, shall pass judgment upon a work like this, to which twenty of the best years of the life of a most able naturalist have been devoted? And who among those naturalists who hold a position that entitles them to pronounce summarily upon the subject, can be expected to divest himself for the nonce of the influence [of] received and favorite systems? In fact, the controversy now opened is not likely to be settled in an off-hand way, nor is it desirable that it should be. A spirited conflict among opinions of every grade must ensue, which,—to borrow an illustration from the doctrine of the book before us—may be likened to the conflict in nature among races in the struggle for life, which Mr. Darwin describes; through which the views most favored by facts will be developed and tested by 'Natural Selection,' the weaker ones be destroyed in the process, and the strongest in the long run alone survive.

The duty of reviewing this volume in the American Journal of Science would naturally devolve upon the principal Editor,[2] whose wide observation and profound knowledge of various departments of natural history, as well as of geology, particularly qualify him for the task. But he has been obliged to lay aside his pen, and to seek in distant lands the entire repose from scientific labor so essential to the restoration of his health,—a consummation devoutly to be wished, and confidently to be expected. Interested as Mr. Dana would be in this volume, he could not be expected to accept its doctrine. Views so idealistic as those upon which his "Thoughts upon Species"

[1] *On the Origin of Species by means of Natural Selection, or the Preservation of Favored Races in the Struggle for Life;* by CHARLES DARWIN, M.A., Fellow of the Royal, Geological, Linnæan, etc. Societies, Author of "Journal of Researches during H.M.S. Beagle's Voyage round the World." London: John Murray. 1859. pp. 502, post 8vo. [*Footnote in original.*]

[2] James Dwight Dana of Yale.

are grounded, will not harmonize readily with a doctrine so thoroughly naturalistic as that of Mr. Darwin. Though it is just possible that one who regards the kinds of elementary matter, such as oxygen and hydrogen, and the definite compounds of these elementary matters, and their compounds again, in the mineral kingdom, as constituting species, in the same sense, fundamentally, as that of animal and vegetable species, might admit an evolution of one species from another in the latter as well as the former case.

Between the doctrines of this volume and those of the other great Naturalist whose name adorns the title-page of this Journal,[3] the widest divergence appears. It is interesting to contrast the two, and, indeed, is necessary to our purpose; for this contrast brings out most prominently, and sets in strongest light and shade the main features of the theory of the origination of species by means of Natural Selection.

The ordinary and generally received view assumes the independent, specific creation of each kind of plant and animal in a primitive stock, which reproduces its like from generation to generation, and so continues the species. Taking the idea of species from this perennial succession of essentially similar individuals, the chain is logically traceable back to a local origin in a single stock, a single pair, or a single individual, from which all the individuals composing the species have proceeded by natural generation. Although the similarity of progeny to parent is fundamental in the conception of species, yet the likeness is by no means absolute: all species vary more or less, and some vary remarkably—partly from the influence of altered circumstances, and partly (and more really) from unknown constitutional causes which altered conditions favor rather than originate. But these variations are supposed to be mere oscillations from a normal state, and in Nature to be limited if not transitory; so that the primordial differences between species and species at their beginning have not been effaced, nor largely obscured, by blending through variation. Consequently, whenever two reputed species are found to blend in nature through a series of intermediate forms, community of origin is inferred, and all the forms, however diverse, are held to belong to one species. Moreover, since bisexuality is the rule in nature (which is practically carried out, in the long run, far more generally than has been suspected), and the heritable qualities of two distinct individuals are mingled in the offspring, it is supposed that the general sterility of hybrid progeny, interposes an effectual barrier against the blending of the original species by crossing.

From this generally accepted view the well-known theory of Agassiz and the recent one of Darwin diverge in exactly opposite directions.

That of Agassiz differs fundamentally from the ordinary view only in this, that it discards the idea of a common descent as the real bond of union among the individuals of a species, and also the idea of a local origin,—supposing, instead, that each species originated simultaneously, generally

[3] That is, Louis Agassiz; see below in this selection and also next selection.

speaking over the whole geographical area it now occupies or has occupied, and in perhaps as many individuals as it numbered at any subsequent period.

Mr. Darwin, on the other hand, holds the orthodox view of the descent of all the individuals of a species not only from a local birth-place, but from a single ancestor or pair; and that each species has extended and established itself, through natural agencies, wherever it could; so that the actual geographical distribution of any species is by no means a primordial arrangement, but a natural result. He goes farther, and this volume is a protracted argument intended to prove that the species we recognize have not been independently created, as such, but have descended, like varieties, from other species. Varieties on this view, are incipient or possible species: species are varieties of a larger growth and a wider and earlier divergence from the parent stock: the difference is one of degree, not of kind.

The ordinary view—rendering unto Cæsar the things that are Cæsar's—looks to natural agencies for the actual distribution and perpetuation of species, to a supernatural for their origin.

The theory of Agassiz regards the origin of species and their present general distribution over the world as equally primordial, equally supernatural; that of Darwin, as equally derivative, equally natural.

The theory of Agassiz, referring as it does the phenomena both of origin and distribution directly to the Divine will,—thus removing the latter with the former out of the domain of inductive science (in which efficient cause is not the first, but the last word),—may be said to be theistic to excess. The contrasted theory is not open to this objection. Studying the facts and phenomena in reference to proximate causes, and endeavoring to trace back the series of cause and effect as far as possible, Darwin's aim and processes are strictly scientific, and his endeavor, whether successful or futile, must be regarded as a legitimate attempt to extend the domain of natural or physical science. For though it well may be that "organic forms have no physical or secondary cause," yet this can be proved only indirectly, by the failure of every attempt to refer the phenomena in question to causal laws. But, however originated, and whatever be thought of Mr. Darwin's arduous undertaking in this respect, it is certain that plants and animals are subject from their birth to physical influences, to which they have to accommodate themselves as they can. How literally they are "born to trouble," and how incessant and severe the struggle for life generally is, the present volume graphically describes. Few will deny that such influences must have gravely affected the range and the association of individuals and species on the earth's surface. Mr. Darwin thinks that, acting upon an inherent predisposition to vary, they have sufficed even to modify the species themselves and produce the present diversity. Mr. Agassiz believes that they have not even affected the geographical range and the actual association of species, still less their forms; but that every adaptation of species to climate and of species to species is as aboriginal, and therefore as inexplicable, as are the organic forms themselves.

Who shall decide between such extreme views so ably maintained on either hand, and say how much of truth there may be in each? The present reviewer has not the presumption to undertake such a task. Having no prepossession in favor of naturalistic theories, but struck with the eminent ability of Mr. Darwin's work, and charmed with its fairness, our humbler duty will be performed if, laying aside prejudice as much as we can, we shall succeed in giving a fair account of its method and argument, offering by the way a few suggestions, such as might occur to any naturalist of an inquiring mind. An editorial character for this article must in justice be disclaimed. The plural pronoun is employed not to give editorial weight, but to avoid even the appearance of egotism, and also the circumlocution which attends a rigorous adherence to the impersonal style.

We have contrasted these two extremely divergent theories, in their broad statements. It must not be inferred that they have no points nor ultimate results in common.

In the first place they practically agree in upsetting, each in its own way, the generally received definition of species, and in sweeping away the ground of their objective existence in Nature. The orthodox conception of species is that of lineal descent: all the descendants of a common parent, and no other, constitute a species; they have a certain identity because of their descent, by which they are supposed to be recognizable. So naturalists had a distinct idea of what they meant by the term species, and a practical rule, which was hardly the less useful because difficult to apply in many cases, and because its application was indirect,—that is, the community of origin had to be inferred from the likeness; that degree of similarity, and that only, being held to be conspecific[4] which could be shown or reasonably inferred to be compatible with a common origin. And the usual concurrence of the whole body of naturalists (having the same data before them) as to what forms are species attests the value of the rule, and also indicates some real foundation for it in nature. But if species were created in numberless individuals over broad spaces of territory, these individuals are connected only in idea, and species differ from varieties on the one hand and from genera, tribes, &c. on the other only in degree; and no obvious natural reason remains for fixing upon this or that degree as specific, at least no natural standard, by which the opinions of different naturalists may be correlated. Species upon this view are enduring, but subjective and ideal. Any three or more of the human races, for example, are species or not species, according to the bent of the naturalist's mind. Darwin's theory brings us the other way to the same result. In his view, not only all the individuals of a species are descendants of a common parent but of all the related species also. Affinity, relationship,

[4] That is, belonging to the same species.

all the terms which naturalists use figuratively to express an underived, unexplained resemblance among species, have a literal meaning upon Darwin's system, which they little suspected, namely, that of inheritance. Varieties are the latest offshoots of the genealogical tree in "an unlineal" order; species, those of an earlier date, but of no definite distinction; genera, more ancient species, and so on. The human races, upon this view likewise may or may not be species according to the notions of each naturalist as to what differences are specific: but, if not species already, those races that last long enough are sure to become so. It is only a question of time.

How well the simile of a genealogical tree illustrates the main ideas of Darwin's theory the following extract from the summary of the fourth chapter shows.

> "It is a truly wonder fact,—the wonder of which we are apt to overlook from familiarity—that all animals and all plants throughout all time and space should be related to each other in group subordinate to group, in the manner which we everywhere behold—namely, varieties of the same species most closely related together, species of the same genus less closely and unequally related together, forming sections and sub-genera, species of distinct genera much less closely related, and genera related in different degrees, forming sub-families, families, orders, sub-classes, and classes. The several subordinate groups in any class cannot be ranked in a single file, but seem rather to be clustered round points, and these round other points, and so on in almost endless cycles. On the view that each species has been independently created, I can see no explanation of this great fact in the classification of all organic beings; but, to the best of my judgment, it is explained through inheritance and the complex action of natural selection, entailing extinction and divergence of character, as we have seen illustrated in the diagram.
>
> "The affinities of all the beings of the same class have sometimes been represented by a great tree. I believe this simile largely speaks the truth. The green and budding twigs may represent existing species; and those produced during each former year may represent the long succession of extinct species. At each period of growth all the growing twigs have tried to branch out on all sides, and overtop and kill the surrounding twigs and branches, in the same manner as species and groups of species have tried to overmaster other species in the great battle for life. The limbs divided into great branches, and these into lesser and lesser branches, were themselves once, when the tree was small, budding twigs; and this connexion of the former and present buds by ramifying branches may well represent the classification of all extinct and living species in groups subordinate to groups. Of the many twigs which flourished when the tree was a mere bush, only two or three, now grown into great branches, yet survive and bear all the other branches; so with the species which lived during long-past geological periods, very few now have living and modified descendants. From the first growth of the tree, many a limb and branch has decayed and dropped off; and these lost branches of various sizes may represent those whole orders, families, and genera which have now no living representatives, and which are known to us only from having been found in a fossil state. As we here and there see a thin straggling

branch springing from a fork low down in a tree, and which by some chance has been favored and is still alive on its summit, so we occasionally see an animal like the Ornithorhynchus or Lepidosiren,[5] which in some small degree connects by its affinities two large branches of life, and which has apparently been saved from fatal competition by having inhabited a protected station as buds give rise by growth to fresh buds, and these, if vigorous, branch out and overtop on all sides many a feebler branch, so by generations I believe it has been with the great Tree of Life, which fills with its dead and broken branches the crust of earth, and covers the surface with its ever branching and beautiful ramifications."

It may also be noted that there is a significant correspondence between the rival theories as to the main facts employed. Apparently every capital fact in the one view is a capital fact in the other. The difference is in the interpretation. To run the parallel ready made to our hands:[6]

"The simultaneous existence of the most diversified types under identical circumstances, . . . the repetition of similar types under the most diversified circumstances, . . . the unity of plan in otherwise highly diversified types of animals, . . . the correspondence, now generally known as special homologies, in the details of structure otherwise entirely disconnected, down to the most minute peculiarities, . . . the various degrees and different kinds of relationship among animals which [apparently] can have no genealogical connection, . . . the simultaneous existence in the earliest geological periods . . . of representatives of all the great types of the animal kingdom, . . . the gradation based upon complications of structure which may be traced among animals built upon the same plan; the distribution of some types over the most extensive range of surface of the globe, . . . the definite relations in which individuals of the same species stand to one another, . . . the limitation of the range of changes which animals undergo during their growth, . . . the return to a definite norm of animals which multiply in various ways, . . . the order of succession of the different types of animals and plants characteristic of the different geological epochs, . . . the localization of some types of animals upon the same points of the surface of the globe during several successive geological periods; . . . the parallelism between the order of succession of animals and plants in geological times, and the changes their living representatives undergo during their embryological growth, . . . *the combination in many extinct types of characters which in later ages appear disconnected in different types,* . . . the parallelism between the gradation among animals and the changes they undergo during their growth,[7] . . . the relations existing between these different series and the geographical distribution of animals, . . . the connection of all the known features of nature into one system,—"

[5] The duck-billed platypus and the lung fish (the latter the most highly organized fish and one resembling the amphibians).

[6] Agassiz, Essay on Classification; Contrib. to Nat. Hist. i, p. 132, et seq. [*Footnote in original.*]

[7] This was a reference to Agassiz's observation that the stages of embryonic development appear to resemble at first the lowest forms of life and then progressively higher ones; Ernst Haeckel later formulated this observation into the dictum, "ontogeny recapitulates phylogeny."

In a word, the whole relations of animals, &c. to surrounding nature and to each other, are regarded under the one view as ultimate facts, or in their ultimate aspect, and interpreted theologically;—under the other as complex facts, to be analyzed and interpreted scientifically. The one naturalist, perhaps too largely assuming the scientifically unexplained to be inexplicable, views the phenomena only in their supposed relation to the Divine mind. The other, naturally expecting many of these phenomena to be resolvable under investigation, views them in their relations to one another, and endeavors to explain them as far as he can (and perhaps farther) through natural causes.

But does the one really exclude the other? Does the investigation of physical causes stand opposed to the theological view and the study of the harmonies between mind and Nature?[8] More than this, is it not most presumable that an intellectual conception realized in nature would be realized through natural agencies? Mr. Agassiz answers these questions affirmatively when he declares that "the task of science is to investigate what has been done, to enquire if possible *how it has been done,* rather than to ask what is possible for the Deity, since *we can know that only by what actually exists;*" and also when he extends the argument for the intervention in nature of a creative mind to its legitimate application in the inorganic world; which, he remarks, "considered in the same light, would not fail also to exhibit unexpected evidence of thought, in the character of laws regulating the chemical combinations, the action of physical forces, etc., etc." Mr. Agassiz, however, pronounces that "the connection between the facts is *only intellectual;*"—an opinion which the analogy of the inorganic world, just referred to, does not confirm, for there a material connection between the facts is justly held to be consistent with an intellectual,—and which the most analogous cases we can think of in the organic world do not favor; for there is a material connection between the grub, the pupa, and the butterfly, between the tadpole and the frog, or, still better, between those distinct animals which succeed each other in alternate and very dissimilar generations. So that mere analogy might rather suggest a natural connection than the contrary; and the contrary cannot be demonstrated until the possibilities of nature under the Diety are fathomed.

But the intellectual connection being undoubted, Mr. Agassiz properly refers the whole to "the agency of Intellect[9] as its first cause." In doing so, however, he is not supposed to be offering a scientific explanation of the phenomena. Evidently he is considering only the ultimate *why,* not the proximate why or *how.*

Now the latter is just what Mr. Darwin is considering. He conceives of a physical connection between allied species: but we suppose he does not deny their intellectual connection, as related to a Supreme Intelligence.

[8] That is, the creative mind of God.

[9] That is, God's creative intellect.

Certainly we see no reason why he should, and many reasons why he should not. Indeed, as we contemplate the actual direction of investigation and speculation in the physical and natural sciences, we dimly apprehend a probable synthesis of those divergent theories, and in it the ground for a strong stand against mere naturalism. Even if the doctrine of the origin of species through natural selection should prevail in our day, we shall not despair; being confident that the genius of an Agassiz will be found equal to the work of constructing, upon the mental and material foundations combined, a theory of nature as theistic and as scientific, as that which he has so eloquently expounded.

To conceive the possibility of "the descent of species from species by insensibly fine gradations" during a long course of time, and to demonstrate its compatibility with a strictly theistic view of the universe, is one thing: to substantiate the theory itself or show its likelihood is quite another thing. This brings us to consider what Darwin's theory actually is, and how he supports it.

That the existing kinds of animals and plants, or many of them, may be derived from other and earlier kinds, in the lapse of time, is by no means a novel proposition. Not to speak of ancient speculations of the sort, it is the well-known Lamarckian theory. The first difficulty which such theories meet with is that, in the present age, with all its own and its inherited prejudgments, the whole burden of proof is naturally, and indeed properly, laid upon the shoulders of the propounders; and thus far the burden has been more than they could bear. From the very nature of the case, substantive proof of specific creation is not attainable; but that of derivation or transmutation of species may be. He who affirms the latter view is bound to do one or both of two things. Either, 1, to assign real and adequate causes, the natural or necessary result of which must be to produce the present diversity of species and their actual relations; or, 2, to show the general conformity of the whole body of facts to such assumption, and also to adduce instances explicable by it and inexplicable by the received view,—so perhaps winning our assent to the doctrine, through its competency to harmonize all the facts, even though the cause of the assumed variation remain as occult as that of the transformation of tadpoles into frogs, or that of *Coryne* into *Sarzia*.[10]

The first line of proof, successfully carried out, would establish derivation as a true physical theory; the second, as a sufficient hypothesis.

Lamarck mainly undertook the first line, in a theory which has been so assailed by ridicule that it rarely receives the credit for ability to which in its day it was entitled. But he assigned partly unreal, partly insufficient causes; and the attempt to account for a progressive change in species through the direct influence of physical agencies, and through the appetencies and habits of animals reacting upon their structure, thus causing the production and the successive modifications of organs, is a conceded and total failure. The shad-

[10] This refers to the stages in the life cycle of a small jellyfish.

owy author of the Vestiges of the Natural History of Creation[11] can hardly be said to have undertaken either line, in a scientific way. He would explain the whole progressive evolution of nature by virtue of an inherent tendency to development,—thus giving us an idea or a word in place of a natural cause, a restatement of the proposition instead of an explanation. Mr. Darwin attempts both lines of proof, and in a strictly scientific spirit; but the stress falls mainly upon the first; for, as he does assign real causes, he is bound to prove their adequacy.

It should be kept in mind that, while all direct proof of independent origination is unattainable from the nature of the case, the overthrow of particular schemes of derivation has not established the opposite proposition. The futility of each hypothesis thus far proposed to account for the derivation may be made apparent, or unanswerable objections may be urged against it; and each victory of the kind may render derivation more improbable, and therefore specific creation more probable, without settling the question either way. New facts, or new arguments and a new mode of viewing the question may some day change the whole aspect of the case. It is with the latter that Mr. Darwin now reopens the discussion.

Having conceived the idea that varieties are incipient species, he is led to study variation in the field where it shows itself most strikingly and affords the greatest facilities to investigation. Thoughtful naturalists have had increasing grounds to suspect that a re-examination of the question of species in zoology and botany, commencing with those races which man knows most about, viz. the domesticated and cultivated races, would be likely somewhat to modify the received idea of the entire fixity of species. This field, rich with various but unsystematized stores of knowledge accumulated by cultivators and breeders, has been generally neglected by naturalists, because these races are not in a state of nature; whereas they deserve particular attention on this very account, as experiments, or the materials for experiments, ready to our hand. In domestication we vary some of the natural conditions of a species, and thus learn experimentally what changes are within the reach of varying conditions in nature. We separate and protect a favorite race against its foes or its competitors, and thus learn what it might become if nature ever afforded it equal opportunities. Even when, to subserve human uses, we modify a domesticated race to the detriment of its native vigor, or to the extent of practical monstrosity, although we secure forms which would not be originated and could not be perpetuated in free nature, yet we attain wider and juster views of the possible degree of variation. We perceive that some species are more variable than others, but that no species subjected to the experiment persistently refuses to vary; and that when it has once begun

[11] In 1844 Robert Chambers, a Scotsman, published anonymously *Vestiges of the Natural History of Creation,* in which, mostly on the basis of advances in geology up to that time, he put forth the idea of evolution but without suggesting the natural means by which development of life forms proceeded. The book occasioned an unbelievable amount of controversy and vituperation.

to vary, its varieties are not the less but the more subject to variation. "No case is on record of a variable being ceasing to be variable under cultivation." It is fair to conclude, from the observation of plants and animals in a wild as well as domesticated state, that the tendency to vary is general, and even universal. Mr. Darwin does "not believe that variability is an inherent and necessary contingency, under all circumstances, with all organic beings, as some authors have thought." No one supposes variation could occur under all circumstances; but the facts on the whole imply an universal tendency, ready to be manifested under favorable circumstances. In reply to the assumption that man has chosen for domestication animals and plants having an extraordinary inherent tendency to vary, and likewise to withstand diverse climates, it is asked:

> "How could a savage possibly know, when he first tamed an animal, whether it would vary in succeeding generations, and whether it would endure other climates? Has the little variability of the ass or guinea-fowl, or the small power of endurance of warmth by the rein-deer, or of cold by the common camel, prevented their domestication? I cannot doubt that if other animals and plants, equal in number to our domesticated productions, and belonging to equally diverse classes and countries, were taken from a state of nature, and could be made to breed for an equal number of generations under domestication, they would vary on an average as largely as the parent species of our existing domesticated productions have varied."

As to amount of variation, there is the common remark of naturalists that the varieties of domesticated plants or animals often differ more widely than do the individuals of distinct species in a wild state: and even in nature the individuals of some species are known to vary to a degree sensibly wider than that which separates related species. In his instructive section on the breeds of the domestic pigeon, our author remarks that:—"at least a score of pigeons might be chosen, which if shown to an ornithologist, and he were told that they were wild birds, would certainly be ranked by him as well defined species. Moreover, I do not believe that any ornithologist would place the English carrier, the short-faced tumbler, the runt, the barb, pouter, and fantail in the same genus; more especially as in each of these breeds several truly inherited sub-breeds, or species as he might have called them, could be shown him." That this is not a case like that of dogs, in which probably the blood of more than one species is mingled, Mr. Darwin proceeds to show, adducing cogent reasons for the common opinion that all have descended from the wild rock-pigeon. Then follow some suggestive remarks:—

> "I have discussed the probable origin of domestic pigeons at some, yet quite insufficient, length; because when I first kept pigeons and watched the several kinds, knowing well how true they bred, I felt fully as much difficulty in believing that they could ever have descended from a common parent, as any naturalist could in coming to a similar conclusion in regard to many species of finches, or other large groups of birds, in nature. One circumstance has

struck me much; namely, that all the breeders of the various domestic animals and the cultivators of plants, with whom I have ever conversed, or whose treatises I have read, are firmly convinced that the several breeds to which each has attended, are descended from so many aboriginally distinct species. Ask, as I have asked, a celebrated raiser of Hereford cattle, whether his cattle might not have descended from long-horns, and he will laugh you to scorn. I have never met a pigeon, or poultry, or duck, or rabbit fancier, who was not fully convinced that each main breed was descended from a distinct species. Van Mons, in his treatise on pears and apples, shows how utterly he disbelieves that the several sorts, for instance a Ribston-pippin or Codlin-apple, could ever have proceeded from the seeds of the same tree. Innumerable other examples could be given. The explanation, I think, is simple: from long-continued study they are strongly impressed with the differences between the several races; and though they well know that each race varies slightly, for they win their prizes by selecting such slight differences, yet they ignore all general arguments, and refuse to sum up in their minds slight differences accumulated during many successive generations. May not those naturalists who, knowing far less of the laws of inheritance than does the breeder, and knowing no more than he does of the intermediate links in the long lines of descent, yet admit that many of our domestic races have descended from the same parents—may they not learn a lesson of caution, when they deride the idea of species in a state of nature being lineal descendants of other species?"

The actual causes of variation are unknown. Mr. Darwin favors the opinion of the late Mr. Knight, the great philosopher of horticulture, that variability under domestication is somehow connected with excess of food. He also regards the unknown cause as acting chiefly upon the reproductive system of the parents, which system, judging from the effect of confinement or cultivation upon its functions, he concludes to be more susceptible than any other to the action of changed conditions of life. The tendency to vary certainly appears to be much stronger under domestication than in free nature. But we are not sure that the greater variableness of cultivated races is not mainly owing to the far greater opportunities for manifestation and accumulation—a view seemingly all the more favorable to Mr. Darwin's theory. The actual amount of certain changes, such as size or abundance of fruit, size of udder, stands of course in obvious relation to supply of food.

Really, we no more know the reason why the progeny occasionally deviates from the parent than we do why it usually resembles it. Though the laws and conditions governing variations are known to a certain extent, while those governing inheritance are apparently inscrutable. "Perhaps," Darwin remarks, "the correct way of viewing the whole subject would be, to look at the inheritance of every character whatever as the rule, and non-inheritance as the anomaly." This, from general and obvious considerations, we have long been accustomed to do. Now, as exceptional instances are expected to be capable of explanation, while ultimate laws are not, it is quite possible that variation may be accounted for, while the great primary law of inheritance remains a mysterious fact.

The common proposition is, that *species reproduce their like*; this is a sort of general inference, only a degree closer to fact than the statement that genera reproduce their like. The true proposition, the fact incapable of further analysis is, that *individuals reproduce their like,*— that characteristics are inheritable. So varieties, or deviations once originated, are perpetuable, like species. Not so likely to be perpetuated, at the outset; for the new form tends to resemble a grand-parent and a long line of similar ancestors, as well as to resemble its immediate progenitors. Two forces which coincide in the ordinary case, where the offspring resembles its parent, act in different directions when it does not, and it is uncertain which will prevail. If the remoter, but very potent ancestral influence predominates, the variation disappears with the life of the individual. If that of the immediate parent—feebler no doubt, but closer—the variety survives in the offspring; whose progeny now has a redoubled tendency to produce its own like; whose progeny again is almost sure to produce its like, since it is much the same whether it takes after its mother or its grandmother.

In this way races arise, which under favorable conditions may be as hereditary as species. In following these indications, watching opportunities, and breeding only from those individuals which vary most in a desirable direction, man leads the course of variation as he leads a streamlet,—apparently at will, but never against the force of gravitation,—to a long distance from its source, and makes it more subservient to his use or fancy. He unconsciously strengthens those variations which he prizes when he plants the seed of a favorite fruit, preserves a favorite domestic animal, drowns the uglier kittens of a litter, and allows only the handsomest or the best mousers to propagate. Still more, by methodical selection, in recent times almost marvellous results have been produced in new breeds of cattle, sheep, and poultry, and new varieties of fruit of greater and greater size or excellence.

It is said that all domestic varieties if left to run wild, would revert to their aboriginal stocks. Probably they would wherever various races of one species were left to commingle. At least the abnormal or exaggerated characteristics induced by high feeding, or high cultivation, and prolonged close breeding would promptly disappear, and the surviving stock would soon blend into a homogeneous result (in a way presently explained), which would naturally be taken for the original form; but we could seldom know if it were so. It is by no means certain that the result would be the same if the races ran wild each in a separate region. Dr. Hooker doubts if there is a true reversion in the case of plants. Mr. Darwin's observations rather favor it in the animal kingdom. With mingled races reversion seems well made out in the case of pigeons. The common opinion upon this subject therefore probably has some foundation. But even if we regard varieties as oscillations around a primitive centre or type, still it appears from the readiness with which such varieties originate, that a certain amount of disturbance would carry them beyond the influence of the primordial attraction, where they may become new centres of variation.

Some suppose that races cannot be perpetuated indefinitely even by keeping up the conditions under which they were fixed: but the high antiquity of several, and the actual fixity of many of them, negative this assumption. "To assert that we could not breed our cart and race horses, long and short-horned cattle, and poultry of various breeds, for almost an infinite number of generations would be opposed to all experience."

Why varieties develope so readily and deviate so widely under domestication, while they are apparently so rare or so transient in free nature, may easily be shown. In nature, even with hermaphrodite plants, there is a vast amount of cross fertilization among various individuals of the same species. The inevitable result of this . . . is to repress variation, to keep the mass of a species comparatively homogeneous over any area in which it abounds in individuals. Starting from a suggestion of the late Mr. Knight, now so familiar, that close interbreeding diminishes vigor and fertility; and perceiving that bisexuality is ever aimed at in nature,—being attained physiologically in numerous cases where it is not structurally,—Mr. Darwin has worked out the subject in detail, and shown how general is the concurrence, either habitual or occasional, of two hermaphrodite individuals in the reproduction of their kind; and has drawn the philosophical inference that probably no organic being self-fertilizes indefinitely; but that a cross with another individual is occasionally—perhaps at very long intervals—indispensable. . . .

In domestication, this intercrossing may be prevented; and in this prevention lies the art of producing varieties. But "the art itself is Nature," since the whole art consists in allowing the most universal of all natural tendencies in organic things (inheritance) to operate uncontrolled by other and obviously incidental tendencies. No new power, no artificial force is brought into play either by separating the stock of a desirable variety so as to prevent mixture, or by selecting for breeders those individuals which most largely partake of the peculiarities for which the breed is valued.[12]

We see everywhere around us the remarkable results which Nature may be said to have brought about under artificial selection and separation. Could she accomplish similar results when left to herself? Variations might begin, we know they do begin, in a wild state. But would any of them be preserved and carried to an equal degree of deviation? Is there anything in nature which in the long run may answer to artificial selection? Mr. Darwin thinks that there is; and *Natural Selection* is the key-note of his discourse.

As a preliminary, he has a short chapter to show that there is variation in nature, and therefore something for natural selection to act upon. He

[12] The rules and processes of breeders of animals, and their results, are so familiar that they need not be particularized. Less is popularly known about the production of vegetable races. . . . [*Footnote in original.*]

readily shows that such mere variations as may be directly referred to physical conditions (like the depauperation of plants in a sterile soil, or their dwarfing as they approach an alpine summit, the thicker fur of an animal from far northward, &c.), and also those individual differences which we everywhere recognize but do not pretend to account for, are not separable by any assignable line from more strongly marked varieties; likewise that there is no clear demarcation between the latter and subspecies, or varieties of the highest grade (distinguished from species not by any known inconstancy, but by the supposed lower importance of their characteristics); nor between these and recognized species. "These differences blend into each other in an insensible series, and the series impresses the mind with an idea of an actual passage."

This gradation from species downward is well made out. To carry it one step farther upwards, our author presents in a strong light the differences which prevail among naturalists as to what forms should be admitted to the rank of species. Some genera (and these in some countries) give rise to far more discrepancy than others; and it is concluded that the large or dominant genera are usually the most variable. In a flora so small as the British, 182 plants generally reckoned as varieties, have been ranked by some botanists as species. Selecting the British genera which include the most polymorphous forms, it appears that Babington's Flora gives them 251 species, Bentham's only 112, a difference of 139 doubtful forms. These are nearly the extreme views; but they are the views of two most capable and most experienced judges, in respect to one of the best known floras of the world. The fact is suggestive, that the best known countries furnish the greatest number of such doubtful cases. Illustrations of this kind may be multiplied to a great extent. They make it plain that, whether species in nature are aboriginal and definite or not, our practical conclusions about them, as embodied in systematic works, are not *facts* but *judgments,* and largely fallible judgments.

How much of the actual coincidence of authorities is owing to imperfect or restricted observation, and to one naturalist's adopting the conclusions of another without independent observation, this is not the place to consider. It is our impression that species of animals are more definitely marked than those of plants; this may arise from our somewhat extended acquaintance with the latter, and our ignorance of the former. But we are constrained by our experience to admit the strong likelihood, in botany, that varieties on the one hand and what are called closely related species on the other do not differ except in degree. Whenever the wider difference separating the latter can be spanned by intermediate forms, as it sometimes is, no botanist long resists the inevitable conclusion. Whenever, therefore, this wider difference can be shown to be compatible with community of origin, and explained through natural selection or in any other way, we are ready to adopt the *probable* conclusion; and we see beforehand how strikingly the actual geographical association of related species favors the broader view. Whether we should continue to regard the forms in question as distinct species, depends upon what meaning we shall finally attach to that term; and that depends upon how far the doctrine of derivation can be carried back and how well it can be supported.

In applying his principle of natural selection to the work in hand, Mr. Darwin assumes, as we have seen: 1, some variability of animals and plants in nature; 2, the absence of any definite distinction between slight variations, and varieties of the highest grade; 3, the fact that naturalists do not practically agree and do not increasingly tend to agree, as to what forms are species and what are strong varieties, thus rendering it probable that there may be no essential and original difference, or no possibility of ascertaining it, at least in many cases; also, 4, that the most flourishing and dominant species of the larger genera on an average vary most (a proposition which can be substantiated only by extensive comparisons, the details of which are not given);—and, 5, that in large genera the species are apt to be closely but unequally allied together, forming little clusters round certain species,—just such clusters as would be formed if we suppose their members once to have been satellites or varieties of a central or parent species, but to have attained at length a wider divergence and a specific character. The fact of such association is undeniable; and the use which Mr. Darwin makes of it seems fair and natural.

The gist of Mr. Darwin's work is to show that such varieties are gradually diverged into species and genera through *natural selection*; that natural selection is the inevitable result of the *struggle for existence* which all living things are engaged in; and that this struggle is an unavoidable consequence of several natural causes, but mainly of the high rate at which all organic beings tend to increase.

Curiously enough, Mr. Darwin's theory is grounded upon the doctrine of Malthus and the doctrine of Hobbes. The elder DeCandolle had conceived the idea of the struggle for existence, and in a passage which would have delighted the cynical philosopher of Malmesbury, had declared that all nature is at war, one organism with another or with external nature; and Lyell and Herbert had made considerable use of it. But Hobbes in his theory of society and Darwin in his theory of natural history, alone have built their systems upon it. However moralists and political economists may regard these doctrines in their original application to human society and the relation of population to subsistence, their thorough applicability to the great society of the organic world in general is now undeniable. And to Mr. Darwin belongs the credit of making this extended application, and of working out the immensely diversified results with rare sagacity and untiring patience. He has brought to view *real causes* which have been largely operative in the establishment of the actual association and geographical distribution of plants and animals. In this he must be allowed to have made a very important contribution to an interesting department of science, even if his theory fails in the endeavor to explain the origin or diversity of species.

> "Nothing is easier," says our author, "than to admit in words the truth of the universal struggle for life, or more difficult—at least I have found it so—than constantly to bear this conclusion in mind. Yet unless it be thoroughly engrained in the mind, I am convinced that the whole economy of nature, with every fact on distribution, rarity, abundance, extinction, and variation, will be dimly seen or quite misunderstood. We behold the face of nature bright with gladness, we often see superabundance of food; we do not see, or we forget, that the birds which are idly singing round us mostly live on insects or seeds, and are thus constantly destroying life; or we forget how largely these songsters, or their eggs, or their nestlings, are destroyed by birds and beasts of prey; we do not always bear in mind, that though food may be now superabundant, it is not so at all seasons of each recurring year."—p. 62
>
> "There is no exception to the rule that every organic being naturally increases at so high a rate, that if not destroyed, the earth would soon be covered by the progeny of a single pair. Even slow-breeding man has doubled in twenty-five years, and at this rate, in a few thousand years, there would literally not be standing room for his progeny. Linnæus has calculated that if an annual plant produced only two seeds—and there is no plant so unproductive as this—and their seedlings next year produced two, and so on, then in twenty years there would be a million plants. The elephant is reckoned to be the slowest breeder of all known animals, and I have taken some pains to estimate its probable minimum rate of natural increase: it will be under the mark to assume that it breeds when thirty years old, and goes on breeding till ninety years old, bringing forth three pairs of young in this interval; if this be so, at the end of the fifth century there would be alive fifteen million elephants, descended from the first pair.
>
> "But we have better evidence on this subject than mere theoretical calculations, namely, the numerous recorded cases of the astonishingly rapid increase of various animals in a state of nature, when circumstances have been favorable to them during two or three following seasons. Still more striking is the evidence from our domestic animals of many kinds which have run wild in several parts of the world; if the statements of the rate of increase of slow-breeding cattle and horses in South America, and latterly in Australia, had not been well authenticated, they would have been quite incredible. So it is with plants: cases could be given of introduced plants which have become common throughout whole islands in a period of less than ten years. . . .
>
> "All plants and animals are tending to increase at a geometrical ratio; all would most rapidly stock any station in which they could anyhow exist; the increase must be checked by destruction at some period of life."—p. 65.

The difference between the most and the least prolific species is of no account.

> "The condor lays a couple of eggs, and the ostrich a score; and yet in the same country the condor may be the more numerous of the two. The Fulmar

petrel lays but one egg, yet it is believed to be the most numerous bird in the world."—p. 68.

"The amount of food gives the extreme limit to which each species can increase; but very frequently it is not the obtaining of food, but the serving as prey to other animals, which determines the average numbers of a species." —p. 68.

"Climate plays an important part in determining the average numbers of a species, and periodical seasons of extreme cold or drought, I believe to be the most effective of all checks. I estimated that the winter of 1854–55 destroyed four-fifths of the birds in my own grounds; and this is a tremendous destruction, when we remember that ten per cent is an extraordinarily severe mortality from epidemics with man. The action of climate seems at first sight to be quite independent of the struggle for existence; but in so far as climate chiefly acts in reducing food, it brings on the most severe struggle between the individuals, whether of the same or of distinct species, which subsist on the same kind of food. Even when climate, for instance extreme cold, acts directly, it will be the least vigorous, or those which have got least food through the advancing winter, which will suffer most. When we travel from south to north, or from a damp region to a dry, we invariably see some species gradually getting rarer and rarer, and finally disappearing; and the change of climate being conspicuous, we are tempted to attribute the whole effect to its direct action. But this is a very false view: we forget that each species, even where it most abounds, is constantly suffering enormous destruction at some period of its life, from enemies or from competitors for the same place and food; and if these enemies or competitors be in the least degree favored by any slight change of climate, they will increase in numbers, and, as each area is already stocked with inhabitants, the other species will decrease. . . . When we reach the Arctic regions, or snowcapped summits, or absolute deserts, the struggle for life is almost exclusively with the elements.

"That climate acts in main part indirectly by favoring other species, we may clearly see in the prodigious number of plants in our gardens which can perfectly well endure our climate, but which never become naturalized, for they cannot compete with our native plants, nor resist destruction by our native animals."—pp. 68, 69.

After an instructive instance in which "cattle absolutely determine the existence of the Scotch Fir," we are referred to cases in which insects determine the existence of cattle.

"Perhaps Paraguay offers the most curious instance of this; for here neither cattle nor horses nor dogs have ever run wild, though they swarm southward and northward in a feral state; and Azara and Rengger have shown that this is caused by the greater number in Paraguay of a certain fly, which lays its eggs in the navels of these animals when first born. The increase of these flies, numerous as they are, must be habitually checked by some means, probably by birds. Hence, if certain insectivorous birds (whose numbers are probably regulated by hawks or beasts of prey) were to increase in Paraguay, the flies would decrease—then cattle and horses would become feral, and this would certainly greatly alter (as indeed I have observed in parts of South America) the vegetation: this again would largely affect the insects; and this, as we just have seen in Staffordshire, the insectivorous birds, and so onwards in ever-increasing circles of complexity. We began this series by insectivorous birds, and we had ended with them. Not that in nature the relations can ever be as simple as this. Battle within battle must ever be recurring with varying success; and yet in the long run the forces are so nicely balanced, that the face of nature remains uniform for long periods of time, though assuredly the merest trifle would often give the victory to one organic being over another. Nevertheless so profound is our ignorance, and so high our presumption, that we marvel when we hear of the extinction of an organic being; and as we do not see the cause, we invoke cataclysms to desolate the world, or invent laws on the duration of the forms of life!"—pp. 72, 73.

"When we look at the plants and bushes clothing an entangled bank, we are tempted to attribute their proportional numbers and kinds to what we call chance. But how false a view is this! Every one has heard that when an American forest is cut down, a very different vegetation springs up; but it has been observed that the trees now growing on the ancient Indian mounds, in the Southern United States, display the same beautiful diversity and proportion of kinds as in the surrounding virgin forests. What a struggle between the several kinds of trees must have gone on during long centuries, each annually scattering its seeds by the thousands; what war between insect and insect—between insects, snails, and other animals with birds and beasts of prey—all striving to increase, and all feeding on each other or on the trees or their seeds and seedlings, or on the other plants which first clothed the ground and thus checked the growth of the trees! Throw up a handful of feathers, and all must fall to the ground according to definite laws; but how simple is this problem compared to the action and reaction of the innumerable plants and animals which have determined, in the course of centuries, the proportional numbers and kinds of trees now growing on the old Indian ruins!"—pp. 74, 75.

For reasons obvious upon reflection the competition is often, if not generally, most severe between nearly related species when they are in contact, so that one drives the other before it, as the Hanoverian, the old English rat, the small Asiatic cockroach in Russia, its greater congener, &c.: and this, when duly considered, explains many curious results;—such, for instance, as the considerable number of different genera of plants and animals which are generally found to inhabit any limited area. . . .

The abundance of some forms, the rarity and final extinction of many others, and the consequent divergence of character or increase of difference among the surviving representatives are other consequences. As favored forms increase, the less favored must diminish in number, for there is not room for all; and the slightest advantage, at first probably inappreciable to human observation, must decide which shall prevail and which must perish, or be driven to another and for it more favorable locality.

We cannot do justice to the interesting chapter upon natural selection by separated extracts. The following must serve to show how the principle is supposed to work.

> "If during the long course of ages and under varying conditions of life, organic beings vary at all in the several parts of their organization, and I think this cannot be disputed; if there be, owing to the high geometrical powers of increase of each species, at some age, season, or year, a severe struggle for life, and this certainly cannot be disputed; then, considering the infinite complexity of the relations of all organic beings to each other and to their conditions of existence, causing an infinite diversity in structure, constitution, and habits, to be advantageous to them, I think it would be a most extraordinary fact if no variation ever had occurred useful to each being's own welfare, in the same way as so many variations have occurred useful to man. But if variations useful to any organic being do occur, assuredly individuals thus characterized will have the best chance of being preserved in the struggle for life; and from the strong principle of inheritance they will tend to produce offspring similarly characterized. This principle of preservation, I have called, for the sake of brevity, Natural Selection."—pp. 126, 127
>
> "In order to make it clear how, as I believe, natural selection acts, I must beg permission to give one or two imaginary illustrations. Let us take the case of a wolf, which preys on various animals, securing some by craft, some by strength, and some by fleetness; and let us suppose that the fleetest prey, a deer for instance, had from any change in the country increased in numbers, or that other prey had decreased in numbers, during that season of the year when the wolf is hardest pressed for food. I can under such circumstances see no reason to doubt that the swiftest and slimmest wolves would have the best chance of surviving, and so be preserved or selected,—provided always that they retained strength to master their prey at this or at some other period of the year, when they might be compelled to prey on other animals. I can see no more reason to doubt this, than that man can improve the fleetness of his greyhounds by careful and methodical selection, or by that unconscious selection which results from each man trying to keep the best dogs without any thought of modifying the breed. . . .
>
> "Now, if any slight innate change of habit or structure benefited an individual wolf, it would have the best chance of surviving and of leaving offspring. Some of its young would probably inherit the same habits or structure, and by the repetition of this process, a new variety might be formed which would either supplant or coexist with the parent-form of wolf. Or, again, the wolves inhabiting a mountainous district, and those frequenting the lowlands, would naturally be forced to hunt different prey; and from the continued preservation of the individuals best fitted for the two sites, two varieties might slowly be formed. These varieties would cross and blend where they met; but to this subject of intercrossing we shall soon have to return. I may add, that, according to Mr. Pierce, there are two varieties of the wolf inhabiting the Catskill Mountains in the United States, one with a light greyhound-like form, which pursues deer, and the other more bulky, with shorter legs, which more frequently attacks the shepherd's flocks."—pp. 90, 91.

We eke out the illustration here with a counterpart instance, viz., the remark of Dr. Bachman that "The deer that reside permanently in the swamps of Carolina are taller and longer-legged than those in the higher grounds."

The limits allotted to this article are nearly reached, yet only four of the fourteen chapters of the volume have been touched. These, however, contain the fundamental principles of the theory and most of those applications of it which are capable of something like verification, relating as they do to phenomena now occurring. Some of our extracts also show how these principles are thought to have operated through the long lapse of the ages. The chapters from the sixth to the ninth inclusive are designed to obviate difficulties and objections, "some of them so grave that to this day," the author frankly says, he "can never reflect on them without being staggered." We do not wonder at it. After drawing what comfort he can from the "imperfection of the geological record," . . . which we suspect is scarcely exaggerated, the author considers the geological succession of organic beings, . . . to see whether they better accord with the common view of the immutability of species, or with that of their slow and gradual modification. Geologists must settle that question. Then follow two most interesting and able chapters on the geographical distribution of plants and animals, the summary of which we should be glad to cite; then a fitting chapter upon classification, morphology, embryology, &c., as viewed in the light of this theory, closes the argument; the fourteenth chapter being a recapitulation.

The interest for the general reader heightens as the author advances on his perilous way and grapples manfully with the most formidable difficulties.

To account, upon these principles, for the gradual elimination and segregation of nearly allied forms,—such as varieties, subspecies, and closely related or representative species,—also in a general way for their geographical association and present range, is comparatively easy, is apparently within the bounds of possibility, and even of probability. Could we stop here we should be fairly contented. But, to complete the system, to carry out the principles to their ultimate conclusion, and to explain by them many facts in geographical distribution which would still remain anomalous, Mr. Darwin is equally bound to account for the formation of genera, families, orders, and even classes, by natural selection. He does "not doubt that the theory of descent with modification embraces all the members of the same class," and he concedes that analogy would press the conclusion still farther; while he admits that "the more distinct the forms are, the more the arguments fall away in force." To command assent we naturally require decreasing probability to be overbalanced by an increased weight of evidence. An opponent might plausibly, and perhaps quite fairly, urge that the links in the chain of argument are weakest just where the greatest stress falls upon them.

To which Mr. Darwin's answer is, that the best parts of the testimony have been lost. He is confident that intermediate forms must have existed; that in the olden times when the genera, the families and the orders diverged from their parent stocks, graduations existed as fine as those which now connect closely related species with varieties. But they have passed and left no sign. The geological record, even if all displayed to view, is a book from which not only many pages, but even whole alternate chapters have been lost out, or rather which were never printed from the autographs of nature. The record was actually made in fossil lithography only at certain times and under certain conditions (i.e., at periods of slow subsidence and places of abundant sediment); and of these records all but the last volume is out of print; and of its pages only local glimpses have been obtained. Geologists, except Lyell, will object to this,—some of them moderately, others with vehemence. Mr. Darwin himself admits, with a candor rarely displayed on such occasions, that he should have expected more geological evidence of transition than he finds, and that all the most eminent palæontologists maintain the immutability of species.

The general fact, however, that the fossil fauna of each period as a whole is nearly intermediate in character between the preceding and the succeeding faunas, is much relied on. We are brought one step nearer to the desired inference by the similar "fact, insisted on by all palæontologists, that fossils from two consecutive formations are far more closely related to each other, than are the fossils of two remote formations. Pictet gives a well-known instance,—the general resemblance of the organic remains from the several stages of the chalk formations, though the species are distinct at each stage. This fact alone, from its generality seems to have shaken Professor Pictet in his firm belief in the immutability of species." . . . What Mr. Darwin now particularly wants to complete his inferential evidence is a proof that the same gradation may be traced in later periods, say in the tertiary,[13] and between that period and the present; also that the later gradations are finer, so as to leave it doubtful whether the succession is one of species,—believed on the one theory to be independent, on the other, derivative,—or of varieties, which are confessedly derivative. The proof of the finer gradation appears to be forthcoming. Des Hayes and Lyell have concluded that many of the middle tertiary, and a large proportion of the later tertiary mollusca are specifically identical with living species; and this is still the almost universally prevalent view. But Mr. Agassiz states that, "in every instance where he had sufficient materials, he had found that the species of the two epochs supposed to be identical by Des Hayes and Lyell were in reality distinct, although closely allied species." Moreover he is now satisfied, as we understand, that the same gradation is traceable not merely in each great division of the tertiary, but in particular deposits or successive beds, each answering to a great number of years; where what have passed unquestioned as members of one species, upon closer examination of numerous specimens exhibit differences which in his opinion entitle them to be distinguished into

[13] The most recent of the three great geological epochs distinguishable at this time, but although most recent still not including a short period of the most recent and present.

two, three, or more species. It is plain, therefore, that whatever conclusions can be fairly drawn from the present animal and vegetable kingdoms in favor of a gradation of varieties into species, or into what may be regarded as such, the same may be extended to the tertiary period. In both cases, what some call species others call varieties; and in the later tertiary shells this difference in judgment affects almost half of the species!

We pass to a second difficulty in the way of Mr. Darwin's theory; to a case where we are perhaps entitled to demand of him evidence of gradation like that which connects the present with the tertiary mollusca. Wide, very wide is the gap, anatomically and physiologically (we do not speak of the intellectual) between the highest quadrumana and man; and comparatively recent, if ever, must the line have bifurcated. But where is there the slightest evidence of a common progenitor? Perhaps Mr. Darwin would reply by another question: where are the fossil remains of the men who made the flint knives and arrow-heads of the Somme valley?

We have a third objection, one, fortunately, which has nothing to do with geology. We can only state it here, in brief terms. The chapter on hybridism is most ingenious, able, and instructive. If sterility of crosses is a special, original arrangement to prevent the confusion of species by mingling, as is generally assumed, then, since varieties cross readily and their offspring is fertile *inter se,* there is a fundamental distinction between varieties and species. Mr. Darwin therefore labors to show that it is not a special endowment, but an incidental acquirement. He does show that the sterility of crosses is of all degrees;—upon which we have only to say, *Natura non facit saltum,*[14] here any more than elsewhere. But, upon his theory he is bound to show how sterility might be acquired, through natural selection or through something else. And the difficulty is, that, whereas individuals of the very same blood tend to be sterile, and somewhat remoter unions diminish this tendency, and when they have diverged into two varieties the cross-breeds between the two are more fertile than either pure stock,—yet when they have diverged only one degree more the whole tendency is reversed, and the mongrel is sterile, either absolutely or relatively. He who explains the genesis of species through purely natural agencies should assign a natural cause for this remarkable result; and this Mr. Darwin has not done. Whether original or derived, however, this arrangement to keep apart those forms which have, or have acquired (as the case may be) a certain moderate amount of difference, looks to us as much designed for the purpose, as does a ratchet to prevent reverse motion in a wheel. If species have originated by divergence, this keeps them apart.

Here let us suggest a possibly attainable test of the theory of derivation, a kind of instance which Mr. Darwin may be fairly asked to produce,—viz., an instance of two varieties, or what may be assumed as such, which have

[14] Nature does not proceed by leaps.

diverged enough to reverse the movement, to bring out some sterility in the crosses. The best marked human races might offer the most likely case. If mulattoes are sterile or tend to sterility, as some naturalists confidently assert, they afford Mr. Darwin a case in point. If, as others think, no such tendency is made out, the required evidence is wanting.

A fourth and the most formidable difficulty is that of the production and specialization of organs.

It is well said that all organic beings have been formed on two great laws; Unity of type, and Adaptation to the conditions of existence. The special teleologists, such as Paley,[15] occupy themselves with the latter only; they refer particular facts to special design, but leave an overwhelming array of the widest facts inexplicable. The morphologists build on unity of type, or that fundamental agreement in the structure of each great class of beings, which is quite independent of their habits or conditions of life; which requires each individual "to go through a certain formality," and to accept, at least for a time, certain organs whether they are of any use to him or not. Philosophical minds form various conceptions for harmonizing the two views theoretically. Mr. Darwin harmonizes and explains them naturally. Adaptation to the conditions of existence is the result of Natural Selection; Unity of type, of unity of descent. Accordingly, as he puts his theory, he is bound to account for the origination of new organs, and for their diversity in each great type, for their specialization, and every adaptation of organ to function and of structure to condition, through natural agencies. Whenever he attempts this he reminds us of Lamarck, and shows how little light the science of a century devoted to structural investigation has thrown upon the mystery of organization. Here purely natural explanations fail. The organs being given, natural selection may account for some improvement; if given of a variety of sorts or grades, natural selection might determine which should survive and where it should prevail.

On all this ground the only line for the theory to take is to make the most of gradation and adherence to type as suggestive of derivation, and unaccountable upon any other scientific view,—deferring all attempts to explain *how* such a metamorphosis was effected, until naturalists have explained *how* the tadpole is metamorphosed into a frog, or one sort of polyp into another. As to *why* it is so, the philosophy of efficient cause, and even the whole argument from design, would stand, upon the administration of such a theory of derivation, precisely where they stand without it. At least there is, or need be, no ground of difference here between Darwin and Agassiz. The latter will admit, with Owen and every morphologist, that hopeless is the attempt to explain the similarity of pattern in members of the same class by utility or the doctrine of final causes. "On the ordinary view

[15] William Paley, an English clergyman, and many other writers of the early nineteenth century explained the wonders of nature in terms of God's designs. All natural phenomena, they held, have a purpose.

of the independent creation of each being, we can only say that so it is, that it has so pleased the Creator to construct each animal and plant." Mr. Darwin in proposing a theory which suggests a *how* that harmonizes these facts into a system, we trust implies that all was done wisely, in the largest sense designedly, and by an Intelligent First Cause. The contemplation of the subject on the intellectual side, the amplest exposition of the Unity of Plan in Creation, considered irrespective of natural agencies, leads to no other conclusion.

We are thus, at last, brought to the question; what would happen if the derivation of species were to be substantiated, either as a true physical theory, or as a sufficient hypothesis? What would come of it? The enquiry is a pertinent one, just now. For, of those who agree with us in thinking that Darwin has not established his theory of derivation, many will admit with us that he has rendered a theory of derivation much less improbable than before; that such a theory chimes in with the established doctrines of physical science, and is not unlikely to be largely accepted long before it can be proved. Moreover, the various notions that prevail,—equally among the most and the least religious,—as to the relations between natural agéncies or phenomena and Efficient Cause, are seemingly more crude, obscure, and discordant than they need be.

It is not surprising that the doctrine of the book should be denounced as atheistical. What does surprise and concern us is, that it should be so denounced by a scientific man, on the broad assumptions that a material connection between the members of a series of organized beings is inconsistent with the idea of their being intellectually connected with one another through the Deity, i.e., as products of one mind, as indicating and realizing a preconceived plan. An assumption the rebound of which is somewhat fearful to contemplate, but fortunately one which every natural birth protests against.

It would be more correct to say, that the theory in itself is perfectly compatible with an atheistic view of the universe. That is true; but it is equally true of physical theories generally. Indeed, it is more true of the theory of gravitation, and of the nebular hypothesis, than of the hypothesis in question. The latter merely takes up *a particular, proximate cause,* or set of such causes, from which, it is argued, the present diversity of species has or may have *contingently resulted.* The author does not say *necessarily* resulted; that the actual results in mode and measure, and none other must have taken place. On the other hand the theory of gravitation, and its extension in the nebular hypothesis, assume a *universal and ultimate* physical cause, from which the effects in nature must *necessarily* have resulted. Now it is not thought, at least at the present day, that the establishment of the Newtonian theory was a step towards atheism or pantheism. Yet the great achievement of Newton consisted in proving that certain forces, (blind forces, so far as the theory is concerned,) acting upon matter in certain directions, must *necessarily* produce planetary orbits of the exact measure and form in which observation shows them to exist;—a view which is just as consistent with eternal necessity, either in the atheistic or the pantheistic form, as it is with theism.

Nor is the theory of derivation particularly exposed to the charge of the atheism of fortuity; since it undertakes to assign real causes for harmonious and systematic results. But of this, a word at the close.

The value of such objections to the theory of derivation may be tested by one or two analogous cases. The common scientific as well as popular belief is that of the original, independent creation of oxygen and hydrogen, iron, gold, and the like. Is the speculative opinion, now increasingly held, that some or all of the supposed elementary bodies are derivative or compound, developed from some preceding forms of matter, irreligious? Were the old alchemist atheists as well as dreamers in their attempts to transmute earth into gold? Or, to take an instance from force (power),—which stands one step nearer to efficient cause than form—was the attempt to prove that heat, light, electricity, magnetism, and even mechanical power are variations or transmutations of one force, atheistical in its tendency? The supposed establishment of this view is reckoned as one of the greatest scientific triumphs of this century.

Perhaps, however, the objection is brought, not so much against the speculation itself, as against the attempt to show how derivation might have been brought about. Then the same objection applies to a recent ingenious hypothesis made to account for the genesis of the chemical elements out of the etherial medium, and to explain their several atomic weights and some other characteristics by their successive complexity,—hydrogen consisting of so many atoms of etherial substance united in a particular order, and so on. The speculation interested the philosophers of the British Association, and was thought innocent, but unsupported by facts. Surely Mr. Darwin's theory is none the worse, morally, for having some foundation in fact.

In our opinion, then, it is far easier to vindicate a theistic character for the derivative theory, than to establish the theory itself upon adequate scientific evidence. Perhaps scarcely any philosophical objection can be urged against the former to which the nebular hypothesis is not equally exposed. Yet the nebular hypothesis finds general scientific acceptance, and is adopted as the basis of an extended and recondite illustration in Mr. Aggasiz's great work.

How the author of this book [Darwin] harmonizes his scientific theory with his philosophy and theology, he has not informed us. Paley, in his celebrated analogy with the watch, insists that if the time-piece were so constructed as to produce other similar watches, after the manner of generation in animals, the argument from design would be all the stronger. What is to hinder Mr. Darwin from giving Paley's argument a further *a-fortiori* extension to the supposed case of a watch which sometimes produces better watches, and contrivances adapted to successive conditions, and so at length turns out a chronometer, a townclock, or a series of organisms of the same type? From certain incidental expressions at the close of the volume, taken in connection with the motto adopted from Whewell, we judge it probable that our author regards the whole system of nature as one which had received at its first formation the impress of the will of its Author, foreseeing the varied yet necessary laws of its action throughout the whole of its existence, ordaining when and how each particular of the stupendous plan should be realized in effect, and—with Him to whom to will is to do—in ordaining doing it. Whether profoundly philosophical or not, a view maintained by eminent philosophical physicists and thologians, such as Babbage on the one hand and Jowett on the other, will hardly be denounced as atheism. Perhaps Mr. Darwin would prefer to express his ideas in a more general way, by adopting the thoughtful words of one of the most eminent naturalists of this or any age,[16] substituting the word *action* for 'thought,' since it is the former (from which alone the latter can be inferred) that he has been considering. "Taking nature as exhibiting thought for my guide, it appears to me that while human thought is consecutive, Divine thought is simultaneous, embracing at the same time and forever, in the past, the present and the future, the most diversified relations among hundreds of thousands of organized beings, each of which may present complications again, which, to study and understand even imperfectly,— as for instance man himself—mankind has already spent thousands of years." In thus conceiving of the Divine Power in act as cöetaneous with Divine Thought, and of both as far as may be apart from the human element of time, our author may regard the intervention of the Creator either as, humanly speaking, *done from all time,* or else as *doing through all time.* In the ultimate analysis we suppose that every philosophical theist must adopt one or the other conception.

A perversion of the first view leads towards atheism, the notion of an eternal sequence of cause and effect, for which there is no first cause,—a view which few sane persons can long rest in. The danger which may threaten the second view is pantheism. We feel safe from either error, in our profound conviction that there is order in the universe; that order presupposes mind; design, will; and mind or will, personality. Thus guarded, we much prefer the second of the two conceptions of causation, as the more philosophical as well as Christian view,—a view which leaves us with the same difficulties and the same mysteries in Nature as in Providence, and no other. Natural law, upon this view, is the human conception of continued and orderly Divine action.

We do not suppose that less power, or other power, is required to sustain the universe and carry on its operations, than to bring it into being. So, while conceiving no improbability of "interventions of Creative mind in nature, if by such is meant the bringing to pass of new and fitting events at

[16] That is, Agassiz.

fitting times, we leave it for profounder minds to establish, if they can, a rational distinction in kind between His working in nature carrying on operations, and in initiating those operations.

We wished under the light of such views, to examine more critically the doctrine of this book, especially of some questionable parts;—for instance, its explanation of the natural development of organs, and its implication of a "necessary acquirement of mental power" in the ascending scale of gradation. But there is room only for the general declaration that we cannot think the Cosmos a series which began with chaos and ends with mind, or of which mind is a result: that if by the successive origination of species and organs through natural agencies, the author means a series of events which succeed each other irrespective of a continued directing intelligence,—events which mind does not order and shape to destined ends,—then he has not established that doctrine, nor advanced towards its establishment, but has accumulated improbabilities beyond all belief. Take the formation and the origination of the successive degrees of complexity of eyes as a specimen. The treatment of this subject, . . . upon one interpretation is open to all the objections referred to; but if, on the other hand, we may rightly compare the eye "to a telescope, perfected by the long continued efforts of the highest human intellects," we could carry out the analogy and draw satisfactory illustrations and inferences from it. The essential, the directly intellectual thing is the making of the improvements in the telescope or the steam-engine. Whether the successive improvements, being small at each step, and consistent with the general type of instrument, are applied to some of the individual machines, or entire new machines are constructed for each, is a minor matter. Though if machines could engender, the adaptive method would be most economical; and economy is said to be a paramount law in nature. The origination of the improvements, and the successive adaptations to meet new conditions or subserve other ends, are what answer to the supernatural, and therefore remain inexplicable. As to bringing them into use, though wisdom foresees the result, the circumstances and the natural competition will take care of that, in the long run. The old ones will go out of use fast enough, except where an old and simple machine remains still best adapted to a particular purpose or condition,—as, for instance, the old Newcomen engine for pumping out coal-pits. If there's a Divinity that shapes these ends, the whole is intelligible and reasonable; otherwise, not.

We regret that the necessity of discussing philosophical questions has prevented a fuller examination of the theory itself, and of the interesting scientific points which are brought to bear in its favor. One of its neatest points, certainly a very strong one for the local origination of species, and their gradual diffusion under natural agencies, we must reserve for some other convenient opportunity.

The work is a scientific one, rigidly restricted to its direct object; and by its science it must stand or fall. Its aim is, probably not to deny creative intervention in nature,—for the admission of the independent origination of certain types does away with all antecedent improbability of as much intervention as may be required,—but to maintain that Natural Selection in explaining the facts, explains also many classes of facts which thousand-fold repeated independent acts of creation do not explain, but leave more mysterious than ever. How far the author has succeeded, the scientific world will in due time be able to pronounce.

As these sheets are passing through the press a copy of the second edition has reached us. We notice with pleasure the insertion of an additional motto on the reverse of the title-page, directly claiming the theistic view which we have vindicated for the doctrine. Indeed these pertinent words of the eminently wise Bishop Butler, comprise, in their simplest expression, the whole substance of our latter pages:—

"The only distinct meaning of the word 'natural' is *stated, fixed,* or *settled;* since what is natural as much requires and presupposes an intelligent mind to render it so, i.e., to effect it continually or at stated times, as what is supernatural or miraculous does to effect it for once."

13. An Orthodox Zoologist Attacks Darwin on Religious and Scientific Grounds. 1860

Asa Gray and LOUIS AGASSIZ (1807–1873) had already met each other in public debate over the issue of Darwinism in encounters in the Boston area when this classic confrontation took place in the pages of *The American Journal of Science.* Although writing a purely descriptive account of Acalephs—jellyfishes and their relations—Agassiz could not restrain himself from repeating his opinions on evolution. He was a worthy combatant, and he raised many important and embarrassing questions about Darwin's theory: How does one explain the discontinuities in the paleontological record? Is animal and plant breeding evidence for occurrences in nature? What is the tendency of variation in a species? Do the social speculations of Malthus constitute scientific evidence? What power or force does the selecting in natural selection? In making his own explanations Agassiz invokes the handiwork of God in several specific ways which raises a serious question of logic as Gray (see previous selection) recognized fully.

Agassiz was born in French Switzerland. He studied medicine at Zurich and natural history at Heidelberg, Munich, and Paris, specializing in ichthyology. He was professor at the Academy at Neuchâtel and while there produced his great work on glaciation. In 1846 he came to the United States, where he had a professorship at Harvard and became the most eminent man in natural history in the country and one of the great naturalists of the world.

See references for the previous section and see also, for a study of Agassiz, Edward Lurie, *Louis Agassiz, A Life in Science* (Chicago: University of Chicago Press, 1960).

PROF. AGASSIZ ON THE ORIGIN OF SPECIES.*

1. *Contributions to the Natural History of the United States;* by L. Agassiz.——The third volume of this work, now in the press, will appear shortly. We copy from the advance sheets the following paragraphs relating to the origin of species, which has lately attracted much attention, in consequence of the publication of Darwin's book on that subject.

INDIVIDUALITY AND SPECIFIC DIFFERENCES AMONG ACALEPHS

The morphological phenomena discussed in the preceding section naturally lead to a consideration of individuality and of the extent and importance of specific differences among the Acalephs. A few years ago the prevailing opinion among naturalists was that, while genera, families, orders, classes, and any other more or less comprehensive divisions among animals were artificial devices of science to facilitate our studies, species alone had a real existence in nature. Whether the views I have presented in the first volume of this work . . . where I showed that species do not exist in any different sense from genera, families, etc., have had anything to do with the change which seems to have been brought about upon this point among scientific men, is not for me to say; but whatever be the cause, it is certainly true that, at the present day, the number of naturalists who deny the real existence of species is greatly increased. Darwin in his recent work on the "Origin of Species," has also done much to shake the belief in the real existence of species, but the views he advocates are entirely at variance with those I have attempted to establish. For many years past I have lost no opportunity of urging the idea that while species have no material existence, they yet exist as categories of thought, in the same way as genera, families, orders, classes, and branches of the animal kingdom. Darwin's fundamental idea, on the contrary, is that species, genera, families, orders, classes, and any other kind of more or less comprehensive divisions among animals do not exist at all, and are altogether artificial, differing from one another only

* *The American Journal of Science and Arts,* 80 (1860), 142–154. Slightly abridged.

in degree, all having originated from a successive differentiation of a primordial organic form, undergoing successively such changes as would at first produce a variety of species; then genera, as the difference became more extensive and deeper; then families, as the gap widened still farther between the groups, until in the end all that diversity was produced which has existed or exists now. Far from agreeing with these views, I have, on the contrary, taken the ground that all the natural divisions in the animal kingdom are primarily distinct, founded upon different categories of characters, and that all exist in the same way, that is, as categories of thought,[1] embodied in individual living forms. I have attempted to show that branches in the animal kingdom are founded upon different plans of structure, and for that very reason have embraced from the beginning representatives between which there could be no community of origin; that classes are founded upon different modes of execution of these plans, and therefore they also embrace representatives which could have no community of origin; that orders represent the different degrees of complication in the mode of execution of each class, and therefore embrace representatives which could not have a community of origin any more than the members of different classes or branches; that families are founded upon different patterns of form, and embrace representatives equally independent in their origin; that genera are founded upon ultimate peculiarities of structure, embracing representatives, which, from the very nature of their peculiarities could have no community of origin; and that finally, species are based upon relations and proportions that exclude, as much as all the preceding distinctions, the idea of a common descent.

As the community of characters among the beings belonging to these different categories arises from the intellectual connection which shows them to be categories of thought, they cannot be the result of a gradual material differentiation of the objects themselves. The argument on which these views are founded may be summed up in the following few words: Species, genera, families, &c. exist as thoughts, individuals as facts. It is presented at full length in the first volume of this work . . . , where I have shown that individuals alone have a definite material existence, and that they are, for the time being, the bearers not only of specific characteristics, but of all the natural features in which animal life is displayed in all its diversity; individuality being, in fact, the great mystery of organic life.

Since the arguments presented by Darwin in favor of a universal derivation from one primary form, of all the peculiarities existing now among living beings have not made the slightest impression on my mind, nor modified in any way the views I have already propounded, I may fairly refer the reader to the paragraphs alluded to above as containing sufficient evidence of their correctness, and I will here only add a single argument, which seems to leave the question where I have placed it.

[1] That is, God's thought.

It seems to me that there is much confusion of ideas in the general statement of the variability of species so often repeated lately. If species do not exist at all, as the supporters of the transmutation theory maintain, how can they vary? and if individuals alone exist, how can the differences which may be observed among them prove the variability of species? The fact seems to me to be that while species are based upon definite relations among individuals which differ in various ways among themselves, each individual, as a distinct being, has a definite course to run from the time of its first formation to the end of its existence, during which it never loses its identity nor changes its individuality, nor its relations to other individuals belonging to the same species, but preserves all the categories of relationship which constitute specific or generic or family affinity, or any other kind or degree of affinity. *To prove that species vary it should be proved that individuals born from common ancestors change the different categories of relationship which they bore primitively to one another.* While all that has thus far been shown is, that there exists a considerable difference among individuals of one and the same species. This may be new to those who have looked upon every individual picked up at random, as affording the means of describing satisfactorily any species; but no naturalist who has studied carefully any of the species now best known, can have failed to perceive that it requires extensive series of specimens accurately to describe a species, and that the more complete such series are, the more precise appear the limits which separate species. Surely the aim of science cannot be to furnish amateur zoölogists or collectors, a *recipe* for a ready identification of any chance specimen that may fall into their hands. And the difficulties with which we may meet in attempting to characterize species do not afford the first indication that species do not exist at all, as long as most of them can be distinguished, as such, almost at first sight. I foresee that some convert to the transmutation creed will at once object that the facility with which species may be distinguished is no evidence that they were not derived from other species. It may be so. But as long as no fact is adduced to show that any one well known species among the many thousands that are buried in the whole series of fossiliferous rocks, is actually the parent of any one of the species now living, such arguments can have no weight; and thus far the supporters of the transmutation theory have failed to produce any such facts. Instead of facts we are treated with marvelous bear, cuckoo, and other stories. Credat Judaeus Apella![2]

Had Mr. Darwin or his followers furnished a single fact to show that individuals change, in the course of time, in such a manner as to produce at least species different from those known before, the state of the case might be different. But it stands recorded now as before, that the animals known to the ancients are still in existence, exhibiting to this day the characters they exhibited of old. The geological record, even with all its imperfections,

[2] Tell the crazed Jews, meaning too improbable to obtain general credence; customarily used, as here, disdainfully.

exaggerated to distortion, tells now, what it has told from the beginning, that the supposed intermediate forms between the species of different geological periods are imaginary beings, called up merely in support of a fanciful theory. The origin of all the diversity among living beings remains a mystery as totally unexplained as if the book of Mr. Darwin had never been written, for no theory unsupported by fact, however plausible it may appear, can be admitted in science.

It seems generally admitted that the work of Darwin is particularly remarkable for the fairness with which he presents the facts adverse to his views. It may be so; but I confess that it has made a very different impression upon me. I have been more forcibly struck by his inability to perceive when the facts are fatal to his argument, than by anything else in the whole work. His chapter on the Geological Record, in particular, appears to me, from beginning to end, as a series of illogical deductions and misrepresentations of the modern results of Geology and Palæontology. I do not intend to argue here, one by one, the question he has discussed. Such arguments end too often in special pleading, and any one familiar with the subject may readily perceive where the truth lies by confronting his assertions with the geological record itself. But since the question at issue is chiefly to be settled by palæontological evidence, and I have devoted the greater part of my life to the special study of the fossils, I wish to record my protest against his mode of treating this part of the subject. Not only does Darwin never perceive when the facts are fatal to his views, but when he has succeeded by an ingenious circumlocution in overleaping the facts, he would have us believe that he has lessened their importance or changed their meaning. He would thus have us believe that there have been periods during which all that had taken place during other periods was destroyed, and this solely to explain the absence of intermediate forms between the fossils found in successive deposits, for the origin of which he looks to those missing links; whilst every recent progress in Geology shows more and more fully how gradual and successive all the deposits have been which form the crust of our earth.—He would have us believe that entire faunæ have disappeared before those were preserved, the remains of which are found in the lowest fossiliferous strata; when we find everywhere non-fossiliferous strata below those that contain the oldest fossils now known. It is true, he explains their absence by the supposition that they were too delicate to be preserved; but any animals from which Crinoids, Brachiopods, Cephalopods, and Trilobites[3] could arise, must have been sufficiently similar to them to have left, at least, traces of their presence in the lowest non-fossiliferous rocks, had they ever existed at all.—He would have us believe that the oldest organisms that existed were simple cells, or some-

[3] Crinoids are mostly fossil lily-shaped echinoderms related to the starfish. Brachiopods are a class of mollusks. Cephalopods are cuttlefishes, octopuses, and their relatives—the most highly developed invertebrates. Trilobites were a widely distributed group of crustaceans whose fossil remains have been extremely important in correlating and dating geological formations.

thing like the lowest living beings now in existence; when such highly organized animals as Trilobites and Orthoceratites[4] are among the oldest known.—He would have us believe that these lowest first-born became extinct in consequence of the gradual advantage some of their more favored descendants gained over the majority of their predecessors; when there exist now, and have existed at all periods in past history, as large a proportion of more simply organized beings, as of more favored types, and when such types as Lingula were among the lowest Silurian[5] fossils, and are alive at the present day.—He would have us believe that each new species originated in consequence of some slight change in those that preceded; when every geological formation teems with types that did not exist before.—He would have us believe that animals and plants became gradually more and more numerous; when most species appear in myriads of individuals, in the first bed in which they are found. He would have us believe that animals disappear gradually; when they are as common in the uppermost bed in which they occur as in the lowest, or any intermediate bed. Species appear suddenly and disappear suddenly in successive strata. That is the fact proclaimed by Palæontology; they neither increase successively in number, nor do they gradually dwindle down; none of the fossil remains thus far observed show signs of a gradual improvement or of a slow decay.—He would have us believe that geological deposits took place during the periods of subsidence; when it can be proved that the whole continent of North America is formed of beds which were deposited during a series of successive upheavals. I quote North America in preference to any other part of the world, because the evidence is so complete here that it can only be overlooked by those who may mistake subsidence for the general shrinkage of the earth's surface in consequence of the cooling of its mass. In this part of the globe, fossils are as common along the successive shores of the rising deposits of the Silurian system, as anywhere along our beaches; and each of these successive shores extends from the Atlantic States to the foot of the Rocky Mountains. The evidence goes even further; each of these successive sets of beds of the Silurian system contains peculiar fossils, neither found in the beds above nor in the beds below, and between them there are no intermediate forms. And yet Darwin affirms that "the littoral and sub-littoral deposits are continually worn away, as soon as they are brought up by the slow and gradual rising of the land within the grinding action of the coast waves." Origin of Species, p. 290.—He would have us believe that the most perfect organs of the body of animals are the product of gradual improvement, when eyes as perfect as those of the Trilobites are preserved with the remains of these oldest animals.—He would have us believe that it required millions of years to effect any one of these changes; when far more extraordinary trans-

[4] Orthoceratites are cephalopods with horn-shaped shells.

[5] Lingula are a type of marine shellfish, and the Silurian was one of the most ancient geological periods from which fossil remains were known at this time.

formations are daily going on, under our eyes, in the shortest periods of time, during the growth of animals.—He would have us believe that animals acquire their instincts gradually; when even those that never see their parents, perform at birth the same acts, in the same way, as their progenitors.—He would have us believe that the geographical distribution of animals is the result of accidental transfers; when most species are so narrowly confined within the limits of their natural range, that even slight changes in their external relations may cause their death. And all these, and many other calls upon our credulity, are coolly made in the face of an amount of precise information, readily accessible, which would overwhelm any one who does not place his opinions above the records of an age eminently characterized for its industry, and during which, that information was laboriously accumulated by crowds of faithful laborers.

It would be superfluous to discuss in detail the arguments by which Mr. Darwin attempts to explain the diversity among animals. Suffice it to say, he has lost sight of the most striking of the features, and the one which pervades the whole, namely, that there runs throughout Nature unmistakable evidence of thought, corresponding to the mental operations of our own mind, and therefore intelligible to us as thinking beings, and unaccountable on any other basis than that they owe their existence to the working of intelligence; and no theory that overlooks this element can be true to nature.

There are naturalists who seem to look upon the idea of creation, that is, a manifestation of an intellectual power by material means, as a kind of bigotry; forgetting, no doubt, that whenever they carry out a thought of their own, they do something akin to creating, unless they look upon their own elucubrations as something in which their individuality is not concerned, but arising without an intervention of their mind, in consequence of the working of some "bundles of forces," about which they know nothing themselves. And yet such men are ready to admit that matter is omnipotent, and consider a disbelief in the omnipotence of matter as tantamount to imbecility; for, what is the assumed power of matter to produce all finite beings, but omnipotence? And what is the outcry raised against those who cannot admit it, but an insinuation that they are *non-compos*?[6] The book of Mr. Darwin is free of all such uncharitable sentiments towards his fellow-laborers in the field of science; nevertheless his mistake lies in a similar assumption that the most complicated system of combined thoughts can be the result of accidental causes; for he ought to know, as every physicist will concede, that all the influences to which he would ascribe the origin of species are accidental in their very nature, and he must know, as every naturalist familiar with the modern progress of science does know, that the organized beings which live now, and have lived in former geological periods, constitute an organic whole, intelligibly and methodically combined in all its parts. As a zoölogist

[6] That is, not mentally competent.

he must know in particular, that the animal kingdom is built upon four different plans of structure, and that the reproduction and growth of animals takes place according to four different modes of development, and that unless it is shown that these four plans of structure, and these four modes of development, are transmutable one into the other, no transmutation theory can account for the origin of species. The fallacy of Mr. Darwin's theory of the origin of species by means of natural selection, may be traced in the first few pages of his book, where he overlooks the difference between the voluntary and deliberate acts of selection applied methodically by man to the breeding of domesticated animals and the growing of cultivated plants, and the chance influences which may effect [*sic*] animals and plants in the state of nature. To call these influences "natural selection," is a misnomer which will not alter the conditions under which they may produce the desired results. Selection implies design; the powers to which Darwin refers the order of species, can design nothing. Selection is no doubt the essential principle on which the raising of breeds is founded, and the subject of breeds is presented in its true light by Mr. Darwin; but this process of raising breeds by the selection of favorable subjects, is in no way similar to that which regulates specific differences. Nothing is more remote from the truth than the attempted parallelism between the breeds of domesticated animals and the species of wild ones. Did there exist such a parallelism, as Darwin maintains, the difference among the domesticated breeds should be akin to the differences among wild species, and afford a clue to determine their relative degree of affinity by a comparison with the pedigrees of well-known domesticated races. Again, if there were any such parallelism, the distinctive characteristics of different breeds should be akin to the differences which exist between fossil species of earlier periods and those of the same genera now living. Now let any one familiar with the fossil species of the genera Bos and Canis,[7] compare them with the races of our cattle and of our dogs, and he will find no correspondence whatever between them; for the simple reason that they do not owe their existence to the same causes. It must therefore be distinctly stated that Mr. Darwin has failed to establish a connection between the mode of raising domesticated breeds, and the cause or causes to which wild animals owe their specific differences.

It is true, Mr. Darwin states that the close affinity existing among animals can only be explained by a community of descent, and he goes so far as to represent these affinities as evidence of such a genealogical relationship; but I apprehend that the meaning of the words he uses has misled him into the belief that he had found the clue to phenomena which he does not even seem correctly to understand. There is nothing parallel between the relations of animals belonging to the same genus or the same family, and the relations between the progeny of common ancestors. In the one case we have the result of a physiological law regulating reproduction, and in other affinities

[7] The genera of cattle and dogs respectively.

which no observation has thus far shown to be in any way connected with reproduction. The most closely allied species of the same genus or the different species of closely allied genera, or the different genera of one and the same natural family, embrace representatives which at some period or other of their growth resemble one another more closely than the nearest blood relations; and yet we know that they are only stages of development of different species distinct from one another at every period of their life. The embryo of our common fresh water turtle, *Chrysemis picta,* and the embryo of our snapping turtle, *Chelydra serpentina,* resemble one another far more than the different species of Chrysemis in their adult state, and yet not a single fact can be adduced to show that any one egg of an animal has ever produced an individual of any species but its own. A young snake resembles a young turtle or a young bird much more than any two species of snakes resemble one another; and yet they go on reproducing their kinds, and nothing but their kinds. So that no degree of affinity, however close, can, in the present state of our science, be urged as exhibiting any evidence of community of descent, while the power that imparted all their peculiarities to the primitive eggs of all the species now living side by side, could also impart similar peculiarities with similar relations, and all degrees of relationship, to any number of other species that have existed. Until, therefore it can be shown that any one species has the ability to delegate such specified peculiarities and relations to any other species or set of species, it is not logical to assume that such a power is inherent in any animal, or that it constitutes part of its nature.[8] We must look to the original power that imparted life to the first being for the origin of all other beings, however mysterious and inaccessible the modes by which all this diversity has been produced may remain for us. The production of a plausible explanation is no explanation at all, if it does not cover the whole ground.

All attempts to explain the origin of species may be brought under two categories: viz. 1st, some naturalists admitting that all organized beings are created, that is to say, endowed from the beginning of their existence with all their characteristics, while 2d, others assume that they arise spontaneously. This classification of the different theories of the origin of species, may appear objectionable to the supporters of the transmutation theory; but I can per-

[8] The difficulty of ascertaining the natural limits of some species, and the mistakes made by naturalists when describing individual peculiarities as specific, has nothing to do with the question of the origin of species, and yet Darwin places great weight, in support of his theory, upon the differences which exist among naturalists in their view of species. Some of the metals are difficult to distinguish, and have frequently been mistaken, and the specific differences of some may be questioned; but what could that have to do with the question of the origin of metals, in the minds of those who may doubt the original difference of metals? Nothing more than the blunders of some naturalists in identifying species with the origin of species of animals and plants. The great mischief in our science now lies in the self-complacent confidence with which certain zoologists look upon a few insignificant lines, called diagnoses, which they have the presumption to offer as characteristics of species, or, what is still worse, as checks upon others to secure to themselves a nominal priority. Such a treatment of scientific subjects is unworthy of our age. [*Footnote in original.*]

ceive no essential difference between their views and the old idea that animals may have arisen spontaneously. They differ only in the modes by which the spontaneous appearance is assumed to be effected; some believe that physical agents may so influence organized beings as to modify them—this is the view of DeMaillet and the Vestiges of Creation;[9] others believe that the organized beings themselves change in consequence of their own acts, by changing their mode of life, etc., this is the view of Lamarck; others still assume that animals and plants tend necessarily to improve, in consequence of the struggle for life, in which the favored races are supposed to survive; this is the view lately propounded by Darwin. I believe these theories will, in the end, all share the fate of the theory of spontaneous generations so called, as the facts of nature shall be confronted more closely with the theoretical assumptions. The theories of DeMaillet, Oken, and Lamarck are already abandoned by those who have adopted the transmutation theory of Darwin; and unless Darwin and his followers succeed in showing that the struggle for life tends to something beyond favoring the existence of certain individuals over that of other individuals, they will soon find that they are following a shadow. The assertion of Darwin, which has crept into the title of his work, is, that favored *races* are preserved, while all his facts go only to substantiate the assertion, that favored *individuals* have a better chance in the struggle for life than others. But who has ever overlooked the fact that myriads of individuals of every species constantly die before coming to maturity? What ought to be shown, if the transmutation theory is to stand, is that these favored individuals diverge from their specific type, and neither Darwin nor any body else has furnished a single fact to show that they go on diverging. The criterion of a true theory consists in the facility with which it accounts for facts accumulated in the course of long-continued investigations and for which the existing theories afforded no explanation. It can certainly not be said that Darwin's theory will stand by that test. It would be easy to invent other theories that might account for the diversity of species quite as well, if not better than Darwin's preservation of favored races. The difficulty would only be to prove that they agree with the facts of Nature. It might be assumed, for instance, that any one primary being contained the possibilities of all those that have followed, in the same manner as the egg of any animal possesses all the elements of the full-grown individual; but this would only remove the difficulty one step further back. It would tell us nothing about the nature of the operation by which the change is introduced. Since the knowledge we now have, that similar metamorphoses go on in the eggs of all living beings has not yet put us on the track of the forces by which the changes they undergo are brought about, it is not likely that by mere guesses we shall arrive at any satisfactory explanation of the very origin of these beings themselves.

[9] See footnote 11, previous selection.

Whatever views are correct concerning the origin of species, one thing is certain, that as long as they exist they continue to produce generation after generation, individuals which differ from one another only in such peculiarities as relate to their individuality. The great defect in Darwin's treatment of the subject of species lies in the total absence of any statement respecting the features that constitute individuality. Surely, if individuals may vary within the limits assumed by Darwin, he was bound first to show that individuality does not consist of a sum of hereditary characteristics, combined with variable elements, not necessarily transmitted in their integrity, but only of variable elements. That the latter is not the case, stands recorded in every accurate monograph of all the types of the animal kingdom upon which minute embryological investigations have been made. It is known, that every individual egg undergoes a series of definite changes before it reaches its mature condition; that every germ formed in the egg passes through a series of metamorphoses before it assumes the structural features of the adult; that in this development the differences of sex may very early become distinct; and that all this is accomplished in a comparatively very short time, extremely short, indeed, in comparison to the immeasurable periods required by Darwin's theory to produce any change among species; and yet all this takes place without any deviation from the original type of the species, though under circumstances which would seem most unfavorable to the maintenance of the type. Whatever minor differences may exist between the products of this succession of generations are all *individual peculiarities,* in no way connected with the essential features of the species, and therefore as transient as the individuals; while the specific characters are forever fixed. A single example will prove this. All the robins of North America now living have been for a short time in existence; not one of them was alive a century ago, when Linnæus for the first time made known that species under the name of Turdus migratorius, and not one of the specimens observed by Linnæus and his cotemporaries was alive when the Pilgrims of the Mayflower first set foot upon the Rock of Plymouth. Where was the species at these different periods, and where is it now? Certainly nowhere but in the individuals alive for the time being; but not in any single one of them, for that one must be either a male or a female, and not the species; not in a pair of them, for the species exhibits its peculiarities in its mode of breeding, in its nest, in its eggs, in its young, as much as in the appearance of the adult; not in all the individuals of any particular district, for the geographical distribution of a species over its whole area, forms also part of its specific characters.[10] A species is

[10] We are so much accustomed to see animals reproducing themselves, generation after generation, that the fact no longer attracts our attention, and the mystery involved in it no longer excites our admiration. But there is certainly no more marvellous law in all nature than that which regulates this regular succession. And upon this law the maintenance of species depends; for observation teaches us that all that is not individual peculiarity is unceasingly and integrally reproduced while all that constitutes individuality, as such, constantly disappears. [*Footnote in original.*]

only known when its whole history has been ascertained, and that history is recorded in the life of individuals through successive generations. The same kind of argument might be adduced from every existing species, and with still greater force by a reference to those species already known to the ancients.

Let it not be objected that the individuals of successive generations have presented marked differences among themselves; for these differences, with all the monstrosities that may have occurred, during these countless generations, have passed away with the individuals, as individual peculiarities, and the specific characteristics alone have been preserved, together with all that distinguishes the genus, the family, the order, the class, and the branch to which the individual belonged. Moreover all this has been maintained through a succession of repeated changes, amounting in each individual to the whole range of transformations, through which an individual passes, from the time it is individualized as an egg, to the time it is itself capable of reproducing its kind, and, perhaps, with all the intervening phases of an unequal production of males and females, of sterile individuals, of dwarfs, of giants, etc., etc., during which there were millions of chances for a deviation from the type. Does this not prove that while individuals are perishable, they transmit, generation after generation, all that is specific or generic, or, in one word, *typical* in them, to the exclusion of every *individual peculiarity* which passes away with them, and that, therefore, while individuals alone have a material existence, species, genera, families, orders, classes, and branches of the animal kingdom exist only as categories of thought in the Supreme Intelligence, but as such have as truly an independent existence and are as unvarying as thought itself after it has once been expressed.

Returning, after this digression, to the question of individuality among Acalephs, we meet here phenomena far more complicated than among higher animals. Individuality, as far as it depends upon material isolation, is complete and absolute in all the higher animals, and there maintained by genetic transmission, generation after generation. Individuality, in that sense, exists only in comparatively few of the Radiates. Among Acalephs it is ascertained only for the Ctenophoræ and some Discophoræ. In others, the individuals born from eggs end by dividing into a number of distinct individuals. In others still, the successive individuals derived from a primary one, remain connected to form compound communities. We must therefore, distinguish different kinds and different degrees of individuality, and may call *hereditary individuality* that kind of independent existence manifested in the successive evolutions of a single egg, producing a single individual, as is observed in all the higher animals. We may call *derivative* or *consecutive individuality* that kind of independence resulting from an individualization of parts of the product of a single egg. We have derivative individuals among the Nudibranchiate Mollusks, whose eggs produce singly, by a process of complete segmentation, several independent individuals. We observe a similar phenomenon among those Acalephs the young of which (Scyphostoma) ends in producing, by transverse division (Strobila), a number of independent free Medusæ (Ephyræ). We have it also among the Hydroids which produce free Medusæ. Next, we must distinguish *secondary individuality,* which is inherent to those individuals arising as buds from other individuals, and remaining connected with them. This condition prevails in all the immovable Polyparia and Hydraria, and I say intentionally in the immovable ones; for, in the movable communities, such as Renilla, Pennatula, etc., among Polyps, and all the Siphonophoræ among Acalephs, we must still further distinguish another kind of individuality, which I know not how to call properly, unless the name of *complex individuality* may be applied to it. In complex individuality a new element is introduced, that is not noticeable in the former case. The individuals of the community are not only connected together, but, under given circumstances, they act together as if they were one individual, while at the same time each individual may perform acts of its own.

As to the specific differences observed among Acalephs, there is as great a diversity between them as between their individuals. In some types of this class the species are very uniform; all the individuals belonging to one and the same species resembling one another very closely, and exhibiting hardly any difference among themselves, except such as arises from age. This identity of the individuals of one and the same species is particularly striking among the Ctenophoræ. In this order there are not even sexual differences among the individuals, as they are all hermaphrodites. In the Discophoræ proper a somewhat greater diversity prevails. In the first place we notice male and female individuals, and the difference between the sexes is quite striking in some genera, as, for instance, in Aurelia. Next there occur frequent deviations, among them, in the normal number of their parts; their body consisting frequently of one or two spheromeres more than usual, sometimes, even, of double normal number, or of a few less. And yet, year after year, the same Discophoræ reappear upon our shores, with the same range of differences among their individuals. Among Hydroids polymorphism[11] prevails to a greater or less extent, besides the differences arising from sex. Few species have only one kind of individuals. Mostly the cycle of individual differences embraces two distinct types of individuals, one recalling the peculiarities of common Hydræ, the other those of Medusæ; but even the Hydra type of one and the same species may exhibit more or less diversity, there being frequently two kinds of Hydræ united in one and the same community, and sometimes even a larger number of heterogeneous Hydræ. And this is equally true, though to a less extent, of the Medusa type. Yet among Siphonophoræ there are generally at least two kinds of Medusæ in one and the same community. But notwithstanding this polymorphism among the individuals of one and the same community, genetically connected together, each successive generation reproduces the same kinds of heterogeneous individuals, and nothing

[11] That is, a wide range of variation of form and appearance among individuals of the same species.

but individuals linked together in the same way. Surely we have here a much greater diversity of individuals, born one from the other, than is exhibited by the most diversified breeds of our domesticated animals; and yet all these heterogeneous individuals remain true to their species, in one case as in the other, and do not afford the slightest evidence of a transmutation of species.

Would the supporters of the fanciful theories lately propounded, only extend their studies a little beyond the range of domesticated animals, would they investigate the alternate generations of the Acalephs, the extraordinary modes of development of the Helminth, the reproduction of the Salpæ,[12] etc., etc., they would soon learn that there are, in the world, far more astonishing phenomena, strictly circumscribed between the natural limits of unvarying species, than the slight differences produced by the intervention of men, among domesticated animals, and, perhaps, cease to be so confident as they seem to be, that these differences are trustworthy indications of the variability of species. For my own part I must emphatically declare that I do not know a single fact tending to show that species do vary in any way, while it is true that the individuals of one and the same species are more or less polymorphous. The circumstance, that naturalists may find it difficult to trace the natural limits of any one particular species, or the mistakes they may make in their attempts to distinguish them, has nothing whatsoever to do with the question of their origin. . . .

When considering Individuality and Specific Differences, as manifested in the class of Acalephs, I have taken an opportunity of showing, upon general grounds how futile the arguments are upon which the theory of transmutation of species is founded. Having now shown that class is circumscribed within definite limits, I may be permitted to add here a few more objections to that theory, based chiefly upon special grounds, connected with the characteristics of classes. If there is any thing striking in the features which distinguish classes, it is the definiteness of their structural peculiarities; and this definiteness goes on increasing, with new and additional qualifications, as we pass from the class characters to those which mark the orders, the families, the genera, and the species. Granting, for the sake of argument, that organized beings living at a later period may have originated by a gradual change of those of earlier periods, one of the most characteristic features of all organized beings remains totally unexplained by the various theories brought forward to explain that change; the definiteness of their respective groups, be they ever so comprehensive, or ever so limited, combined with the greatest inequality in their numeric relations. There exist a few thousand Mammalia and Reptiles, and at least three times their number of Birds and Fishes. There may be twenty thousand Mollusks; but there are over a hundred thousand Insects, and only a few thousand Radiates. And yet the limits of the class of Insects are as well defined as those of any other class, with the only excep-

[12] Helminth and Salpæ, respectively worms and a type of primitive vertebrate ocean fish, sometimes free-swimming and sometimes living in colonies.

tion of the class of Birds which is unquestionably the most definite in its natural boundaries. Now the supporters of the transmutation theory may shape their views in whatever way they please to suit the requirements of the theory, instead of building the theory upon the facts of Nature, they never can make it appear that the definiteness of the characters of the class of Birds is the result of a common descent of all Birds, for the first Bird must have been brother or cousin to some other animal that was not a Bird, since there are other animals besides Birds in this world, to no one of which any bird bears as close a relation as it bears to its own class. The same argument applies to every other class; and as to the facts, they are fatal to such an assumption, for Geology teaches us that among the oldest inhabitants of our globe known, there are representatives of nine distinct classes of animals, which by no possibility can be descendants of one another, since they are cotemporaries.

The same line of argument and the same class of facts forbid the assumption that either the representatives of one and the same order, or those of one of the same family, or those of one of the same genus should be considered as lineal descendants of a common stock; for orders, families and genera are based upon different categories of characters, and not upon more or less extensive characters of the same kind, as I have shown years ago . . . , and numbers of different kinds of representatives of these various groups, make their appearance simultaneously in all the successive geological periods. There appear together Corals and Echinoderms of different families and of different genera in each successive geological formation, and this is equally true for Bryozoa, Brachiopods and Lamellibranchiata, for Trilobites and the other Crustacea, in fact for the representatives of all the classes of the animal kingdom, making due allowance for the period of the first appearance of each; and at all times and in all classes the representatives of these different kinds of groups are found to present the same definiteness in their characteristics and limitation. Were the transmutation theory true, the geological record should exhibit an uninterrupted succession of types blending gradually into one another. The fact is that throughout all geological times each period is characterized by definite specific types, belonging to definite genera, and these to definite families, referable to definite orders, constituting definite classes and definite branches, built upon definite plans. Until the facts of Nature are shown to have been mistaken by those who have collected them, and that they have a different meaning from that now generally assigned to them, I shall therefore consider the transmutation theory as a scientific mistake, untrue in its facts, unscientific in its method, and mischievous in its tendency.

14. A Young Physicist Surveys Science as It Existed in One Year. 1869

This summary of science in one year is far more revealing than the simple record put down was intended to be. What science meant to that day, and the place of America in that science, is made unwittingly obvious. This is a unique glimpse of how Americans saw the world of science and functioned in it ten years after Darwin.

JOHN TROWBRIDGE (1843–1923) was a young physicist at the Massachusetts Institute of Technology when he wrote this. He was born in Boston and had been graduated from the Lawrence Scientific School of Harvard University. Shortly after he wrote this article he was called to teach physics at Harvard, where he introduced modern laboratory instruction in physics. He eventually held the Rumford professorship there.

His biography appears in *Dictionary of American Biography,* XVII, 654–655.

NOTES BY THE EDITOR, on the Progress of Science for the Year 1869.*

John Trowbridge

The opening of the Pacific Railway and of the Suez Canal, and the completion of the laying of the French Cable, are tempting subjects to dwell upon.

It is not fitting to indulge in national boasting, at the completion of our line to the Pacific, before we learn the exact condition of the road, and the thoroughness of the work; although the rapidity of its execution, and its magnitude, might excuse any display of national egotism. The opening, however, of our great territories to the enterprise of both Atlantic and Pacific coasts, and to the cheap labor of Asia, is a result clearly to be seen.

We shall soon be called upon to chronicle other Pacific Railways; a northern, and possibly a southern one. As in the case of the French Atlantic Cable, the success of later attempts will be received as a matter of course, and the Pacific Railroad, whose completion we note to-day, will lose its prestige among the coming number of routes to the Pacific. In the present volume

* *Annual of Scientific Discovery, Or, Year-Book of Facts in Science and Art, For 1870* (Boston: Gould and Lincoln, 1870), pp. iii–xxii, with a few editorial deletions.

will be found accounts of the coal-fields of the territories. Apprehensions of lack of fuel for our great railway, by the discovery of these deposits, are seen to be ill-founded. It is felt that the Pacific Railway, with all its great realities and possibilities, is inadequate as a means of communication between our lines of coast, and attention has been redirected to the Darien Ship Canal.[1] An appropriation has been made by Congress to pay the expenses of a new survey for this work, and an expedition has already sailed.

The completion of the Suez Canal undoubtedly had its share in directing public attention in the United States to the possibility of this enterprise. This canal has been opened with impressive ceremonies; the reports are somewhat contradictory in regard to the work.

Shallow iron steam-ships are being built on the Tyne, for the navigation of the canal. Mr. Ashbury, who sailed through the canal in his yacht, Cambria, writes that after taking careful soundings, he is of the opinion that no vessel drawing over nineteen feet of water can pass through the canal.

The "New York Tribune" states: "Two of the steamers of the Messageries Imperiales (French Company), of 2,400 tons burden, have safely passed through the Suez Canal. Steamers drawing fifteen feet can navigate the canal from Port Said to Suez, with ease, in fifteen hours. The water does not wash away the banks as much as apprehended. The complete success of the great work exceeds all expectations."

The Suez Canal Company has issued regulations for the navigation of the canal. Article I. states that the navigation of the Suez Maritime Canal will be open to all ships without distinction of nationality, provided their draught of water does not exceed 7½ metres, the depth of the canal being 8 metres, equal to 26 English feet.

To-day we witness a return to old routes of commerce. In early times, the track of commerce between the West and the East was by the way of Egypt and the Red Sea; from this commerce Alexandria rose to opulence, and Venice became a first-rate power. Afterwards, by the discovery of the Cape route, trade was diverted into new channels, and Venice and Alexandria sank in wealth and importance. The opening of the Suez Canal brings commerce back into its old channel.

We are called upon to chronicle the successful laying of the French Atlantic Cable.

A project to extend telegraphic communication from Cuba (already in connection with Florida) by Porto Rico, through the West India Island, is favorably entertained. Prussia, too, we hear, is beginning to think of securing more direct communication with America. It has been suggested that if a cable were laid from a point on her seaboard round by the north of Scotland, and by the western shore of Ireland, to join the Anglo-American cables at Valentia, Prussia would send all the North of Europe messages by this route.

[1] That is, a canal through the Isthmus of Panama.

It is understood that the Prussian Government have had the subject recently before them, and that a concession has been granted to carry out an Atlantic cable, having North Germany for its terminus. The old project of the North Atlantic is being again mooted. That route was to go by Iceland, Greenland, and so on to Canada and the United States, Denmark being the assumed starting-point. The cable to India by the Red Sea is going on satisfactorily, and an auxiliary line, one between Marseilles and Malta, is spoken of.

All of these projects indicate increased convenience and gain to the public. At present the use of the ocean telegraph is confined to the commercial community; but ere long, when the tariff is reduced from Europe to America, and to India, the general public will send messages as freely as they do by the land wires. We may reasonably hope, too, that the cost of submarine cables will be reduced by and by, and this will do more to cheapen messages than anything else.

In Northern Russia the construction of a land line is far advanced to connect St. Petersburg with the mouth of the Amour River, on completion of which only a submarine link will be wanting to complete the telegraphic girdle round the earth.

Electricity and steam are the great agents of civilization. The introduction of telegraphic lines and railways in Russia and Asia is destined to revolutionize this part of the globe. We Americans are apt to think ourselves the most progressive nation, and point with especial pride to our Pacific Railway. Russia, however, is making great strides; and the English railways in India compete in difficulty of execution and magnitude with the Pacific Railway.

During the past year several improvements in railway carriages have been brought before the public.

Mr. Robert F. Fairlie has invented a steam carriage which will round curves of 50 feet radius at 20 miles an hour, with, it is alleged, perfect safety. The carriage, instead of seating the usual complement of 100 passengers (English car), seats only 66. The English papers are enthusiastic in regard to this carriage.

The Portmadoc and Festiniog Railway, in Wales, has also attracted much attention, from the narrowness of its gauge—two feet only. The Fairlie carriage and the narrow-gauge railway will undoubtedly come into play in difficult countries.[2]

We are certainly far from perfection in the construction of our railways in America. The fearful catastrophes that have taken place from cars taking fire have reawakened an interest in new methods of heating them. No method has yet been devised to meet the difficulty satisfactorily. In this volume will

[2] This Welsh railway was originally built and used as a tram line in 1836 for the hauling of slate. In 1869 it was converted to passenger use and became known as the "Toy Railway." Narrow gauge railways are appropriate for mountainous terrain where very sharp curves are necessary, or at least economical.

be found the description of an electro-heating apparatus. The introduction of steel rails promises to make accidents from defective rails rarer.

The English have lately turned their attention to the American system of constructing railroads. They have found, to their surprise, that in India they must adopt American ideas.[3] Notwithstanding its defects, it has been found that our system is likely to prove the best for their colonies. A commission of English engineers are now investigating our system with a view to the railways of India.

The brake power on several of the French and Spanish railways has been greatly increased by an ingenious arrangement conceived by Monsieur Chatelier, of applying what has been termed "contre vapeur" to the engine, converting it, for the time being, into a pump forcing steam and water into the boiler.

At a meeting of the American Academy of Arts and Sciences, held in Boston, U.S., the Rumford medals were presented to Mr. George H. Corliss, of Providence, R. I., for his improvements in the steam-engine. The presentation was made by Dr. Asa Gray, the President of the Academy. We make the following extract from his remarks:

"It appears that within the twenty years since this machinery was perfected, more than 1,000 engines of the kind have been built in the United States, and several hundred in other countries, giving an aggregate of not less than 250,000 horse-power; that as to economy of fuel, evidence has been afforded to the Rumford Committee, showing a saving over older forms of engines of about one-third. As to its other crowning excellence, uniformity of velocity, the purchasers of one of the engines, now in its eighteenth year of service, certify that, with the power varying from 60 to 360 horse-power within a minute, the speed of the engine is not perceptibly affected."

While we chronicle the great works in engineering, the improvements of the past year in making steel promise still greater achievements. The Bessemer process has already done much; the later discovery of Bessemer, the high-pressure furnace, by which the melting of ores is accomplished much more speedily and economically than by the old processes, is destined, it is thought, to further cheapen steel. It is stated that Bessemer was led to this discovery by meditation on the cause of the heat of the sun, and the influence that the force of gravity, 27 times greater than that upon our earth, must have upon the intensity of that heat.

The Siemens regenerating furnaces are being rapidly introduced into this country.

These processes tend towards cheapening a very first-class material, which will undoubtedly supersede iron for almost all structural purposes. Engineers hesitate at present to use this material, since no adequate experiments have been made in regard to the limits to which steel structures can

[3] American railway construction practice included building much sharper curves and steeper grades, for example, than would have been permissible in British practice.

be loaded with safety. Experimental researches have been carried on for some time in England, at Woolwich, under a committee appointed by the Institution of Civil Engineers, which promise to supply this want.

The results of Mr. Whitworth's experiments, tending to supersede the hammer and rolls by forcing cast steel, while in a semi-fluid state, into strong iron moulds by hydraulic pressure, are regarded with great interest.

The use of pulverized fuel, experiments on which are now being conducted, promises, by surrounding each particle with just the amount of oxygen which it needs for perfect combustion, to utilize fuel to greater advantage.

"The Bulletin of the American Iron and Steel Association" states that 65 new blast furnaces have been erected in this country during the last 18 months. It adds that it has a record of 58 more in contemplation, the greater number at the West, nearly all of which will be built the coming year, if those engaged can be assured of the stability of the tariff.

The "Bulletin" computes the total product of Pig Iron in this country during 1869 at more than 1,900,000 tons. In 1865 (the first year after the war), it was but 931,000 tons—an increase without a parallel in the history of any country.

Steel rails are being largely adopted both at home and abroad. The results of the experiments made are not merely satisfactory in regard to the increased durability of the new material. They demonstrate that the section might be materially reduced. The Northern Railway Company, of Austria, was one of the earliest to experiment upon rails of Bessemer steel, and exhibited specimens of its rails at the exposition of 1867. With a weight per yard of only 45 pounds, the company obtained a steel rail having double the strength of the iron rails of a larger section previously employed by them; the cost to the company per ton of iron rails having been from 60 dollars to 70 dollars, and that of steel rails being from 90 dollars to 100 dollars. The expense per running mile is still kept nearly within its original limits, with a very great improvement in regard to strength and durability.

The French Railway Companies are also extensively introducing rails of Bessemer steel upon their roads. These rails, as manufactured at the principal French works, cost from 60 to 70 dollars per ton.

There is a growing feeling among engineers and steel makers, that the compound rail, made wholly or partly of steel, will prove more safe and economical than any solid rail, for, if the same durability of track can be obtained with a steel cap as with an all-steel rail, the first cost will be greatly decreased. A rail made in two or three continuous parts, breaking joints, is also a practical insurance against disaster from broken rails.

It is estimated that in the United States from 40,000 to 50,000 tons of steel rails are in use on our various railways.

The Lehigh and Susquehannah is entirely built of steel. Other railways are using them largely, the Hudson River, Erie, and Pennsylvania Railways using 10,000 tons or more each. The last report of the New Jersey Railway and Transportation Company says: "It is probable that steel rails will be gradually laid the entire length of the road, the greater durability of these rails overcoming the objection to their increased cost."

The use of steel rails will guarantee greater safety of life and limb, and their introduction, therefore, should be hailed with delight, for the term American rails has become a synonym for the cheapest and least durable rails manufactured.

Our late war taught us much in regard to ordnance and iron ships. The great advances in the manufacture of steel, and the discovery of new explosives, are destined to materially further our knowledge.

The most noteworthy improvement of this year in fortifications is Captain Moncrief's system. By an ingenious device he lowers his gun upon its rocking carriage after firing, and thereby does away with embrasures (the weak places in protecting works), while he gains the advantage of reloading his gun in comparative . . . safety.

What influences the new explosives, picrates, dynamite, and ammonia powder will have on warlike operations, remains to be seen.

Attention has lately been turned to gas as a calorific agent. Profs. Silliman and Wurtz, by their researches, promise to increase our knowledge of its illuminating power. . . .

Prof. Tyndall says that the superiority of gas for light-houses over oil is rendered very manifest by the experiments lately instituted at Howth Baily and Wicklow Head.

One cannot fail to notice the impulse which the completion of the Pacific Railway, the Suez Canal, and the French Atlantic Cable have given to the desire implanted in the human breast to overcome natural obstacles. M. Lesseps advocates flooding the desert of Sahara by means of a canal, and thus afford communication with the interior of Africa.

Among the projects that have been re-agitated the past year, are the project of a canal around the Falls of Niagara; a re-enlargement of the Erie for vessels of 1,000 tons; one across the Alleghanies in Virginia; one through the Isthmus of Darien, the expedition for surveying which has already started, and one from Huron to Ontario. In tunnels we have that of Mt. Cenis, 8 miles, and the Hoosac, 5 miles, in length, both in rapid progress; one of wrought-iron tubes at London, and another at Chicago; tunnels proposed under the East and North Rivers at New York; under the Ganges at Calcutta, and under the Straits of Dover.

In view of past achievements, it is not safe to pronounce any of these projects not feasible.

In physical science, Tyndall commenced the year with a picturesque account of a discovery of the peculiar action of light upon vapors.

In electricity we have no startling discoveries to chronicle. M. Jamin, it is said, has ascertained that magnetism can be condensed for a short period in the same way as electricity. Prof. LeRoy Cooley, of Albany, has discovered a way of registering vibrations by means of electricity. . . .

He dispenses with the sirene, and obtains a direct registration, the vibrating body itself opening and closing a circuit.

We have about the usual number of new batteries to chronicle. The combination of elements to produce currents seems unlimited.

The energies, however, of most of our physicists, both at home and abroad, have been directed to the field of spectrum analysis.

The late eclipse undoubtedly awakened greater interest in this new branch of science.[4]

Prof. Magnus has lately published a research upon heat spectra.

Angström and Thalen have also lately published laborious and accurate tables of the wave lengths of the different metals.

Roscoe's work on Spectrum Analysis, published this year, presents the subject in a very lucid manner.

We incorporate herewith the notes of Mr. Nichols on the progress in the field of chemistry.

In Chemistry no startling discovery has been made during the past year. Yet each year marks progress, especially in the contributions to the history of the compounds of carbon, and each year adds to the number of those complex bodies which, a short time ago, were found only in the bodies of animals or in plants, but which now can be prepared at will, in the laboratory. Especial attention may be called to the production during the past year, by artificial means, of *alizarine,* the coloring matter of the madder root. . . . Such discoveries extend our views of the domain of chemistry and cause us to have less apprehension in regard to the limited supply of many substances, the demand for which is continually increasing.

The publication by Professor Bunsen, of Heidelberg, of a paper on the "Washing of Precipitates" . . . has wrought a great change in the manner of conducting in the laboratory an operation of constant occurrence, that of filtration. By his method, as contrasted with that formerly employed, the saving of time amounts in certain cases to many hundred per cent.,—an advantage which, at the present day, we cannot afford to overlook.

The researches of Graham on the metallic character of hydrogen as deduced from the deportment of the alloys of palladium and hydrogen show how close is the relation between mechanical force and chemical affinity. It seems as if he were "led not only to manifest the metallic character of hydrogen, but also to seize the very moment at which the phenomenon of the mechanical condensation of a gas by a porous body changes into a truly chemical combination."[5]

[4] There was an eclipse of the sun in 1869; see below in this essay.

[5] Whether or not hydrogen ought to be considered a metal was a long-standing problem in chemistry. Graham believed that it is and was studying one of the basic questions of chemical inquiry, what makes for chemical combination? Palladium metal has the ability to absorb extremely large quantities of hydrogen gas—very considerably more than the porosity of the metal would indicate. Graham came to believe that the absorption of the hydrogen was not mechanical force but chemical affinity—the capacity of one substance to interact with another in a definite ratio.

The council of the Chemical Society (London) having determined to found an annual lectureship in honor of Faraday, the inaugural lecture was delivered this year (June 18) by the French chemist, Dumas, who was eminently fitted to perform this duty, not only on account of his having been an intimate friend of Faraday, but also on account of his great eloquence, and on account of his eminent position among the chemists of his own country. He began with an admirable eulogy of him whom his discourse commemorated, and then reviewed, from the stand-point of the present day, the progress of chemistry from its first beginnings.

He paid tribute to the labors of Lavoisier, Dalton, and Prout, and, pointing out the analogies existing between the elements of mineral chemistry and the compound radicals of organic chemistry (so called), and at the same time the relations between the atomic weights of those bodies which are now accepted as elements, he argued the probability of their being themselves proved to be complex.

The limits of chemistry he defines in these words: "The existing chemistry is, therefore, all powerful in the circle of mineral nature, even when its processes are carried on in the heart of the tissues of plants or of animals and at their expense; and she has advanced no further than the chemistry of the ancients in the knowledge of life and in the exact study of living matter; like them she is ignorant of the mode of generation.

"The ancients were mistaken when they confounded, under the name of *organic matter,* sugar and alcohol, which have never lived with the living tissue of plants, or in the flesh of animals. Sugar and alcohol have no more share of life than bone-earth, or salts contained in the various liquids. The chemist has never manufactured anything which, near or distant, was susceptible even of the appearance of life. Everything he has made in his laboratories belongs to 'brute' matter; as soon as he approaches life and organization, he is disarmed."

The medal which accompanies the Faraday lectureship is struck in palladium, and, in addition to this medal, Dumas carries back to France with him a medal struck in the alloy of palladium and hydrogen to commemorate the discovery of the alloy by Graham.

The subject of the disposal of the sewage of towns becomes daily of more importance. At the meeting of the British Association[6] at Exeter, a report was presented, in which were collected statistics showing the various methods adopted in towns and cities on the continent for utilizing the sewage, and that committee has issued circulars to the town authorities throughout England asking for aid in collecting information and in making experiments in regard to this matter. The earth-closet is finding favor, and, no doubt, will eventually supersede the water-closet in rural districts, and in towns where a supply of water cannot readily be obtained. The fact that it has been made

[6] That is, the British Association for the Advancement of Science.

the subject of a patent adds to its cost, and will retard somewhat its adoption in this country, but now that attention has been called to the matter, use will be made, and with advantage in a sanitary aspect, of the principle which is involved in it,—the disinfecting power of dry earth.

"Like some other valuable discoveries, it seems surprising that nobody thought of it or applied it before. But the simple fact is, that the privy may be made as inoffensive as the corn-barn by the application of about a pint and a half of *dry earth* every time it is used. There are one or two things about it important to remember: (1.) It should be *earth* (not sand or gravel), and should be thoroughly dried by exposure to the sun or otherwise; (2.) The privy-vault should be kept free from rain, from slops, and from excessive moisture of any sort. The more fluid thrown in, the more dry earth required to absorb it. How often it happens that a country hotel or boarding-house, crowded with people, becomes late in the season disgusting and unhealthy from decomposing material when a few shovelsful of dried earth thrown into the privy once a day would remove all offence."[7]

"It has been proposed for dried earth to substitute charcoal, which would be regenerated by burning. It is stated that one hundred weight of charcoal per month would be sufficient for a closet used by six persons daily. It is not likely, however, that this modification will find extensive adoption, except in localities peculiarly situated."

We incorporate herewith the notes of Dr. Kneeland on the progress in biology:

"The theory of Darwin is steadily progressing in the estimation of naturalists; indeed it may be said to be no longer simply a theory, as it has been demonstrated, in a few instances at least, both in the vegetable and animal kingdom, that 'natural selection,' or the survival of the fittest, is one of the causes of the existing varieties and so-called species among animals and plants. No naturalist can now presume to sneer at or ignore this and kindred theories, when such men as Lyell, Hooker, Huxley, and Owen reject utterly the doctrine of innumerable special acts of creation, and accept in variously modified forms the development of living things by the operation of laws impressed upon them at the beginning. The 'derivative hypothesis' of Owen . . . apparently meets the approval of naturalists more generally than any other; this maintains the incessant new development of living beings out of non-living material, and sees the grandeur of creative power, not in the exceptional miracle of one or few original forms of life, but in the daily and hourly calling into existence many forms by conversion of chemical and physical into vital modes of force; his conclusion is that, from the magnet which chooses between steel and zinc, to the philosopher who chooses between good and evil, the difference is one of degree, not of kind, and that there is no need of assuming a special miracle to account for mental pheno-

[7] Report of Massachusetts State Board of Health, 1870. [*Footnote in original.*]

mena. 'Natural selection' also is operative in the case of men, among whom there is a perpetual survival of the fittest; in the most barbarous conditions of mankind the struggle is almost entirely between individuals; in proportion as civilization has increased among men, it is easy to trace the transference of a great part of the struggle, little by little, from individuals to tribes, nations, leagues, guilds, corporations, societies, and similar combinations; and accompanying this transference has been undeniably the development of the moral qualities and of social virtues.

"The Social Science Associations are actively working out the great problems of moral and physical evils incident to civilization, especially those pertaining to hygienic or sanitary reform. The first step in the moral elevation of a community has been found to be the diffusion of knowledge of sanitary laws; cleanliness and good health are recognized as the best foundations of public prosperity. Hence science is constantly progressing in attempts to secure for the masses of the people cheap and wholesome food, pure air and pure water, ventilation of public buildings and the crowded dwellings of the poor; the removal of sewage, so as not only not to contaminate the earth, air, and water, but to convert it, even in our own houses, into an inodorous and valuable fertilizer, has been successfully accomplished. Fire-extinguishing and life-saving apparatus, both on land and sea, have reached a high degree of efficiency; man is gradually obtaining the mastery over the epidemic diseases which have for ages decimated the human race; and the return from human to vaccine lymph direct from the cow will restore the wavering faith of the public in the efficacy of vaccination, and eventually put a stop to the ravages of small-pox.

The recent successful employment of chloral as an anæsthetic, by the stomach instead of the lungs, and its undoubted efficacy as a sedative in nervous diseases and insanity, has drawn the attention of physiological chemists to the nearly unexplored field of the action of medicines by decomposition within the inmost recesses of the body.

Deep-sea dredgings have revealed an extensive and varied range of life at depths heretofore deemed untenanted, and have proved that there is a band of organisms encircling the globe at the bottom of the ocean,—these organisms, too, resembling those found in the immensely remote cretaceous epoch. The amœba . . . seems to be one of the links which connect the inorganic with the organic world, its organless tissue being capable of combining physical forces so as to assume organic functions."

Great advances have been made in celestial chemistry during the year, through the medium of spectrum analysis.

The observations of Huggins by means of this delicate method have proved that the star Sirius is receding from the earth at the rate of 29.4 miles per second; the observations of Huggins have been confirmed by Father Secchi, made at Rome. It is thought that the results of these and similar observations may one day lead to a determination of the motion of the solar system in space. By the same method of analysis, traces of aqueous vapor have been discovered in some of the planets.

The President of the British Association, in his address at Exeter, thus details Lockyer's discovery:

"After having observed the remarkable spectrum of the prominences[8] during the total eclipse, it occurred to M. Janssen that the same method might allow the prominences to be detected at any time; and on trial he succeeded in detecting them the very day after the eclipse. The results of his observations were sent by post, and were received shortly after the account of Mr. Lockyer's discovery had been communicated by Mr. De La Rue to the French Academy. In the way hitherto described a prominence is not seen as a whole, but the observer knows when its image is intercepted by the slit; and by varying a little the position of the slit, a series of sections of the prominence are obtained, by putting which together the form of the prominence is deduced. Shortly after Mr. Lockyer's communication of his discovery, Mr. Huggins, who had been independently engaged in the attempt to render the prominences visible by the aid of the spectroscope, succeeded in seeing a prominence as a whole by somewhat widening the slit, and using a red glass to diminish the glare of the light, admitted by the slit, the prominence being seen by means of the C line in the red. Mr. Lockyer had a design for seeing the prominences as a whole by giving the slit a rapid motion of small extent, but this proved to be superfluous, and they are now habitually seen with their actual forms. Nor is our power of observing them restricted to those which are so situated that they are seen by projection outside the sun's limb; such is the power of the spectroscopic method of observation, that it has enabled Mr. Lockyer and others to observe them right on the disc of the sun,—an important step for connecting them with other solar phenomena. One of the most striking results of the habitual study of these prominences is the evidence they afford of the stupendous changes which are going on in the central body of our system. Prominences, the heights of which are to be measured by thousands and tens of thousands of miles, appear and disappear in the course of some minutes. And a study of certain minute changes of position in the bright line F, which receive a simple and natural explanation by referring them to proper motion in the glowing gas by which that line is produced, and which we see no other way of accounting for, have led Mr. Lockyer to conclude that the gas in question is sometimes travelling with velocities comparable with that of the earth in its orbit. Moreover, these exhibitions of intense action are frequently found to be intimately connected with the spots, and can hardly fail to throw light on the disputed question of their formation. Nor are chemical composition and proper motion the only physical conditions of the gas which are accessible to spectral analysis. By comparing the breadth of the bright bands (for though narrow they are not mere lines) seen in the prominences, with those observed in the spectrum

[8] That is, large red gas clouds extending beyond the edge of the sun.

of hydrogen, rendered incandescent under different physical conditions, Dr. Frankland and Mr. Lockyer have deduced conclusions respecting the pressure to which the gas is subject in the neighborhood of the sun."

Since the discovery of Lockyer's, Janssen's, and Huggins' method of viewing the prominences, Zöllner has discovered a way of seeing them as a whole. . . . He makes use also of the C line, and likens it to looking at the sunset sky through a chink in the window. It is thought that this method may be used to advantage in the coming transit of Venus. The total eclipse of August 7th, 1869, was very fully observed.

The "American Journal of Arts and Sciences" [*sic*] thus speaks of the arrangements made for observing the phenomenon:

"Few astronomical phenomena have probably ever called out a more thoroughly organized system of observation than that arranged for the recent eclipse. The line of total obscuration crossed the North American continent diagonally, entering the territory of the United States at Behring's Straits, in about the 65th degree of latitude, and longitude 90° west of Washington, while it left our shore at the latitude of 34° and the meridian of Washington itself. It traversed a central belt of well-populated territory, yet there seems to have been scarcely a town of any considerable magnitude along the entire line which was not garrisoned by observers having some special astronomical problem in view.

An appropriation was made by Congress, at its last session, for carrying out a series of observations under the direction of the Superintendent of the Nautical Almanac, and Prof. Coffin has succeeded, by the liberal aid of the Navy Department, and the very generous and extensive facilities contributed by some of the principal railroads, in providing for an amount of work which for magnitude, variety, and thoroughness, seems large beyond all proportion to the sum placed at his disposal. Three cities in Iowa, Burlington, Mount Pleasant, and Ottumwa, were occupied by astronomical, photographic, and physical observers under his direction, and special observers, provided with telescopes and instruments for determining geographical positions, were sent by him to the North and South, to fix the limits of the belt of total obscuration.

The Navy Department, besides making other provisions, sent observers to the western shore of Behring's Straits; and the War Department detailed Dr. Curtis to make special photographic observations at Des Moines, Iowa.

The Coast Survey established parties on the Yaken [*sic*] River, in Alaska, at Des Moines in Iowa, Springfield in Illinois, and Abingdon in West Virginia, and perhaps at still other stations—that at Springfield being amply provided with photographic observers and apparatus. Most of the principal observatories likewise organized expeditions of greater or less magnitude. From Washington, the several observers arranged independent series of investigations, stellar, spectroscopic, physical, and meteorological. From Cambridge, a large party went to Shelbyville, Ky., with large photographic outfit, and spectroscopic equipments. From Albany, a similar party went to Mattoon,

Illinois; others, from Clinton and Chicago, went to Des Moines, from Cincinnati to Sioux City; and the number of private astronomers who established themselves along the central line with telescopes and other apparatus of investigation must have been exceedingly large.

The beginning and end of the eclipse seem to have been observed a few seconds later, and the beginning and end of the totality about fifteen seconds later than the predictions of the American Nautical Almanac. As regards the exact position of the central line, and of the limits of the total belt, we have as yet insufficient information to determine the degree of accordance with computation. There can be no doubt that materials have been collected capable of improving the adopted values of the moon's diameter and horizontal parallax. One of the most interesting results is the introduction of a new and accurate method of determining the time of first contact, by observing with a spectroscope the gradual occultation of the bright lines of the chromosphere. This we owe to Prof. Young, of Dartmouth College, who formed one of Prof. Coffin's Nautical Almanac party at Burlington. By keeping the centre of the slit directed to the point at which the contact is to take place, the observer is forewarned of the approach of the moon's limb, by the shortening of the bright lines belonging to the chromosphere. The line C is well adapted to this purpose, and is seen to grow steadily shorter, until it is totally extinguished. The moment of disappearance of the last bright ray is of course that of the first contact, which is thus observed with the same care and accuracy as any other appulsive phenomenon. Although the first contact, as determined in this way by Prof. Young, was noted some five seconds before its recognition by any other observer, it was subsequently found by Prof. Mayer to accord within a small fraction of a second with the time as determined by measurement of a series of photographs taken during the first minute.

Prof. Harkness, of Washington Observatory, observed at Des Moines the spectra of five protuberances, no two of which gave the same lines. In the corona spectrum he found no absorptive lines, and but one bright line. Measures of the protuberances were made by Prof. Rogers, at Des Moines, who found the largest to be nearly a minute and a half high, and observed a peculiar honeycombed or cellular appearance in all of them. Special search was made for intra-mercurial planets by Prof. Newcomb, at Des Moines, according to the plan suggested by him in the April number of the "American Journal of Science and Arts," with two 6-inch object-glasses, having a field of about 20° each, and previously clamped to the desired position. A similar scrutiny of the ecliptic near the sun was made by Dr. Gould, at Burlington, in connection with Prof. Coffin's party, using a Tolles' telescope of five inches' aperture and a field of nearly 2°, provided with occulting discs at the focus. But neither of these observers, nor any others engaged in similar research, found any indications of planets nearer than Mercury. . . .

The report of Com. B. F. Sands, U.S.N., Superintendent of the U.S. Naval Observatory, on the late eclipse, just published, is an exhaustive one, and compares favorably with the best efforts of a similar nature on the other side of the Atlantic.

Prof. Kirkwood, of Bloomington, Indiana, has lately published two able papers; one upon the periodicity of the solar spots, and another on comets and meteors. In the first-named paper he discusses the disturbing action of the planets on the sun's envelope, and suggests the hypothesis that a particular portion of the sun's surface is more favorable to spot formation than other portions. From his discussions he concludes:

1. A connection between the behavior of sun-spots and the configuration of certain planets has been placed beyond reasonable doubt.

2. The theory, however, of spot formation by planetary influence is encumbered with anomalies and even inconsistencies, unless we admit the co-operation of a modifying cause.

3. The hypothesis that a particular part of the solar surface is more susceptible than others to planetary disturbance is rendered probable by the observations of different astronomers.

4. The 11-year cycle of spot variation is mainly dependent on the influence of Mercury.

5. The marked irregularity of this period from 1822 to 1867 is in a great measure due to the disturbing action of Venus.

6. Wolf's 56-year cycle is determined by the joint action of Mercury and the earth; and, finally, the hypothesis proposed accounts for all the well-defined cycles of spot-variations.

In the paper on comets and meteors, Prof. Kirkwood considers the probable consequences of the sun's motion through regions of space in which cosmical matter is widely diffused, and compares these theoretical deductions with the observed phenomena of comets, aerolites, and falling stars.

From the variation in the number of observed comets and the periodicity of shooting stars, it is concluded that during the interval from 700 to 1200 the solar system was passing through, or near, a meteoric cloud of very great extent; that from 1200 to 1700 it was traversing a region comparatively destitute of such matter; and that about the commencement of the 18th century it again entered a similar nebula of unknown extent.

The present Earl of Rosse has been engaged upon the determination of the radiation of heat from the moon. It appears from his research that the greater part of the moon's heat which reaches the earth appears to have been first absorbed by the lunar surface. The amount of lunar heat appears to indicate an elevation of temperature for the moon's surface at full moon of 500°F.

Full arrangements have been made in France and England to observe the coming transit of Venus. Some constants in astronomical science will be tested by these observations.

The new facts in geography may be thus summarized:

The explorations and discoveries in South-eastern and East Equatorial Africa.

The additional and conclusive evidence now brought to light of a climate in the ice-bound region of the Arctic, at a past and remote period of time, resembling that of the countries lying near the equator.

The marvelous results of the deep-sea dredgings of Profs. Thompson and Carpenter, revealing the existence of animal life at immense depths in the ocean, where it has been supposed to have been impossible.

The very general disturbance throughout this year of the earth's surface by earthquakes, distinguishable not so much for the effects in particular localities as for the wide distribution of the phenomena over the globe, and its appearance in parts of the world where such disturbances have never been previously witnessed within the memory of man.

The attractive power of mountains, discovered in the pendulum experiments made during the past year at the observing stations upon the Himalayas, in India.

The discovery of trees of enormous size in Australia, one of which was found to be 69 feet in circumference; of great deposits of valuable coal throughout the whole of New Zealand, and the finding of coal upon the borders of the Caspian, verifying in the last particular a prediction of Humboldt, made forty years ago; both of which discoveries are of the highest importance to commerce.

The anthropological researches in Europe, Asia, and Africa, revealing the structure, mode of life, and customs of the earliest inhabitants of the earth.

The assembling at Copenhagen, last August, of the International Congress of Prehistoric Archæology, under the auspices of the King of Denmark, interesting in the circumstance that it brought into communication with each other learned men from all parts of Europe, and for the valuable information the papers and descriptions elicited in respect to the three successive periods of man's early history, known as the stone, the bronze, and the iron.

The return of Capt. Hall from the Arctic regions with valuable information respecting that mysterious country.

The exploration by Dr. Hayes of the remains of the early settlements made on the south-eastern shore of Greenland. The return of captain Adams and his men from the exploration of the Colorado and its tributaries.

The completion of the French explorations of the river Cambodia to the province of Tunan in China, the official details of which have not yet appeared.

The expedition of Sir Samuel Baker into the interior of Africa, which started last October.

The escape of Captain Livingston, of the American ship Congress, through a cyclone of extraordinary intensity and force, and the gaining of valuable information thereby.

The expedition of the Russian Merchant Soidorow, in his own steamer, around the coast of Norway, and through the polar ocean, to the mouth of the Pitschora.

A dispatch from Bombay, Oct. 6, states: A letter has just been received here from Dr. Livingstone, the great African traveller. He was at Lake Bangweolo at time of writing (in July, 1868), and was in excellent health and spirits. He mentioned that he believed he had at last found the true source of the Nile.

A caravan arrived at Zanzibar, Oct. 14, 1869, bringing the news that Dr. Livingstone had arrived at Nigi alive and well.

A later report, at our time of writing, Feb. 5th, 1870, states that he has been burnt as a wizard, by a native chief; it is trusted that time will contradict this.

In a letter to the Earl of Clarendon, he says: "I think that I may safely assert that the chief sources of the Nile arise between 10° and 12° south latitude, or nearly in the position assigned to them by Ptolemy. . . .

"The springs of the Nile have hitherto been searched for very much too far to the north. They rise some 400 miles south of the most southerly portion of the Victoria Nyanza, and, indeed, south of all the lakes except Bangweolo.

An International Exhibition of select works of fine and industrial art, and scientific inventions, is to be held in 1871, at South Kensington, England. This is the first of a series of annual exhibitions.

The movement which established the South Kensington Museum is having its parallel in Massachusetts and New York. It is proposed to establish a museum of the fine arts in New York and Boston. At the last session of the Legislature of Massachusetts, the following resolve was passed:

"*Resolved,* That the Board of Education be directed to consider the expediency of making provision by law for giving free instruction to men, women, and children, in mechanical drawing, either in existing schools, or in those to be established for that purpose, in all the towns in the Commonwealth having more than five thousand inhabitants, and report a definite plan therefor to the next General Court. [Approved June 12, 1869.]"

It is felt that our common schools do not give the right training to the industrial classes, and that if we are to have skilled mechanics, we must educate them. In view of the great natural advantages of the West, we at the East can hold our ground only by skilled labor; and the proper education of the lower classes has become a question of vital importance.

15. A Physician Explores the Physical Correlates of Human Thinking. 1870

Here a mature thinker, OLIVER WENDELL HOLMES (1809–1894), gently carries his audience along to a firm confrontation with the materialistic mechanism that developed in the wake of discoveries in neurophysiology. Science had revealed not only that the brain is the seat of the mind and that changes in the brain cause changes in thinking and behavior but that a model of the nervous system with impulses automatically going and sensations automatically coming has physiological justification. Holmes's reservations about—and his acceptance of—this mechanical model are instructive. By the 1870s scientists were well acquainted with reflex action, automatic control of bodily processes by the autonomic nervous system, and other nonvolitional activities of the nervous system. Holmes explores the inferences possible from the patterns of mental symptoms of physiological activity. His solution to the problem of mechanical versus free will—some decades before William James formulated pragmatism—suggests the sophistication of Holmes's thinking.

Holmes, born into the New England aristocracy, has had an enduring reputation as an American man of letters. In addition, he was a celebrated physician and teacher at Harvard Medical School. He took his M.D. from Harvard in 1836, but he had also studied medicine abroad and all of his life he kept up with the latest scientific advances.

For an interesting study of Holmes's life and work, see Eleanor M. Tilton, *Amiable Autocrat, A Biography of Dr. Oliver Wendell Holmes* (New York: H. Schuman, 1947).

MECHANISM IN THOUGHT AND MORALS*

Oliver Wendell Holmes

As the midnight train rolls into an intermediate station, the conductor's voice is heard announcing, "Cars stop ten minutes for refreshments." The passengers snatch a brief repast, and go back, refreshed, we will hope, to their places. But, while they are at the tables, one may be seen going round among the cars with a lantern and a hammer, intent upon a graver business.

* Oliver Wendell Holmes, *Mechanism in Thought and Morals. An Address Delivered Before the Phi Beta Kappa Society of Harvard University, June 29, 1870. With Notes and After-Thoughts* (Boston: J. R. Osgood & Co., 1871), pp. 5–13, 16–17, 20–30, 33–42, 44–53, 56–58, 62–99. Somewhat abridged.

He is clinking the wheels to try if they are sound. His task is a humble and simple one: he is no machinist, very probably; but he can cast a ray of light from his lantern, and bring out the ring of iron with a tap of his hammer.

Our literary train is stopping for a very brief time at its annual station. . . . It is not unlikely the passengers may stand much in need of refreshment before I have done with them: for I am the one with the hammer and the lantern; and I am going to clink some of the wheels of this intellectual machinery, on the soundness of which we all depend. The slenderest glimmer I can lend, the lightest blow I can strike, may at least call the attention of abler and better-equipped inspectors.

I ask your attention to some considerations on the true mechanical relations of the thinking principle, and to a few hints as to the false mechanical relations which have intruded themselves into the sphere of moral self-determination.

I call that part of mental and bodily life mechanical which is independent of our volition. The beating of our hearts and the secretions of our internal organs will go on, without and in spite of any voluntary effort of ours, as long as we live. Respiration is partially under our control: we can change the rate and special mode of breathing, and even hold our breath for a time; but the most determined suicide cannot strangle himself without the aid of a noose or other contrivance which shall effect what his mere will cannot do. The flow of thought is, like breathing, essentially mechanical and necessary, but incidentally capable of being modified to a greater or less extent by conscious effort. Our natural instincts and tastes have a basis which can no more be reached by the will than the sense of light and darkness, or that of heat and cold. All these things we feel justified in referring to the great First Cause: they belong to the "laws of Nature," as we call them, for which we are not accountable.

Whatever may be our opinions as to the relations between "mind" and "matter," our observation only extends to thought and emotion as connected with the living body, and, according to the general verdict of consciousness, more especially with certain parts of the body; namely, the central organs of the nervous system. The bold language of certain speculative men of science has frightened some more cautious persons away from a subject as much belonging to natural history as the study of any other function in connection with its special organ. If Mr. Huxley maintains that his thoughts and ours are "the expression of molecular changes in that matter of life which is the source of our other vital phenomena;" if the Rev. Prof. Haughton suggests, though in the most guarded way, that "our successors may even dare to speculate on the changes that converted a crust of bread, or a bottle of wine, in the brain of Swift, Molière, or Shakspeare, into the conception of the gentle Glumdalclitch, the rascally Sganarelle, or the immortal Falstaff,"—all this need not frighten us from studying the conditions of the thinking organ in connection with thought, just as we study the eye in its relations to sight. The brain is an instrument, necessary, so far as our direct observation extends, to thought. The "materialist" believes it to be wound up by the ordinary cosmic forces, and to give them out again as mental products: the "spiritualist" believes in a conscious entity, not interchangeable with motive force, which plays upon this instrument. But the instrument must be studied by the one as much as by the other: the piano which the master touches must be as thoroughly understood as the musical box or clock which goes of itself by a spring or weight. A slight congestion or softening of the brain shows the least materialistic of philosophers that he must recognize the strict dependence of mind upon its organ in the only condition of life with which we are experimentally acquainted. And what all recognize as soon as disease forces it upon their attention, all thinkers should recognize, without waiting for such an irresistible demonstration. They should see that the study of the organ of thought, microscopically, chemically, experimentally, on the lower animals, in individuals and races, in health and in disease, in every aspect of external observation, as well as by internal consciousness, is just as necessary as if mind were known to be nothing more than a function of the brain, in the same way as digestion is of the stomach.

These explanations are simply a concession to the timidity of those who assume that they who study the material conditions of the thinking centre necessarily confine the sphere of intelligence to the changes in those conditions; that they consider these changes constitute thought; whereas all that is held may be, that they accompany thought. It is a well-ascertained fact, for instance, that certain sulphates and phosphates are separated from the blood that goes to the brain in increased quantity after severe mental labor. But this chemical change may be only one of the factors of intellectual action. So, also, it *may* be true that the brain is inscribed with material records of thought; but what that is which reads any such records, remains still an open question. I have meant to leave absolutely untouched the endless discussion as to the distinctions between "mind" and "matter," and confine myself chiefly to some results of observation in the sphere of thought, and some suggestions as to the mental confusion which seems to me a common fact in the sphere of morals.

The central thinking organ is made up of a vast number of little starlike bodies embedded in fine granular matter, connected with each other by raylike branches in the form of pellucid threads; the same which, wrapped in bundles, become nerves,—the telegraphic cords of the system. The brain proper is a double organ, like that of vision; its two halves being connected by a strong transverse band, which unites them like the Siamese twins. The most fastidious lover of knowledge may study its general aspect as an after-dinner amusement upon an English walnut, splitting it through its natural sphere, and examining either half. . . .

The brain must be fed, or it cannot work. Four great vessels flood every part of it with hot scarlet blood, which carries at once fire and fuel to each of its atoms. Stop this supply, and we drop senseless. Inhale a few whiffs of ether, and we cross over into the unknown world of death with a return-ticket; or we prefer chloroform, and perhaps get no return-ticket.[1] Infuse a few drachms of another fluid into the system, and, when it mounts from the stomach to the brain, the pessimist becomes an optimist; the despairing wretch finds a new heaven and a new earth, and laughs and weeps by turns in his brief ecstasy. But, so long as a sound brain is supplied with fresh blood, it perceives, thinks, wills. . . .[2]

Such is the aspect, seen in a brief glance of the great nervous centre. It is constantly receiving messages from the senses, and transmitting orders to the different organs by the "up and down trains" of the nervous influence. It is traversed by continuous lines of thought, linked together in sequences which are classified under the name of "laws of association." The movement of these successions of thought is so far a result of mechanism, that, though we may modify them by an exertion of will, we cannot stop them, and remain vacant of all ideas.

My bucolic friends tell me that our horned cattle always keep a cud in their mouths: when they swallow one, another immediately replaces it. If the creature happens to lose its cud, it must have an artificial one given it, or, they assure me, it will pine, and perhaps die. Without committing myself to the exactness of the interpretation of the statement, I may use it as an illustration. Just in the same way, one thought replaces another; and in the same way the mental cud is sometimes lost while one is talking, and he must ask his companion to supply its place. "What was I saying?" we ask; and our friend furnishes us with the lost word or its equivalent, and the jaws of conversation begin grinding again. . . .

Our conscious mental action, aside from immediate impressions on the senses, is mainly pictured, worded, or modulated, as in remembered music; all, more or less, under the influence of the will. In a general way, we refer the seat of thinking to the anterior part of the head. *Pictured* thought is in relation with the field of vision, which I perceive—as others do, no doubt—as a transverse ellipse; its vertical to its horizontal diameter about as one to three. We shut our eyes to recall a visible object: we see visions by night. The bright ellipse becomes a black ground, on which ideal images show more distinctly than on the illuminated one. The form of the mental field of vision is illustrated by the fact, that we can follow in our idea a ship sailing, or a horse running, much farther, without a sense of effort, than we can a balloon rising. In seeing persons, this field of mental vision seems to be a little in front of the eyes. Dr. Howe kindly answers a letter of inquiry as follows:—

[1] Because the range of safe dosage is relatively narrow, chloroform in the early years produced a high mortality rate when ignorantly used as an anesthetic.

[2] That is, acts as the immediate instrument through which these phenomena are manifested. So a good watch, in good order and wound up, tells us the time of day. The making and winding-up forces remain to be accounted for. [*Footnote in original.*]

"Most congenitally-blind persons, when asked with what part of the brain they think, answer, that they are not conscious of having any brain.

"I have asked several of the most thoughtful and intelligent among our pupils to designate, as nearly as they can, the seat of sensation in thought; and they do so by placing the hand upon the *anterior* and *upper* part of the cranium."

Worded thought is attended with a distinct impulse towards the organs of speech: in fact, the effort often goes so far, that we "think aloud," as we say. The seat of this form of mental action seems to me to be beneath that of pictured thought; indeed, to follow certain nerves downward: so that as we say, "My heart was in my mouth," we could almost say, "My brain is my mouth." A particular spot has been of late pointed out by pathologists, not phrenologists,[3] as the seat of the faculty of speech. I do [not] know that our sensations ever point to it. . . .

The seat of the *will* seems to vary with the organ through which it is manifested; to transport itself to different parts of the brain, as we may wish to recall a picture, a phrase, or a melody; to throw its force on the muscles or the intellectual processes. Like the general-in-chief, its place is anywhere in the field of action. It is the least like an instrument of any of our faculties; the farthest removed from our conceptions of mechanism and matter, as we commonly define them.

This is my parsimonious contribution to our knowledge of the relations existing between mental action and space. Others may have had a different experience; the great apostle did not know at one time whether he was in the body or out of the body: but my system of phrenology extends little beyond this rudimentary testimony of consciousness.

When it comes to the relation of mental action and *time,* we can say with Leibnitz, *"Calculemus;"* for here we can reach quantitative results. The "personal equation," or difference in rapidity of recording the same occurrence, has been recognized in astronomical records since the time of Maskelyne, the royal astronomer. . . .[4] More recently, the time required in mental processes and in the transmission of sensation and the motor impulse along nerves has been carefully studied by Helmholtz, Fizeau, Marey, Donders, and others.[5] From forty to eighty, a hundred or more feet a second are estimates of different observers: so that, as the newspapers have been repeating, it would take a whale a second, more or less, to feel the stroke of a harpoon in his tail. Com-

[3] This is a reference to the by then discredited science of phrenology, practitioners of which purported to demonstrate the seats of various mental functions and propensities in the brain on the basis of skull configurations.

[4] Astronomers had discovered that in order to achieve exactitude in correlation of their observations they had by comparison with each other to determine the fraction of a second that it took each of them to observe and record the passing of a heavenly body past a certain point, for the reaction times differed significantly from one astronomer to another.

[5] As early as 1850 in a classic experiment Helmholtz had determined that nerve impulses travel at the leisurely rate of 25 to 40 meters per second.

pare this with the velocity of galvanic signals, which Dr. Gould has found to be from fourteen to eighteen thousand miles a second through iron wire on poles, and about sixty-seven hundred miles a second through the submarine cable. The brain, according to Fizeau, takes one-tenth of a second to transmit an order to the muscles; and the muscles take one-hundredth of a second in getting into motion. These results, such as they are, have been arrived at by experiments on single individuals with a very delicate chronometric apparatus. I have myself instituted a good many experiments with a more extensive and expensive machinery than I think has ever been employed,—namely, two classes, each of ten intelligent students, who with joined hands represented a nervous circle of about sixty-six feet: so that a hand-pressure transmitted ten times round the circle traversed six hundred and sixty feet, besides involving one hundred perceptions and volitions. My chronometer was a "horse-timer," marking quarter-seconds. After some practice, my second class gradually reduced the time of transmission ten times round, which, like that of the first class, had stood at fourteen and fifteen seconds, down to ten seconds; that is, one-tenth of a second for the passage through the nerves and brain of each individual,—less than the least time I have ever seen assigned for the whole operation; no more than Fizeau has assigned to the action of the brain alone. The mental process of judgment between colors (red, white, and green counters), between rough and smooth (common paper and sand-paper), between smells (camphor, cloves, and assafœtida), took about three and a half tenths of a second each; taste, twice or three times as long, on account of the time required to reach the true sentient portion of the tongue. These few results of my numerous experiments show the rate of working of the different parts of the machinery of consciousness. Nothing could be easier than to calculate the whole number of perceptions and ideas a man could have in the course of a lifetime. But as we think the same thing over many millions of times, and as many persons keep up their social relations by the aid of a vocabulary of only a few hundred words, or, in the case of some very fashionable people, a few scores only, a very limited amount of thinking material may correspond to a full set of organs of sense, and a good development of the muscular system. . . .

Do we ever think without knowing that we are thinking? The question may be disguised so as to look a little less paradoxical. Are there any mental processes of which we are unconscious at the time, but which we recognize as having taken place by finding certain results in our minds?

That there are such unconscious mental actions is laid down in the strongest terms by Leibnitz. . . . The existence of unconscious thought is maintained by him in terms we might fairly call audacious, and illustrated by some of the most striking facts bearing upon it. . . . It does not follow, he says . . . that, because we do not perceive thought, it does not exist.—Something goes on in the mind which answers to the circulation of the blood and all the internal movements of the viscera.—In one word, it is a great source of error to believe that there is no perception in the mind but those of which it is conscious.—

This is surely a sufficiently explicit and peremptory statement of the doctrine, which, under the names of "latent consciousness," "obscure perceptions," "the hidden soul," "unconscious cerebration," "reflex action of the brain," has been of late years emerging into general recognition in treatises of psychology and physiology. . . .

Unconscious activity is the rule with the actions most important to life. The lout who lies stretched on the tavern-bench, with just mental activity enough to keep his pipe from going out, is the unconscious tenant of a laboratory where such combinations are being constantly made as never Wöhler or Berthelot could put together; where such fabrics are woven, such colors dyed, such problems of mechanism solved, such a commerce carried on with the elements and forces of the outer universe, that the industries of all the factories and trading establishments in the world are mere indolence and awkwardness and unproductiveness compared to the miraculous activities of which his lazy bulk is the unheeding centre. All these unconscious or reflex actions take place by a mechanism never more simply stated than in the words of Hartley, as "*vibrations* which ascend up the sensory nerves first, and then are detached down the motory nerves, which communicate with these by some common trunk, plexus, or ganglion." The doctrine of Leibnitz, that the brain may sometimes act without our taking cognizance of it, as the heart commonly does, as many internal organs always do, seems almost to belong to our time. The readers of Hamilton and Mill, of Abercrombie, Laycock, and Maudsley, of Sir John Herschel, of Carpenter, of Lecky, of Dallas, will find many variations on the text of Leibnitz, some new illustrations, a new classification and nomenclature of the facts; but the root of the matter is all to be found in his writings.

I will give some instances of work done in the underground workshop of thought,—some of them familiar to the readers of the authors just mentioned.

We wish to remember something in the course of conversation. No effort of the will can reach it; but we say, "Wait a minute, and it will come to me," and go on talking. Presently, perhaps some minutes later, the idea we are in search of comes all at once into the mind, delivered like a prepaid bundle, laid at the door of consciousness like a foundling in a basket. How it came there we know not. The mind must have been at work groping and feeling for it in the dark; it cannot have come of itself. Yet, all the while, our consciousness, so far as we are conscious of our consciousness, was busy with other thoughts.

In old persons, there is sometimes a long interval of obscure mental action before the answer to a question is evolved. I remember making an inquiry, of an ancient man whom I met on the road in a wagon with his daughter, about a certain old burial-ground which I was visiting. He seemed to listen attentively; but I got no answer. "Wait half a minute or so," the daughter said, "and he will tell you." And sure enough, after a little time, he answered me, and to the point. The delay here, probably, corresponded to what machinists call "lost time," or "back lash," in turning an old screw, the thread of which is worn. But, within a fortnight, I examined a young man for his degree, in whom I noticed a certain regular interval, and a pretty long one, between every question and its answer. Yet the answer was, in almost every instance, correct, when at last it did come. It was an idiosyncrasy, I found, which his previous instructors had noticed. I do not think the mind knows what it is doing in the interval, in such cases. This latent period, during which the brain is obscurely at work, may, perhaps, belong to mathematicians more than others. Swift said of Sir Isaac Newton, that, if one were to ask him a question, "he would revolve it in a circle in his brain, round and round" (the narrator here describing a circle on his own forehead), "before he could produce an answer."

I have often spoken of the same trait in a distinguished friend of my own, remarkable for his mathematical genius, and compared his sometimes long-deferred answer to a question, with half a dozen others stratified over it, to the thawing-out of the frozen words as told of by Baron Munchausen and Rabelais, and nobody knows how many others before them.

I was told, within a week, of a business-man in Boston, who, having an important question under consideration, had given it up for the time as too much for him. But he was conscious of an action going on in his brain which was so unusual and painful as to excite his apprehensions that he was threatened with palsy, or something of that sort. After some hours of this uneasiness, his perplexity was all at once cleared up by the natural solution of his doubt coming to him,—worked out, as he believed, in that obscure and troubled interval.

The cases are numerous where questions have been answered, or problems solved, in dreams, or during unconscious sleep. . . . Somnambulism and double-consciousness offer another series of illustrations.[6] Many of my audience remember a murder-case, where the accused was successfully defended, on the ground of somnambulism, by one of the most brilliant of American lawyers. In the year 1686, a brother of Lord Culpeper was indicted at the Old Bailey for shooting one of the guards, and acquitted on the same ground of somnambulism; that is, an unconscious, and therefore irresponsible, state of activity.

A more familiar instance of unconscious action is to be found in what we call "absent" persons,—those who, while wide awake, act with an apparent purpose, but without really knowing what they are doing; as in La Bruyère's character, who threw his glass of wine into the backgammon-board, and swallowed the dice.

[6] That is, complex activity while still apparently asleep, on the one hand, or, on the other hand, completely under the influence of a second consciousness (a second personality).

There are a vast number of movements which we perform with perfect regularity while we are thinking of something quite different,—"automatic actions of the secondary kind," as Hartley calls them, and of which he gives various examples. The old woman knits; the young woman stitches, or perhaps plays her piano, and yet talks away as if nothing but her tongue was busy. Two lovers stroll along side by side, just born into the rosy morning of their new life, prattling the sweet follies worth all the wisdom that years will ever bring them. How much do they think about that wonderful problem of balanced progression which they solve anew at every step?

. . . There are thoughts that never emerge into consciousness, which yet make their influence felt among the perceptible mental currents, just as the unseen planets sway the movements of those which are watched and mapped by the astronomer. Old prejudices, that are ashamed to confess themselves, nudge our talking thought to utter their magisterial veto. In hours of languor, as Mr. Lecky has remarked, the beliefs and fancies of obsolete conditions are apt to take advantage of us. We know very little of the contents of our minds until some sudden jar brings them to light, as an earthquake that shakes down a miser's house brings out the old stockings full of gold, and all the hoards that have hid away in holes and crannies.

We not rarely find our personality doubled in our dreams, and do battle with ourselves, unconscious that we are our own antagonists. Dr. Johnson dreamed that he had a contest of wit with an opponent, and got the worst of it: of course, he furnished the wit for both. Tartini heard the Devil play a wonderful sonata, and set it down on awaking. Who was the Devil but Tartini himself? I remember, in my youth, reading verses in a dream, written, as I thought, by a rival fledgling of the Muse. They were so far beyond my powers, that I despaired of equalling them; yet I must have made them unconsciously as I read them. Could I only have remembered them waking!

But I must here add another personal experience, of which I will say beforehand,—somewhat as honest Izaak Walton said of his pike, "This dish of meat is too good for any but anglers or very honest men,"—this story is good only for philosophers and very small children. I will merely hint to the former class of thinkers, that its moral bears on two points: first, the value of our self-estimate, sleeping,—possibly, also waking; secondly, the significance of general formulæ when looked at in certain exalted mental conditions.

I once inhaled a pretty full dose of ether, with the determination to put on record, at the earliest moment of regaining consciousness, the thought I should find uppermost in my mind. The mighty music of the triumphal march into nothingness reverberated through my brain, and filled me with a sense of infinite possibilities, which made me an archangel for the moment. The veil of eternity was lifted. The one great truth which underlies all human experience, and is the key to all the mysteries that philosophy has sought in vain to solve, flashed upon me in a sudden revelation. Henceforth all was clear: a few words had lifted my intelligence to the level of the knowledge of the cherubim. As my natural condition returned, I remembered my resolution; and, staggering to my desk, I wrote, in ill-shaped, straggling characters, the all-embracing truth still glimmering in my consciousness. The words were these (children may smile; the wise will ponder): *"A strong smell of turpentine prevails throughout."*

My digression has served at least to illustrate the radical change which a slight material cause may produce in our thoughts, and the way we think about them. If the state just described were prolonged, it would be called insanity. I have no doubt that there are many ill-organized, perhaps over-organized, human brains, to which the common air is what the vapor of ether was to mine: it is madness to them to drink in this terrible burning oxygen at every breath; and the atmosphere that infolds them is like the flaming shirt of Nessus.[7]

The more we examine the mechanism of thought, the more we shall see that the automatic, unconscious action of the mind enters largely into all its processes. Our definite ideas are stepping-stones; how we get from one to the other, we do not know: something carries us; we do not take the step. A creating and informing spirit which is with us, and not of us, is recognized everywhere in real and in storied life. It is the Zeus that kindled the rage of Achilles; it is the Muse of Homer; it is the Daimon of Socrates; it is the inspiration of the seer; it is the mocking devil that whispers to Margaret as she kneels at the altar; and the hobgoblin that cried, "Sell him, sell him!" in the ear of John Bunyan: it shaped the forms that filled the soul of Michael Angelo when he saw the figure of the great Lawgiver in the yet unhewn marble, and the dome of the world's yet unbuilt basilica against the blank horizon; it comes to the least of us, as a voice that will be heard; it tells us what we must believe; it frames our sentences; it lends a sudden gleam of sense or eloquence to the dullest of us all, so that . . . we wonder at ourselves, or rather not at ourselves, but at this divine visitor, who chooses our brain as his dwelling-place, and invests our naked thought with the purple of the kings of speech or song.

After all, the mystery of unconscious mental action is exemplified, as I have said, in every act of mental association. What happens when one idea brings up another? Some internal movement, of which we are wholly unconscious, and which we only know by its effect. What is this action, which in Dame Quickly agglutinates contiguous circumstances by their surfaces; in men of wit and fancy, connects remote ideas by partial resemblances; in men of imagination, by the vital identity which underlies phenomenal diversity; in the man of science, groups the objects of thought in sequences of maximum resemblance? Not one of them can answer. . . .

[7] When Hercules put this shirt on, it stuck to his flesh and tore the flesh off with it.

The poet sits down to his desk with an odd conceit in his brain; and presently his eyes fill with tears, his thought slides into the minor key, and his heart is full of sad and plaintive melodies. Or he goes to his work, saying, "To-night I would have tears;" and, before he rises from his table, he has written a burlesque, such as he might think fit to send to one of the comic papers, if these were not so commonly cemeteries of hilarity interspersed with cenotaphs of wit and humor. These strange hysterics of the intelligence, which make us pass from weeping to laughter, and from laughter back again to weeping, must be familiar to every impressible nature; and all is as automatic, involuntary, as entirely self-evolved by a hidden organic process, as are the changing moods of the laughing and crying woman. The poet always recognizes a dictation *ab extra;*[8] and we hardly think it a figure of speech when we talk of his inspiration.

The mental attitude of the poet while writing, if I may venture to define it, is that of the "nun, breathless with adoration." Mental stillness is the first condition of the listening state; and I think my friends the poets will recognize that the sense of effort, which is often felt, accompanies the mental spasm by which the mind is maintained in a state at once passive to the influx from without, and active in seizing only that which will serve its purpose. It is not strange that remembered ideas should often take advantage of the crowd of thoughts, and smuggle themselves in as original. Honest thinkers are always stealing unconsciously from each other. Our minds are full of waifs and estrays which we think are our own. Innocent plagiarism turns up everywhere. Our best musical critic tells me that a few notes of the air of "Shoo Fly" are borrowed from a movement in one of the magnificent harmonies of Beethoven. . . .

Persons who talk most do not always think most. I question whether persons who think most—that is, have most conscious thought pass through their minds—necessarily do most mental work. The tree you are sticking in "will be growing when you are sleeping." So with every new idea that is planted in a real thinker's mind: it will be growing when he is least conscious of it. An idea in the brain is not a legend carved on a marble slab; it is an impression made on a living tissue, which is the seat of active nutritive processes. Shall the initials I carved in bark increase from year to year with the tree? and shall not my recorded thought develop into new forms and relations with my growing brain? Mr. Webster told one of our greatest scholars that he had to change the size of his hat every few years. His head grew larger as his intellect expanded. Illustrations of this same fact were shown me many years ago by Mr. Deville, the famous phrenologist, in London. But organic mental changes may take place in shorter spaces of time. A single night of sleep has often brought a sober second-thought, which was a surprise to the hasty conclusion of the day before. Lord Polkommet's

[8] That is, from beyond, or external to oneself.

description of the way he prepared himself for a judicial decision is in point, except for the alcoholic fertilizer he employed in planting his ideas: "Ye see, I first read a' the pleadings; and then, after letting them wamble in my wame wi' the toddy two or three days, I gie my ain interlocutor."

. . . The vast amount of blood sent to the brain implies a corresponding amount of material activity in the organ. In point of fact, numerous experiments have shown (and I may refer particularly to those of our own countrymen,—Professors Flint, Hammond, and Lombard) that the brain is the seat of constant nutritive changes, which are greatly increased by mental exertion.

The mechanical co-efficient of mental action may be therefore considered a molecular movement in the nervous centres, attended with waste of material conveyed thither in the form of blood,—not a mere tremor like the quiver of a bell, but a process more like combustion; the blood carrying off the oxidated particles, and bringing in fresh matter to take their place.

This part of the complex must, of course, enter into the category of the correlated forces. The brain must be fed in order to work; and according to the amount of waste of material will be that of the food required to repair losses. So much logic, so much beef; so much poetry, so much pudding; and, as we all know that all growing things are but sponges soaked full of old sunshine, Apollo[9] becomes as important in the world of letters as ever.[10]

But the intellectual product does not belong to the category of force at all, as defined by physicists. It does not answer their definition as "that which is expended in producing or resisting motion." It is not reconvertible into other forms of force. One cannot lift a weight with a logical demonstration, or make a tea-kettle boil by writing an ode to it. A given amount of molecular action in two brains represents a certain equivalent of food, but by no means an equivalent of intellectual product. . . . It may be doubted whether the present Laureate of England[11] consumed more oxidable material in the shape of nourishment for every page of "Maud" or of "In Memoriam" than his predecessor Nahum Tate, whose masterpiece gets no better eulogy than that it is "the least miserable of his productions," in eliminating an equal amount of verse.

As mental labor, in distinction from the passive flow of thought, implies an exercise of will, and as mental labor is shown to be attended by an increased waste, the presumption is that this waste is in some degree referable to the material requirements of the act of volition. We see why the latter should be attended by a sense of effort, and followed by a feeling of fatigue.

A question is suggested by the definition of the physicists. What is that which changes the form of force? Electricity leaves what we call magnetism in iron, after passing through it: what name shall we give to that virtue in

[9] Greek god of sunlight, music, and poetry.

[10] It is curious to compare the Laputan idea of extracting sunbeams from cucumbers with George Stephenson's famous saying about coal. [*Footnote in original.*]

[11] Alfred Lord Tennyson.

iron which causes the force we know as electricity thus to manifest itself by a precipitate, so to speak, of new properties? Why may we not speak of a *vis ferrea*[12] as causing the change in consequence of which a bar through which an electrical current has flowed becomes capable of attracting iron and of magnetizing a million other bars? And so why may not a particular brain, through which certain nutritious currents have flowed, fix a force derived from these currents in virtue of a *vis Platonica* or a *vis Baconica,* and thus become a magnet in the universe of thought, exercising and imparting an influence which is not expended, in addition to that accounted for by the series of molecular changes in the thinking organ?

We must not forget that force-equivalent is one thing, and quality of force-product is quite a different thing. The same outlay of muscular exertion turns the winch of a coffee-mill and of a hand-organ. It has been said that thought cannot be a physical force, because it cannot be measured. An attempt has been made to measure thought as we measure force. I have two tables, one from the "Annales Encyclopédiques," and another, earlier and less minute, by the poet Akenside, in which the poets are classified according to their distinctive qualities; each quality and the total average being marked on a scale of twenty as a maximum. I am not sure that mental qualities are not as susceptible of measurement as the aurora borealis or the changes of the weather. But even measurable *quality* has no more to do with the correlation of forces than the color of a horse with his power of draught; and it is with quality we more especially deal in intellect and morals.

I have spoken of the material or physiological co-efficient of thought as being indispensable for its exercise during the only condition of existence of which, apart from any alleged spiritualistic experience, we have any personal knowledge. We know our dependence too well from seeing so many gallant and well-freighted minds towed in helpless after a certain time of service,—razees at sixty, dismantled at seventy, going to pieces and sinking at fourscore. We recognize in ourselves the loss of mental power, slight or serious, from grave or trifling causes. "Good God," said Swift, "what a genius I had when I wrote that book!" And I remember that an ingenious tailor of the neighboring city, on seeing a customer leave his shop without purchasing, exclaimed, smiting his forehead, "If it had not been for this—emphatically characterized—headache, I'd have had a coat on that man before he'd got out over my doorstep." Such is the delicate adjustment of the intellectual apparatus by the aid of which we clothe our neighbor, whether he will or no, with our thoughts if we are writers of books, with our garments if we are artificers of habiliments.

The problem of memory is closely connected with the question of the mechanical relation between thought and structure. How intimate is the alliance of memory with the material condition of the brain, is shown by the

[12] Iron force, the power of iron.

effect of age, of disease, of a blow, of intoxication. I have known an aged person repeat the same question five, six, or seven times during the same brief visit. I was once asked to see to a woman who had just been injured in the street. On coming to herself, "Where am I? what has happened?" she asked. "Knocked down by a horse, ma'am; stunned a little: that is all." A pause, "while one with moderate haste might count a hundred;" and then again, "Where am I? what has happened?"—"Knocked down by a horse, ma'am; stunned a little: that is all." Another pause and the same question again; and so on during the whole time I was by her. The same tendency to repeat a question indefinitely has been observed in returning members of those worshipping assemblies whose favorite hymn is, "We won't go home till morning."

Is memory, then, a material record? Is the brain, like the rocks of the Sinaitic Valley, written all over with inscriptions left by the long caravans of thought, as they have passed year after year through its mysterious recesses?

When we see a distant railway-train sliding by us in the same line, day after day, we infer the existence of a track which guides it. So, when some dear old friend begins that story we remember so well; switching off at the accustomed point of digression; coming to a dead stop at the puzzling question of chronology; off the track on the matter of its being first or second cousin of somebody's aunt; set on it again by the patient, listening wife who knows it all as she knows her well-worn wedding-ring,—how can we doubt that there is a track laid down for the story in some permanent disposition of the thinking-marrow?

I need not say that no microscope can find the tablet inscribed with the names of early loves, the stains left by tears of sorrow or contrition, the rent where the thunderbolt of passion has fallen, or any legible token that such experiences have formed a part of the life of the mortal, the vacant temple of whose thought it is exploring. It is only as an inference, aided by an illustration which I will presently offer, that I suggest the possible existence, in the very substance of the brain-tissue, of those inscriptions which Shakspeare must have thought of when he wrote,—

> "Pluck from the memory a rooted sorrow;
> Raze out the written troubles of the brain."

The objection to the existence of such a material record—that we renew our bodies many scores of times, and yet retain our earliest recollections—is entirely met by the fact, that a scar of any kind holds its own pretty nearly through life in spite of all these same changes, as we have not far to look to find instances.

It must be remembered that a billion of the starry brain-cells could be packed in a cubic inch, and that the convolutions contain one hundred and thirty-four cubic inches, according to the estimate already given. My illustration is derived from microscopic photography. I have a glass slide on which is a minute photographic picture, which is exactly covered when the head of a small pin is laid upon it. In that little speck are clearly to be seen, by a proper magnifying power, the following objects: the Declaration of Independence, with easily-recognized facsimile autographs of all the signers; the arms of all the original thirteen States; the Capitol at Washington; and very good portraits of all the Presidents of the United States from Washington to Mr. James K. Polk. These objects are all distinguishable as a group with a power of fifty diameters: with a power of three hundred, any one of them becomes a sizable picture. You may see, if you will, the majesty of Washington on his noble features, or the will of Jackson in those hard lines of the long face, crowned with that bristling head of hair in a perpetual state of electrical divergence and centrifugal self-assertion. Remember that each of these faces is the record of a life.

Now recollect that there was an interval between the exposure of the negative in the camera and its development by pouring a wash over it, when all these pictured objects existed potentially, but absolutely invisible, and incapable of recognition, in a speck of collodion-film, which a pin's head would cover, and then think what Alexandrian libraries, what Congressional document-loads of positively intelligible characters,—such as one look of the recording angel would bring out . . . might be held in those convolutions of the brain which wrap the talent intrusted to us, too often as the folded napkin of the slothful servant hid the treasure his master had lent him!

Three facts, so familiar that I need only allude to them, show how much more is recorded in the memory than we may ever take cognizance of. The first is the conviction of having been in the same precise circumstances once or many times before. . . . The second is the panorama of their past lives, said, by people rescued from drowning, to have flashed before them. I had it once myself, accompanied by an ignoble ducking and scrambling self-rescue. The third is the revival of apparently obsolete impressions, of which many strange cases are related in nervous young women and in dying persons. . . . It is possible, therefore, and I have tried to show that it is not improbable, that memory is a material record; that the brain is scarred and seamed with infinitesimal hieroglyphics, as the features are engraved with the traces of thought and passion. And, if this is so, must not the record, we ask, perish with the organ? Alas! how often do we see it perish *before* the organ!—the mighty satirist tamed into oblivious imbecility; the great scholar wandering without sense of time or place among his alcoves, taking his books one by one from the shelves, and fondly patting them; a child once more among his toys, but a child whose to-morrows come hungry, and not full-handed,—come as birds of prey in the place of the sweet singers of morning. We must all become as little children if we live long enough; but how blank an existence the wrinkled infant must carry into the kingdom of heaven, if the Power that gave him memory does not repeat the miracle by restoring it!

The connection between thought and the structure and condition of the brain is evidently so close, that all we have to do is study it. It is not in this direction that materialism is to be feared: we do not find Hamlet and Faust, right and wrong, the valor of men and the purity of women, by testing for albumen, or examining fibres in microscopes.

It is in the moral world that materialism has worked the strangest confusion. In various forms, under imposing names and aspects, it has thrust itself into the moral relations, until one hardly knows where to look for any first principles without upsetting every thing in searching for them.

The moral universe includes nothing but the exercise of choice: all else is machinery. What we can help and what we cannot help are on two sides of a line which separates the sphere of human responsibility from that of the Being who has arranged and controls the order of things.

The question of the freedom of the will has been an open one, from the days of Milton's demons in conclave. . . . It still hangs suspended between the seemingly exhaustive strongest motive argument and certain residual convictions. The sense that we are, to a limited extent, self-determining; the sense of effort in willing; the sense of responsibility in view of the future, and the verdict of conscience in review of the past,—all of these are open to the accusation of fallacy; but they all leave a certain undischarged balance in most minds. . . .

It is one thing to prove a proposition like the doctrine of necessity in terms, and another thing to accept it as an article of faith. . . . Even Mr. Huxley, who throws quite as much responsibility on protoplasm as it will bear, allows that "our volition counts for something as a condition of the course of events."

I reject, therefore, the mechanical doctrine which makes me the slave of outside influences, whether it work with the logic of Edwards, or the averages of Buckle; whether it come in the shape of the Greek's destiny, or the Mahometan's fatalism. . . .

But I claim the right to eliminate all mechanical ideas which have crowded into the sphere of intelligent choice between right and wrong. The pound of flesh I will grant to Nemesis; but, in the name of human nature, not one drop of blood,—not one drop.

Moral chaos began with the idea of transmissible responsibility. It seems the stalest of truisms to say that every moral act, depending as it does on choice, is in its nature exclusively personal; that its penalty, if it have any, is payable, not to bearer, not to order, but only to the creditor himself. To treat a mal-volition, which is inseparably involved with an internal condition, as capable of external transfer from one person to another, is simply to materialize it. When we can take the dimensions of virtue by triangulation; when we can literally weigh Justice in her own scales; when we can speak of the specific gravity of truth, or the square root of honesty; when we can send a statesman his integrity in a package to Washington, if he happen to have left it behind,—then we may begin to speak of the moral character of inherited tendencies, which belong to the machinery for which the Sovereign Power alone is responsible. The misfortune of perverse instincts, which adhere to us as congenital inheritances, should go to our side of the account, if the books of heaven are kept, as the great Church of Christendom maintains they are, by double entry. But the absurdity which has been held up to ridicule in the nursery has been enforced as the highest reason upon older children. Did our forefathers tolerate Æsop among them? "I cannot trouble the water where you are," says the lamb to the wolf: "don't you see that I am farther down the stream?"—"But a year ago you called me ill names."—"O sir! a year ago I was not born."—"Sirrah," replies the wolf, "if it was not you, it was your father, and that is all one;" and finishes with the usual practical application. . . . If we suffer for any thing except our own wrong-doing, to call it punishment is like speaking of a yard of veracity or a square inch of magnanimity.

So to rate the gravity of a mal-volition by its consequences is the merest sensation materialism. A little child takes a prohibited friction-match: it kindles a conflagration with it, which burns down the house, and perishes itself in the flames. Mechanically, this child was an incendiary and a suicide; morally, neither. Shall we hesitate to speak as charitably of multitudes of weak and ignorant grown-up children, moving about on a planet whose air is a deadly poison, which kills all that breathe it four or five scores of years?

. . . The next movement in moral materialism is to establish a kind of scale of equivalents between perverse moral choice and physical suffering. Pain often cures *ignorance,* as we know,—as when a child learns not to handle fire by burning its fingers,—but it does not change the moral nature. Children may be whipped into obedience, but not into virtue; and it is not pretended that the penal colony of heaven has sent back a single reformed criminal. We hang men for our convenience or safety; sometimes shoot them for revenge. Thus we come to associate the infliction of injury with offences as their satisfactory settlement,—a kind of neutralization of them, as of an acid with an alkali: so that we feel as if a jarring moral universe would be all right if only suffering enough were added to it. This scheme of chemical equivalents seems to me, I confess, a worse materialism than making protoplasm master of arts, and doctor of divinity.

Another mechanical notion is that which treats moral evil as bodily disease has so long been treated,—as being a distinct entity, a demon to be expelled, a load to be got rid of, instead of a condition, or the result of a condition. But what is most singular in the case of moral disease is, that it has been forgotten that it is a living creature in which it occurs, and that all living creatures are the subjects of natural and spontaneous healing processes. A broken vase cannot mend itself; but a broken bone can. Nature, that is, the Divinity, in his every-day working methods, will soon make it as strong as ever.

Suppose the beneficent self-healing process to have repaired the wound in the moral nature: is it never to become an honest scar, but always liable to be re-opened? Is there no outlawry of an obsolete self-determination? If the President of the Society for the Prevention of Cruelty to Animals impaled a fly on a pin when he was ten years old, is it to stand against him, crying for a stake through his body . . .? So it is that a condition of a conscious being has been materialized into a purely inorganic brute fact,—not merely dehumanized, but deanimalized and devitalized.

Here is was that Swedenborg, whose whole secret I will not pretend to have fully opened, though I have tried with the key of a thinker whom I love and honor,—that Swedenborg, I say, seems to have come in, if not with a new revelation, at least infusing new life into the earlier ones. *What we are* will determine the company we are to keep, and not the avoirdupois weight of our moral exuviæ, strapped on our shoulders like a porter's burden.

Having once materialized the whole province of self-determination and its consequences, the next thing is, of course, to materialize the methods of avoiding these consequences. We are all, more or less, idolaters, and believers in quackery. We love specifics better than regimen, and observances better than self-government. The moment our belief divorces itself from character, the mechanical element begins to gain upon it, and tends to its logical conclusion in the Japanese prayer-mill.

Brothers of the Phi Beta Kappa Society, my slight task is finished. I have always regarded these occasions as giving an opportunity of furnishing hints for future study, rather than of exhibiting the detailed results of thought. I cannot but hope that I have thrown some ray of suggestion, or brought out some clink of questionable soundness, which will justify me for appearing with the lantern and the hammer. . . .

Does not the man of science who accepts with true manly reverence the facts of Nature, in the face of all his venerated traditions, offer a more acceptable service than he who repeats the formulæ, and copies the gestures, derived from the language and customs of despots and their subjects? The attitude of modern Science is erect, her aspect serene, her determination inexorable, her onward movement unflinching; because she believes herself, in the order of Providence, the true successor of the men of old who brought down the light of heaven to men. She has reclaimed astronomy and cosmogony, and is already laying a firm hand on anthropology, over which another battle must be fought, with the usual result, to come sooner or later. Humility may be taken for granted as existing in every sane human being; but it may be that it most truly manifests itself to-day in the readiness with which we bow to new truths as they come from the scholars, the teachers, to whom the inspiration of the Almighty giveth understanding. If a man should try to show it in the way good men did of old,—by covering himself with tow-cloth, sitting on an ash-heap, and disfiguring his person,—we should send him straightway to Worcester or Somerville; and, if he began to "rend his garments," it would suggest the need of a straight-jacket. . . .

We must bestir ourselves; for the new generation is upon us,—the marrow-bone-splitting descendants of the old cannibal troglodytes. Civilized as well as savage races live upon their parents and grandparents. Each generation strangles and devours its predecessor. The young Feejeean carries a cord in his girdle for his father's neck; the young American, a string of propositions or syllogisms in his brain to finish the same relative. The old man says, "Son, I have swallowed and digested the wisdom of the past." The young man says, "Sire, I proceed to swallow and digest thee with all thou knowest." There never was a sand-glass, nor a clepsydra, nor a horologe, that counted the hours and days and years with such terrible significance as this academic chronograph which has completed a revolution. The prologue of life is finished here at twenty: then come five acts of a decade each, and the play is over, with now and then a pleasant or a tedious afterpiece, when half the lights are put out, and half the orchestra is gone.

16. An Astronomer Bemoans the Condition of the Exact Sciences in America. 1874

"Exact Science in America" is the classic statement of the late nineteenth century concerning the obvious backwardness of American science in terms of basic research and discovery. Here are the familiar themes of the superiority of Europe, the neglect of pure science in the United States. Of the greatest interest are the several types of factors that account for the neglect of the exact sciences, the specific institutions that were at fault. This indictment raises questions as to its accuracy in general and also whether the inclusion of the biological sciences would have altered the picture the author paints or modified the point that he makes.

SIMON NEWCOMB (1835–1909) was a brilliant exception to his depressing assessment of American science. Born in poverty in Nova Scotia, he was apprenticed to a quack herb doctor. He ran away to the United States, where he taught school and educated himself until, in 1857, he was appointed computer in the Nautical Almanac Office. In 1858 he received a B.Sc. from Harvard's Lawrence Scientific School. The United States Navy in 1861 appointed him professor of mathematics, and in 1877 he became the superintendent of the Nautical Almanac Office. He worked primarily in mathematical astronomy and was best known for his work on the motions of the moon. He was celebrated throughout the word for his scientific achievements. He also wrote in the field of political economy and once produced a novel.

See Simon Newcomb, *The Reminiscences of an Astronomer* (Boston: Houghton Mifflin, 1903); W. W. Campbell, "Simon Newcomb," National Academy of Sciences, *Biographical Memoirs,* 10 (1924), 1–69; Paul F. Boller, Jr., "New Men and New Ideas: Science and the American Mind," in H. Wayne Morgan, ed., *The Gilded Age, A Reappraisal* (Syracuse: Syracuse University Press, 1963), 221–243; Richard H. Shryock, "American Indifference to Basic Research during the Nineteenth Century," *Archives Internationales d'Histoire des Sciences,* 2 (1948), 50–65; Daniel J. Kevles, "The Study of Physics in America, 1865–1916" (doctoral dissertation, Princeton University, 1964).

EXACT SCIENCE IN AMERICA*

Simon Newcomb

However strong may be the faith of Americans in the future greatness of their country, their opinion of its present intellectual development is modest in the extreme. To the average intelligent citizen, the idea of this country producing such a mathematician as Le Verrier, or such a physicist as Tyndall, is simply incredible. All he has known of the scientific world leads him to believe that such men are necessarily of transatlantic origin. Our citizen has also a theory which explains the deficiency to his entire satisfaction, and, indeed, reconciles him to it, namely, that the activities and energies of this country are all directed toward material development, and that the atmosphere produced by this development is unfavorable to the production of the highest intellectual qualities. He therefore waits for the intellectual development of his country with the same philosophic patience with which he looks forward to the day when her infant manufactures will no longer need protection against European labor and capital.

Now, while, as we shall presently see, our citizen is quite right in the general belief that we are a generation behind the age in nearly every branch of exact science, he is entirely wrong in his view of the form which that state of backwardness assumes, and of the causes which have led to it. As a matter of fact, this country does produce abler mathematicians than Le Verrier, and abler physicists than Tyndall; only he has never heard of them, and would not believe in them, if, by any accident, he were to hear. Scientific investigators have to be born; and there is no sufficient evidence that they are born in any smaller numbers in proportion to population here than in Europe. What little real science we have, hidden in out-of-the-way corners, is fully equal to the average of European science. A comparison of census returns

* *North American Review,* 119 (1874), 286–308.

would probably show that the proportion of men engaged in intellectual and professional pursuits is nearly as great in this as in the average of other civilized countries. Nor are the facilities for prosecuting science here so much less than in Europe as to excuse our backwardness. And yet, that the amount of published investigation, in nearly every branch of exact science, is small in an extraordinary degree, can be seen by a comparison of our contributions to science with those of Europe.

In the present survey we confine ourselves to the exact or physical sciences in contradistinction to biological science, and to pure in contradistinction to applied science. It is quite true that we thus bring to light our weakest points, and that, if we extended our survey into other fields, we could find things more pleasing to our national pride than those we expect to find in the field we have chosen. There is no objection to our knowing where we are weak, because then we may know where our efforts must be applied in order that we may become strong. The division between our strong and our weak points is not, however, by any means that between the exact and the biological sciences, but that which separates the highest purely intellectual efforts from every other field of activity. Were we to look at the philosophic side of natural history, as exemplified by the works of such men as Darwin and Huxley, we might find here nearly as great a dearth of activity as in the exact sciences. On the other hand, were we concerned with the application of even the exact sciences to the arts of life, we should find our country in the front ranks of progress. We omit the consideration of these, because our object is, not to treat of the efforts to which we are impelled by our daily physical wants, but those to which we are impelled by the purely intellectual wants of our nature.

In such a survey as that proposed, we naturally begin with mathematics, not merely as standing at the head of the exact sciences, but as the key with which the portals of every science must be opened before it can properly be called exact. It is probable that few, even among intelligent men, understand that hardly any physical science can attain its highest development until its propositions are expressed in mathematical language, and its deductions are effected by algebraic formulæ. Now, when we seek for published mathematical investigation in this country, we find hardly anything but an utter blank. Of mathematical journals designed for original investigations, such as we find in nearly every country in Europe, we have none and never have had any. There have been a number of short-lived attempts to establish mathematical periodicals suited to the state of the science here, some of them worthy of all praise; but the necessity of adapting their contents to the capacity of their readers prevented them from containing anything of importance in the way of original investigation. Since the discontinuance of the "Mathematical Monthly," in 1861, we have not had, until the present year, even one of these popular mathematical journals. Quite lately an indication that some cultivators or students of mathematics must still exist has been given by the appearance from Iowa of several numbers of "The Analyst."

With no wish to disparage the praiseworthy effort to start a mathematical journal of any kind, it must be confessed that this publication is of the same kind with its predecessors, being principally devoted to the solution of problems, and to discussions too elementary to have much interest for mathematicians. The only place in which we can search for anything in the shape of original contributions to mathematics is in the transactions of our learned societies; and here we find since the Declaration of Independence a score or two of papers professedly of this character, but it is not likely that more than one or two of them contain anything worthy of quotation or remark. The whole of them together would not amount to so much as the mathematical journals of Europe publish in a month. Of late years, even these solitary contributions appear to have ceased, and the German *Jahrbuch der Mathematik* has not, since its first appearance, in 1868, found a mathematical paper published in America, though it gives a list of ninety-six European serials containing such papers.

The prospect of mathematics here is about as discouraging as the retrospect. To become a mathematician it is necessary to study the writings of the great mathematicians, and these are entirely inaccessible, except by private purchase, outside of New York, the Eastern States, and the city of Baltimore. The writer has made diligent but vain inquiry to find in a single public library of the second city on the continent[1] any of the writings of the great mathematicians of the present century. The National Library at Washington is almost equally deficient. Its collection of such books as a mathematical student would wish to refer to is poorer than that of many a private individual. The probability that there would be no one to consult them affords a reason for not purchasing them. If the reader will visit the Royal Library at Berlin, he will find among the dictionaries and other works kept for constant reference in the public reading-room a set of Crelle's Mathematical Journal, the volumes of which are among the best thumbed there; and if, as he looks at them, he will reflect that our National Library does not possess the volumes at all, he will need no further illustration of the relative state of mathematical science in the two countries.

When we pass to the physical sciences, the prospect is a little more encouraging. We find the names of Franklin and Henry in the history of electrical science. We have two or three active workers in the line of experimental physics who deserve to be ranked among the intellectual ornaments of our country. But the smallness of the sum total of their published productions may be inferred from the circumstance that there is but a single journal[2] to publish their researches, and that this journal covers the whole range of biological and physical science. Here, as in every other science, we find our

[1] Philadelphia.

[2] *The American Journal of Science* [*and Arts*], also known as *Silliman's Journal*. See below, in this essay.

deficiency to increase just in proportion as the science becomes exact. Many branches of physics have attained, and nearly all the remaining branches are rapidly attaining, the mathematical stage of development. As they enter this stage we find our American cultivators all dropping off.

In astronomy we can make as good, perhaps a better, showing than in physics. In exact astronomy we have the names of Bowditch and Peirce.[3] We have observatories quite comparable with those of Europe, well supplied with astronomers fully equal in ability to the ablest of Europe. In no country is the astronomical work that is done better done. Yet we do not find our astronomers engaging in investigations of the utmost delicacy; and the first determination of the parallax of a fixed star by an American astronomer has yet to appear. Like all other scientific investigators, our astronomers are crippled by the want of a medium of prompt and certain publication. Measured by the quantity of matter published, we fall far behind Germany, France, and England.

The general deficiency alluded to will be brought out with statistical exactness by an enumeration of our scientific journals and transactions. We have but a solitary scientific journal of the first class. This was established half a century since by Professors Silliman and Dana, during which time it has honorably represented American science. It served the purposes of all the sciences when it was founded, and it serves them still. The only other medium of publication of original investigations in exact science is to be looked for in the proceedings and transactions of our learned societies. Of these, the former generally contain only short notices of papers, while the latter appear at such long intervals that they afford only a very tardy means of publication. Our two most active societies have been the Philosophical Society of Philadelphia and the American Academy of Arts and Sciences, each of which has brought out about a dozen volumes of transactions since the beginning of the century. Excluding societies whose publications are purely biological, we are not aware that half a dozen other volumes of transactions have appeared within the interval alluded to. Add the eighteen volumes published by the Smithsonian Institution, itself founded by a foreigner, and we shall have a total of between forty-five and fifty volumes in three fourths of a century. This total combined product of the Smithsonian Institution and all the scientific societies of the country is about equal to what either the Royal Society or the French Academy of Sciences publishes in one third the time.

The great mass of scientific papers in Europe do not, however, appear in transactions, but in scientific journals. Here we stand at a much more striking

[3] Nathaniel Bowditch was a New England navigator turned astronomer and mathematician who perfected the work of others and was easily the leading figure in American mathematics in the early years of the nineteenth century. Benjamin Peirce was a professor at Harvard, a competent mathematician of international reputation and the outstanding man in mathematics in America in the mid-nineteenth century; he was, however, not as distinguished as many capable Europeans.

disadvantage. Against a hundred and fifty or two hundred pages annually on astronomy and physics in Silliman's Journal, Germany can show us two journals of pure mathematics, publishing together three or four large volumes of matter every year, two more of mathematics and physics, one of astronomy, and one of physics and chemistry. Altogether these journals issue ten or eleven volumes annually, half of them quarto and half octavo.

The comparison with England is more difficult, because, for reasons which we shall hereafter explain, the great mass of English research appears in the journals, proceedings, and transactions of societies. Were we to compare these, bulk for bulk, with our own publications, the comparison might be as much to our disadvantage as in the case of Germany. But this comparison would not be fair, for the reason that a great deal of the matter published by these societies is of semi-popular character, hardly entitled to be classed among original contributions to science. Making allowance for this, it is probable that, instead of finding in England, as we do in Germany, thirty or forty times as much publication of original research in exact science as in America, we should find only five or ten times as much. A comparison with France would probably be more to our disadvantage than that with England, as the *Comptes Rendus* of the French Academy alone contain ten times more matter pertaining to exact science than Silliman's Journal does.

It may be asked whether this comparison of gross quantities of scientific publication affords a fair and accurate criterion of the scientific status of the several countries, and whether there is really thirty or forty times as much scientific investigation in Germany as here. We reply that, in order to make the comparison entirely fair, there are two circumstances to be allowed for. One is, that much of our work is published in a very condensed form, and therefore does not occupy so much space as corresponding work might where the means of publication are more ample. But results lose so much in value by condensation, that the amount of allowance to be made for this circumstance is not very large. The other circumstance is that a great deal of our work is published in foreign journals. Our observatories, for instance, have no medium of publication except the *Astronomische Nachrichten.*

Making every possible allowance, and viewing the facts from every standpoint, we shall be able to make only the most beggarly and humiliating showing. What is yet worse, we cannot claim to be improving our relative position, but are rather falling back, scientific activity increasing more rapidly in Europe than here. If we find improvement in one place or one direction, we find decadence in another to counterbalance it. In what was, in times past, one of our great scientific centres we now find, so far as the world can learn, not a solitary mathematician, astronomer, or physicist.

We have here a picture presented to our view which should command the serious attention of all who desire the intellectual progress of this country. Why, with our numerous educational institutions, and our great crowd of professors, should our contributions to the exact sciences be so nearly zero? If the popular idea of the cause of this state of things were correct, we should have neither literature, art, nor liberal education. However low an opinion we may form of our education, literature and art, we cannot assign them to the contemptible position alongside the education, literature, and art of the world which our science holds in the science of the world. We conceive that the true cause is not to be found in any lack of ability among our scientific men, nor in anything in our intellectual atmosphere which is positively unfavorable to the prosecution of science. The lack of a proper educational system does indeed prevent the training of a proper number of investigators in the higher branches; but this does not account for the comparative inactivity of our admittedly first-class men. The real proximate cause is found in the lack of any sufficient incentive to the activity which characterizes the scientific men of other nations, and of any sufficient inducement to make young men of the highest talents engage in scientific pursuits. The reason that so much more scientific investigating is done in Germany than in this country, is simply that the inducements to do it are there so much more powerful. A glance at two features of the scientific development of the present day will enable us more clearly to appreciate the state of things in the various countries referred to. The spirit of modern science and its law of development are largely expressed in two antithetical and, to the superficial glance, inconsistent propositions.

The first proposition is, that the methods and objects of our scientific investigation are distinguished by their purely practical character, using the word "practical" in its best sense. Indeed, the most marked characteristic of the science of the present day, and that which distinguishes it both from the science of the ancients and from the speculations of untrained minds on scientific subjects at the present time, is its entire rejection of all speculation on propositions which do not admit of being brought to the test of experience. Let us take an astronomical prediction as an example; on July 29, 1878, at half past three in the afternoon, the dark shadow of the moon will pass over the city of Denver in the Territory of Colorado; and during three minutes the gloom of night will take the place of day. This prediction is complete with respect to the phenomena and to everything connected with it which can influence the material interest of mankind, yet it is entirely independent of the question, What causes the moon to gravitate toward the earth and sun? It is founded on certain laws of the motion of the moon; but in investigating those laws the questions why the moon gravitates toward the earth and sun; whether she moves by virtue of the immediate will of the Deity, or of some quality inherent in the matter of which she is composed; and whether such a quality was impressed upon her matter by the Deity, or is self-existent in the matter itself,—are entirely ignored. It is by ignoring them, and confining the attention to the purely phenomenal aspect of the problem, that the laws in question have been discovered.

The antithetical proposition is, that, as an almost or quite universal rule, scientific discoveries are never made by men having any practical object in view. To take a single instance among the long series of discoveries in electricity which finally made the telegraph possible, not one was made with any other object than the increase of knowledge of natural laws; and no practical telegrapher was ever known to discover a new law of electricity. The same remark applies to all those scientific discoveries which have made modern society what it is. It seems as if Nature persistently refuses the knowledge of her secrets to those who seek them from any other motive than the love of truth. In this case, as in many others, the highest utility is most effectually reached by not seeking utility at all. Nothing is more certain than that, if we are to make any further advance in our knowledge of nature, it must be through the labors of men who make the study of nature the principal business of their lives, and who are trained accordingly.

At the same time, such men must have some inducement for entering upon this field of labor. The question may here arise, whether the claim that any worldly inducement will really lead to successful scientific investigation is not invalidated by the very principle just laid down. If knowledge must be sought for its own sake, how can we be asked to hold out other rewards for seeking it? If men will not pursue it without such reward, will they really pursue it successfully under the influence of the reward?

It might be a sufficient reply to these questions to say, that the simple acquisition of knowledge by investigators is not sufficient; the knowledge acquired must be published to the world before it can be of any value to others. Now, investigating and publishing are two very different things. The latter requires the work to be put into such a form that it can pass the closest criticism. One may study Nature most industriously for her own sake, and yet refuse to read proof-sheets, without a strong inducement of a much lower order. It is in the quantity of published matter that the deficiencies we have pointed out are seen. If esoteric science were of any public importance, we might find a great deal of it in this country. The difficulty is, not that our scientific men are indifferent to knowledge, but that they do not go through the laborious and thankless process of digesting and elaborating their knowledge and publishing it to the world.

There is yet another aspect of the case. While the motives which may induce one to spend one's life in a particular line of activity seldom admit of accurate analysis and definition, there are two propositions which cover the ground now under discussion. Let us admit that pursuing knowledge with no higher object than the applause of mankind, or the acquisition of wealth, will never lead to any result. It must also be admitted that to assert that the good opinion of his fellow-men should not be a motive to exertion, is simply saying that one should be entirely indifferent to that opinion,—a proposition which none will directly maintain. And however derogatory to science it may be to pursue it with a view to wealth, no one will directly maintain that scientists as a class should be above supporting their families and paying their debts. The distinction between love of notoriety and honorable ambition, and between the love of wealth and the desire to pay one's way in the world, is too obvious to make it necessary for us to draw it here.

Let us now consider more closely the relative inducements to scientific research here and in Europe, and see whether they are not nearly or quite sufficient to account for the backwardness of research here. In Germany the seats of scientific activity are the universities; in France and England, the learned societies. We have, therefore, to compare our universities and learned societies with those of England and the Continent. When we examine the universities of Germany, and compare them with our own, one of the most obvious points of contrast is that, while here the universities make the professors, there it is the professors who make the university. Students flock to Berlin, not because the university is an old, celebrated, and good one, but to hear Helmholtz and Virchow.[4] If all the men like these should leave the university, the students would follow them, and the university would at once sink to the second or third rank. But in this country students are not attracted to Harvard and Yale by the names of individual professors, but by the reputation and organization of the colleges. They choose one or other of these institutions because they know that a certain system of instruction has been organized, and that certain facilities are afforded by which they can be well instructed. If the professor is a good teacher, they demand nothing further of him.

This is by no means a simple difference of organization between the universities of the two countries, but a difference arising out of a fundamental peculiarity of the German mind, namely, a desire to be taught by a great man. In the American student we can hardly see any evidence of the existence of this trait. But in the German it is most strongly marked, and to it, combined with that love of thoroughness, and that wholesome contempt of "rule of thumb" systems which in this country pass as "practical," is largely due, not only German pre-eminence in every branch of intellectual activity, but very largely the present position of the German nation.

It is quite possible that the difference of which we speak would be found, on analysis, to be not so much a difference in the amount of respect entertained for intellectual greatness as in the standard of greatness. It is quite likely that, on the whole, the genus Professor is held in as high respect here as in Europe. But here the respect is called forth by the title, the position, and the learning, without which it is supposed the position could not have been gained, while in Germany it is dependent on what the professor adds to knowledge. The simple knowledge of what others have done in any special branch of research commands no more respect there than here, but probably a great deal less. Whether in the departments of science, philosophy, or the

[4] Helmholtz was one of the greatest innovators in both physics and physiology; Virchow was founder of modern pathology.

humanities, the question is, not what does he know, but what has he discovered that is new. What doubts has he cleared up? What fallacies has he exposed? What increase of precision has he given to the subject he has studied? The professor for whom satisfactory answers to these questions cannot be given, who is unknown as an investigator or thinker, and who can impart to his hearers no ideas but such as he has acquired from others, loses caste, gradually perhaps, but as certainly as, in mercantile life, the merchant who does not study the course of the market loses wealth. His students drop off, and, if his fellow-professors are like him, nothing will prevent his university from rapidly sinking to a low grade.

On the other hand, the man who has made really important additions to knowledge, and who has something fresh to make known to his hearers, needs neither the title nor position of professor to secure himself an appreciative and paying audience, and an honorable standing at the university. The humblest *Privat-docent* can compete with the oldest professor, certain that his future success will not depend on the favor of a board, or on any system of promotion by seniority, but on his real merit as an advancer and expounder of knowledge. Under this system there are no favors to deserving young men as such, and no positions for men who have failed to make a living in other spheres of activity. Instead of these, we have the same fierce struggle for existence which is at the root of human development, and which has full play in most of the affairs of men,—a struggle in which the weak are pushed to the wall without mercy, and prizes are gained in proportion to strength.

In striking contrast to this system is that which we find to prevail here. Among the large number of our so-called universities, in fact at all outside of the Eastern States, nothing more is expected of a professor than acquaintance with a certain defined curriculum, and ability to carry the student through it. He has nothing to do but satisfy the appointing power that he understands what is found in a certain text-book, and that he can teach what he knows to others. Even at our highest institutions of learning, Harvard and Yale for instance, we find nothing like the German standard. However great the knowledge of the subject which may be expected in a professor, he is not for a moment expected to be an original investigator, and the labor of becoming such, so far as his professional position is concerned, is entirely gratuitous. He may thereby add to his reputation in the world, but will scarcely gain a dollar or a hearer at the university.

If the immediate necessity of originality is so much greater on the part of a German professor, so, in a certain way, is the reward of entire success. We do not refer mainly to a pecuniary reward, but if we did, the proposition would be partly true and the state of things would forcibly illustrate the difference of the two systems which we are trying to bring out. We have said that probably the genus Professor is held in as high estimation here as in Germany. We may add, that the average professional income is much higher here, even when we make allowance for the greater cost of the necessaries of life. But there the greater part of the income is earned by being a great man and a great expounder, and there being no definite limit to what may thus be gained, the highest incomes, and, consequently, the pecuniary rewards held out for becoming a first-class man, are higher than any which a board of trustees can vote in this country. Of course the third and fourth rate men are much worse off than here; so much worse off, in fact, that they are compelled to betake themselves to some other occupation for a livelihood. The reward to which we principally refer is, however, that of reputation and esteem. The successful German professor, as he grows old, secures a position in the affections of a large body of educated men corresponding to which we have nothing at all to show, at least in the case of one who is merely an investigator and teacher. If one wishes to fully realize this, he need only witness the *Jubiläum* of any eminent German professor or scientist, and see the congratulations showered upon him by individuals and organizations from every part of the country, and then try to remember when he ever heard of any one taking any notice of an American professor becoming an "Emeritus," or even giving the occasion more than a passing thought.

When we pass from Germany to England, we find the state of things entirely different. The love of being taught by a great man, to which we have traced some of the peculiarities of German universities, is as completely unknown in England as here, and, in respect to the features we have been describing, the English university is much more like the American than like the German. If, therefore, the English investigator depended on the universities for support and encouragement, we should expect to find scientific investigation as backward in England as here. But the part which is played by the educational system of Germany is in England played by the great scientific societies. The general organization of these societies is substantially the same as in this country. On the European Continent each considerable nation has its Academy of Sciences, either supported or patronized by the government. In England and America, however, it is considered no part of the functions of the government to support such a body; the scientific societies of these countries must therefore be supported by the contributions of their members and by private endowments. Now, it is a fact which we have to face, and which it would be folly to disguise, that our scientific societies do not compare with those of England in wealth and power. There are a score or two of English societies which regularly publish transactions. The membership in the leading societies ranges from four hundred to a thousand, or even several thousand, while the annual contribution of each member is from two to four guineas. The annual income from the members alone may therefore range from four thousand to twenty thousand dollars, while the older societies have invested their surplus income from time to time in such amounts as to now receive a considerable annual interest.

Of the corresponding data for American societies we have no exact statistics available. We cannot, however, be seriously in error in saying that the

Geographical Society of New York and the American Association for the Advancement of Science are the only bodies in the country, of which exact or physical science forms one of the objects, the paying membership in which amounts to two hundred. We believe also that the maximum annual contribution in any of our societies is five dollars, while in a great many it is much less. In the American Association it is only three dollars. The societies we have mentioned are probably the only ones of which the annual income from the contributions of their members amounts to a thousand dollars. So far as is known, it is only two or three of the older societies that have any other source of income than this. By this lack of wealth, not only are our societies deprived of the means of publishing papers, but they are deficient in one of the great elements of interest and strength. However out of place the love of wealth may be in such a body, it cannot be denied that every one will take much more interest in an organization in which he has a good deal of money invested than in one in which he has none.

The great weakness of most of our scientific organizations does not, however, consist in the want of financial means, but in something much more difficult to determine and define. We can only say that, with a few exceptions, they exhibit a total lack of cohesive power, vitality, and that undefinable something which may be called weight and importance. However eminent may be the men who compose them, most of them are, as organizations, insignificant, and exhibit the same liability to die from slight causes that weak and sickly individuals do. A history of all the attempts to organize learned societies in this country would afford an instructive study in human nature, and might show that they died by causes as uniform as those which cause the decay and death of individuals. The principal features of the decadence would be, first, a constant enlargement of the range of membership, and consequent lowering of the tone of society; then a gradual and unnoticed falling off of the abler members, until the society is reduced to a state which a physician would describe as great constitutional weakness. This change may require a few months or many years. The society is then attacked by a parasite in the form of a charlatan desirous of using it as a means of securing a temporary notoriety for himself. He secures the management and control of it, and furnishes most of its discussions, till he makes it ridiculous, and then society and charlatan disappear together from public view.

If strength is relative, so that power in charlatanry is the same thing with weakness in integrity, then a curious instance of the weakness of real science is afforded by an attempt to organize a national scientific society a few years since. Many of our readers may remember "The American Union Academy of Literature, Science, and Art,'" organized in Washington in 1869. It had no known sponsor, unless the messenger who carried round a paper to get the signatures of prominent men should be considered such. But it soon began to fill a prominent place in the local papers. It had an opening meeting, and an inaugural address from a president. A few months later it had another inaugural address from another president. It got scientific questions referred to it by the Commissioner of Internal Revenue, and made reports on them. It applied to Congress for a charter recognizing it as a scientific adviser of the government; and an act of incorporation so recognizing it passed the House of Representatives without opposition, and was favorably reported from the proper committee of the Senate.

During all this time not a solitary man eminent in literature or science was known to be responsible for the society. The list of corporators named in the charter included judges, lawyers, bankers, officers of the army and navy, and a few civil officers of the government of every grade, but only a single eminent scientist, and he resided in a distant State, and probably knew nothing of the real character and objects of the organization. In fact, it is not likely that the large majority of the corporators themselves knew anything more than that they had been asked to join a very modest association for the mutual improvement of its members by scientific discussions. Why such an association should want national recognition from Congress, no one ever explained.

As we have said, the charter was reported to the Senate from the Committee on the District of Columbia, with recommendation that it pass. But there happened to be a single senator who saw the absurdity of choosing such a set of corporators to report on scientific "questions of importance to the public interest, which may be officially submitted to the Academy or its branches by the officers of the general or State governments." This was Mr. Sumner.[5] After a vigorous debate, in which, however, the provision we have just quoted was strangely overlooked, the bill was laid upon the table, and the society gradually sank into oblivion.

The most important function of the English societies is to take the place of the university system of Germany in encouraging scientific research. This they do by prompt publication of papers, by favorable notices of small works, and by the public award of prizes to great works. The details of the way in which this encouragement is given are, however, not important. The important fact which we wish to impress on the mind of the reader is, that when an Englishman makes any scientific investigation or discovery of merit and importance, he is considered a valuable member of society, and society takes pains publicly to indicate its appreciation of his value. When we say that in this country one may devote his life to science, and may gratuitously give to scientific investigation an amount of labor and talent which would secure him both wealth and distinction in any other profession, without receiving therefor a solitary public mark or expression of appreciation from any source whatever, or the slightest additional consideration from the public, hardly anything more is necessary to show that there is here comparatively little incentive to such work. One fact connected with our governmental administration will

[5] Senator Sumner was known for his incisive thinking, sharp tongue, and powerful rhetoric—and sometimes unpopular opinions.

illustrate the small amount of weight which the public assign to science. There are several government establishments of a scientific character, the best administration of which would require high scientific attainments, to the head of which scientific men would naturally be appointed if the public deemed them of sufficient importance. Yet not one of these establishments has now a head appointed on the score of eminence as an investigator in the sciences with which the operations of the establishment are allied, and there is probably not more than one of which the head could make pretensions to an acquaintance with any science whatever. This does not arise from those defects of the civil service of which the public now complain.[6] If science carried with it here the same social and personal influence as abroad, the scientific clique would under our present system divide the scientific officers among its members just as the politicians do the political offices. They fail to do so, not because they are not politicians, but because they are weak.

The great drawback to American science which we have been considering is, not lack of facilities, but lack of incentive. In fields of research requiring good libraries, or expensive appliances in the way of investments and apparatus, our facilities are, it is true, confined to the East. In other respects they are as good as in Europe, and in some points better. For one thing, it may be doubted whether any other government is as ready as our own to appropriate money for scientific objects. However contracted many of the views of the "average Congressman" may be, he has intelligence enough to know that the knowledge of nature forms an important element of our civilization, and is quite willing to afford the means of increasing that knowledge on being satisfied that the money will be judiciously expended. It is not always easy to insure the fulfilment of the last condition under our present system, owing partly to the want of some permanent scientific body to which the expenditure of moneys for scientific objects might be intrusted, and which could be held responsible for its doings. This drawback has not generally proved a serious one, and we believe that scientific appropriations have been, without exception, honestly expended, if not always in the most judicious manner.

On the whole, we may say that if any one possessing the qualities of sound judgment, accurate thinking powers, and quick perception of the relations of cause and effect, chooses to devote them to the increase of knowledge, and to forego the wordly advantages they would secure for him in the ordinary fields of human activity, his opportunities will be as good here as elsewhere. But he must fight his way with the same persistence that he would in any other profession, with the probability that he will be a great deal longer in winning even the modest position which society here assigns to the eminent investigator. He can adopt another profession with the assurance of gaining employment, wealth, and position as fast as he can satisfy the public of his ability; while in scientific as in military life, age and seniority are still indis-

[6] The widespread use of the spoils system at that time often made any government job the plaything of politics.

pensable to any high recognition. In political life one may become representative in Congress, senator, chairman of a congressional committee, or even Cabinet officer, at an age when his youth would be considered a serious if not a fatal objection to the highest recognition of his scientific claims. Some well-known examples from the French Academy of Sciences, during the period when it was in the prime of its activity, and the leading scientific body of the world, will afford an instructive contrast to our system. To this day the Academy is reproached by its historians for its unpardonable neglect of the claims of young La Place, whom, notwithstanding the talent and activity displayed in his memoirs, it persisted in considering a neophyte long after he had proved himself a master; and La Place[7] himself is complimented for not being discouraged by this neglect. Yet he was elected a member of the Academy when he was between twenty-eight and thirty years of age,—an age at which it would be an act of hardihood even to propose a candidate for membership to our National Academy of Sciences, no matter how high his claims.

In 1809, Poisson and Arago were candidates for the vacancy in the section of Astronomy caused by the death of La Lande. The former, though he had never made a distinctively astronomical investigation, was earnestly supported by La Place on the score of age and position; but Arago was elected by a large majority, in consideration of his work in the measurement of an arc of the meridian. Yet the man of age and position was only twenty-eight, and his successful competitor was only twenty-three. The former was not elected till he was thirty-one, when the Academy, no longer willing to sustain the reproach of overlooking so eminent a man, elected him to a vacancy in the section of Physics, although it was only as a mathematician that he had acquired eminence.

We do not propose to enter into any discussion of the relative merits of simple age and social position on the one hand, and talent and industry on the other. But it is undoubtedly true that the latter must be recognized in preference to the former, just in proportion to our earnestness in promoting works which can be performed only by talent and industry. For instance, in our military and naval service, in time of peace, promotion and command go strictly by seniority,—a system which has the advantage of avoiding every suspicion of favoritism, and saving the appointing power the responsibility of deciding between the merits of a large number of candidates. But in the selection of leaders in time of war, the system breaks down, and ability necessarily takes the place of seniority in determining the choice.

So in scientific affairs: by making recognition the reward of age and personal influence rather than of talent and industry, we avoid a great deal of heartburning, bitterness, and strife. So far as the avoidance of these is an important end in society, so far is our system a good one. But it is none the less true that our country will never contribute its share to human knowledge

[7] Laplace (the usual spelling) presented his first paper and was elected an associate of the Academy at the age of twenty-four.

under this system. Science forms no exception to that law of nature which proportions awards to merit and exertions, and gives the strong the advantage over the weak; and no amount of sentiment will do away with the fact that it is to the operation of this law that the progress of the world is due.

What we have just said refers to the scientific recognition of the young man of science among his fellow-workers. But respect and reputation are awarded by the intelligent public as tardily as by scientific men. We refer, of course, to respect for scientific work, not to that for the popularization of science. If our young man is a good popular writer and lecturer, he can hew his own way to fame with nearly as much rapidity in this as in any other country. But popular lecturing is something entirely different from scientific research,—so different that few succeed in both fields of activity; and it is the worker in the latter field whom the public are prone to treat with indifference. If it is urged that reputation with the public is too low a motive to be set before a young man, we reply, that the very reason it is low is the comparative lowness of the means by which the end may be secured. Apart from this, it cannot be denied that a wide and solid reputation among educated and influential men is a valuable thing to any one; and the more rapidly a young man can obtain it in any pursuit, the more encouragement he has to enter upon that pursuit. If such a reputation in the scientific line can be more readily obtained by showy works than by those which are really solid and valuable, the remedy is, not to treat it with contempt, but to educate that small portion of the public whose good opinion is of any moment into an appreciation of the real state of things.

Now, our instrumentalities for communicating to the educated public a knowledge of the doings of the scientific world have, until very lately, been nearly as defective as our means of scientific publication, and, notwithstanding certain recent improvements, are still far behind those of other nations. In England, France, and Germany weekly, monthly, and quarterly journals of popular science are too numerous to be recounted; while, previous to the establishment of the "Popular Science Monthly" by the Appletons, we had not in this country a single journal designed to diffuse the knowledge either of general or exact science. The "American Naturalist," as its name implies, is devoted entirely to biology. One of our principal scientific wants has been a publication which should serve as a medium of communication between scientific men and the educated public as well as between the various classes of scientific workers. To fulfil the latter object it need not be too technical for the former object, for a specialist in one branch may know as little about the work of a specialist in another branch as the general public does. A geological theory, for instance, must be explained to an astronomer just as it would be explained to any other intelligent reader, and *vice versa.*

In the absence of any such publication, nearly all the scientific information attainable by the public has been derived directly or indirectly from English sources, generally from the proceedings of the English scientific societies. The latter, being organized for the promotion and encouragement of the science of their own country, naturally give prominence to the labors of their own members; and we may say that English scientific writers generally adopt the same system. The consequence is, that in acquiring an American reputation the American stands at the same disadvantage alongside the Englishman that he does in acquiring an English reputation. Another cause conspires to make his disadvantage really greater, namely, the general incredulity respecting the possibility of native talent which seems to be inherent in the American mind. The result of both causes combined is, that an American must obtain a European reputation before his own countrymen begin to believe in him.

Of course there are exceptions to this rule. One occupying so high an official position that he must be treated with respect, and taking care to keep himself prominently before the public, may secure even more public consideration than he is entitled to. But neither the holding of office nor public prominence constitutes scientific position, so that it is doubtful whether even this should be considered an exception. The main fact with which we have to deal is, that original scientific research does not by itself command the public consideration which the same talent would if directed in other ways, nor which it would if exercised in the same way by a European. The proposition that an original investigator can receive consideration in his own country only after acquiring a European reputation, is not invalidated by showing that if he be something else than an investigator the proposition will not apply.

Within the past three or four years there has been a large increase in the amount of popular scientific publication in this country, which is seen in the establishment of a scientific magazine, and in the appearance of "scientific columns" and "scientific departments" in many of our newspapers and magazines. But the great object of educating the intelligent public in scientific matters is very imperfectly fulfilled by these publications. A considerable portion of the matter they present to us consists of fugitive items, hardly more interesting or important than the column of daily clippings of one short sentence each which has become a feature of our newspapers. The most notable exceptions have been the science department of the "Atlantic Monthly," while it lasted; the Editors' Table of the "Popular Science Monthly"; and, of late, the Science Record of "Harper's Magazine." Here we have found original discussions of scientific questions, and reviews of the progress of science by competent writers. For the rest of the "Popular Science Monthly" so much cannot be said. When first started it was mostly made up of extracts from English publications and of essays which could hardly have found a place in any other publication. Of late it has gradually improved by including more original matter, and that of a better class. But it has never attempted to supply the great want to which we have referred, namely, that of making known the progress of science in this country; and the reader who wishes to learn what our scientific men are doing here will find far more copious accounts of it in "Nature," an English periodical, than he will in the American magazine referred to.

The various deficiencies in the incentives to scientific research which we have described may be summed up in the single proposition, that the American public has no adequate appreciation of the superiority of original research to simple knowledge. It is too prone to look upon great intellectual efforts as mere *tours de force,* worthy of more admiration than the feats of the gymnast, but not half so amusing, and no more in need of public support. The remedy is to educate the intelligent public into an appreciation of the importance of scientific investigation, and of the necessity of bestowing upon those who are successfully engaged in it something in the way of consideration which may partially compensate them for devoting their energies to tasks which, from their very nature, can bring them no pecuniary compensation. This reward must be proportioned to merit, and the absence of commanding personal qualities must not be an obstacle to obtaining it. A great man can just as well be produced here as in Europe, if the public will only become disposed to encourage intellectual greatness here as there. If we had an equally rigorous system of intellectual natural selection and equal public encouragement for talent of the highest class, America would rapidly take a leading position among the scientific nations of the world.

17. A Chemist Suggests That Science Is Unraveling the Riddle of Life. 1875

As the science of chemistry advanced in the nineteenth century, the mystery of life receded. So successful were the chemists in analyzing and synthesizing in the laboratory even complex organic substances that it became reasonable—but not mandatory—to infer that there was no substance, even living matter, that would not some day be reduced to its chemical elements and created at will by the scientist. Here IRA REMSEN (1846–1927) expresses some of the excitement that he and his contemporaries felt as they explored the implications of organic chemistry. It is also possible from his words to infer the kind of opposition such thinking engendered. This famous nineteenth century controversy over whether life could be reproduced by the scientist, or only by God, was known as the vitalism controversy.

Remsen was a distinguished chemist who was born in New York and educated in the City College and the College of Physicians and Surgeons there. After taking his M.D., he went on to the University of Göttingen, where he earned his Ph.D. When he wrote this article, Remsen was teaching chemistry at Williams College. A year later he was appointed the first professor of chemistry at Johns Hopkins University. Later he became president of Johns Hopkins.

For more about Remsen and chemists in America see Edward H. Beardsley, *The Rise of the American Chemistry Profession, 1850–1900* (Gainesville, Fla.: University of Florida Press, 1964) and Frederick H. Getman, *The Life of Ira Remsen* (Easton, Pa.: Journal of Chemical Education, 1940).

THE ARTIFICIAL PREPARATION OF ORGANIC BODIES*

Ira Remsen

The "good, old" foundation upon which our fathers stood has been sadly shaken. Its complete overthrow has at times seemed inevitable. "Scientific men" have led the attacking army, and thus gradually brought themselves into disgrace with a portion of the community. The fight has been carried on to a great extent blindly, and most attempts to establish order have only succeeded in increasing the confusion. Sorties from the camp of "the fathers" have been made, and weapons have been carried back; but, alas! the weapons were useless, or, if used, they injured the user. The conflict is still waging, and it will continue to wage. Occasionally faint promises of a better understanding are given, but some misguided enthusiast, on one side or the other, hastens to destroy the hopes of a happy issue. The frequent shocks received by "the fathers" have unduly excited them, and they look upon each advance of science as something dangerous. Often they do not stop to examine whether the movement of the hostile party is, or is not, antagonistic to their position, but blindly throw their whole force against it, and anxiously look for the results of the crash. It sometimes happens that they thus waste their force, and weaken themselves for future necessary encounters.

Dropping the figure, we may safely assert that those who are avowedly the opponents of science, though their objects may be the highest—though they may be actuated by only noble desires—have, unfortunately, from time to time brought ridicule upon themselves by upholding views which were not tenable, and which a careful examination and thorough knowledge of the subject would show to be unnecessary for the support of their theories. These somewhat trite remarks lead to a consideration of the subject embraced in the title of this paper.

There are certain chemical substances known to us which only occur in the organs of plants or animals. The number of these substances at present known is very great, and new ones are being rapidly added to the list. They consist often of but three elements—carbon, oxygen, and hydrogen; some-

* *The Popular Science Monthly*, 7 (1875), 726–732.

times nitrogen is added to these, and, rarely, phosphorous or sulphur. Notwithstanding the fact that they are made up of few constituents, they are usually of complicated structure; indeed, the complication in some of them is so great that, with our present means of analysis, we are unable to express their composition by means of satisfactory formulæ. The substances referred to have been known by the name *organic bodies.*

Up to within a few years chemists were, to a certain extent, justified in drawing a line of division between two classes of bodies, both occurring in Nature: 1. Those which can be prepared in the laboratory; 2. Those which cannot be prepared in the laboratory. The second class included the so-called organic bodies. These were known to occur only in the organs of plants or animals. The two facts, taken together, were significant, and but little surprise can be expressed that a connection was traced between them. The simplest conclusion that could be drawn from the premises *was* drawn, and the scientific world, buoyed up by certain preconceived notions in regard to life, tacitly accepted it. Chemical substances which are produced under the mysterious influence of life, in the dark, unfathomed cavities of living organisms, cannot be produced by the hand of the mortal chemist. This was the conclusion which grew to be a dogma, and was used as a kind of *ex post facto* argument in favor of certain views in regard to the so-called "vital force."

But its influence did not cease here. Having worked so beneficially as an important link in a chain of retrograde logical sequences, it was afterward made a starting-point for other lines of argument. It was employed in religious and purely philosophical discussions, and assisted in the establishment of subsequent illogical conclusions.

As these discussions were taking place, the chemist quietly continued his strange dealings with the elements. Discovery followed discovery, until the fact could no longer be doubted that the dogma must fall. Its fall was, however, not the matter of a moment. It received repeated blows before it gave up its existence. Its place has been taken by an hypothetical statement founded upon a large array of facts, viz.: every chemical body, no matter of how complicated a structure, or what its nature may be, will probably, in good time, be prepared artificially in the chemist's laboratory. And this statement becomes more and more probable every day. Already a large number of the compounds, the formation of which was formerly supposed to be dependent upon the action of the vital force, have been reproduced entirely independently of any suspicion of the action of this force; and thousands of other analogous compounds which have never been found in plant nor animal are now known to us. Let us look briefly at some of the steps that were taken in this advance of opinion.

In the year 1828 Wöhler made the first observation bearing directly upon this subject. A few years earlier he had discovered cyanic acid, and he was now engaged in the thorough investigation of this acid. He prepared its ammonium salt, and, on evaporating the aqueous solution of the salt, he noticed the formation of large, well-developed crystals that in every respect resembled *urea.* Urea was well known, but had, up to that time, only been found among the products formed in animal bodies. Its existence was, in accordance with the then prevalent views, supposed to be due to the inexplicable action of the vital force. A careful examination failed to disclose any points of difference between the two bodies, and Wöhler was forced to the conclusion that at least one organic substance could be prepared outside of the organism.

But this by no means brought about a change of views. The upholders of the old dogma immediately found relief which was apparently satisfactory. Cyanogen compounds, of course including cyanic acid, had only been prepared from substances which had had their origin in the organs of animals, and, although these original substances had been subjected to purely chemical influences, and thus another animal substance produced, the vital force had nevertheless played its part as an essential agent in the formative process. This argument seemed plausible, and could hardly be objected to. Other and more decisive experiments were necessary.

In 1841 Fownes succeeded for the first time in preparing cyanogen directly from its elements. He passed nitrogen-gas over a mixture of charcoal and hydrate or carbonate of potassium at a red heat, and obtained a salt of cyanogen and potassium—cyanide of potassium. From this salt it was a comparatively simple matter to prepare all the other cyanogen compounds, and, finally, urea. Thus, then, there could be no doubt that a direct construction of some organic bodies from their elements was possible.

But urea, in comparison with most animal or vegetable products, is of simple structure. It contains but one atom of carbon in each molecule, whereas many others contain a very large number of carbon-atoms. The transformation of cyanate of ammonia into urea, which took place so readily, was a very simple one, if we consider merely the relation of the two bodies to each other. They have exactly the same composition. They contain the same percentages of carbon, hydrogen, nitrogen, and oxygen. A change of the arrangement of the atoms was necessary, but this was all—no addition of material, no building up, no binding together of a large number of atoms into one compound. There was still left something which could be ascribed to the influence of vital force, and this fact was seized upon and made to do service. It was now stated that, although it might be possible to prepare artificially some of the simpler organic bodies, the vital force was necessary to bring about the complicated form of union found in the greater number of the products of the life-process. This statement held its own for a number of years. In the mean time a series of brilliant experiments by Berthelot had established the fact that a large number of organic bodies could with ease be prepared artificially. In 1856 this chemist published the first results of his investigations. He had effected a direct union of carbonic oxide with hydrate of potassium, and thus obtained the potassium salt of formic acid. Later he showed that a direct union of carbon with hydrogen was possible; using carbon-poles, he passed a current of electricity through them in an atmosphere of hydrogen; he thus obtained acetylene, a hydrocarbon made up of two atoms of carbon and two of hydrogen. He also produced marsh-gas, ethylene, and a number of other hydrocarbons from inorganic materials which, in their turn, could be obtained from the elements. It was shown that marsh-gas could be converted into methyl alcohol; ethylene into ethyl alcohol; and from these alcohols it was an easy step to formic and acetic acid; to the aldehydes, amines, acetines, etc., etc. These results, although startling when viewed from the oldest stand-point, could still be reconciled to these views as necessarily modified subsequently to Wöhler's discovery. The compounds thus formed artificially were still of comparatively simple structure; and, in the numerous transformations effected by Berthelot, in no case was the passage from a compound of a lower to one of a higher order. Marsh-gas, methyl alcohol, and formic acid, each contained but one atom of carbon; ethylene, ethyl alcohol, and acetic acid, each contained two atoms of carbon. Surely the vital force alone could build up more complicated bodies.

Not so. The series of advances in the new doctrine, thus so propitiously begun, did not stop. New methods of investigation were introduced. Questions of a different character were put to chemical substances, and answers were not wanting. The interest in chemical science increased, and the army of those who were to carry it forward also increased. The growth of the science became proverbially rapid, and, during the excitement attendant upon this development, the last of the old landmarks between inorganic and organic bodies was swept away; vital force, as far as it was directly concerned in the formation of organic bodies, lost prestige. Both classes of bodies were found to be subject to the same fixed laws. A chemical substance is a chemical substance, look at it as we will. Its constituents, in one case as in the other, are bound together by chemical affinity, simply and alone. Whatever the conditions may be which surround the formation of organic bodies in the animal or vegetable organism, the final combination of the atoms, necessary to the formation, is brought about by chemical affinity. Although we cannot reproduce these conditions outside of the body, we can in so far imitate them that the same kind of combination will take place. We have at our command at present many means for the building up of the most complicated organic bodies from the simplest. Some of these are easily understood, and were discovered as the result of strict logical deduction; others are still inexplicable, and were discovered by accident. We can pass readily from one hydrocarbon to another, adding carbon-atoms to an extent which, theoretically at least, is unlimited; from one acid to another of higher order; from alcohol to alcohol; from alcohol to acid; from acid to hydrocarbon; from hydrocarbon to acid, through all the normal series of organic compounds. So great is our power in this direction, that it is possible to produce any member of any regular series of organic compounds from marsh-gas as a starting-point, or from any other member whatsoever. But marsh-gas can be indirectly produced from its elements, carbon and hydrogen; hence, we have the possibility given of preparing artificially by far the greater number of organic compounds. This number includes many of those substances which are formed in the animal or vegetable organism.

The formation of urea and formic acid has been alluded to. Without reference to the historical order, a few of the achievements of chemists, which have from time to time astonished and delighted the world, may here be briefly noted. Among vegetable products are oxalic acid, which was formed directly from carbonic acid, a combination of two carbon-atoms being necessary in the process; valeric acid, containing five carbon-atoms; malic acid, with four carbon-atoms, one of the most widely-distributed acids of the vegetable kingdom, being contained in a large number of unripe fruits; cinnamic acid, containing nine atoms of carbon; tartaric acid, the acid of grape-juice. Wintergreen oil, obtained from *Gaultheria procumbens,* has been found to consist mainly of an organic ether, which can be, and has been prepared artificially. The oil of garlic *Allium sativum* contains carbon, hydrogen, and sulphur. It can be prepared with all its properties without the plant. The oil of mustard, with its peculiar arrangement of carbon, hydrogen, nitrogen, and sulphur, is now manufactured on the large scale by a patented process, the mustard-plant being outrivaled by the chemist. The deadly poison conine [*sic*], and the beautiful colors alizarine and indigo, finally, belong in the same list. In regard to alizarine or Turkey-red, it may be remarked that the discovery of the methods for its artificial preparation has led to the establishment of an important branch of industry of far-reaching influence. It is doubtful, however, whether as much will ever be said concerning the preparation of indigo. Among animal products that have yielded up the secrets of their internal structure to the chemist are the simple fats and the lactic acids. In a great many portions of the animal organism, as the brain, pancreas, liver, lungs, the thyroid and thymoid glands, is found a substance, containing six atoms of carbon, which has been called leucine. This substance is also a frequent product of the decomposition of organic bodies. Leucine is obtained more readily by artificial means than it can be extracted from the tissues in which it exists ready formed. A constant ingredient of the juice of flesh is creatine; and one of the products of decomposition of creatine is sarcosine. Both creatine and sarcosine can be constructed from the elements by purely chemical processes. Taurine, which occurs in the bile, in the contents of the alimentary canal, in the lung-tissue and the kidneys, and contains carbon, hydrogen, nitrogen, oxygen, and sulphur, can be prepared by a very simple process.

These examples suffice to indicate the character of the results already achieved, and furnish justification for the hope now entertained by chemists that in good time it will be possible to produce all chemical substances in the laboratory. No one who has given the subject a sufficient amount of attention to enable him to form an opinion can for a moment feel a doubt on this subject.

The old dogma no longer exists. There are those who sigh at its death; who consider that the sacrilegious step of Science which annihilated it has, in some way, tended to lessen the mystery of life, and to embolden the votaries of science to look for, work for, further disclosures which may threaten some favorite view—it may be one of more importance than that which we have considered. But others, supporting themselves on the basis that truth can never be dangerous to the right, see no cause for alarm in such advances. They hail them with pleasure, and encourage the spirit which hastens their arrival. In regard to the special question treated in this paper, arguments are hardly necessary to show that the results of investigation, as we have stated them, could not materially modify any time-honored, fundamental views. Is life less of a mystery? Has the question concerning the nature of life been even approached in these researches? We think not. That chemical substances of peculiar structure are found in the living organism is true. That these substances are formed by the action of the force called chemical affinity is just as surely a truth. Do these two truths mutually detract from the importance of each other? When the active agent in the formation of the so-called organic bodies became known, a thousand questions could be proposed to one that could be proposed previously. The conditions for its working became subjects of inquiry, and an almost endless series of possibilities presented itself. From what substances have the new ones been formed? What chemical processes have brought about the final formation? Years—ages must elapse before our knowledge on these points can begin to be exhaustive. And then what? Is the mystery solved? No.

We are ascending a mountain of great light. Our views are becoming more and more extended as we reach higher and higher positions. Should we ever be enabled to reach the summit, there would be found a pleasant harmony in the broad panorama, and our eyes would rest in delight upon it; but the most extensive view has its horizon, the barrier between the visible and the invisible.

18. A Teacher Blames Higher Education for the Backwardness of American Science. 1876

In this article FRANK WIGGLESWORTH CLARKE (1847–1931) explores in detail one of the factors that scientists of the late nineteenth century blamed for what they considered to be the backward state of science in America. He suggests in a striking way that apparent interest in science did not go very deep and could indeed turn into covert opposition. Some of Clarke's strictures and suggestions still had validity nearly a century later. It should be emphasized, of course, that there were in the 1870s colleges and universities in which the best science was supported generously.

Clarke was a native of Boston and a graduate of Harvard's Lawrence Scientific School. From 1874 to 1883 he was professor of physics and chemistry at the University of Cincinnati. Afterward he became chief chemist of the U.S. Geological Survey and had a distinguished career as a government scientist.

See L. M. Dennis, "Biographical Memoir of Frank Wigglesworth Clarke, 1847–1931," National Academy of Sciences, *Biographical Memoirs,* 15 (1934), 137–165; Frederick Rudolph, *The American College and University, A History* (New York: Alfred A. Knopf, 1962); and Edward H. Beardsley, *The Rise of the American Chemistry Profession, 1850–1900* (Gainesville, Fla.: University of Florida Press, 1964).

"AMERICAN COLLEGES versus AMERICAN SCIENCE"*
F. W. Clarke

America, when compared with other first-class nations, occupies a low position in science. For every research published in our country, at least fifty appear elsewhere. England, France, Germany, Austria, Russia, Italy, and Sweden, outrank us as producers of knowledge. Our original investigators in any department of learning may almost be counted on the fingers. Fifteen or twenty chemists and physicists, as many mathematicians and astronomers, and a somewhat larger number of zoölogists, entomologists, botanists, and geologists, would fill out our meagre catalogue. Among these few discoverers a

* *The Popular Science Monthly,* 9 (1876), 467–479.

comparatively small proportion are of high rank. There may be in the United States, all told, twenty men of really notable scientific standing, although there is no one to compare in actual achievements with Sir William Thomson, Helmholtz, or Regnault. In geology we make a pretty fair showing, perhaps, because of the great facilities for research offered by our surveys and exploring expeditions. The newness of our country has also been of advantage to our zoölogists who have not failed to improve their opportunities. But in chemistry and physics, the two sciences most intimately connected with the greater industries, we have accomplished very little.

Several causes have combined to bring about this state of affairs. There is native ability enough in America to carry on work of the highest order, but inducements and opportunities have been lacking. The labor of developing new regions, of building up commerce, manufactures, and agriculture, of constructing railroads, bridges, and telegraphs, has diverted public attention from matters apparently of a more abstract and less immediately practical character. Material necessities have taken a natural precedence of intellectual wants. Now, having laid our foundations, we begin to think seriously about the future superstructure.

But apart from all these drawbacks to American scientific growth, there is yet another of almost equal magnitude. This is to be found in the system (or rather lack of system) which has shaped our higher education. Our country is dotted over with a multitude of so-called colleges and universities, which have sprung up, not in response to any well-defined necessity, not under the developing influence of broad and clear ideas, generous culture, and wise motives, but because of personal ambition, sectarian jealousy, or petty local pride. States have conferred charters almost indiscriminately, without reason or forethought. Any body of trustees, no matter how ignorant or how foolish, has had but to ask for university powers, and the request has been granted. Incapacity on their part, or injudiciousness in the plans, has seemed to offer no impediments. This policy may be democratic, but it certainly is not wise. Its chief result must invariably be to degrade the standard of education. A college or university charter should be issued only with extreme care, and to fully responsible persons. It ought to demand compliance with certain rigid conditions, and should be forfeited whenever the institution holding it falls below the proper standards. But the mischief has been done, and science has suffered. Let us see how.

In order that science may flourish in any community, several things are needful. There must be a general appreciation of its true value to the world, clear understanding by men of culture as to the best means for its promotion, facilities for both study and research, and suitable inducements to attract intellectual labor. No matter how able and enthusiastic an investigator may be, he can do little without apparatus or specimens, encouragement, and the means of support. Indeed, the last-named, or bread-and-butter element, is a very important feature of the problem. The human brain is a marketable commodity, at the service of the best-paying master. Payment may come partly in the shape of fame, but something of a decidedly material nature is demanded also. A man may love science devotedly, and yet be starved into adopting some more lucrative profession.

Suppose, now, that a young man of culture, genius, and enthusiasm, wishes to devote his life to science. He has received the necessary training in his favorite branch, and simply asks for an opportunity to apply his attainments both to bodily support and to the extension of human knowledge. At the very start the chances are against him. Many such men are annually driven by necessity out of the field of science, and forced to seek a maintenance in trade, manufactures, or some other department of industry. That a great deal of valuable talent is thus wasted, and turned into channels unsuited to its development, there can be no doubt. That so much good work has been done in a society where so much is lost, speaks well for the human intellect, and shows that real ability is commoner than the majority of people suppose. If seed never fell by the wayside, but only in fruitful places, our views of human nature would soon undergo a wonderful change.

But in the case of our particular novice, employment is at last secured as "Professor of Natural Science" in an average American college. In fact, scarcely any other career would be open to him. Now, how many of the requisites for success are likely to be at his command?

To begin with, he encounters a board of trustees among whom not one has the remotest idea of what science is, or what is essential to its growth. He is called upon by these gentlemen to "teach" chemistry, physics, astronomy, botany, zoölogy, mineralogy, geology, physiology, and perhaps Paley's evidences[1] on top of all. For study and research he has neither time, books, nor apparatus. For study, indeed, he is not supposed to need any time; and if he should press this necessity upon his employers he would probably be told that he ought to know his lessons before attempting to teach. His students come to him miserably prepared, caring little for what he considers important, and regarding his instruction as so much of an impediment between them and their degrees. And for all this he may receive less than a thousand dollars a year, and that with a feeling of precariousness and uncertainty. At last one of three things happens: he is either called to a chair in some respectable institution, gives up teaching altogether for another less annoying occupation, or else, his enthusiasm quenched and his aspirations gone, settles down into a dreary rut, to rust out the remainder of his days.

This picture may seem exaggerated, and yet it is wholly within bounds. Many men have been ground through the mill of an unendowed country college professorship, and know how hard and thankless were the tasks as-

[1] This refers to a famous book by an English clergyman, William Paley, *Natural Theology, or, Evidences of the Existence and Attributes of the Deity Collected from the Appearances of Nature* (1802). In this preevolutionary work Paley attempted to prove the existence of God and verity of the Scriptures by showing the necessity of God's hand to create the wonders of nature, which Paley asserts could have been conceived only by God.

signed for them to do. In such a position the true man of science can very rarely find either appreciation, encouragement, facilities, or pecuniary reward. Discouragement of the most wearing kind will, in nine cases out of ten, be his lot.

The American college system, then, is clearly an impediment in the way of American science. It acts adversely in several modes, and these I purpose tracing.

There are to-day in America over five hundred institutions claiming the name of college or university. Of these more than forty are in the single State of Ohio. Some are exclusively for male students, others receive only young ladies, the majority are arranged for the co-education of the sexes. Every religious sect, or fragment of a sect, is represented: Baptists, Free-will Baptists, Seventh-day Baptists, Presbyterians, United Presbyterians, Cumberland Presbyterians, Episcopalians, both High-Church and Low-Church, Methodists of divers complexions, Adventists, Swedenborgians, Friends, Unitarians, and Universalists: all control special institutions, equipped and endowed with due reference to the perpetuation of sound faith, and, incidentally, to the encouragement of what is supposed to be learning. Among Catholics, who now control seventy-four colleges, the inter-sectarian character is strongly marked, and institutions are recognized as especially Jesuit, or Franciscan, or Benedictine, or managed by the Christian Brothers, or by the Congregation of the Sacred Heart.

Now, there are several ways by which this sectarianism in education works mischief to science. The very fact that a college has been established for theological purposes, or for ecclesiastical aggrandizement, is adverse to good scientific research. Even though the teacher of science may not be directly hindered, the studies which are of especial value to theological students will be given undue prominence. In fact, nearly every American college emphasizes the classics and literary studies, and looks upon natural science as something of minor importance, often as a dangerous accessory, which must be tolerated but not encouraged. A college catalogue which now lies open before me, after announcing that full provision has been made in its course for the inculcation of religion and morality, asserts that "scientific culture is of value only in so far as it is based on a true conception of God, and our relation to him." Such a statement as this, viewed from the standpoint of any particular sect, will usually be found to mean more than the mere words indicate.

But the great injury to science is done by the unnecessary subdivision of forces. Forty institutions spring up where only one is needed, and nearly all of them are necessarily weaklings. Libraries, cabinets, apparatus, buildings, and faculties, are foolishly duplicated. Each college lives in a continual struggle for existence, doing inferior work, and paying miserable salaries to an inadequate corps of teachers. If there were such things as Presbyterian mathematics, Baptist chemistry, Episcopalian classics, and Methodist geology, such a scattering of educational forces would be pardonable; but, as matters really stand, it is a nuisance for which no valid excuse can be found. Here there seems to be a real conflict, not between religion and science, but between the injudiciousness of religious people and the requirements of scientific research. Where one good laboratory should exist, we have forty small and inferior sets of apparatus, each fit only for elementary instruction, and wholly unsuited to purposes of investigation. Thus the very institutions which we should naturally expect to advance science have been made by sectarian spirit incapable of yielding solid results. Other branches of learning suffer also, only science is most impeded of all. The classics, mathematics, philosophy, or literature, demand few appliances. Give the professors a fair library, perhaps some maps or charts, and a recitation or lecture room apiece, and all is provided for. But science, to be properly taught, demands much more. There must be not only laboratories and apparatus, but material and specimens; and these all cost much money. No wonder, then, that a poor institution cramps its scientific teachers, and offers meagre opportunities for the prosecution of their best and most valuable work.

Going a step beyond this curtailment of material means, we shall find that the division of forces again operates contrary to science in the selection of professors. In the first place, poverty compels a college to demand more work from a professor than any man can well do. A teacher who is called upon to instruct elementary students in half a dozen distinct branches cannot accomplish much real work in any one. Every branch of science is vigorously growing, and can be properly taught only by one who has the time to keep abreast of its growth. A large majority of American college professors are now incompetent, because the policy of college management keeps them so. Let us glance at a few of the professorships which some country colleges have established. Here, for example, is McCorkle College, situated in Eastern Ohio, whose ministerial president is "Professor of Hebrew, Natural, Mental, and Moral Science." Surely this gentleman, if his professions are honest, must be the most learned scholar in the world. His "moral science" would, of course, prevent him from undertaking any work which he was incompetent to do. We cannot suspect a "reverend" of hypocrisy in such a matter as this. In Maryland, [a] . . . college, St. John's, rejoices in a "Professor of Natural Philosophy, Chemistry, Mineralogy, and Geology, and Lecturer on Zoölogy and Botany." Penn College, in Iowa, has a "Professor of Natural Science and Political Economy"; and Eminence College, Kentucky, a "Professor of Biblical Literature, Mental Philosophy, and Chemistry." Even in New York State there is Hobart College, with its "Professor of Civil Engineering and Chemistry, and acting Professor of Mathematics and Modern Languages." Professorships like these are by no means rare; they are the rule rather than the exception. A very large majority of our so-called "institutions of learning" employ Jacks-of-all-trades to do the work of instruction, and how well that work is likely to be done we can easily imagine; indeed, it is difficult to understand how a conscientious man can undertake such tasks. Every teacher who is competent to teach at all must know that he is unable to cover so much ground, and should refuse to be a party to such fraudulent teaching. Fraudulent is not too strong a word to use in this connection. An institution which receives money from its students in payment for an education such as it cannot give, is certainly guilty of fraud. These frauds are the natural outgrowth of improperly-granted charters, incompetent or ignorant boards of trustees, and reckless sectarian pride. Every denomination seems to be imbued with the characteristic American anxiety for display, and the establishment of a new college is a convenient piece of clap-trap to resort to. Surely the advancement of religion ought not to render necessary such sacrifices of true principle! If false pretensions are to be thus directly encouraged by the churches, what can we expect from the people at large?

The smaller colleges, however, are not the only ones to blame in this matter of professorships. They are perforce compelled to employ smatterers, because of their inability to pay the proper number of specialists. But institutions of considerable wealth often injure science in their selection of teachers by introducing false issues into the question. Every year professors are chosen, not on account of scientific ability, but for reasons of a theological or sectarian character. If two men, one a Baptist, and the other a Unitarian, were candidates for the same professorship in a Baptist university, the former, even if very much inferior to his rival, would almost certainly be elected. There may be exceptions to this general rule, but they are very rare. Even at Princeton issues of this sort are frequently raised, and the ablest candidates have been rejected on purely dogmatic grounds. Theological soundness in such an institution far outranks scientific ability. If Laplace[2] had lived in America, no college would have tolerated him for an instant. Almost any decayed minister, seeking an asylum, would have beaten him in the race for a professorship. Not many years ago, the ablest chemist America has ever produced was a candidate for the chair of chemistry in a very prominent Eastern college. He did not believe in the Trinity, and for that reason alone failed of an election. The immorality of such a system is manifest. When success or failure is made to depend upon a mere *profession* of belief, a direct premium is put upon hypocrisy. Incompetent men are not unlikely to be unscrupulous also. Science cannot really flourish in America until, in this respect, the colleges mend their ways. Men must be chosen professors because of their fitness to teach specified subjects, and not on account of their notions, real or professed, concerning abstract theological dogmas. Moral character ought, of course, to be considered; but mere speculative belief, never.

Another objectionable result of college scattering is the under-payment of professors. Even our best universities have shortcomings in this respect.

[2] Laplace was famous for a conversation with Napoleon in which the Emperor taxed the astronomer for having written an enormous book on the system of the universe without having mentioned its Creator. "Sir," Laplace is supposed to have replied, "I have no need of that hypothesis."

A teacher upon small salary is naturally somewhat unsettled in his mind, is apt to be looking about for better employment, and is liable to feel a constantly diminishing interest in his work. Stability of place and freedom from pecuniary anxiety are very important to an investigator; and just these requisites few American colleges are able to supply. A large salary is not absolutely necessary to a scholar, but a certain means of comfortable subsistence is. At present, when wholly inadequate payment is offered, there is scarcely any inducement to attract a young man into the scientific life. A professorship or tutorship may be accepted for a year or two, perhaps, just as a stepping-stone to something more lucrative, but how rarely is the teacher's vocation taken up as a career! Almost every other important occupation yields surer rewards, and a fairer prospect of attaining to a competency. A young lawyer, doctor, or merchant, if careful and industrious, may reasonably look forward to possessing at some time a home of his own, with the means of sustaining and properly educating his children. The young devotee of science, however, has rarely any such possibilities before him. His labor is as arduous as, and demands even more talent than, that of the attorney or physician, but the recompense is vastly less. If, as he ought, he gives his leisure months to the advancement of learning, he will find his salary insufficient for the maintenance of a family. In order really to live, he must constantly be doing outside work. He will thus struggle along, year after year, in constant danger of being discharged or supplanted, and in his old age, weary and broken down, will find himself little more than a pauper. Is it strange, then, that the best intellectual talent of America is repelled from professorial positions, and attracted into other fields of labor? Can science be expected to flourish under such a system? We pay mere popular lecturers well enough; and surely the real workers, who create science, ought to be fairly recompensed also. But we can hope for little improvement until the number of colleges is reduced, and the means of those remaining suitably enlarged. Science must offer careers to men of ability, with the rewards which capacity, skill, and faithful industry, always ought to receive.

But, after tracing all the effects produced by the division of educational forces, we shall still find other points in which our college system is prejudicial to science. Glance over the curriculum laid down in almost any college catalogue, and see how the scientific instruction is arranged. In nearly every instance there will be found an enormous disproportion between linguistic studies and science. As a rule, over one-half of a student's time for four years is assigned to language; the remaining half being divided between mathematics, English literature, history, philosophy, and "natural science." Chemistry, for example, is generally taught through a single term (one-third or one-half, as the case may be) of the junior year. Thus a study, extremely important both practically and as means of culture, is pursued by a student for perhaps three hours a week during one-eighth or one-twelfth of his college course. In some institutions, undoubtedly, more time is given to chemistry;

but such cases are comparatively rare. A youth will enter college with at least a year's preparation in Greek, and then will follow that study for the greater part of his four years' course; but the science from whose applications he derives direct benefit every day of his life is crowded out into an obscure corner of the curriculum, and made to seem of little value. Physics is treated like chemistry; while geology, botany, zoölogy, and astronomy, are pushed even closer to the wall.

Now, what effect has this unfair distribution of studies produced upon American science? Plainly, a very bad effect. Our scientific men must be recruited mainly from among the ranks of our college graduates, and hence the latter ought to be imbued with something of the scientific spirit. That spirit is not likely to be very strongly aroused by the present policy of make-believe teaching. In fact, an enthusiasm for science is dampened rather than encouraged in the majority of American universities. The student sees men of fair training employed to teach the classics, while the work in scientific branches is done by wholly-untrained or imperfectly-trained instructors. Frequently it happens that Latin and Greek are taught by separate professors, while a single teacher is called upon to cover all science outside of mathematics. It is easy to see what effect such a state of affairs is liable to produce upon the mind of an average pupil. He becomes accustomed to regard the sciences as comparatively unimportant. He learns almost nothing of their true relations to life, and the little which he does happen to pick up is gleaned from a few superficial lectures and two or three trivial text-books. If he fails in these studies at examination, the failure counts practically nothing against him from graduating. In short, the college deliberately carries out a policy of scientific smattering, and the student is influenced about as might be expected. He graduates in complete ignorance both of the methods and of the aims of science, having learned only a few disconnected facts concerning the great world about him.

Very many American colleges, however, now provide what claim to be "scientific courses," running for four years parallel with those in classics, and leading to bachelor of science degrees. This fact illustrates only a sham deference to the public demand for less Latin and Greek, and amounts to very little in favor of science. A striking case in point is furnished by McCorkle College, the learned president of which we have already referred to. Let us analyze the course laid down in the catalogue. There are three terms per annum for four years, or twelve terms in all, and in the regular classical course the studies run as follows: Latin is taught during ten terms; Greek, through eight terms; mathematics, five; history, four; Hebrew, three; natural philosophy, two; chemistry, two; geology and astronomy, one each; other studies, mainly philosophical but none scientific, seven. The modern languages seem to be omitted altogether! Then, following the schedule from which this abstract was made, comes the announcement that "the scientific department will embrace all the above course except the classics." Could a more contemptible sham be invented? Would it be possible to do more in the way of belittling science? The total omission of scientific studies would be more honest and more truly in the spirit of science. And yet this institution is empowered to grant degrees, and has the same legal authority as Harvard, Yale, or Cornell. This is, to be sure, an extreme case, but it is not much worse than a host of others. As a general rule, the "scientific course" in a Western college is the classical course, *plus* a little mathematics, and with French and German substituted for Latin and Greek. Less preparation on the part of the student is required to enter it, and every applicant is given to understand that it does not rank quite equally with its older rival. In both courses the natural sciences are similarly arranged, so that the graduated bachelor of science knows really no more chemistry, physics, botany, zoölogy, geology, or astronomy, than the supposably less scientific bachelor of arts. In fact, the great majority of so-called "scientific courses" are mere makeshifts, intended to accommodate those students who are too dull, or too imperfectly prepared for taking the more thoroughly-equipped line of study in the classics. Here, again, American colleges oppose the development of the scientific spirit, and hinder seriously the growth of American science.

It would be possible to multiply indefinitely these illustrations of weakness on the part of our college system. Institution after institution might be cited in which not science only, but all culture, is at the lowest possible ebb. Just the bare facts concerning some Western and Southern colleges would, if published here, seem like incredible exaggerations or distortions of the truth. I have beside me college catalogues which are positively grotesque in their absurdities; no satire could do justice to them. One institution in particular, situated in Tennessee, has fairly reached the point at which the sublime and the ridiculous meet. In respect to science, even some of our oldest and best universities are open to criticism. Some apply theological tests in the election of professors, and in a mild way act toward modern science as some of the Spanish universities once acted toward the discoveries of Newton. Many others make lower standards for scientific than for classical students, seemingly upon the idea that a bachelor of science is expected to know less than a bachelor of arts. Perhaps the scientific spirit is now best represented in this country by the Sheffield Scientific School at New Haven. Here the policy of the institution seems to have been entirely shaped and guided by the Faculty rather than by the trustees. The Lawrence Scientific School did stand higher before the abolition of its special laboratory, and approximated closely to the German idea; but of late its Connecticut rival has passed it in the race. As a university, taken for all in all, Harvard is probably far ahead of Yale, but in training scientific students the latter can at present claim superiority.[3] The

[3] The Sheffield Scientific School at Yale and the Lawrence Scientific School at Harvard represented pre-Civil War attempts to include science within the colleges in special schools rather than to integrate it into the curriculum. Both schools, as Clarke indicates, had varying fortunes over the years before their parent institutions were transformed into real universities on the German model, at which point the schools were absorbed.

Columbia College School of Mines is also a good institution, but it errs in the direction of over-thoroughness. The students have so much routine and detail work to do that no time is left for originality. The instructors, too, are overworked, so that they can accomplish little in the way of research, and they are, moreover, in many cases, underpaid. This latter evil the trustees can and should remedy. It also occurs at Cornell University, and has lost to that institution the services of several valuable men. These points are mentioned now, not hypercritically, but because they serve to illustrate certain discouragements which our scientific men have to encounter.

Now, having recognized some of the weaknesses in our American mode of conducting the work of higher education, we may reasonably ask how they are to be remedied. How shall reform be brought about, and by whom?

It is quite evident that improvement must come partly from within and partly from without. The internal management of each college must modify itself for the better, and its efforts should be strengthened and encouraged by exterior influences. From the latter, however, we have most to hope. As long as our colleges are controlled by men who do not appreciate thoroughness in scientific culture, we can expect but little from within. An incompetent Faculty is not likely to become suddenly conscientious and resign, neither are average boards of trustees prone to confess their incapacity. External pressure must be brought to bear both upon trustees and upon professors before they can be made fully to realize the responsibilities resting upon them. This pressure may come, partly from public sentiment, and partly, though later, through legislation.

But how shall public sentiment be properly shaped and made available for service? How is its natural though slow growth to be fostered and directed? Mainly by the efforts, organized and individual, of scientific men. Personally, every worker in science should strive to awaken in the community about him a comprehension of the value and the purposes of his particular branch. In other words, the real investigators ought to do more toward popularizing their discoveries, instead of leaving the task to amateurs or charlatans. At present, unfortunately, too many able scientific men depreciate popular work and hold aloof from it. They do nothing themselves to interest the general public, and then lament the fact that the public does not become interested. Yet just here is where the beginning must be made. With a wider public interest in science will come a deeper public appreciation, and this will develop the tendencies necessary for the improvement of our colleges and schools. Until the people see and recognize the difference between true investigators and mere collectors of specimens, between original workers and text-book amateurs, little real progress can be made.

Organized effort is also needed. Just as lawyers or physicians band themselves together, so also men of science should combine for mutual self-protection against quackery. A man who had never been admitted to the bar could scarcely be chosen to a law professorship, neither could any one but a regular graduate be elected to teach in a respectable medical school. Why should not organization among chemists, geologists, or naturalists, produce in the long-run a similar state of affairs? Such an effective organization might be difficult to bring about, and still something could be done. Even a very little improvement would be better than no improvement at all. Local scientific societies might do good in two ways: 1. By preventing, or at least opposing bad appointments in colleges; 2. By furnishing the means for popular lectures and field-excursions. They could also, perhaps, do something toward breaking up the present vicious and absurd mode of teaching science by mere text-book recitations, and so help forward the adoption of correct methods. An attempt to teach drawing or music by lectures only, would be universally recognized as nonsensical; the same system of instruction applied to any one of the natural sciences is equally ridiculous. Nature must be studied at first hand to be properly understood.

Through legislation also something may be accomplished. This something may be very little, but a good many littles taken together aggregate much. Just as a single dollar may be the beginning of a great fortune, so one apparently trifling measure can become the starting-point of a sweeping reform. The first step to take in this direction is to prevent the issue of more charters. Inflation is as bad in education as it is in finance. No State which already contains more than one fair college or university should permit another to be established. Let the millionaires who wish to help learning give their money to institutions already in existence, or else not give at all. No benefaction is better than a mischievous benefaction. It is not long since Massachusetts lost a splendid opportunity to inaugurate the policy here recommended. The Methodist denomination of that State were discussing the foundation of a new educational institution in or near Boston. Harvard University at once made a very liberal offer; namely, that if the Methodists chose to establish merely a theological school, and to place the same in Cambridge, it would give them rent free the use of a lot of land for their building, and would permit their students to have access to the great library, and to attend, without expense, fifty courses of lectures. This magnificent offer was foolishly declined, and the Methodists founded, only four miles away, the Boston University—a school for which there was no real demand, and which signified merely sectarian folly. If at that time the Massachusetts Legislature had refused to grant a charter, a good move would have been made. The money bequeathed by Isaac Rich might perhaps have gone to the Wesleyan University at Middletown, making that comparatively weak institution really strong. As it was, the Methodist denomination, with more zeal than discretion, divided its forces in New England, started a college within half a dozen miles of at least three others, and contributed heavily toward the perpetuation of the present vicious policy. Tufts College is another wealthy institution close to Harvard, doing little save to adorn a high hill with brick and mortar, and wholly unable to compete with its great rival. All over the country there are to be found similar examples of what is at once multiplication of means and division of forces. Galesburg, Illinois, has two colleges: one Presbyterian, the other Universalist. Nashville rejoices in four: one Methodist Episcopal, another Methodist Episcopal South, a third for colored people, and the fourth vaguely described as "non-sectarian." This senseless scattering of appliances ought never to have been permitted. The true policy is, to establish great central universities, around which as nuclei the theological schools may cluster. A plan of consolidation among existing colleges would be difficult to carry out, but to some such plan we must eventually look for reform.

Perhaps at some future time it may also become possible to regulate colleges by law, and to compel them to maintain certain standards of scholarship. If a few institutions which are now doing sham work should be summarily deprived of their charters, and so rendered unable to confer degrees, much good would result. No Legislature, however, could as yet be induced to take such a step, even supposing it to be perfectly legal. A policy of this kind must follow after the awakening of public sentiment. But the principle that every institution of learning ought to be what it pretends to be, is unquestionable. No kind of fraud is more objectionable than fraud in education.

As a matter of course, legislation upon the college problem would have to be different in different States. Neither Rhode Island nor New Hampshire need act at all upon the question; but Ohio, Indiana, and Illinois, ought to move vigorously. In these and other Western States, especially the States which sustain universities at public expense, a healthy and judicious system of taxation might be desirable. If every college controlled by a private corporation was energetically taxed, the weaklings would soon be either suppressed entirely or forced to consolidate with other stronger institutions. Ohio alone has at least a dozen colleges which taxation would affect in this way. At present, they are public nuisances; united, they might become a source of public good.

19. A Scientific Administrator Asserts the Adequacy of Positivistic Science. 1896

In the last decade or two of the nineteenth century, scientists became more secure in their position in American society. Science had won a surprising degree of recognition in higher education, in the press, and even in industry. The fight had been a hard one, and many of those who supported science came to think of it as not only a way of life but a philosophy in and of itself. This faith in science was no crude materialism, but it did represent a belief that science could sooner or later solve all of man's problems. A necessary corollary was the assertion that any questions that science cannot answer are not worth asking. This essay provides a striking example of a professional who not only defended science but carried the offensive to the enemy camp. He even mentions by name as his symbolic adversaries the two philosophers most influential among educated Americans: Herbert Spencer, an Englishman who was no enemy of science directly but who believed that everything could be classified into a closed system; and G. W. F. Hegel, whose doctrine of reality has been parodied as "pure thought thinking about pure thought."

The author, JOHN WESLEY POWELL (1834–1902), Major Powell, was a veteran fighter who had lost his right arm in the Civil War. He was born in upper New York and before the war had studied in several colleges and, finally, because of his devotion to natural science, had been elected secretary of the Illinois Society of Natural History. After the war he taught briefly at Illinois Wesleyan College and Illinois Normal University. Later he devoted himself to exploring the West, mostly with the aid of small government appropriations. His most famous exploit was navigating the Colorado River in 1869. He was associated with the federal surveys in the West, and from 1880 to 1894 director of the United States Geological Survey. He was also head of the Bureau of American Ethnology. He wrote on other subjects, too, including social philosophy. A colorful figure, Powell's greatest contribution came out of his work as a scientific administrator.

See W. C. Darrah, *Powell of the Colorado* (Princeton: Princeton University Press, 1951); Wallace Stegner, *Beyond the Hundredth Meridian: John Wesley Powell and the Second Opening of the West* (Boston: Houghton Mifflin, 1954); and John W. Powell, *Truth and Error, Or, The Science of Intellection* (Chicago: The Open Court Publishing Company, 1898), of which this paper constituted chap. 1.

CERTITUDES AND ILLUSIONS; Chuar's Illusion*

J. W. Powell

In the fall of 1880 I was encamped on the Kaibab plateau at the edge of the forest above the canyon gorge of a little stream. White men and Indians composed the party with me. Our task was to make a trail down this side canyon into the depths of the Grand Canyon of the Colorado. While in camp after the day's work was done, both Indians and white men engaged in throwing stones across the little canyon, which was many hundreds of feet in depth. The distance from the brink of the wall on which we were camped to the brink of the opposite wall seemed not very great, yet no man could throw

* *Science,* n.s. 3 (1896), 263–271. Slightly abridged.

a stone across the chasm, though Chuar, the Indian chief, could strike the opposite wall very near its brink. The stones thrown by others fell into the depths of the canyon. I discussed these feats with Chuar and led him on to an explanation of gravity. Now Chuar believed that he could throw a stone much farther along the level of the plateau than over the canyon. His first illusion was thus one very common among mountain travelers—an underestimate of the distance of towering and massive rocks when the eye has no intervening objects to divide the space into parts as measures of the whole.

I did not venture to correct Chuar's judgment, but simply sought to discover his method of reasoning. As our conversation proceeded he explained to me that the stone could not go far over the canyon, for it was so deep that it would make the stone fall before reaching the opposite bank; and he explained to me with great care that the hollow or empty space pulled the stone down. He discoursed on this point at length, and illustrated it in many ways: "If you stand on the edge of the cliff you are likely to fall; the hollow pulls you down, so that you are compelled to brace yourself against the force and lean back. Any one can make such an experiment and see that the void pulls him down. If you climb a tree the higher you reach the harder the pull; if you are at the very top of a tall pine you must cling with your might lest the void below pull you off."

Thus my dusky philosopher interpreted a subjective fear of falling as an objective force; but more, he reified void and imputed to it the force of pull. I afterward found these ideas common among other wise men of the dusky race, and once held a similar conversation with an Indian of the Wintun on Mount Shasta, the sheen of whose snowclad summit seems almost to merge into the firmament. On these dizzy heights my Wintun friend expounded the same philosophy of gravity.

Now in the language of Chuar's people, a wise man is said to be a traveler, for such is the metaphor by which they express great wisdom, as they suppose that a man must learn by journeying much. So in the moonlight of the last evening's sojourn in the camp on the brink of the canyon, I told Chuar that he was a great traveler, and that I knew of two other great travelers among the white men of the East, one by the name of Hegel[1] and another by the name of Spencer, and that I should ever remember these three wise men, Chuar, Hegel and Spencer, who spoke like words of wisdom, for it passed through my mind that all three of these philosophers had reified void and founded a philosophy thereon.

In the history of philosophy an illusion is discovered concerning matter and each of the constituents or categories of matter, which are number, extension, motion, duration and judgment; and as bodies are related elements of

[1] See introduction. Exposition (or defense) of Hegel's and Spencer's ideas beyond Powell's treatment is not necessary for the purposes of seeing the place of this essay in American science.

matter, relation itself comes to be the object of illusion. Matter is the substrate of all bodies; bodies thus have a substrate, and the illusion of matter arises from supposing that matter, which is the substrate, has also its substrate, which is sometimes called essence. Classes are orders of number; the illusion of number relates to class or kind, and this is also usually called essence. Extensions combined have figure and structure, which produce form, and the illusion of extension is an illusion in relation to forms which are derived from extensions, and is called space. Motions through collisions are forces, and the illusion relating to motion is also called force. Duration is persistence and change, which give rise to time, and the illusion of duration is called time. Judgment is consciousness and inference, which give rise to comprehension of ideas, and the illusion of idea is called ghost. Bodies are related to one another, hence numbers, extensions, motions, durations and judgments are related. Certain of the relations of these things are called cause, and the illusion of relation also is called cause.

Now it must be clearly understood that the terms substrate, essence, space, force, time, ghost and cause refer sometimes to real things, as when properly used in science, and sometimes to illusions, when they are improperly used, as they often are in metaphysics; but usually the word ghost is now used only in reference to an illusion, and this is the sole case where we have a term for an illusion which is commonly understood in that sense, but the term spirit is used in both senses, for the certitude and for the illusion.

The seven illusions here enumerated are perhaps the most fundamental and far-reaching of the vast multitude of illusions which appear in the history of error. The words substrate, essence, space, force, time, ghost and cause are terms of universal use and their synonyms appear in all civilized languages, and perhaps in all lower languages. They have always stood for certitudes and illusions; here they require definitions both as certitudes and as illusions, in so far as we are able to define them.

SUBSTRATE

Substrate is matter, matter is the substrate of all bodies. Essence is any collocation of units into a unit of a higher order which makes it a kind or one of a class. Space is any extension or any collocation of extensions; force is any collocation of motions that are related by collisions; time is any duration or collocation of durations; mind or spirit or ghost is any cognition or collocation of cognitions; cause is any related antecedent or collocation of such antecedents of a change. Such are the fundamental meanings of the words when used to designate realities. We shall hereafter see what they mean when they are used to desginate illusions. Matter is the substrate of body and has no substrate for itself. All matter has four factors or constituents, number, extension, motion and duration, and some matter at least has a fifth factor, namely judgment. Matter is not a substrate for these factors, but exists in these constituents which are never dissociated, but constitute matter, or are the moments of matter; and this matter is the substrate of all bodies.

ESSENCE

The term essence as used in philosophy is employed in a double manner and is thus often ambiguous. It is sometimes used as a synonym for substrate of matter, at other times it is used to designate the occult substrate of class. In this latter sense it is here used. Essence, then, is the number essential to make an order or kind of a class. As the whole number is essential, every one is essential; they are severally and conjointly essential, so that it is possible correctly to speak of them all as being essential and to speak of every one severally as being essential. All of the particles which make up a body are conjointly and severally essential to that body, and the essence of a body is the hierarchy of particles of which it is composed. The term essence, therefore, is a general term or pronoun for all collocations of number, and its special meaning is derived from the context. As an illusion, essence is the name of an unknown something which produces a kind or class, and is a property of an unknown or unknowable substrate of matter.

If, as the chemist believes, with much good reason, the ultimate chemical particles are alike, they are alike only in number, extension, motion and duration; they are unlike in association, position, direction or motion and the duration of association, so that likeness and unlikeness is inherent in matter itself. In bodies innumerable combinations of number, extension, motion and duration are found, and out of these are developed innumerable likenesses and unlikenesses, so that one body is like another in many respects and unlike that other in many other respects. The science of classification takes these likenesses and unlikenesses and discovers degrees among them which are of profound importance in the study of the world, and upon which a large share of knowledge rests. All knowledge does not rest upon likeness and unlikeness; but likeness is founded upon number, and men have discovered that what is true of a body is true of any other body of like kind, under the axiom that whatever is true of anything is true of its identity in so far only as it is a constant property or an absolute, and not in so far as it is a variable or relative. These are all simple, self-evident propositions, but in the compounding and recompounding of matter it is not always possible to disentangle the constants from the variables. Men lost in the meaning of words, forever wandering in linguistic jungles, have engaged in discussions about essences and have at last reified the word as something which is not number associated with extension, motion and duration, but as some occult existence unknown and unknowable, which gives to bodies their likeness or unlikeness. Having reached the conclusion that matter is something more than its constituents, with an occult, unknown and unknowable substrate, they take the next step that the essence of class or likeness and unlikeness exists not in the fundamental properties of body or the fundamental constituents of matter, but in their substrate.

All known things are classified either properly or improperly. The characters upon which they are classed are thus innumerable. These characters which constitute class are all the bodies embraced in the class and all the properties embraced in all the bodies of the class. The term essence, then, used in this sense, means all of these things. Therefore it is a general name for everything in the universe, but obtaining its particular meaning in any case by the context. What is the meaning of the word *this*? It may be applied to any constituent of matter, to matter itself, to any body or to any property, relation or quality in the material world, and to any idea in the mental world, and its meaning is derived from the context; it has no definite meaning in itself. *Essence*, as a word used by philosophers, is a pronoun of like character without specific meaning, and attains its specific meaning only by the context; it has one meaning at one time, and another at another, and thus it seems to be illusive. As the substrate of matter, a reified nothing, is entertained in the minds of some as an entity, so some thinkers make essence a property of this substrate—a nonenity of a nonenity [*sic*]. Chuar, Hegel and Spencer reason in this manner. Essence as connoting the essential characters of a class is a word the meaning of which scientific men clearly understand; it is never ambiguous, although naturalists may sometimes disagree about the essentiality of a particular character, but the essence of which the philosopher thinks is nonexistent, the opinions of the three wise men to the contrary notwithstanding.

SPACE

The word space is the pronoun of all extensions, figures and structures of extensions in the multitudinous bodies of the world. There are many extensions, and every known body is a constituent of some other body, and this synthesis may be continued until the mind is lost in immensity. The space occupied by a body is its extension in structure and figure. This desk before me has extension, or we say that it occupies space; the space which it occupies is its extension, from which it excludes other bodies. Remove the furniture from the room, it is said to be empty, yet it is full of air; remove the air from the room, yet it is full of ether;[2] remove the ether; may be, we know not, all is removed; then the wall encloses void—nothing—but the walls of the room yet have extension, and we can measure this by measuring the walls, but void cannot be measured; there is nothing to be measured. Thus it is that space is the pronoun of all dimensions of all bodies, severally and conjointly, and as they are variable, space seems to be illusive, and it comes

[2] The ether was conceived to be an extremely fine fluid or substance that was supposed to fill all space, even a vacuum, and account for the propagation properties of light and in a less important way field phenomena such as magnetism and gravity.

at last in the minds of careless thinkers to mean something more than extension, an unknown and unknowable thing that like essence, belongs to the unknown and unknowable substrate of matter. The word is useful when its use is understood as a pronoun or general word whose meaning is given by the context.

FORCE

Force is the pronoun for combinations of motion. It thus may be applied to numerous things now existing, or which have existed in the past or may exist in the future. It is the general word for all collisions and all combinations of collisions; collisions of particles of ether in light and heat, collisions of particles of air in sound, collisions of particles of water in stress, collisions of particles of matter in all solids exhibited in the structure and strength of those materials. It thus stands for the action of two or more bodies as they come in collision, and thus influence each other's motions. It is not an occult, unknown or unknowable something which belongs to an occult, unknown and unknowable substrate. The term has no particular or determined meaning in itself, but derives its meaning from the context. It is a word of universal use, whose meaning must be determined by its application; it is the general term or pronoun to denote any or all actions and reactions.

TIME

Time is the pronoun of all durations. It means any duration to which the term is applied, all durations or any collocation of durations the mind may entertain. When reified it comes to be thought of as applying to an existence independent of the things which have duration. Then time, like essence, space and force, becomes a property of the substrate of matter, an illusion about an illusion.

GHOST

Spirit is the general term or pronoun for all judgments in the infinite variety of sensations, perceptions, understandings, acceptions and reflections. It is a name for all ideation. It is known to us only in its association or connection with the universal constituents of matter, which are number, extension, motion and duration. There is no spirit which is not a unity of many and one. There is no spirit which has not force. There is no spirit which has not duration; in so far all are agreed; and it is here affirmed that there is no spirit which has not extension, for without extension all the other constituents would vanish, become nothing, absolutely unimaginable or unthinkable. When spirit is considered to be something which is not number or many in one, which has not extension with figure and structure without force, or the power of action and reaction and without duration as persistence or persistence and change, that it, without time, it becomes a nonentity, a nothing, and it is then an illusion and is usually called ghost.

CAUSE

We use the word cause as we use the words *this* and *that,* as a general term or pronoun for anything that stands in relation to any other thing in the production of a change. The multitudinous bodies and particles of the universe coöperate with one another in the production of changes. The condition before a particular change is considered in respect to the condition after the change, and the condition which coöperated in the production of the change, is called a cause, and the condition after the change is called an effect. It is thus that the term cause may be applied to any body, to any property, or to any relation; it is a term for any of these things, any collocation of these things or any part of these things, and just what its meaning may be can be discovered only by the context in which the word is used. In the multitude of bodies, properties and relations which coöperate in the production of the change whose result is called an effect, we may stop to consider any one and call that the cause. Failing to appreciate the variable significance of the word, men are led into the illusion that there is some entity, some separate existence called cause.

Metaphorically, essence is sometimes used for space, sometimes for force, sometimes for time, sometimes for spirit, and sometimes for cause, and interchangeably all of these terms may be used as metaphors for one another.

Thus it is that we have a family of chimeras in substrate, essence, space, force, time, ghost and cause that are not bodies or the properties of bodies, but things nonexistent—mysteries that are at the foundation of all philosophies of the unknowable and all philosophies of the contradictory, and the ground of all antinomies. They constitute the substrate, the essence, the space, the time, the cause of the philosophies of the three wise men, Chuar, Hegel and Spencer.

We shall hereafter see more clearly how these illusions have been developed and how other illusions have gathered about them. Here we simply call attention to the fundamental illusions to indicate somewhat the purposes of this argument.

It is within the experience of every human being, and has been through all generations, that man is forever discovering number, extension, motion, duration and judgment. He learns something of number in infancy and adds to his knowledge daily and extends his knowledge to an indefinite multiplicity. He adds to his knowledge the extension of one body and another still embodied in a higher order; and thus his knowledge of extension increases to an indefinite extent. He is forever discovering new motions and new combinations of motions as forces and finds that he is able thus to add more and more of like motions and forces to his knowledge. Ever he is discovering durations—the durations of coexistent things and the durations of past things, extending to high antiquity, and he prophesies durations to come, and many do come, until his mind is led into the illimitable future. Mind is then trained by constant experience to expect a further enlargement of knowledge and to consider the possibilities into which it may expand, until it dwells upon endless number, endless extension, endless force, endless duration. Man contemplates multiples and submultiples of the things of which he already has knowledge, and then invents implements of research by which submultiples are discovered, and other implements by which multiples in higher orders are discovered. Finding that he has explored but a small part of the universe, and that within the universe wherever bodies are to be met they have been resolved into numbers, extensions, motions and durations, he grasps the idea of infinity not as something other than that of which he knows, but as more of that which he best knows. The experience of men through countless generations has organized the concepts of number, extension, motion and duration as the universal factors of matter, and never has any mind discovered any other things saving only those which are included in the terms of mind. Of matter without mind, man has absolutely no vestige of knowledge which is not included under the terms number, extension, motion and duration. These terms absorb them all. Therefore matter is number, extension, motion and duration, and at least some matter has judgment.

The mind discovers another factor or category in the universe—judgment, which develops into cognition of the constituents of matter, of their relations, and also a cognition of cognitions and the relations of cognitions. It is thus that the universe is resolved into material elements and judgments, the five things best known, and science in dealing with the universe explains them by resolving them into these best known things. Science does not lead to mystery, but to knowledge, and the mind rests satisfied with the knowledge thus gained when the analysis is complete—when any newly discovered body is resolved into its constituents or any new idea into its judgments.

Concepts of number, extension, motion, duration and judgment are developed by all minds; from that of the lowest animal to that of the highest human genius. Through the evolution of animal life, these concepts have been growing as they have been inherited down the stream of time in the flood of generations. It is thus that an experience has been developed, combined with the experience of all the generations of life for all the time of life, so that it is impossible to expunge from human mind these five concepts. They can never be cancelled while sanity remains. Things having something more than number, extension, motion, duration and judgment cannot even be invented; it is not possible for the human mind to conceive anything else, but semblances of such ideas may be produced by mummification of language.

Ideas are expressed in words which are symbols, and the word may be divested of all meaning in terms of number, extension, motion, duration and judgment and still remain, and it may be claimed that it still means something unknown and unknowable; this is the origin of reification. There are many things unknown at one stage of experience which are known at another, so man comes to believe in the unknown by constant daily experience; but has by further converse with the universe known things previously unknown, and they invariably become known in terms of number, extension, motion, duration and judgment, and are found to be only combinations of these things. It is thus that something unknown may be imagined, but something unknowable cannot be imagined.

No man imagines reified substrate, reified essence, reified space, reified force, reified time, reified ghost, or reified cause. Words are blank checks on the bank of thought, to be filled with meaning by the past and future earnings of the intellect. But these words are coin signs of the unknowable and no one can acquire the currency for which they call.

Things little known are named and man speculates about these little known things and erroneously imputes properties or attributes to them until he comes to think of their possessing such unknown and mistaken attributes. At last he discovers the facts; then all that he discovers is expressed in the terms of number, extension, motion, duration and cognition. Still the word for the little known thing may remain to express something unknown and mystical, and by simple and easily understood processes he reifies what is not, and reasons in terms which have no meaning as used by him. Terms thus used without meaning are terms of reification.

Such terms and such methods of reasoning become very dear to those immersed in thaumaturgy[3] and who love the wonderful and cling to the mysterious, and, in the revelry developed by the hashish of mystery, the pure water of truth is insipid. The dream of intellectual intoxication seems more real and more worthy of the human mind than the simple truths discovered by science. There is a fascination in mystery and there has ever been a school of intellects delighting to revel therein, and yet, in the grand aggregate, there is a spirit of sanity extant among mankind which loves the true and simple.

Often the eloquence of the dreamer has even subverted the sanity of science, and clear-headed, simple-minded scientific men have been willing to affirm that science deals with trivialities, and that only metaphysics deals with the profound and significant things of the universe. In a late great text-book on physics, which is a science of simple certitudes, it is affirmed:

To us the question, *What is matter*?—What is, assuming it to have a real existence outside ourselves, the essential basis of the phenomena with which we may as physicists make ourselves acquainted?—appears absolutely insoluble. Even if we become perfectly and certainly acquainted with the intimate structure of what we call Matter, we would but have made a further step in the study of its properties; and as physicists we are forced to say that while somewhat has been learned as to the properties of Matter, its essential nature is quite unknown to us.

[3] The performance of miracles or magic.

As though its properties did not constitute its essential nature.

So, under the spell of metaphysics, the physicist turns from his spectroscope to exclaim that all his researches may be dealing with phantasms.

Science deals with realities. These are bodies with their properties. All the facts embraced in this vast field of research are expressed in terms of number, extension, motion, duration and judgment; no other terms are needed and no other terms are coined, but by a process well known in philology as a disease of language, sometimes these terms lapse into meanings which connote illusions. The human intellect is of such a nature that it has notions or ideas which may be certitudes or illusions. All the processes of reasoning, including sensation and perception, proceed by inference; the inference may be correct or erroneous, and certitudes are reached by verifying opinions. This is the sole and only process of gaining certitudes. The certitudes are truths which properly represent noumena, the illusions are errors which misrepresent noumena. All knowledge is the knowledge of noumena, and all illusion is erroneous opinion about noumena. The human mind knows nothing but realities and deals with nothing but realities, but in this dealing with the realities—the noumena of the universe—it reaches some conclusions that are correct and others that are incorrect. The correct conclusions are certitudes about realities; the incorrect conclusions are illusions about realities. Science is the name which mankind has agreed to call this knowledge of realities, and error is the name which mankind has agreed to give to all illusions. Thus it is that certitudes are directly founded upon realities; and illusions as they are always about realities, are thus indirectly, though incorrectly, founded upon realities, but certitudes and illusions alike all refer to realities. In this sense then it may be stated that all error as well as knowledge testifies to reality, and that all our knowledge is certitude based upon reality, and that illusions would not be possible were there not realities about which inferences are made.

Known realities are those about which mankind has knowledge; unknown things are those things about which man has not yet attained knowledge. Scientific research is the endeavor to increase knowledge, and its methods are observation, experience and verification. Illusions are erroneous inferences in relation to known things. All certitudes are described in terms of number, extension, motion, duration and judgment; nothing else has yet been discovered and nothing else can be discovered with the faculties with which man is possessed.

In the material world we have no knowledge of something which is not a unity of itself or a unity of a plurality; of something which is not an extension of figure or an extension of figure and structure; of something which has not motion or a combination of motions as force; of something which has not duration as persistence or duration with persistence and change.

In the mental world we have no knowledge of something which is not a judgment of consciousness and inference; of a judgment which is not a judgment of a body with number, extension, motion and duration. Every notion of something in the material world devoid of one or more of the constituents of matter is an illusion; every notion of something in the spiritual world devoid of the factors of matter and judgment is an illusion. . . .

In the intoxication of illusion facts seem cold and colorless, and the wrapt [*sic*] dreamer imagines that he dwells in a realm above science—in a world which as he thinks absorbs truth as the ocean the shower, and transforms it into a flood of philosophy. Feverish dreams are supposed to be glimpses of the unknown and unknowable, and the highest and dearest aspiration is to be absorbed in this sea of speculation. Nothing is worthy of contemplation but the mysterious. Yet the simple and the true remain. The history of science is the history of the discovery of the simple and the true; in its progress illusions are dispelled and certitudes remain.

Part 3

THE EARLY TWENTIETH CENTURY: PRESTIGE AND MATURITY

During the first decades of the twentieth century, scientists in the United States built upon the substantial base that they had established during the previous eighty years. They had a well defined profession, well developed institutions, and a substantial amount of funding. Above all, they had aspirations. The most marked of these aspirations by 1900 was somehow to cause "pure" science to flourish. "A wonderful spirit of research has literally *seized* us," noted physicist Louis Bauer in 1909. But the scientists' frantic quest for funds to keep their laboratories going and expanding led them to talk publicly not only about "pure" science but about the utilitarian justification for science, even "pure" science. This public stance involved the scientists in other historical developments of the time.

In the first third of the century the history of American science was marked by two general trends. First, it prospered and grew. Where there had been 4,000 names listed in *American Men of Science* in 1906, in 1933 there were 22,000 (the population had increased from 85 to 126 million). The rise of the universities was especially crucial to this growth and prosperity. The graduate school continued to increase its prestige, but such an institution could flourish only because between 1900 and 1934 the income of colleges and universities had increased by a factor of more than ten. By the middle 1930s despite a great depression American universities were spending at least 50 million dollars a year on research, a very substantial fraction of which was, of course, scientific research. Private industry was spending twice that amount on research, and the government even more than that.

The well-being of science was not just a matter of numbers and wealth. American researchers began to have some reason to feel that their achievements were commensurate with the place of the United States in world politics and economics. For the first time, there were glimmerings of the idea that American science was not still purely colonial and derivative from Europe. Between 1907 and the mid-1930s, for example, thirteen Americans won Nobel prizes for scientific accomplishments.

Beyond prosperity and growth, the second general historical trend in American science had to do with the relationship between it and American society. In the nineteenth century, despite the obeisance made to science and technology, in America social forces tended to shape, or at least limit, the development of science. In the early years of the twentieth century, the relationship became more reciprocal, and in important ways: Despite talk of "pure" research, science had a substantial influence upon American social policies. As early as 1898 W J McGee (called by Theodore Roosevelt "the scientific brains" of the conservation movement) realized what had happened: "America has become a nation of science. There is no industry, from agriculture to architecture, that is not shaped by research and its results; there is not one of our fifteen millions of families that does not enjoy the benefits of scientific advancement; there is no law on our statutes, no motive in our conduct, that has not been made juster by the straightforward and unselfish habit of thought fostered by scientific methods."

The first two decades of the century have been characterized as the Progressive era. In government, in industry, in social enterprises of every sort it became a commonplace to appeal to experts and expertise to solve problems. The accepted model of an expert was of course the scientist in his capacity as specialist. (The appeal of science was very great; membership in the American Association for the Advancement of Science, for example, quadrupled between 1900 and 1910 alone.)

As a tool of social change, science had two aspects: as accumulated and systematized factual material suitable to be used as a basis for decision, whether regarding the construction of a bridge or a railroad tariff schedule, and as a method to be applied to difficult unsolved problems, either technical or social. Leaders in the Progessive period tended to believe that education, by bringing information to bear on both personal situation and social issues, could change the course of events. Institutions of higher learning were often characterized as "the university in the service of society." Significantly, Progressive leaders believed that the successes of science were based upon the scientific method which, again, could be applied to all human problems. As political writer William E. Walling put it in 1914: "Now, every day, science is becoming more consciously pragmatic, more consciously concerned with the service of man. . . . Science is being rapidly endowed, absorbed, and directed by government and is being applied more and more exclusively to work of a practical nature and of the highest value, though it gives no immediate profits. And this is the science which now has the unqualified support and respect of the most able and advanced of the scientists of the time."

People of the Progressive period had a strong sense of their own power—power to control both nature and man. This sense of power derived directly from the achievements of science, applied science, and technology. Technology, especially, fired Americans' imaginations with what might lie ahead. The steam engines and railways, despite their enormous impact on thought in the nineteenth century, did nothing to inspire visions of the future compared with what was done by the everyday harnessing of electricity and the use of the internal combustion engine, along with the conquest of the air by the Wright brothers. The social effects of technology were real enough; and while the psychological impact can only be inferred, it was clearly equally great. Where once the hero of boys' fiction had been a youthful business entrepreneur in the Horatio Alger tradition, by the 1910s the ideal for American youngsters became Tom Swift, scientist and inventor.

Motivated by the prospects of what science could do, Americans now applied it to problems on a scale unthinkable a generation before. Those who knew the harsher side of life repeatedly saw children rescued from certain death by the miracle of diphtheria antitoxin. The public rightly celebrated the conquest of yellow fever and hookworm. Such marvels as these reprieves from sickness and death inspired social action. Not only was the construction of water and sewage systems accelerated but a pure food and drug law was enacted. Mass changes individually vended, too, came from the laboratories, now even more than from the inventor's bench—new materials like rayon, gadgets like the high vacuum radio tube, and vast increments to productivity such as the new gasoline cracking process.

It was just such concrete evidences of the potency of science that increased the authority—both within the scientific community and without—of anyone who spoke in the name of science. Many Americans were prepared to listen with credulity to the great physiologist, Jacques Loeb, who believed that he had found the mechanism of the life process itself and proposed to draw conclusions from the discovery. Adherents of the eugenics movement, which was based largely in a sizable fraction of the national scientific community, believed that knowledge about genetics would permit a change in the physical inheritance of man himself, thus carrying reform to an extreme that not even the most sanguine educators had envisaged.

The cataclysm of World War I destroyed some of the hopefulness of the Progessive period. The Progressives had endeavored to change not only the material but the cultural and moral environment of Americans and, by this means, to bring into existence a new society with changed men in it. Now, after World War I, Americans tended to turn away from great social programs. At the same time, the spokesmen for the public redoubled their emphases upon scientific and technological progress. In the age of the engineer, as the 1920s were called, it was widely believed that correct organization of industry and society and intelligent use of scientific and technical knowledge would bring about a utopian material abundance.

It would be a grave error to think that this dedication of scientists and public alike to increasing productivity was mere materialism. Rather it was a particularly open expression of a traditional American belief: that material plenty solves problems and brings utopia. People with leisure are also, so it was believed, more cultured. Ending poverty brings social stability and reduces the motivation for evil. This doctrine of plenty lay behind much of the Progressive impulse. The actual prosperity of large parts of the population in the 1920s, when income could be used for technological products, modern medical care, and the pursuit of the good life, gave tangible evidence that the idealism of the time was not unrealistic.

Nor was the emphasis upon technology viewed as inhibiting the advance of pure science. In the first third of the twentieth century, scientists and nonscientists alike continued after the Progressive years to justify all kinds of scientific activity in terms of its ultimate utility in the form of applied science. Even great industrial institutions, such as the American Telephone and Telegraph Company and the General Electric Company, subsidized a certain amount of pure science research in the laboratories that sprang up to confirm the dependence of the industrial age upon science. Indeed, scientists employed by both of these companies won Nobel prizes in the 1930s. (By 1938 there were 1769 industrial research laboratories employing over 44,000 people.

While the Great War of 1914–1918 destroyed the social idealism that marked the Progressive years, the conflict nevertheless gave renewed impetus to the belief that science held the key to man's almost celestial powers. The application of technology to warfare, in the form of chemicals particularly, convinced the American people as never before of the potential of science to work material miracles and fulfill man's dreams. The scientists, as their spokesman, Robert A. Millikan, shows, were even more than the public stimulated by the work of the war.

It is ironic that just as scientists were sampling power in the form of major social impact, that just as research was suggesting how the possibilities of scientific effort had never really been explored—the picture of nature and the universe that scientists were building suddenly became less certain and clear than had theretofore generally been thought. The first three decades of the twentieth century should have been marked by a culmination of the positivism of the nineteenth century. But by a perverse turn of fate, the more the scientists actually learned, the less they believed they knew. By the 1930s, the new vision of science, to use the phrase of P. W. Bridgman, was that truth was neither simple nor certain. Science had already come a long way from the confident days of John W. Powell.

The various fields of scientific endeavor illustrate how easy and definite answers seemed always to escape the laboratory hunters. While many discoveries and researches in every field ornamented the efforts of scientists both in the United States and in the world in general, developments in two areas, biology and physics, had a general significance that touched the nature of science itself.

Life scientists in the early twentieth century found that Darwin had left a legacy that was growing increasingly unmanageable. Such paradoxes as nonadaptive evolutionary changes led to the study of ecology, the life environment of the organism. As early as the 1890s field biology was being overshadowed by experimental work in laboratories. Genetics, the study of the mechanisms of inheritance, burst into existence suddenly at the turn of the century, but as the years passed took on a complexity beyond the early conceptions of even men as well informed as psychologist E. L. Thorndike. The search for what exactly is a trait, and how do traits and genes coincide, led into increasingly intricate research, most notably in the laboratory of Columbia University scientist Thomas Hunt Morgan. In physiology in general, both the functioning of the organism and its cellular constituents came to seem almost infinitely involuted, the mechanisms of physiological control more numerous and subtle than had theretofore been dreamed. Each gland and vitamin and enzyme made knowledge about the nature of life processes appear more and more elusive. The great discovery that bacteria cause disease was complicated not only by physiological knowledge but by the discovery of viruses and the difficulty, in turn, of fitting the latter into bodily processes, on the one hand, and, on the other hand, into evolutionary theory and ideas about the nature of life.

Physics in the twentieth century became the spectacular science. First radioactivity and then further theoretical work began to raise basic questions about the nature of matter and the interrelations between matter and energy. The atom, it appeared, was not the ultimate constituent. The widely publicized relativity theories of Albert Einstein (who moved to the United States in 1933) helped to convey to the public the idea that the basic relationships within the Newtonian world machine were actually not as certain and inflexible as they had once appeared. As a rhymester caught the spirit in 1921:

Relativity

Twinkle, twinkle, little star,
How I wonder where you are;
High above I see you shine,
But, according to Einstein,
You are not where you pretend,
You are just around the bend;
And your sweet seductive ray
Has been leading men astray
All these years—O little star,
Don't you know how bad you are?

The great changes in physics, however, came about not because of relativity but because of quantum theory and, ultimately, quantum mechanics.

The retreat from certainty and simplicity in science grew out of three main trends. First was the increasing use of purely theoretical constructs and mathematical models for natural phenomena regardless of whether they were, except in certain particulars, closely and demonstrably tied to actuality. Much could be learned, it appeared, if one simply made convenient—and changeable—assumptions about the nature of the universe. Second, the quantification of science took on new forms with the introduction of statistical considerations into both physics (quantum theory) and biology (genetics) in 1900. A statistical treatment permits one to be concerned only with the gross, and exceptions do not necessarily disprove a rule. In many areas, perhaps all areas, scientists came to hope for no more certainty than a particular probability, a chance of specified dimensions. As E. L. Rice, the biologist, pointed out, maintaining such science required men with a measure of faith so that they could tolerate the uncertainties. Finally, uncertainty itself appeared to be inherent in modern physical—and perhaps all scientific—research. Retreat from confidence and certainty, and on such an array of fronts, inevitably invited attacks on science as such, attacks that had a certain popular appeal in the 1930s. Not the least of the enemies of science as a way of life, however, were first-rate thinkers, such as John W. Buckham, who were aware of moral and social questions as well as the latest developments in science.

For all of its apparent vulnerability, science in America nevertheless flourished. By the first decades of the twentieth century, the typically American pattern of scientific research support had become clear. The primary agency for research was to be the university. Attempts to set up European-type institutes (such as the Rockefeller Institute for Medical Research) were not widely emulated. During World War I the government attempted to mobilize scientists by calling them into the armed forces and setting up special agencies, particularly the National Research Council, in which researchers might work cooperatively on scientific problems. But despite continuation of the National Research Council after the war, and despite the conviction of many men who had been to war, that specially funded cooperative research would pay off as individual efforts could not, the individualistic university continued to be the main channel through which money came for scientific research and the institution in which it made its home.

As funds available for science increased, an important new resource of support showed up in the early twentieth century: the foundation, which supplemented university and government efforts. Foundation grants were extremely welcome. Despite the large amount of federal money actually used for research of one kind or another, the government, except during the war, held back from giving any significant large-scale support to scientific research as such. The only partial exception was in the field of agriculture, where large sums supported investigations at land grant colleges and the experiment stations closely allied with them, especially after passage of the Adams Act in 1906. Even minor sums that could be utilized by the scientific community, such as those administered through the National Advisory Committee for Aeronautics and a multitude of other agencies, helped give scientists some alternative to complete dependence upon university money. The foundations provided still another alternative.

Aside from the Rumford Fund of the American Academy of Arts and Sciences, money from the Smithsonian Institution, and National Academy of Sciences endowments (totaling almost a hundred thousand dollars by 1895), foundations had been insignificant in funding scientific research in the nineteenth century. Occasionally an individual wealthy person might provide support for a scientific project (like the Lick Observatory), but the endowed or funded foundation was a phenomenon of the twentieth century. The first two most important such sources of scientific research funds were the Carnegie Institution of Washington (1902), which soon had the incredibly large endowment of 22 million dollars, and the Eugenics Record Office, which used money supplied by Mrs. E. H. Harriman to support many areas of biology. By the 1920s Rockefeller money was being distributed by several agencies, and the age of the foundation had begun. Between 1920 and 1934 the number of known research funds doubled.

The fact that certain types of funding were available on a large scale clearly affected the configuration of American scientific research efforts. But a complementary factor, and one equally important in searching for the causes of American excellence in particular fields, has now to be added: Among those fields were there areas in which the Johnny-Come-Lately of world science, America, might shine? U. S. accomplishments in big-instrument telescopy were directly related to the fact that philanthropists such as Charles T. Yerkes and finally the Carnegie Institution were willing to build the telescopes. But mere money cannot explain the preeminence of Americans in genetics (although combinations of health and agriculture funds were vital here), some of the newer fields of physics, and physiology, in which Americans by the twentieth century provided a large part of the world literature and did outstanding work in such fields as those of the quickly popularized vitamins and hormones. Americans, characteristically, did best in untraditional areas of research, and in the early twentieth century science had so changed that many such openings existed.

The coming of the Great Depression of the 1930s raised many doubts about the future of science in the United States. Tied to the vision of prosperity of the 1920s, science was damned in the 1930s as the originator of the productivity that had put so many men out of work. Disillusionment with materialism, the promise of plenty, led also to disillusionment with science. Critics who had questioned the beneficence of science in the service of war could now say that society was out of balance and unable to control its science and technology. (There was a serious proposal in the early 1930s to enforce a holiday on scientific research until social progress had proceeded to the point where the nation could benefit from more science.) Such critics claimed that the promises of the truths to be revealed if only science were given support had led only to more uncertainties. It was out of the ashes of this betrayal of promise and the concurrent repudiation of innovation and materialism that the promise and prestige of science had to rise again.

20. A Physicist Describes Science and His Faith in Its Goodness. 1909

This was a prepared address, delivered at the dedication of the University of Illinois Laboratory of Physics in 1909. To a modern reader the sequence of topics may appear accidental, as if the writer spoke about subjects that just happened to come to mind. Yet in this series of ideas one can see a view of science typical of the opening years of the twentieth century—typical in its grounds for justifying scientific research, in its faith in continuing progress, in its cultural nationalism, and in its sense of the mission of science to make the world a better place in which to live. As the author remarks, in 1909 no area of life was without actual, or at least potential, capability of feeling the impact of science.

ARTHUR GORDON WEBSTER (1863–1923) was a most eminent mathematical physicist. He was born in Massachusetts of an old New England family. After graduating from Harvard he went on to Berlin to take his Ph.D. He spent the rest of his life at Clark University, where he soon became head of the department of physics. He helped found the American Physical Society and was elected its third president in 1903.

Concerning more information on American physics and Webster's life, see Daniel J. Kevles, "The Study of Physics in America, 1865–1916" (doctoral dissertation, Princeton University, 1964), and *Dictionary of American Biography,* XIX, 584–585.

SCIENTIFIC FAITH AND WORKS*

Arthur Gordon Webster

. . . It is not my intention here to consider the history of science, and its development from the small beginnings of the cinquecento through its glorious burst in the eighteenth century to full fruition in the nineteenth. Let us briefly recapitulate some of the changes which the works of science have made in the face of the earth, and of mankind inhabiting it. First and most important is the production of power, by which man's energies are inconceivably multiplied. The discovery of coal at just the right time to be utilized in the invention of the steam engine enabled man to command hitherto undreamed of forces, making the constructions and manufactures of the ancients seem like child's play. The raising of cotton, made practical by the invention of the cotton gin, largely transformed the clothing of the

* *The Popular Science Monthly,* 76 (1910), 108–115, 117–123. Slightly abridged.

world, while the development of the iron and steel industry revolutionized methods of construction. With the command of power in centralized units came the development of the industrial system, and the tendency to crowd together into cities, leading to so many scientific problems yet unsolved. With the tremendous increase in the wants of humanity brought about by the increased power to supply them, the supply of natural energy in the form of coal, which at first seemed inexhaustible, seemed menaced, and other natural resources had to be developed, and more efficient methods of application found. Thus in our day the development of the internal combustion or gas engine, which threatens to crowd the steam engine to the wall, has finally permitted the application of petroleum, which by the aid of chemistry has furnished not only great stores of energy, but numerous useful products. Not the least important aspect of the power development is that part which is applied to transportation. The covering of the whole known world with lines of railway has made possible . . . easy movements from place to place not only of peoples, but of products, so that while a few centuries ago a large proportion of the population never moved more than a few miles from their birthplaces, being as good as fettered to the soil, now even the poorest may be easily displaced from country to country, the seas being no more of a barrier than the land. The increase of education by travel, and the tendency toward peace produced by the increased acquaintance of nations with each other, is not to be overestimated. Perhaps no more impressive example of man's power over nature is to be found than the sight of a great ocean steamship, lying at her dock and towering over the surrounding buildings, or ploughing her way at express speed over the stormy waves, whose power she hardly seems to feel. A notion of the huge demands made by ocean transportation on our resources of energy is obtained when we think that one of these marine monsters is using sixty or eighty thousand horsepower, while an express train uses from a thousand to fifteen hundred only. In view of this depletion of our coal supplies the question of water power has become urgent, and science has succeeded in bridling our rivers and waterfalls for further supplies, while the transmission of this power by electricity has made manufacturing possible where it was not before, and is now being applied to transportation on a large scale. Not to be neglected in connection with the application of power is the question of illumination. When we think of the dark and dismal nights in the cities, not only of antiquity, but even of two centuries ago, making it impossible to go out in safety at night, and encouraging all sorts of crimes of violence, we must consider the successive application of gas, oil and electricity to have had no mean influence on the habits of mankind. The use of modern illuminates, especially electrical, has made possible the performance of more work, under more healthful conditions, and has completely changed the habits of man as regards the hours of darkness. Whether this has been entirely for his advantage we may leave until later.

Almost equally important with transportation is communication, which has in like manner changed the possibilities and habits of mankind. At the time of our revolution it took weeks to get any news to or from Europe, while even as late as the civil war our news was two weeks old when it reached England. What a contrast to the present, when the news of the fall of a cabinet or the overthrow of a sultan last night in any part of the world is put before us at breakfast this morning, and that not only in the centers of population, but in remote country districts. Nations can not now ignore each other's feelings and desires, while those misapprehensions which lead to war are made many times less frequent. The use of the ocean cable and of the telephone has largely transformed methods of doing business. Time is money, and although the increased facility of locomotion has led hosts of business men to circulate from one end of the country to the other, this can now in large measure be saved by the use of the telephone.

More important for the existence of man even than transportation and communication is food. The applications of science have made not one, but thousands of blades of grass grow where one grew before. Chemistry has shown how to fertilize the exhausted soil, engineering has furnished water where none was, and caused the desert to blossom as the rose. Its latest feat, in the anxiety due to the exhaustion of the nitrate beds, has been the fixation of the nitrogen of the air, which in Norway combines the harnessing of the waters with the compulsion, in the electric arc, of the nitrogen to unite with the oxygen, thus yielding unlimited nitrates for the restoration of our exhausted food supplies. Here also transportation comes in, so that the famines which formerly vexed large portions of the earth have now lost their terrors. When we think of the misery of the English agricultural classes before the abolition of the corn laws we may well praise the development of transportation which has enabled her to eat out of our full hand. At the same time the application of thermodynamics to freezing machinery has enabled us to send our meat across the ocean to become the roast beef of old England. The effects of all this upon the farmer can not be passed by. Commanding the markets of the world, ploughing his fields by steam or electricity, grinding his grain by gasoline, feeding his stock from silos, milking his cows by vacuum, cooling his cream by cold-producing machinery, separating it in a centrifugal creamer, making his cheese by the aid of chemistry so that he duplicates the product of any locality in the world, in easy reach of the city by automobile or trolley-car, and in communication with all his neighbors by telephone, he is no longer an object of derision, a hayseed, but an example of the works of science, demanding an equal part of influence in the government of the country, and gladly contributing of his rich store to the endowment of institutions like this for the education of his youth and the further advancement of science.

Again let us consider what science has done for the amelioration of health. When we consider the crowding, the filth, the misery of the greater part of the populace in the cities of antiquity, of the middle ages, and of our own time in many cities of the orient, we can but feel that the application of science to sanitation, to sewerage, water supply, and housing, has been of immense benefit, although it has by no means kept up with the needs of civilization. The discoveries of preventive medicine have removed the terrors from small-pox and yellow fever, and made impossible the wholesale devastation of great cities by plagues which were common only a few centuries ago. In our own days we have seen the work of the microscopist reveal the cause of the most various diseases, from malaria and cholera to the hookworm disease, while the marvelous work of the surgeon's knife fills us with amazement. If it be desirable to live long, science has largely contributed to benefit mankind in this way. With the improvement in the conditions of work has come the possibility for increased amusement. Music is stored up in the phonograph, to be carried to the remotest corners of Asia and Africa, while the kinematograph has rendered all corners of the earth accessible to the multitude, and has vivified the scenes of history.

Not the least important of the works of science is its effect in the promotion of general peace. As the nations are more closely linked together by this means of transportation and communication, their interests become more nearly alike, and they do not so easily plunge into wars. The applications of science to war have at the same time made it more terrible and deadly, so that nations do not dare to expose themselves to the chance of physical or commercial extermination thereby involved. If the development of the aeroplane shall make it possible for a fast cruiser like the *Lusitania* to be sent out equipped with rapid flying-machines which, on catching the strongest battleship shall make it possible to sail over her at too great a height to be shot at, but near enough to drop high explosives that shall destroy her, war will be at an end. The late Edward Atkinson once stated that all that was necessary to end war was the invention of a gun that should pick off generals at headquarters as the Boer sharpshooters picked off the British captains and colonels.

But I have said enough in praise of the works of science. It is no doubt possible to exaggerate their praise. A most judicious and learned observer, his Excellency James Bryce, in a Phi Beta Kappa address at Harvard two years ago, has examined the question, "What is progress," and whether all our modern improvements have constituted real progress from the times of the ancients. His conclusion is somewhat disappointing, and at the end the beam inclines very slightly in the positive direction. He does consider it probable, however, that the advances of science have rendered more tolerable human life, and have lengthened its span. We must not forget, indeed, that with nearly every new advance some disadvantage is connected, that with the development of industrialism there is connected great injustice, that the results of crowding in cities have led to great misery and sickness, problems not yet solved, and that the recent survey of Pittsburg [*sic*] has revealed conditions which could doubtless be paralleled elsewhere, but which cause us to blush for our boasted civilization. At the same time, these defects are not to be charged to science, but to the failure to utilize it. On the other hand the increase of insanity due to the greater strenuousness of life brought on by modern conditions is not so easily explained away.

It is not, however, for all these works of science that I wish to arouse your enthusiasm. . . . What is the object of science, and is it worth our devotion? What are its purposes and methods, and what may we hope from it? Does it consist in building railroads and bridges, laying cables, digging tunnels and canals, and converting coal into ice? I believe it does not. Let us suppose that the advance of science, the adoption of socialism, or what not, has furnished every working man not only with three acres and a cow, but with hot and cold water, sanitary plumbing, steam heating, with cold brine for refrigeration, milk and beer laid on in pipes, with electric lighting, heating and power for the sewing machine, vacuum cleaner and the few remaining domestic necessities, with a telephone for communication and for the enjoyment of contemporary music, a phonograph and automatic piano for that of the past, an automobile and flying machine for transportation and sport, and that the hours of labor have been reduced to four, will universal happiness then reign? I fear not, if this is all. For life does not consist exclusively of eating and drinking, nor yet of pleasure. Unless what we call the soul is improved as well as the body, life is likely to be a poor thing. It is here that we come to the improvement of morals and of taste, and the need for art, literature and science. I mention these together, for their purposes are the same. They elevate the mind, kindle the imagination and give a more lofty outlook on the universe in general. It is the satisfaction of man's legitimate curiosity, his desire to know the how and the why of nature, that is, in my opinion, the true end of science. There are in the world, we are told by the late William Kingdon Clifford, three classes of persons: in the first place, scientific thinkers, secondly, persons who are engaged in work upon what are called scientific subjects, but who in general do not, and are not expected to, think about these subjects in a scientific manner, and lastly those whose work and thoughts are unscientific. Scientific thought is not determined by the subject thought of. The subject of science is the universe, its limitations those of the human mind. When the captain of a ship finds its position by means of observations with the sextant, or when an engineer constructs a dynamo with the aid of a drawing and data known to be correct, he does not engage in scientific thought, although he makes use of experience previously collected. When the computer in the office of the Nautical Almanac computes an eclipse of the moon, foretelling it to a second of time several years before the event, he is not engaged in scientific thought, but is making use of technical skill. When, on the other hand, Adams and Leverrier, computing the positions of the planet Uranus, found them not verified in fact, but by the assumption of a new hypothesis, were able to discover the planet Neptune, they were engaged in scientific thought of a high order. The collection of facts, as one collects postage stamps or coins, does not constitute science. In order to have science the facts must be fitted into a definite system, in accordance with a classification on the basis of what we call laws. It is prerequisite for the existence of any science whatever that we admit that nature is subject to uniformity, that is, that similar circumstances of similar things will be followed by similar results. The belief that the order of nature is reasonable, that is, that there is a correspondence between her ways and our thoughts, and that this correspondence can be found out, is what I have called scientific faith. The method of the inductive sciences, those that concern the facts of nature, is first to observe a class of seemingly related facts in order to find out what they have in common, then if possible to form some hypothesis as to their relation, then to compare the different cases with the hypothesis in order to see whether it is justified. When this process has been successfully carried out, we are able to predict what will occur in given circumstances, although these circumstances have not occurred. This is what we mean by discovering a law of nature, namely, finding a common property of a class of phenomena, such that under all circumstances the phenomena which will ensue can be described. This is what constitutes the difference between scientific and technical thought. Technical knowledge enables us to deal with cases that have occurred before, while scientific knowledge enables us to deal with what has not occurred before.

This is a matter that is not always understood in this country. It is a matter of common knowledge that this country stands very high in technical knowledge, but it is not so often pointed out that her contribution to science has as yet been distressingly small. Numerous examples might be given. We have just been celebrating the anniversary of Fulton's steamboat, with well-deserved enthusiasm. Nevertheless we must remember that Fulton did not invent the steamboat, nor did he construct the first one. He combined knowledge then existing with practical sense and business acumen, and was able to build a boat so large and successful as to convince the world of a new mode of transportation. In recent times the questions of developing the power of Niagara involved the construction of turbines larger than had ever been built. These were built in Philadelphia, by means of the technical skill there existing, but the designs were made in Geneva by the well known engineers Faesch and Piccard. As a matter of fact the Swiss had long since developed the theory of the turbine, and were prepared to design one of any size on the principles already found sound. More recently the steam turbine has come into the field formerly the exclusive possession of the reciprocating steam engine. Curiously, the first successful turbines came from England, then a large number were developed in Germany and France, while at the present time we have one very successful

American turbine. Now the physical principles involved in the turbine are quite different from those of the reciprocating engine, and involve considerable theoretical knowledge of the properties of fluids in rapid motion, some of which were familiar in the case of water, but which were of a different sort for an expansive vapor like steam. It is very noticeable that the best treatises on the steam turbine to-day are German, and begin with a large amount of theory on the properties of rotating discs, then of the thermodynamics of vapors, and finally of the flow of steam through jets, before the technical matters are touched. We are now hoping for the development of the gas-turbine, which shall combine the two advantages of the gas-engine and the turbine, and which will demand for its success all the knowledge of thermodynamics which we possess. As a final example take the case of wireless telegraphy. This country was a pioneer in ordinary telegraphy, having not only Morse to contribute the technical knowledge, but before him Henry with his scientific development of the electromagnet[;] but the wireless telegraph was imported in an advanced state of development, from England, where the scientific acumen of Maxwell had predicted the action of the electric waves. I am sorry to say that I feel that there is a tendency among our engineers or at least among our engineering students to try to do their work with a very small amount of scientific thinking, and it seems to me that this tendency must be overcome if we wish to maintain a successful competition in either science or technology with such a thorough-going scientific nation as Germany.

There is a tendency to-day in some quarters to disparage the use of hypotheses. With this tendency I do not sympathize. It is difficult to see how scientific advances can be made without the use of hypotheses, nor has that been the ordinary custom. The phrase of Newton has been quoted, *"Hypotheses non fingo,"*[1] but certainly that must be interpreted as meaning that he did not form unnecessary explanations of phenomena rather than that he did not proceed by means of working hypotheses, for he did. By making the hypothesis that the earth attracted bodies according to the inverse square of the distance, and calculating whether the fall of the moon toward the earth was of the amount required by this supposition, he was able to predicate the law of gravitation, and by the calculation that the orbit of a body attracted according to this law would be an ellipse he was able to explain the law of planetary motion discovered by Kepler. It is difficult to see how Kepler could have arrived at his law of elliptic motion if he had not first guessed that the orbits of the planets were circles or conic sections, and then verified it by comparison with the observations on their apparent positions.

The chief test of the success of a scientific hypothesis and of a train of reasoning therefrom is found in the ability to make predictions. Of this probably the most striking example in all science is the law of gravitation

[1] Literally, I make no suppositions. That is, I concern myself solely with facts. Isaac Newton gave this statement currency.

just alluded to. All the observations of the last two hundred years have only resulted in confirming Newton's conclusion, while the accuracy of astronomical prediction exceeds that of any part of science. Such is an example of scientific faith. . . .

I will now, with your permission, undertake to make a rough classification of the sciences, and make some remarks on the differences in their methods. Sitting serene at the head as queen of all is mathematics. Ready she is to serve all, and what a servant she can be is witnessed by those other sciences that have most need of her. Mathematics is probably the most misunderstood of all the sciences. Huxley called it "that science which knows nothing of observation, nothing of experiment, nothing of induction, nothing of causation." To this a sufficient answer might be that she does not need to, but a better one is that it is not true. Intuition and induction have a great part in all mathematical discoveries, as all of the great mathematicians agree. Mathematics has no subject matter, but may be applied to anything that has exact relations. To sing the beauties of mathematics to those ignorant of that subject is as futile as to praise music to the tone-deaf, or painting to the color-blind. I have a friend who describes a symphony as a horrid noise. The president of a great eastern university has said that the manipulation of mathematical symbols is a mark of no particular intellectual eminence. Presumably he had never tried it. To the often-repeated charge that mathematics will turn out only what is put in we may reply that while from incorrect assumptions it can not get correct results it has the power of so transforming the data as to reveal to us totally unexpected truths. Witness the magnificent generalizations of Adams and Leverrier, of Hamilton, and of Maxwell [showing this]. There is no doubt that the invention of the infinitesimal calculus has furnished man with the most powerful and elegant instrument of thought ever devised. . . . It is on account of the logical importance of the method, the universality of its applicability, and the intellectual power developed, that I could wish that as a counterpart to Plato's motto should be placed over every college gateway, "Let none depart hence who knows not the calculus,"[2] at least as to what it deals with, and its fundamental principles.

I am glad to say that in some of our colleges are now given courses in what is termed "culture calculus." It seems to me that this subject is more deserving of the name of culture than the familiarity with the immoralities of the Greek gods.

Of the natural sciences there are two fundamental ones, physics and biology. Physics has to do with all the universe, in so far as it possesses energy, and exerts forces one part upon another, and in so far as it does not possess life. Biology deals with all matters possessing this difficultly defined attribute, but so far as we know, even the phenomena of living matter are subject to the laws of physics. I presume that every biologist will admit that

[2] The motto of the Platonic Academy was supposed to have been: "Let no one enter who knows not mathematics."

life does not create energy, but merely directs it. Nevertheless, the question of vitality is to-day far beyond the explanation of the physicist. The subdivisions of physics have been, for convenience only, set off as individual sciences, chiefly because the whole subject would be too large for the treatment of any individual scientist. The most important part of physics is dynamics, which treats of the laws of motion, and the forces which are associated therewith. Of this a great division is celestial mechanics, which, as we have seen in the cases of Galileo and Newton, contributed in great part to the inductive establishment of the laws of motion in general. The remainder of astronomy is now catalogued as astrophysics and is dealt with by purely physical methods and instruments. As a subdivision of astronomy may be reckoned geodesy, which deals with the form of the earth, deduced from astronomical measurements and from its gravitational attraction.

Chemistry is that part of physics which deals with the properties of substances that have individual characteristics by which they may be always distinguished, and which combine with each other in definite proportions. Its methods are those of physics, its main instrument is the physical balance, and it is in recent years concentrating more attention upon those physical relations connected with temperature, pressure, and electrical relations, all of which are now found to yield to mathematical treatment in a manner until recently unsuspected.

The methods of physics and chemistry usually involve the controlling of certain of the circumstances under which phenomena occur, so that the changes in others may be more easily observed. This is usually done in a laboratory furnished with many means of controlling circumstances, for instance, temperature, pressure, electrical or magnetic state, so that the same circumstances may be reproduced again and again. Meteorology, or as it is now somewhat grandiloquently called, cosmical physics, has to do with those phenomena of the atmosphere, the ocean, or the magnetic state of the earth, which are not controllable by man, and which can not therefore, be repeated at pleasure in the laboratory, but must be observed when and where they occur. The same applies to geology, which is the application of physics, chemistry and even biology, or any science whatever, to the earth, in relation to its physical constitution and its history. Geography deals with the face of the earth, and uses the results of geology to study the earth as fit to be the dwelling place for man. There remain the technical applications of physics in all kinds of engineering, civil, mechanical, electrical, chemical or mining, involving the strength of materials, elasticity and the direction of the natural sources of energy to the purposes of man. All these applications of physics need, and are highly susceptible to, mathematical treatment, and for that reason they are the most perfectly developed of all the sciences.

Let us now turn to the biological sciences. The two fundamental divisions, zoology and botany, dealing with animals and plants, seem to run continuously one into the other, like chemistry and physics. Under both we have the subdivisions of morphology for the study of form and physiology for function. Under zoology we put anatomy, and various more specialized sciences which find their technical application in medicine. There still remain anthropology, the study of man and his practises, psychology, which deals with the workings of what we call his mind, or that of animals, sociology, properly a part of anthropology, dealing with man when living with his fellows, and economics striving to teach him how to get along with them still better.

This classification is admittedly rough, but it does not separate closely connected things as some that I have seen do. . . . Of these biological sciences the methods are somewhat different, they are mostly still in the descriptive stage, and have rarely attained sufficient quantitative information to be capable of mathematical treatment. And yet that must be their ultimate object, for without mathematics there is no exact description. That this is not impossible even in biology may be seen from the following example. If a bacterial culture be inoculated into a jelly with the point of a needle, it will be seen under the microscope to grow in all directions from the original center, and if pains are taken to ensure the physical homogeneity of the jelly the shape of the colony will be an almost perfect circle. If the diameter of this circle be measured at regular intervals, I have no doubt that a quantitative law of growth can be deduced, and even a differential equation found, which will turn out to resemble that of certain physical phenomena, say the conduction of heat. We may observe that the instruments and methods of the physiologist and the experimental psychologist are already largely physical, and their researches are carried on in laboratories. In proportion as the various circumstances are rendered more amenable to external control, so the methods of biology will more nearly approach those of physics. Whereas biology was until recently chiefly a science of observation, it has now become in a high degree experimental. The physiologist removes or alters organs, removes eggs from the natural parent and places them in a foster-mother, cuts off the heads and tails of worms and observes the conditions of survival and regeneration. If the force of gravity were removed, in what direction would a plant grow? If an egg be subjected to centrifugal force in which direction will the head of the animal appear? These are the sort of questions that the biologist is now attacking. Nor is he without mathematical statements. The great generalization of Darwin of fifty years ago has ever since concentrated attention on problems of development and heredity. Darwin's conclusions were the results of the observations of a long life. Now the experimental method enables one to hasten and accelerate conclusions. The gentle monk and acute man of science, Gregor Mendel, forty years ago in his cloister at Brünn by his careful experiments on the crossing of thousands of peas, and by comparison of their seeds, flowers and stems, succeeded in unveiling a law which has profoundly influenced ideas on heredity, not only in plants but in animals. He finds that in the process of hybridization there are certain characteristics which are transmitted entire to the offspring, and are termed dominant, others which seem to disappear or become latent in the process, which he terms recessive. When however the hybrids are bred together both qualities reappear in the offspring, and in a definite proportion of three of the dominant to one of the recessive. In the next generation another definite proportion occurs, and so on. We here have a very definite arithmetical relation, which is susceptible of very exact study and confirmation.

The method of Mendel, which we may call that of experimental evolution, is now of wide application, and there are laboratories which do nothing else but breed and cross under very exact control. Among one of the large-scale experimenters in this line may be mentioned Mr. Luther Burbank, who, though a master of method and subsidized by the Carnegie Institution, seems to be devoted rather to practical than to scientific results.

In connection with the laboratory or experimental method in evolution, must be mentioned a most promising application of mathematics to biology in the new science of biometrics, or the application of the methods of probability or statistics to great numbers of similar objects. If the doctrines of evolution or of variation are ever to be accurately proved it must be in this manner. To illustrate, suppose we have a phenomenon in which chance is involved, and that two events are equally likely, such as throwing head or tail with a coin. Suppose we have a vertical board in which are struck horizontal pegs in a regular arrangement of rows and columns. Suppose a shot be dropped over the middle of this array of pegs, and assume that if it strikes a peg it is equally likely to drop to the right or left. The next time it strikes a peg the chances are the same. It is obviously very unlikely that a shot will continually fall on the same side, while the likeliest thing that can happen is that it shall fall in the middle. Hence if a large number of shot are let fall they will be found, if caught where they fall, to be arranged in a form limited by a curve highest in the middle, and gradually falling symmetrically toward both sides, known as the curve of errors. This curve represents graphically the result of an infinite number of causes acting, each as likely to produce a certain effect as its opposite. Let us now take some biological subject of investigation, say the length of a certain kind of shell. Many thousands being measured, it is found that they vary from the average, but in such a way that very few differ very far from the mean. If the number having any given length is plotted vertically corresponding to the deviation from the mean laid off horizontally, we shall obtain a curve which will generally closely resemble the curve of errors. If this is the case we shall conclude that the causes of the variations in length are perfectly at random, but if we find that the curve is unsymmetrical, or for instance has two summits, we shall know that at least two sorts of causes are acting. Thus questions of heredity and variation may be mathematically studied. This method has been greatly developed by the mathematician, Karl Pearson, who has now devoted himself to the study of evolution by mathematical means.

Finally, that apparently most remote of the sciences from the exactness of physical laws, economics, has been brought under the treatment of mathematics, not only by statistical methods like those just described, but by methods of the calculus. The distinguished mathematician and economist Cournot applied to the theory of wealth methods like those used in mechanics to treat of equilibria, so that very complicated economic principles were amenable to treatment by symbols.[3]

I have, I think, said enough to show the power of science to transform the world, and to develop the mind of man. Is not this development of high spiritual value, and is not the pursuit of truth irrespective of prejudice and authority a noble object, worthy of the devotion of a lifetime? Of the moral values of science it would be easy to give arguments. One has but to consider the self sacrifice of many of its devotees, who consider neither toil nor time if only the good of the race be advanced. Galileo was tortured, Giordano Bruno was burned, and to-day the daily papers bring us news of lives lost in the study of the cholera, of the plague, of the sleeping sickness. The spirit of science is well illustrated by the gift to the Pasteur Institute by M. Osiris last summer of thirty millions of francs. He was led to do this by the fact that the director, Doctor Roux, having won a prize of one hundred thousand francs for the discovery of a diphtheria serum, though not a rich man, immediately turned it over to the institute. Feeling that a cause capable of producing such unselfishness must deserve support, M. Osiris made it this large bequest. Lord Rayleigh, in like manner, donated his Nobel prize of forty thousand dollars to the physical laboratory at Cambridge.

In closing, permit me to recommend the scientific career to young men as one of great satisfaction, whether one succeeds in it or not. To be even a soldier in this noble army, to feel oneself the follower of Faraday, of Helmholtz and of Maxwell, to push on the standard of truth, is worth more than to dress in purple and fine linen and to own many automobiles. There are in this country of eighty millions only about five thousand scientists. The country needs you, young men; it is a patriotic duty to put her where she should stand intellectually among the nations. Would that I might reach the rich, and sing to them the praises of this sort of service. In other lands the rich serve the state, why not here? Surpass your less fortunate brothers not in your pleasures, but in your achievements. And then the American college will be exempt from some of the criticism that it meets to-day. Finally let us bear in mind that while we admire the palaces of science like this, they are not necessary for the performance of good work, and that those of us who are obliged to work in less sumptuous abodes may be consoled with the reflection that most of the great discoveries in science were made with simple apparatus, in humble quarters, but by great men. It is the *spirit* that quickeneth. For the true scientific spirit may we ever pray, for the works of the Lord are great, sought out of all them that have pleasure therein.

[3] Cournot (1801–1877) attempted to analyze supply and demand in terms of functional equilibrium and made similar mechanical and mathematical analyses of other economic phenomena. His work turned out to be not in the mainstream of the development of the discipline of economics.

21. A Biologist Offers a Mechanistic Explanation of Life and Ethics. 1911

Between 1890 and 1910 scientists made a number of discoveries, such as mutations and x-rays and radioactivity, that excited the imagination of the literate public. One of the most impressive figures demonstrating the extent to which science could penetrate the mysteries of nature was JACQUES LOEB (1859–1929). Here, on the basis of the latest contemporary advances—including the new genetics—and his own laboratory work, Loeb suggests that science was on the brink of discovering the nature of life itself. As he reveals here, unlike many scientists Loeb was as interested in the questions of mechanism, materialism, free will, and ethics as in the science on which he based his beliefs. This address was delivered at the First International Congress of Monists in 1911.

Loeb was born in Germany and educated at Berlin, Munich, and Strassburg. He received his M.D. in 1884. Thereafter he taught and worked in Germany in the field of experimental biology. In 1891 he was called to Bryn Mawr and the next year to the University of Chicago. In 1902 he went to the University of California, and in 1910 he became a member of the Rockefeller Institute. His life work he describes in this article. Ironically, in the 1920s it became clear that Loeb's work did not lie within a profitable line of scientific development.

See Donald Fleming's "Introduction," in Jacques Loeb, *The Mechanistic Conception of Life* (Cambridge, Mass.: Harvard University Press, 1964), vii-xli; and W. J. V. Osterhout, "Biographical Memoir of Jacques Loeb, 1859–1924," National Academy of Sciences, *Biographical Memoirs*, 13 (1930), 314–401.

THE MECHANISTIC CONCEPTION OF LIFE*
Jacques Loeb

I. INTRODUCTORY

The reader is aware that two conflicting conceptions are held in regard to the nature of life, namely, a vitalistic and a mechanistic. The vitalists deny the possibility of a complete explanation of life in terms of physics and chemistry. The mechanists proceed as though a complete and unequivocal

* *The Popular Science Monthly*, 80 (1912), 5–21. Slightly abridged.

physico-chemical analysis of life were the attainable goal of biology. It should also be stated that whenever a vitalist desires to make a contribution to science which is more substantial and lasting than mere argument or metaphor, he forgets or lays aside his vitalism and proceeds on the premises and methods of the mechanist. It is thus obvious that as far as the progress of biology is concerned the difference of viewpoint between vitalists and mechanists is of no consequence.

The difference between the two opposite views becomes only of importance when the results of biology are applied to ethical and sociological problems. Since applications of this kind present themselves constantly, the biologist may be pardoned if he raises the question whether or not our present state of knowledge justifies the expectation that life phenomena may ultimately be completely explained in terms of physics and chemistry. I intend to put before you a brief survey of some results, in the main recent, of scientific inquiry which I think may be utilized for an answer to this question.

Before going into these data, it may be necessary to allude briefly to a not uncommon misapprehension in regard to the nature of biological "truth" and methods. It is seemingly often taken for granted by laymen that "truth" in biology or science in general is of the same order as "truth" in certain of the mental sciences; that is to say, that everything rests on argument or rhetoric and that what is regarded as true to-day may be expected with some probability to be considered untrue to-morrow. It happens in sciences, especially in the descriptive sciences like paleontology or zoology, that hypotheses are forwarded, discussed and then abandoned. It should, however, be remembered that modern biology is fundamentally an experimental and not a descriptive science; and that its results are not rhetorical, but always assume one of two forms: it is either possible to control a life phenomenon to such an extent that we can produce it at desire at any time (as, *e.g.*, the contraction of an excised muscle); or we succeed in finding the numerical relation between the conditions of the experiment and the biological result (*e.g.*, Mendel's law of heredity). Biology as far as it is based on these two principles can not retrogress, but must advance.

II. THE BEGINNING OF SCIENTIFIC BIOLOGY

Scientific biology, defined in this sense, begins with the attempt made by Lavoisier and Laplace (1780) to show that the quantity of heat which is formed in the body of a warm-blooded animal is equal to that formed in a candle, provided that the quantities of carbon dioxide formed in both cases are identical. This was the first attempt to reduce a life-phenomenon, namely, the formation of animal heat, completely to physico-chemical terms. What these two investigators began with primitive means has been completed by more recent investigators—Pettenkofer and Voit, Rubner and Zuntz. The oxidation of a food-stuff always furnishes the same amount of heat, no matter whether it takes place in the living body or outside.

These investigations left a gap. The substances which undergo oxidations in the animal body—starch, fat and proteins— are substances which at ordinary temperature are not easily oxidized. They require the temperature of the flame in order to undergo rapid oxidation through the oxygen of the air. This discrepancy between the oxidations in the living body and those in the laboratory manifests itself also in other chemical processes, *e.g.*, digestion or hydrolytic reactions, which were at first found to occur outside the living body rapidly only under conditions incompatible with life. This discrepancy was done away with by the physical chemists, who demonstrated that the same acceleration of chemical reactions which is brought about by a high temperature can also be accomplished at a low temperature with the aid of certain specific substances, the so-called catalyzers. This progress is connected preeminently with the names of Berzelius and Wilhelm Ostwald. The specific substances which accelerate the oxidations at body temperature sufficiently to allow the maintenance of life are the so-called ferments of oxidation.

The work of Lavoisier and Laplace not only marks the beginning of scientific biology, it also touches the core of the problem of life; for it seems that oxidations form a part, if not the basis, of all life phenomena in higher organisms.

III. THE "RIDDLE OF LIFE"

By the "riddle of life" not everybody will understand the same thing. We all, however, desire to know how life originates and what death is, since our ethics must be influenced to a large extent through the answer to this question. We are not yet able to give an answer to the question as to how life originated on the earth. We know that every living being is able to transform food-stuffs into living matter; and we also know that not only the compounds which are formed in the animal body can be produced artificially, but that chemical reactions which take place in living organisms can also be repeated at the same rate and temperature in the laboratory. The gap in our knowledge which we feel most keenly is the fact that the chemical character of the catalyzers (the enzymes or ferments) is still unknown. Nothing indicates, however, at present that the artificial production of living matter is beyond the possibilities of science.

This view does not stand in opposition to the idea of Arrhenius that germs of sufficiently small dimensions are driven by radiation-pressure through space; and that these germs if they fall upon new cosmic bodies possessing water, salts and oxygen and the proper temperature give rise to a new evolution of organisms. Biology will certainly retain this idea, but I believe that we must also follow out the other problem: namely, either succeed in producing living matter artificially, or find the reasons why this should be impossible.

IV. THE ACTIVATION OF THE EGG

Although we are not yet able to state how life originated in general, another, more modest problem has been solved, that is, how the egg is caused by the sperm to develop into a new individual. Every animal originates from an egg and in the majority of animals a new individual can only then develop if a male sex-cell, a spermatozoon, enters into the egg. The question as to how a spermatozoon can cause an egg to develop into a new individual was twelve years ago still shrouded in that mystery which to-day surrounds the origin of life in general. But to-day we are able to state that the problem of the activation of the egg is for the most part reduced to physico-chemical terms. The egg is in the unfertilized condition a single cell with only one nucleus. If no spermatozoon enters into it, it perishes after a comparatively short time, in some animals in a few hours, in others in a few days or weeks. If, however, a spermatozoon enters into the egg, the latter begins to develop, *i.e.*, the nucleus begins to divide into two nuclei and the egg which heretofore consisted of one cell is divided into two cells. Subsequently each nucleus and each cell divides again into two, and so on. These cells have in many eggs the tendency to remain at the surface of the egg or to creep to the surface and later such an egg forms a hollow sphere whose shell consists of a large number of cells. On the outer surface of this hollow sphere cilia are formed and egg is now transformed into a free-swimming larva. Then an intestine develops through the growing in of cells in one region of the blastula[1] and gradually the other organs, skeleton, vascular system, etc., originate. Embryologists had noticed that occasionally the unfertilized eggs of certain animals, *e.g.*, sea-urchins, worms, or even birds, show a tendency to a nuclear or even a cell division; and R. Hertwig, Mead and Morgan had succeeded in inducing one or more cell divisions artificially in such eggs. But the cell divisions in these cases never led to the development of a larva, but at the best to the formation of an abnormal mass of cells which soon perished.

I succeeded twelve years ago in causing the unfertilized eggs of the sea-urchin to develop into swimming larvæ by treating them with sea-water, the concentration of which was raised through the addition of a small but definite quantity of a salt or sugar. The eggs were put for two hours into a solution the osmotic pressure of which had been raised to a certain height. When the eggs were put back into normal sea-water they developed into larvæ and a part of these larvæ formed an intestine and a skeleton. The same result was obtained in the eggs of other animals, starfish, worms and mollusks. These experiments proved the possibility of substituting physico-chemical agencies for the action of the living spermatozoon, but did not yet explain how the spermatozoon causes the development of the egg, since in these experiments the action of the spermatozoon upon the egg was very

[1] A first stage in embryonic development when the original fertilized egg has multiplied, typically as described earlier in the paragraph, into a hollow sphere.

incompletely imitated. When a spermatozoon enters into the egg it causes primarily a change in the surface of the egg which results in the formation of the so-called membrane of fertilization. This phenomenon of membrane formation which had always been considered as a phenomenon of minor importance did not occur in my original method of treating the egg with hypertonic[2] sea-water. Six years ago while experimenting on the California sea-urchin, *Strongylocentrotus purpuratus,* I succeeded in finding a method of causing the unfertilized egg to form a membrane without injuring the egg. This method consists in treating the eggs for from one to two minutes with sea-water to which a definite amount of butyric acid (or some other monobasic fatty acid) has been added. If after that time the eggs are brought back into normal sea-water, all form a fertilization membrane in exactly the same way as if a spermatozoon had entered. This membrane formation or rather the modification of the surface of the egg which underlies the membrane formation starts the development. It does not allow it, however, to go very far at room temperature. In order to allow the development to go further it is necessary to submit the eggs after the butyric acid treatment to a second operation. Here we have a choice between two methods. We can either put the eggs for about one half hour into a hypertonic solution (which contains free oxygen); or we can put them for about three hours into sea-water deprived of oxygen. If the eggs are then returned to normal sea-water containing oxygen they all develop; and in a large number the development is as normal as if a spermatozoon had entered.

The essential feature is therefore the fact that the development is caused by two different treatments of the eggs; and that [of] these the treatment resulting in the formation of the membrane is the more important one. This is proved by the fact that in certain forms, as for instance the star-fish, the causation of the artificial membrane formation may suffice for the development of normal larvæ; although here too the second treatment increases not only the number of larvæ, but also improves the appearance of the larvæ, as R. Lillie found.

The question now arises, how the membrane formation can start the development of the egg. An analysis of the process and of the nature of the agencies which cause it yielded the result that the unfertilized egg possesses a superficial cortical layer, which must be destroyed before the egg can develop. It is immaterial by what means this superficial cortical layer is destroyed. All agencies which cause a definite type of cell destruction—the so-called cytolysis—cause also the egg to develop, as long as their action is limited to the surface layer of the cell. The butyric acid treatment of the egg mentioned above only serves to induce the destruction of this cortical layer. In the eggs of some animals this cortical layer can be destroyed mechanically by shaking the egg, as A. P. Mathews found in the case of star-fish eggs

[2] That is, having a relatively greater osmotic pressure than a comparable substance, in this case, sea water.

and I in the case of the eggs of certain worms. In the case of the eggs of the frog it suffices to pierce the cortical layer with a needle, as Bataillon found in his beautiful experiments a year ago.[3] The mechanism by which development is caused is apparently the same in all these cases, namely, the destruction of the cortical layer of the eggs. This can be caused generally by certain chemical means which play a rôle also in bacteriology; but it can also be caused in special cases by mechanical means, such as agitation or piercing of the cortical layer. It may be mentioned parenthetically that foreign blood sera have also a cytolytic effect, and I succeeded in causing membrane formation and in consequence the development of the sea-urchin egg by treating it with the blood of various animals, *e.g.*, of cattle, or the rabbit.

Recently Shearer has succeeded in Plymouth in causing a number of parthenogenetic plutei[4] produced by my method to develop beyond the stage of metamorphosis, and Delage has reported that he raised two larvæ of the sea-urchin produced by artificial parthenogenesis to the stage of sexual maturity. We may, therefore, state that the complete imitation of the developmental effect of the spermatozoon by certain physico-chemical agencies has been accomplished.

I succeeded in showing that the spermatozoon causes the development of the sea-urchin egg in a way similar to that in my method of artificial parthenogenesis; namely, by carrying two substances into the egg, one of which acts like the butyric acid and induces the membrane formation, while the other acts like the treatment with a hypertonic solution and enables the full development of the larvæ. In order to prove this for the sea-urchin egg foreign sperm, *e.g.*, that of the star-fish, must be used. The sperm of the sea-urchin penetrates so rapidly into the sea-urchin egg that almost always both substances get into the egg. If, however, star-fish sperm is used for the fertilization of the sea-urchin egg, in a large number of cases, membrane formation occurs before the spermatozoon has found time to entirely penetrate into the egg. In consequence of the membrane formation the spermatozoon is thrown out. Such eggs behave as if only the membrane formation had been caused by some artificial agency, *e.g.*, butyric acid. They begin to develop, but soon show signs of disintegration. If treated with a hypertonic solution they develop into larvæ. In touching the egg contents the spermatozoon had a chance to give off a substance which liquefied the cortical layer and thereby caused the membrane formation by which the further entrance of the spermatozoon into the egg was prevented. If, however, the starfish sperm enters completely into the egg before the membrane formation begins, the spermatozoon carries also the second substance into the egg, the

[3] This method does not work with the eggs of fish and is apparently as limited in its applicability as the causation of development by mechanical agitation. [*Footnote in original.*]

[4] Parthenogenesis is the development of eggs without fertilization, and plutei are larval forms of sea urchins.

action of which corresponds to the treatment of the egg with the hypertonic solution. In this case the egg can undergo complete development into a larva.

F. Lillie has recently confirmed the same fact in the egg of a worm, *Nereis*. He mixed the sperm and eggs of *Nereis* and centrifuged the mass. In many cases the spermatozoon which had begun to penetrate into the egg were thrown off again. The consequence was that only a membrane formation resulted without the spermatozoon penetrating into the egg. This membrane formation led only to a beginning but not to a complete development. We may, therefore, conclude that the spermatozoon causes the development of the egg in a way similar to that which takes place in the case of artificial parthenogenesis. It carries first a substance into the egg which destroys the cortical layer of the egg in the same way as butyric acid does; and secondly a substance which corresponds in its effect to the influence of the hypertonic solution in the sea-urchin egg after the membrane formation.

The question arises as to how the destruction of the cortical layer can cause the beginning of the development of the egg. This question leads us to the process of oxidation. Years ago I had found that the fertilized sea-urchin egg can only develop in the presence of oxygen; if the oxygen is completely withdrawn the development stops, but begins again promptly as soon as oxygen is again admitted. From this and similar experiments I concluded that the spermatozoon causes the development by accelerating the oxidations in the egg. This conclusion was confirmed by experiments by O. Warburg and by Wasteneys and myself in which it was found that through the process of fertilization the velocity of oxidations in the egg is increased to four or six times its original value. Warburg was able to show that the mere causation of the membrane formation by the butyric acid treatment has the same accelerating effect upon the oxidations as fertilization.

What remains unknown at present is the way in which the destruction of the cortical layer of the egg accelerates the oxidations. It is possible that the cortical layer acts like a solid crust and thus prevents the oxygen from reaching the surface of the egg or from penetrating into the latter sufficiently rapidly. The solution of these problems must be reserved for further investigation.

We, therefore, see that the process of the activation of the egg by the spermatozoon, which twelve years ago was shrouded in complete darkness, to-day is practically completely reduced to a physico-chemical explanation. Considering the youth of experimental biology we have a right to hope that what has been accomplished in this problem will occur in rapid succession in those problems which to-day still appear as riddles.

V. NATURE OF LIFE AND DEATH

The nature of life and of death are questions which occupy the interest of the layman to a greater extent than possibly any other purely theoretical problem; and we can well understand that humanity did not wait for experimental biology to furnish an answer. The answer assumed the anthropomorphic form characteristic of all explanations of nature in the prescientific period. Life was assumed to begin with the entrance of a "life principle" into the body; that individual life begins with the egg was of course unknown to primitive or pre-scientific man. Death was assumed to be due to the departure of this "life principle" from the body.

Scientifically, however, individual life begins (in the case of the sea-urchin and possibly in general) with the acceleration of the rate of oxidation in the egg, and this acceleration begins after the destruction of its cortical layer. Life of warm blooded animals—man included—ends with the cessation of oxidation in the body. As soon as oxidations have ceased for some time, the surface films of the cells, if they contain enough water and if the temperature is sufficiently high, become permeable for bacteria, and the body is destroyed by microorganisms. The problem of the beginning and end of individual life is physico-chemically clear. It is, therefore, unwarranted to continue the statement that in addition to the acceleration of oxidations the beginning of individual life is determined by the entrance of a metaphysical "life principle" into the egg; and that death is determined, aside from the cessation of oxidations, by the departure of this "principle" from the body. In the case of the evaporation of water we are satisfied with the explanation given by the kinetic theory of gases and do not demand that—to repeat a well-known jest of Huxley—the disappearance of the "aquosity" be also taken into consideration.

VI. HEREDITY

It may be stated that the egg is the essential bearer of heredity. We can cause an egg to develop into a larva without sperm, but we can not cause a spermatozoon to develop into a larva without an egg. The spermatozoon can influence the form of the offspring only when the two forms are rather closely related. If the egg of a sea-urchin is fertilized with the sperm from a different species of sea-urchin, the larval form has distinct paternal characters. If, however, the eggs of a sea-urchin are fertilized with the sperm of a more remote species, *e.g.*, a star-fish, the result is a sea-urchin larva which possesses no paternal characters, as I found and as Godlewski, Kupelwieser, Hagedoorn and Baltzer were able to confirm. This fact has some bearing upon the further investigation of heredity, inasmuch as it shows that the egg is the main instrument of heredity, while apparently the spermatozoon is restricted in the transmission of characters to the offspring. If the difference between spermatozoon and egg exceeds a certain limit the hereditary effects of the spermatozoon cease and it acts merely as an activator to the egg.

As far as the transmission of paternal characters is concerned, we can say to-day that the view of those authors was correct who, with Boveri, localized the transmission not only in the cell nucleus, but in a special constituent of the nucleus, the chromosomes. The proof for this was given by facts found along the lines of Mendelian investigations. The essential law of Mendel, the law of segregation, can in its simplest form be expressed in the following way. If we cross two forms which differ in only one character every hybrid resulting from this union forms two kinds of sex-cells in equal numbers; two kinds of eggs if it is a female, two kinds of spermatozoa if it is a male. The one kind corresponds to the pure paternal, the other to the pure maternal type. The investigation of the structure and behavior of the nucleus showed that the possibility for such segregation of the sex-cells in a hybrid can easily be recognized during a given stage in the formation of the sex-cells, if the assumption is made that the chromosomes are the bearers of the paternal characters. The proof for the correctness of this view was furnished through the investigation of the heredity of those qualities which occur mainly in one sex; *e.g.,* color blindness which occurs preeminently in the male members of a family.

Nine years ago McClung published a paper which solved the problem of sex determination, at least in its essential feature. Each animal has a definite number of chromosomes in its cell nucleus. Henking had found that in a certain form of insects *(Pyrrhocoris)* two kinds of spermatozoa exist which differ in the fact that the one possesses a nucleolus while the other does not. Montgomery afterwards showed that Henking's nucleolus was an accessory chromosome. McClung first expressed the idea that this accessory chromosome was connected with the determination of sex. Considering the importance of this idea we may render it in his own words:

> A most significant fact, and one upon which almost all investigators are united in opinion, is that the element is apportioned to but one half of the spermatozoa. Assuming it to be true that the chromatin is the important part of the cell in the matter of heredity, then it follows that we have two kinds of spermatozoa that differ from each other in a vital matter. We expect, therefore, to find in the offspring two sorts of individuals in approximately equal numbers, under normal conditions, that exhibit marked differences in structure. A careful consideration will suggest that nothing but sexual characters thus divides the members of a species into two well-defined groups, and we are logically forced to the conclusion that the peculiar chromosome has some bearing upon the arrangement.
>
> I must here also point out a fact that does not seem to have the recognition it deserves; viz., that if there is a cross division of the chromosomes in the maturation mitoses,[5] there must be two kinds of spermatozoa regardless of the presence of the accessory chromosome. It is thus possible that even in the absence of any specialized element a preponderant maleness would attach to one half of the spermatozoa. . . .

[5] That is, cell division in which the chromosomes also divide.

The researches of the following years, especially the brilliant work of E. B. Wilson, Miss Stevens, T. H. Morgan and others, have amply confirmed the correctness of this ingenious idea and cleared up the problem of sex determination in its main features.

According to McClung each animal forms two kinds of spermatozoa in equal numbers, which differ by one chromosome. One kind of spermatozoa produces male animals, the other female animals. The eggs are all equal in these animals. More recent investigations, especially those of E. B. Wilson, have shown that this view is correct for many animals.

While in many animals there are two kinds of spermatozoa and only one kind of eggs, in other animals two kinds of eggs and only one kind of spermatazoa are formed, *e.g.*, sea-urchins and certain species of birds and of butterflies *(Abraxas)*. In these animals the sex is predetermined in the egg and not in the spermatazoon. It is of interest that, according to Guyer, in the human being two kinds of spermatazoa exist and only one kind of eggs; in man, therefore, sex is determined by the spermatazoon. . . .

The problem of sex determination has, therefore, found a simple solution, and simultaneously Mendel's law of segregation finds also its solution.

In many insects and in man the cells of the female have two sex–chromosomes. In a certain stage of the history of the egg one half of the chromosomes leaves the egg (in the form of the "polar-body") and the egg keeps only half the number of chromosomes. Each egg, therefore, retains only one *X* or sex–chromosome. In the male the cells have from the beginning only one *X*-chromosome and each primordial spermatazoon divides into two new (in reality into two pairs of) spermatazoa, one of which contains an *X*-chromosome while the other is without such a chromosome. What can be observed here directly in the male animal takes place in every hybrid; during the critical, so-called maturation division of the sexual cell in the hybrid a division of the chromosomes occurs whereby only one half of the sex cells receive the hereditary substance in regard to which the two original pure forms differ.

That this is not a mere assumption can be shown in those cases in which the hereditary character appears only, or preeminently, in one sex as, *e.g.*, color blindness which appears mostly in the male. If a color-blind individual is mated with an individual with normal color vision the heredity of color blindness in the next two generations corresponds quantitatively with what we must expect on the assumption that the chemical substances determining color vision are contained in the sex–chromosomes. In the color-blind individual something is lacking which can be found in the individual with normal color perception. The factor for color vision is obviously transmitted through the sex-chromosome. In the next generation color blindness cannot appear since each fertilized egg contains the factor for color perception. In the second generation, however, the theory demands that one half of the males should be color blind. In man these conditions cannot always be verified numerically since the number of children is too small to yield the conditions to be expected according to the calculus of probability. T. H. Morgan has found in a fly *(Drosophila)* a number of similar sex-limited characters which behave like color blindness, *e.g.*, lack of pigment in the eyes. These flies have normally red eyes. Morgan has observed a mutation with white eyes, which occurs in the male. When he crossed a white-eyed with a red-eyed female all flies of the first generation were red-eyed, since all flies had the factor for pigment in their sex-cells; in the second generation all females and exactly one half of the males had red eyes, the other half of the males, however, white eyes, as the theory demands.

From these and numerous similar breeding experiments of Correns, Doncaster, and especially of Morgan, we may conclude with certainty that the sex-chromosomes are the bearers of those hereditary characters which appear preeminently in one sex. We say preeminently since theoretically we can predict cases in which color blindness or white eyes must appear also in the female. Breeding experiments have shown that this theoretical prediction is justified. The riddle of Mendel's law of segregation finds its solution through these experiments and incidentally also the problem of the determination of sex which is only a special case of the law of segregation, as Mendel already intimated.

The main task which is left here for science to accomplish is the determination of the chemical substances in the chromosomes which are responsible for the hereditary transmission of a quality, and the determination of the mechanism by which these substances give rise to the hereditary character. Here the ground has already been broken. It is known that for the formation of a certain black pigment the cooperation of a substance—tyrosin—and of a ferment of oxidation—tyrosinase—is required. The hereditary transmission of the black color through the male animal must occur by substances carried in the chromosomes which determine the formation of tyrosin or tyrosinase or of both. We may, therefore, say that the solution of the riddle of heredity has succeeded to the extent that all further development will take place purely in cytological and physico-chemical terms.

While until twelve years ago the field of heredity was the stamping ground for the rhetorician and metaphysician it is to-day perhaps the most exact and rationalistic part of biology, where facts can not only be predicted qualitatively, but also quantitatively.

VII. THE HARMONIOUS CHARACTER OF THE ORGANISMS

It is not possible to prove in a short address that all life phenomena will yield to a physico-chemical analysis. We have selected only the phenomena of fertilization and heredity, since these phenomena are specific for living organisms and without analogues in inanimate nature; and if we can convince ourselves that these processes can be explained physico-chemically we may safely expect the same of such processes for which there exist *a priori* analogies in inanimate nature, as, *e.g.*, for absorption and secretion.

We must, however, settle a question which offers itself not only to the layman but also to every biologist, namely, how we shall conceive that wonderful "adaptation of each part to the whole" by which an organism becomes possible. In the answer [to] this question the metaphysician finds an opportunity to put above the purely chemical and physical processes something specific which is characteristic of life only: the "Zielstrebigkeit," the "harmony" of the phenomena, or the "dominants" of Reinke and similar things.

With all due personal respect for the authors of such terms I am of the opinion that we are dealing here, as in all cases of metaphysics, with a play on words. That a part is so constructed that it serves the "whole" is only an unclear expression for the fact that a species is only able to live—or to use Roux's expression—is only durable, if it is provided with the automatic mechanism for self-preservation and reproduction. If, for instance, warm-blooded animals should originate without a circulation they could not remain alive, and this is the reason why we never find such forms. The phenomena of "adaptation" cause only apparent difficulties since we rarely or never become aware of the numerous faultily constructed organisms which appear in nature. I will illustrate by a concrete example that the number of species which we observe is only an infinitely small fraction of those which can originate and possibly not rarely do originate, but which we never see since their organization does not allow them to continue to exist long. Moenkhaus found ten years ago that it is possible to fertilize the egg of each marine bony fish with the sperm of practically any other marine bony fish. His embryos apparently lived only a very short time. This year I succeeded in keeping such hybrid embryos between distantly related bony fish alive for over a month. It is, therefore, clear that it is possible to cross practically any marine teleost with any other.

The number of teleosts at present in existence is about 10,000. If we accomplish all possible hybridizations 100,000,000 different crosses will result. Of these teleosts only a very small proportion, namely about one one-hundredth of one per cent., can live. It turned out in my experiments that the heterogeneous hybrids between bony fishes formed eyes, brains, ears, fins and pulsating hearts, blood and blood-vessels, but could live only a limited time because no blood circulation was established at all—in spite of the fact that the heart beat for weeks—or that the circulation, if it was established at all, did not last long.

What prevented these heterogeneous fish embryos from reaching the adult stage? The lack of the proper "dominants"? Scarcely. I succeeded in producing the same type of faulty embryos in the pure breeds of a bony fish

(Fundulus heteroclitus) by raising the eggs in 50 c.c. of sea-water to which was added 2 c.c. one one-hundredth per cent. NaCN. The latter substance retards the velocity of oxidations and I obtained embryos which were in all details identical with the embryos produced by crossing the eggs of the same fish with the sperm of remote teleosts, e.g., *Ctenolabrus* or *Menidia.* These embryos, which lived about a month, showed the peculiarity of possessing a beating heart and blood, but no circulation. This suggests the idea that heterogeneous embryos show a lack of "adaptation" and durability for the reason that in consequence of the chemical difference between heterogeneous sperm and egg the chemical processes in the fertilized egg are abnormal.

The possibility of hybridization goes much farther than we have thus far assumed. We can cause the eggs of echinoderms to develop with the sperm of very distant forms, even mollusks and worms (Kupelwieser); but such hybridizations never lead to the formation of durable organisms.

It is, therefore, no exaggeration to state that the number of species existing to-day is only an infinitely small fraction of those which can and possibly occasionally do originate, but which escape our notice because they can not live and reproduce. Only that limited fraction of species can exist which possesses no coarse disharmonies in its automatic mechanism of preservation and reproduction. Disharmonies and faulty attempts in nature are the rule, the harmonically developed systems the rare exception. But since we only perceive the latter we gain the erroneous impression that the "adaptation of the parts to the plan of the whole" is a general and specific characteristic of animate nature, whereby the latter differs from inanimate nature.

If the structure and the mechanism of the atoms were known to us we should probably also get an insight into a world of wonderful harmonies and apparent adaptations of the parts to the whole. But in this case we should quickly understand that the chemical elements are only the few durable systems among a large number of possible but not durable combinations. Nobody doubts that the durable chemical elements are only a product of blind forces. There is no reason for conceiving otherwise the durable systems in living nature.

VIII. THE CONTENTS OF LIFE

The contents of life from the cradle to the bier are wishes and hopes, efforts and struggles and unfortunately also disappointments and suffering. And this inner life should be amenable to a physico-chemical analysis? In spite of the gulf which separates us to-day from such an aim I believe that it is attainable. As long as a life phenomenon has not yet found a physico-chemical explanation it usually appears inexplicable. If the veil is once lifted we are always surprised that we did not guess from the first what was behind it.

That in the case of our inner life a physico-chemical explanation is not beyond the realm of possibility is proved by the fact that it is already possible for us to explain cases of simple manifestations of animal instinct and will on a physico-chemical basis; namely, the phenomena which I have discussed in former papers under the name of animal tropisms. As the most simple example we may mention the tendency of certain animals to fly or creep to the light. We are dealing in this case with the manifestation of an instinct or impulse which the animals can not resist. It appears as if this blind instinct which these animals must follow, although it may cost them their life might be explained by the same law of Bunsen and Roscoe, which explains the photo-chemical effects in inanimate nature. This law states that within wide limits the photo-chemical effect equals the product of the intensity of light into the duration of illumination. It is not possible to enter here into all the details of the reactions of these animals to light, we only wish to point out in which way the light instinct of the animals may possibly be connected with the Bunsen-Roscoe law.

The positively heliotropic animals—*i.e.*, the animals which go instinctively to a source of light—have in their eyes (and occasionally also in their skin) photosensitive substances which undergo alterations by light. The products formed in this process influence the contraction of the muscles—mostly indirectly, through the central nervous system. If the animal is illuminated on one side only, the mass of photochemical reaction products formed on that side in the unit of time is greater than on the opposite side. Consequently the development of energy in the symmetrical muscles on both sides of the body becomes unequal. As soon as the difference in the masses of the photochemical reaction products on both sides of the animal reaches a certain value the animal, as soon as it moves, is automatically forced to turn towards one side. As soon as it has turned so far that its plane of symmetry is in the direction of the rays, the symmetrical spots of its surface are struck by the light at the same angle and in this case the intensity of light and consequently the velocity of reaction of the photochemical processes on both sides of the animal become equal. There is no more reason for the animal to deviate from the motion in a straight line and the positively heliotropic animal will move in a straight line to the source of light. (It was assumed that in these experiments the animal is under the influence of only one source of light and positively heliotropic.)

In a series of experiments I have shown that the heliotropic reactions of animals are identical with the heliotropic reactions of plants. It was known that sessile heliotropic plants bend their stems to the source of light until the axis of symmetry of their tip is in the direction of the rays of light. I found the same phenomenon in sessile animals, *e.g.*, certain hydroids and worms. Motile plant organs, *e.g.*, the swarm spores of plants, move to the source of light (or if they are negatively heliotropic away from it) and the same is observed in motile animals. In plants only the more refrangible rays from green to blue have these heliotropic effects, while the red and yellow rays are little or less effective; and the same is true for the heliotropic reactions of animals.

It has been shown by Blaauw for the heliotropic curvatures of plants that the product of the intensity of a source of light into the time required to induce a heliotropic curvature is a constant; and the same result was obtained simultaneously by another botanist, Fröschl. It is thus proved that the Bunsen-Roscoe law controls the heliotropic reactions of plants. The same fact had already been proved for the action of light on our retina.

The direct measurements in regard to the applicability of Bunsen's law to the phenomena of animal heliotropism have not yet been made. But a number of data point to the probability that the law holds good here also. The first of these facts is the identity of the light reactions of plants and animals. The second is at least a rough observation which harmonizes with the Bunsen-Roscoe law. As long as the intensity of light or the mass of photochemical substances at the surface of the animal is small, according to the law of Bunsen, it must take a comparatively long time until the animal is automatically oriented by the light, since according to this law the photochemical effect is equal to the product of the intensity of the light into the duration of illumination. If, however, the intensity of the light is strong or the active mass of the photochemical substance great, it will require only a very short time until the difference in the mass of photochemical reaction products on both sides of the animal reaches the value which is necessary for the automatic turning to (or from) the light. The behavior of the animals agrees with this assumption. If the light is sufficiently strong the animals go in an almost straight line to the source of light; if the intensity of light (or the mass of photosensitive substances on the surface of the animal) is small the animals go in irregular lines, but at last they also land at the source of light, since the directing force is not entirely abolished. It will, however, be necessary to ascertain by direct measurements to what extent these phenomena in animals are the expression of Bunsen-Roscoe's law. But we may already safely state that the apparent will or instinct of these animals resolves itself into a modification of the action of the muscles through the influence of light; and for the metaphysical term "will" we may in these instances safely substitute the chemical term "photochemical action of light."

Our wishes and hopes, disappointments and sufferings have their source in instincts which are comparable to the light instinct of the heliotropic animals. The need of and the struggle for food, the sexual instinct with its poetry and its chain of consequences, the maternal instincts with the felicity and the suffering caused by them, the instinct of workmanship and some other instincts are the roots from which our inner life develops. For some of these instincts the chemical basis is at least sufficiently indicated to arouse the hope that their analysis, from the mechanistic point of view, is only a question of time.

IX. ETHICS

If our existence is based on the play of blind forces and only a matter of chance; if we ourselves are only chemical mechanisms—how can there be an ethics for us? The answer is, that our instincts are the root of our ethics and that the instincts are just as hereditary as is the form of our body. We eat, drink and reproduce not because mankind has reached an agreement that this is desirable, but because, machine-like, we are compelled to do so. We are active, because we are compelled to be so by processes in our central nervous system; and as long as human beings are not economic slaves the instinct of successful work or of workmanship determines the direction of their action. The mother loves and cares for her children not because metaphysicians had the idea that this was desirable, but because the instinct of taking care of the young is inherited just as distinctly as the morphological characters of the female body. We seek and enjoy the fellowship of human beings because hereditary conditions compel us to do so. We struggle for justice and truth since we are instinctively compelled to see our fellow beings happy. Economic, social and political conditions or ignorance and superstition may warp and inhibit the inherited instincts and thus create a civilization with a faulty or low development of ethics. Individual mutants may arise in which one or the other desirable instinct is lost, just as individual mutants without pigment may arise in animals; and the offspring of such mutants may, if numerous enough, lower the ethical status of a community. Not only is the mechanistic conception of life compatible with ethics; it seems the only conception of life which can lead to an understanding of the source of ethics.

22. A Psychologist Suggests Breeding Better People on the Basis of the Science of Genetics. 1913

During the Progressive reform period prior to World War I, intellectuals and reform leaders tended to believe that science held the solutions to many social problems. Just as medicine had conquered rabies and the Wright brothers gravity, so the application of science would, the Progressive generation hoped, solve all human problems. The means of the solution was generally to be education—changing people's ideas and behavior (as, for example, in the famous no-spitting campaign against tuberculosis contagion). For those aspects of the human animal that resisted environmental reform—idiocy, for example—selective human breeding, according to the Progressive science of eugenics, would eventually solve many problems. This lecture expresses eloquently many of the reformers' assumptions regarding heredity, their hopes, and some of their sophistication.

EDWARD L. THORNDIKE (1874–1949) was a psychologist. A native of Massachusetts, he graduated from Wesleyan and then took an M.A. at Harvard and a Ph.D. at Columbia (1898). He taught briefly at Western Reserve and then spent the rest of his life as a faculty member of Teachers College, Columbia University. His classic experiments on trial-and-error learning marked him as one of America's most creative scientists.

See Mark H. Haller, *Eugenics, Hereditarian Attitudes in American Thought* (New Brunswick: Rutgers University Press, 1963); Geraldine Joncich, *The Sane Positivist: A Biography of Edward L. Thorndike* (Middletown, Conn.: Weslyan University Press, 1968); and Merle Curti, *The Social Ideas of American Educators* (New York: Charles Scribner's Sons, 1935), chap. XIV.

EUGENICS: With Special Reference to Intellect and Character*
Edward L. Thorndike

By eugenics is meant, as you all know, the improvement of mankind by breeding. It has been decided by those responsible for this lecture—Mrs. Huntington Wilson and the president and trustees of the university—that its topic shall be the intellectual and moral, rather than the physical, improvement of the human stock.

Common observation teaches that individuals of the same sex and age differ widely in intellect, character and achievement. The more systematic and exact observations made by scientific students of human nature emphasize the extent of these differences. Whether we take some trivial function—such as memory for isolated words, or delicacy of discrimination of pitch—or take some broad symptom of man's nature, such as his rate of progress through school, or ability in tests of abstract intellect, or even his general intellectual and moral repute—men differ widely. Samples of the amount and distribution of such differences are given in Charts 1, 2 and 3. Chart 1 relates that of 732 children who had studied arithmetic equally long, one could get over a hundred examples done correctly in fifteen minutes, while others could not get correct answers to five. Even if we leave out of account the top three per cent., covering all the records of 60 or over, we have some children achieving twenty-five times as many correct answers as other children.

* *The Popular Science Monthly,* 83 (1913), 125–138.

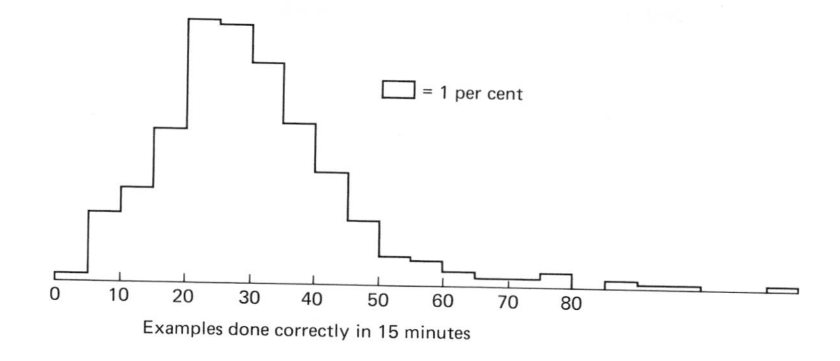

Chart 1. The Relative Frequencies of Different Degrees of Ability in Addition in the Case of Fourth-grade Pupils.

Chart 2 shows that when four hundred children who had had similar school training were given each the same amount of practise in certain work in division, some improve[d] not at all, and others enormously. Chart 3 shows that of children in the same school all of the same year-age (thirteen), some have done the work of the eight grades of the elementary school and of one or two years of the high school, while others have not completed the work of a single year. Still less competence at intellectual tasks could be found by including children from asylums for imbeciles and idiots.

The differences thus found amongst individuals of the same sex and age are due in large measure to original, inborn characteristics of the intellectual

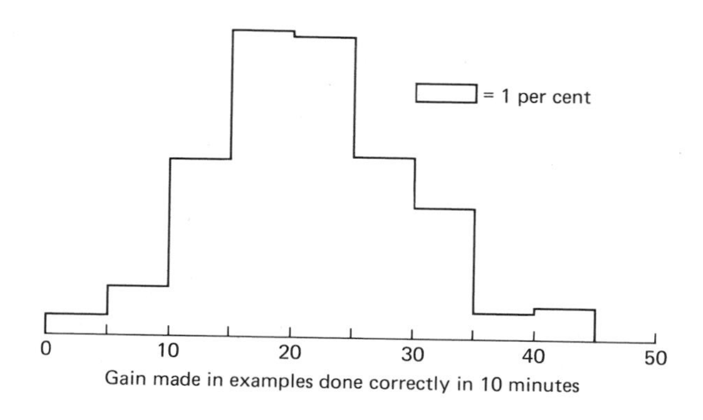

Chart 2. The Relative Frequencies of Different Amounts of Gain from Fifty Minutes of Practise in Division, in the Case of Pupils of the Same School Grade.

and moral constitution of the individuals in question. They are, it is true, in part due to differences in maturity—one thirteen-year-old being further advanced in development than another. They are also due in part to differences in environment, circumstances, training—one sort of home-life being more favorable than another to progress through school, for example. Each advance

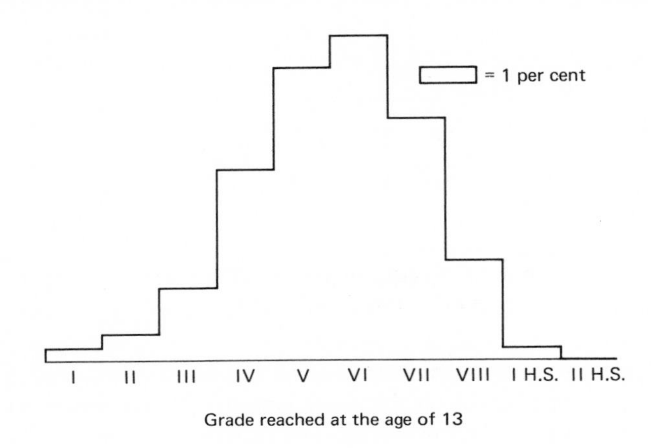

Chart 3. The Relative Frequencies of Different Amounts of Progress in School of Thirteen-year-old Children.

in the study of individual differences, however, shows that differences in maturity and differences in the circumstances of nurture account for only a small fraction of the differences actually found in individuals of the same general environment of an American city in 1900–1912. Long before a child begins his schooling, or a man his work at trade or profession, or a woman her management of a home—long indeed before they are born—their superiority or inferiority to others of the same environmental advantages is determined by the constitution of the germs and ova whence they spring, and which, at the start of their individual lives, they *are*.

Of the score or more of important studies of the causes of individual differences which have been made since Francis Galton led the way, I do not find one that lends any support to the doctrine of human initial equality, total or approximate. On the contrary, every one of them gives evidence that if the thousand babies born this week in New York City were given equal opportunity they would still differ in much the same way and to much the same extent as they will in fact differ.

We find, for instance, that the children of certain families rank very much higher in certain psychological tests of perception, association and the like, than the children of certain other families. Now if this difference were due to the difference between the two groups of families in environment—

in ideals, customs, hygienic conditions and the like—it should increase greatly with the age of the children in some rough proportion to the length of time that they are subject to the beneficent or unfavorable environment. It does not. One family's product differs from another nearly as much at the age of 9 to 11 as at the age of 12 to 14.

Again, if inequalities in the environment produce the greater part of these differences, equalizing opportunity and training should greatly reduce them. Such equalization is found by experiment to reduce them very little, if at all. Chart 4 shows, for example, the result of equal amounts of training applied to two groups of adults whom life in general had previously brought to the conditions shown at the left of the chart. The trait chosen was addition; from life in general one group had gained the ability to do twenty-seven more additions per minute than the other group, accuracy being equal in the two groups. At the end of the special training the superior individuals had gained on the average 28 additions per minute, while the inferior individuals gained only 10 additions per minute. As a result of this partial equalization of opportunity, the superior individuals were farther ahead than ever! If equality of opportunity has no equalizing effect in so easily alterable a trait as rapidity in addition, surely it can have little power in such traits as energy, stability, general intellectual power, courage or kindliness.

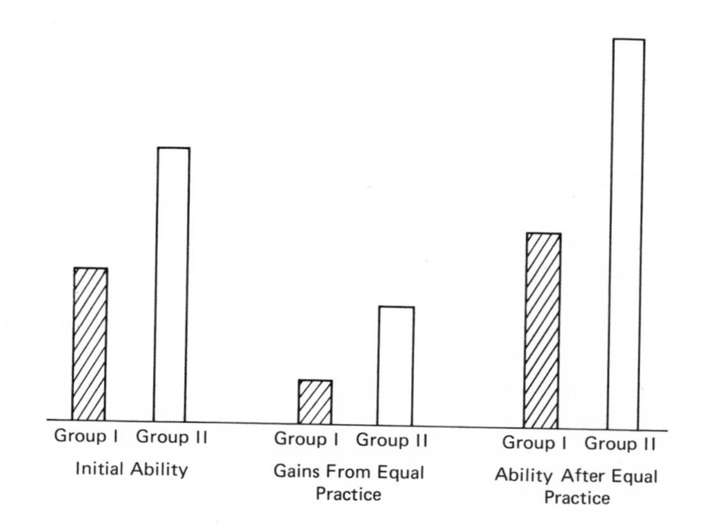

Chart 4. The Relation of the Gains from Equal Amounts of Practise in the Case of Individuals of High and Low Initial Ability.

Men differ by original nature. With equal nurture of an inferior sort they progress unequally to low stations; with equal nurture of a superior sort they progress unequally to high stations. Their absolute achievements, the amounts of progress which they make from zero up, are due largely to the environment which excites and directs their original capacities. Their relative achievements—the amounts of progress which they make, one in comparison with another—are due largely to their variations one from another in original capacities.

The man's original nature, too, has large selective power over his environment. The thousand babies will in large measure each create his own environment by cherishing this feature and neglecting that, amongst those which the circumstances of life offer. As Dr. Woods has well argued, the power of the environment to raise or lower a man is very great only when the environment is unavoidable. We must remember that one of these babies, if of mean and brutal nature, can by enough pains avoid industry, justice and honor, no matter how carefully he is brought up; and that one of them of intellectual gifts can, if he cares enough, seek out and possess adequate stimuli to achievement in art, science, or letters, no matter how poor and sordid his home may be.

If, a hundred years ago, every boy in England could have had as good opportunity—each of the sort fitted to his capacities—as Charles Darwin had, the gain for human welfare would probably have been great; but if every boy then could have had as good inborn capacity for science, art, invention, the management of men—or whatever his strongest capacity was—as Charles Darwin had for science, the gain for welfare would certainly have been enormous.

The original differences in intellect, character, and skill which characterize men are related to the families and races whence the individuals spring. Each man's original mental constitution, which so largely determines how much more or less he will do for the world's good than the average man of his generation, is the product of no fortuity, but of the germs of his parents and the forces which modify the body into which they grow—is the product, as we are accustomed to say, of heredity and variation. The variation within the group of offspring of the same parents is large—a very gifted thinker may have an almost feeble-minded brother—but the variation between families is real. A feeble-minded person's brothers will be feeble-minded hundreds of times as often.

The general average tendency of the original intellectual and moral natures of children to be like the original natures of their ancestry is guaranteed beforehand by the accepted principles of biology. Direct evidence of it is also furnished by investigations of the combination of original and acquired differences which human achievements, as they stand, display. The same studies which find differences of nurture hopelessly inadequate to account for differences of ability and achievement, find that original capacities and interests must be invoked precisely because achievement runs in families, and in a manner or degree which likeness in home training can not explain.

Galton found that the real sons of eminent men had a thousand times the ordinary man's chance of eminence and far excelled the adopted sons of men of equal eminence. Woods has shown that, when each individual is rated for intellect or morals, the achievements of those sons of royal families who succeeded to the throne by paternal death and thus had the special attention given to crown princes and the special unearned opportunities of succession, have, in the estimation of historians, been no greater than those of their younger brothers.

Children of the same parents resemble one another in every mental trait where the issue has been tested, and resemble one another nearly or quite as much in such tests as quickness in marking the A's on a sheet of printed capitals or giving the opposites of words, to which home training has never paid any special attention, as they do in adding or multiplying, where parental ambitions, advice and rewards would be expected to have much more effect, if they have any anywhere.

Mr. Courtis, who has been assiduously studying the details of ability in arithmetic in school children, finds, as one sure principle of explanation, the likeness of children to parents—and this even in subtle traits and relations between traits, of whose very existence the parents were not aware, and which the parents would not have known how to nurture had they known of their existence.

Dr. Keyes has recently made an elaborate study of various possible causes of the rate of progress of a child through the elementary school. He traces the effects of defective vision, of sickness, of moving from one school to another, and so on, but finds nothing of great moment until he happens to trace family relationships. Then it appears that certain families are thick with "accelerates," or pupils who win double promotions, whereas other families are thick with retarded pupils, who require two years to complete a normal year's work. Of 168 families, only 30 contain both an "accelerated" and a "retarded" pupil, whereas 138 show either two or more accelerates or two or more retarded pupils. The differences in home training are here not allowed for, but, in view of what has been found in other cases, it appears certain that the rate at which a child will progress in school in comparison with his fellows is determined in large measure before he is born.

In intellect and morals, as in bodily structure and features, men differ, differ by original nature, and differ by families. There are hereditary bonds by which one kind of intellect or character rather than another is produced. Selective breeding can alter a man's capacity to learn, to keep sane, to cherish justice or to be happy.

Let the lines L_1H_1 and L_2H_2 in Chart 5 be identical scales for the original capacity for intellect, or virtue, or any desirable human trait. Let the surface above line L_1H_1 represent the distribution of this original capacity amongst men to-day. There is every reason to believe that wise selective breeding could change the present state of affairs, at least to that shown above L_2H_2,

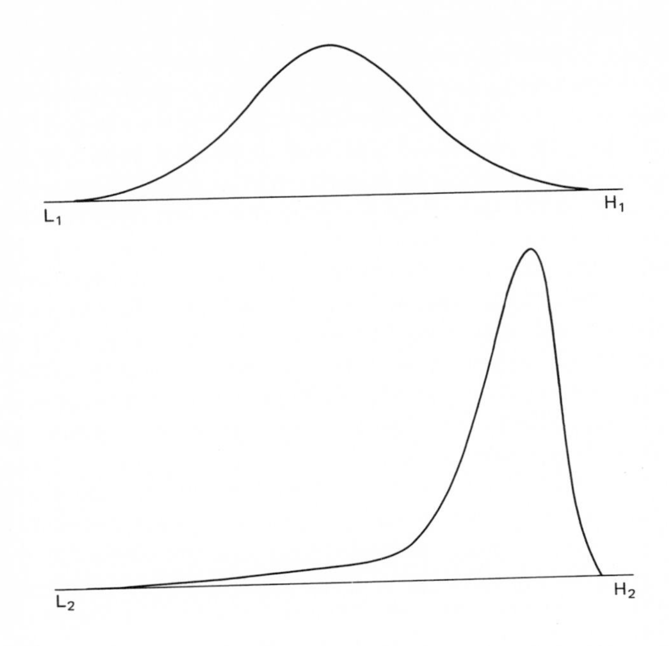

Chart 5. The Improvement Possible by Selective Breeding. The upper surface being taken to represent the existing distribution of intellect, the lower surface represents what might be expected from, say, ten or twenty generations of breeding exclusively from the apparently best tenth of human intellects.

within relatively few generations. Perhaps it could do even more. There is every reason also to believe that each step of improvement in the original nature of man would, in and of itself, improve the environmental conditions in which he lives and learns.

So much for the general possibility of eugenics in the case of intellect, morals and skill—for what should soon be in every primer of psychology, sociology and education, and be accepted as a basis of practise by every wise family, church and state.

The next question concerns the *extrinsic* effects of selective breeding for intellect or for morals, the possibility of injuring the race indirectly by a change in, say, intellect which in and of itself is desirable. If we breed horses for speed, they are likely to lose in strength and vigor; do we run such risks in breeding men for intellect, or for morals, or for skill? This question has been neglected by the hortatory type of enthusiasts for eugenics. It has also not received the attention which it deserves from the real workers for racial improvement, probably because the psychological investigations which answer it are little known. They do, however, give a clear and important answer—that there is practically no chance whatever of injury from selective breeding within a race for intellect, or for morality, or for mental health and balance, or for energy, or for constructive ingenuity and skill—no risk that the improvement of any one of these will cause injury to any other of them, or to physical health or happiness. The investigations have found that, within one racial group, the correlations between the divergences of an individual from the average in different desirable traits are positive, that the man who is above the average of his race in intellect is above rather than below it in decency, sanity, even in bodily health. Chart 6 shows, for example, the average *intellect* of each of the groups, when individuals are graded 1, 2, 3, 4, etc., up to 10 on a scale for *morality,* according to Woods's measurements of royal families. I may add that the effect of chance inaccuracies in Woods's ratings, whereby one individual is rated as 8 or 10 when he should have been rated 9, or is rated 4 or 8 when he should have been rated 6, is to make this obtained and shown relation of intellect to morals *less close* than it really is.

Nature does not balance feeble-mindedness by great manual dexterity, nor semi-insane eccentricities by great courage and kindliness. Correlation of divergences up or down from mediocrity is the rule, not compensation. The child of good reasoning powers has better, not worse, memory than the average; the child superior in observation is superior in inference; scholarship is prophetic of success out of school; a good mind means a better than

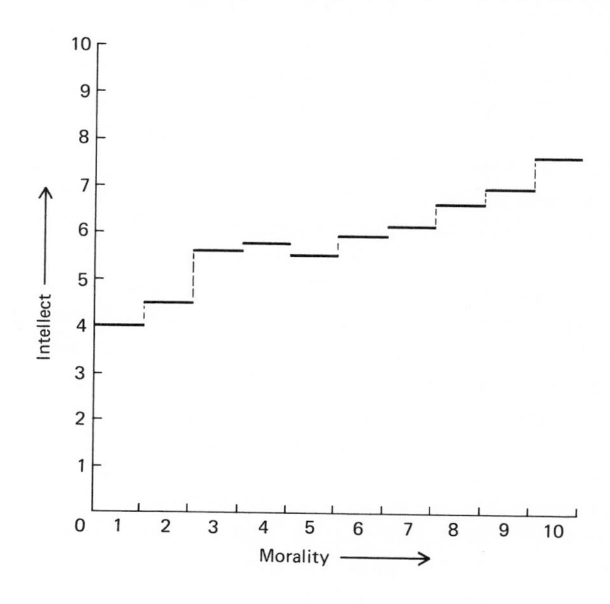

Chart 6. The Relation between Intellect and Morality in European Royal Families. After Woods.

average character. The fifty greatest warriors of the world will be above the average man as poets. The fifty greatest artists of the world will be better scientists than the average. Genius of a certain type does, *via* the nervous temperament, ally itself to eccentricities of a certain type; and very stupid men can not be rated as insane because they are already idiots; but on the average the most intellectual tenth of the population would, under equal conditions of strain, furnish fewer lapses into insanity than its proportional quotum.

Selective breeding for superior intellect and character does not then require great skill to avoid injurious by-products or correlatives of intrinsically good traits. *Intrinsically good traits have also good correlatives.* Any method of selective breeding, then, which increases the productivity of intellectually or morally good stock over that of poor stock, will improve man, with one possible added requirement—that breeding should be for fertility as well, should not be suicidal, should not make the race better, but at the same time put an end to it altogether!

It might be that there was a necessary inverse correlation in human nature between fecundity and high intellectual and moral station whereby, the better men became, the fewer offspring they would have; and whereby, at a certain limit of super-manhood, reproduction would cease. Certain changes of the birth-rate with time, and certain variations in it amongst groups, have given some students the impression that intellect, at least, is, by natural necessity, inversely correlated with fecundity.

It is hard to find the facts by which to either verify or refute the notion, current in superficial discussions of human nature and institutions, that such is the case. Sad testimony to man's neglect of the question which of all questions perhaps concerns him most—the simple question of which men and women produce the men and women of the future—is given by the fact that almost no clear and reliable evidence is available concerning the relations of fecundity to intellect, morality, energy, or balance. The most significant evidence is that collected by Woods in the case of royal families. Woods gives the number of children living till 21 in the case of each individual of the royal families which he studied. From them I have made the summaries noted on Charts 7 and 8. Each of these sets of facts is of course the result of the constitutional fecundity of the women in question plus certain very intricate cooperating circumstances; and neither can be taken at its face-value. What the birth rate would have been had the constitutional capacity of each woman worked under equal conditions, can only be dubiously inferred. My own inference from relevant facts concerning the studies of differentiated birth rates with which I am acquainted is that morality, mental health, energy, and intellect perpetuate a family, and that wherever the *really* better, or saner, or stronger, or more gifted, classes fail to equal the really worse, ill-balanced, feeble or stupid classes, it is a consequence of unfortunate circumstances and customs which are avoidable and which it is the business of human

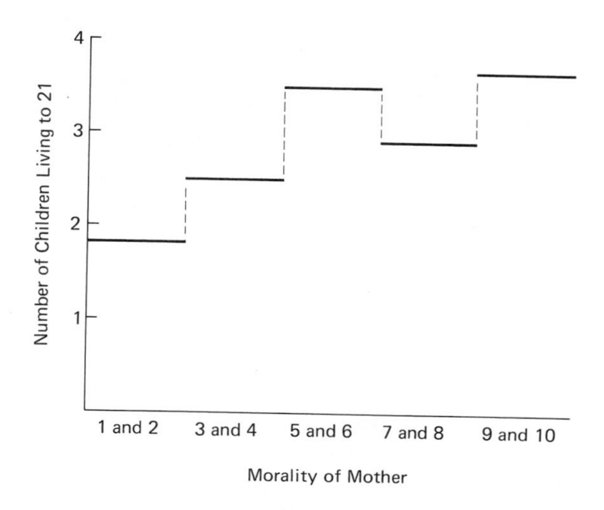

Chart 7. The Relation of Morality of Mother to Number of Children.

policy to avoid. Society may choose to breed from the bottom, but it does not have to.

No great ingenuity or care then seems necessary to make fairly rapid improvement in the human stock. The task is only the usual one of any rational idealism—to teach people to want a certain thing that they ought

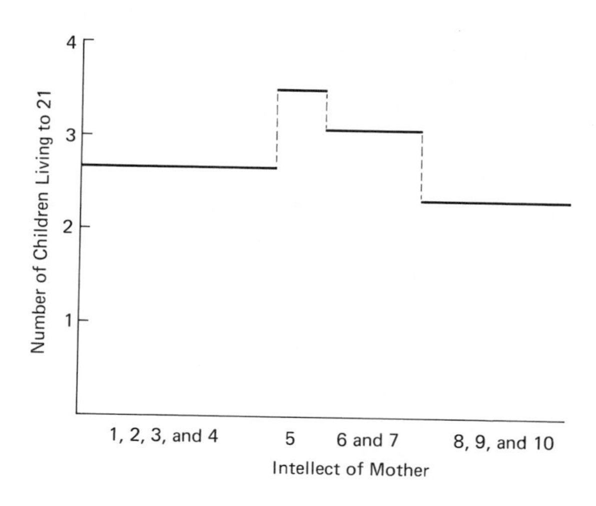

Chart 8. The Relation of Intellect of Mother to Number of Children.

to want, and to change social usages so as to satisfy this new want. The same sort of tuition whereby men are learning to want those who are alive with them to be healthier, nobler and more capable, will serve to teach us to want those who are to live with our children's children to be healthier, nobler and more capable. Provided certain care is taken to favor the sane, balanced type of intellect rather than the neurotic, any selective breeding which increases the fecundity of superior compared to inferior men, and which does not produce deterioration in the physical and social conditions in which men live, will serve.

The danger of deterioration in physical and social conditions from breeding for intellect and morals is trivial. The effect is almost certain to be the opposite—an improvement in physical and social conditions. The more rational the race becomes, the better roads, ships, tools, machines, foods, medicines and the like it will produce to aid itself, though it will need them less. The more sagacious and just and humane the original nature that is bred into man, the better schools, laws, churches, traditions and customs it will fortify itself by. There is no so certain and economical a way to improve man's environment as to improve his nature.

Each generation has of course to use what men it *has* to make the world better for them; but a better world for any future generation is best guaranteed by making better men. Certain worthy customs of present civilization may be endangered by rational control of who is to be born, though this seems to me unlikely. In any case, we may be sure that if the better men are born they will establish better customs in place of those whose violation made their birth possible.

It is not by a timid conservatism sticking to every jot and tittle of the customs which gifted men of the past have taught the world, that we shall prevent backsliding: it is far safer to trust gifted men of the present and future to keep what is good in our traditions, and to improve them. The only safe way to conserve the good wrought by the past is to improve on it.

It is beyond the province of this lecture to devise biologically helpful and socially innocuous schemes of selective breeding, but I may be permitted to record my faith that if mankind to-day really wanted to improve the original nature of its grandchildren as much, say, as it wants to improve the conditions of life for itself and its children, and believed certain facts of biology and psychology as effectively, say, as it believes that wealth gives power or that disease brings misery, appropriate schemes for selective breeding would be devised well within the span of our own lives.

Any form of socially innocuous selective breeding will improve the stock by reproducing from those members of it who have shown, by ancestral and personal achievement, with due allowance for favorable or unfavorable circumstances, the superiority of the germ plasm which they bear. But some forms may be far more effective than others according to the way in which the original components of intellect, character, energy, skill, stability and the like in the germs are constituted. Suppose, for example, that the original germinal basis for human intellect consisted in the presence of a certain constant something, call it "I_n, the determiner for intellect," in the germ or ovum. The fertilized ovum, which is the human life at its beginning, could then have I_n double, if both the germ and ovum had it; I_n single if one or the other had it; or could lack I_n, as it must if neither had it. Suppose that the consequences of these three conditions were that the I_nI_n individuals would tend, with fair conditions in life, to be specially gifted; that the I_n individuals would tend to be of "normal" intellect; that the individuals lacking I_n would tend to be feeble-minded. It is then the case that of the germs produced by the individual who had I_nI_n at the start of his life, each contains I_n, that of the germs produced by the individual who had I_n at the start of his life, half have I_n and half lack it, and that of the germs produced by the individual who lacked I_n at the start of his life, no one has I_n. Consequently, by discovering the individuals who lacked I_n at the start of life and preventing them from breeding, we could rapidly reduce feeble-mindedness. By discovering the individuals who had I_nI_n at the start of life and breeding exclusively from them, we could eradicate feeble-mindedness and ordinariness both, leaving a race of only the specially gifted. The discovery could be made in a few generations of experimental breeding; and the exclusion, of course, could be made one generation after the discovery.

This supposition will be recognized by many of you as a simplified case of Mendelian inheritance of a unit character due to the presence or absence of a single determiner which can either be or not be in a germ or ovum, and which "segregates."

No case quite so simple as this can be true of human intellect, but something approximating it has been suggested as perhaps true.

Suppose, on the other hand, that the germinal basis for intellect consists in the presence, in the germ or ovum, of one or more of four determiners—I_1, I_2, I_3 and I_4—contributing amounts 1, 2, 3 and 4 of intellectual capacity. The fertilized ovum could then have any one of 256 different constitutions ranging from the entire absence of all these determiners to the presence of each one "duplex"—*i.e.*, in both germ and ovum. If such duplex presence meant that the two contributions combined additively, the original intellect of the individual could range from 0 to 20. Individuals, all of one same original intellect—10—might be of very different germinal constitutions, and so of very different possibilities in breeding. If two individuals, each of original intellect 10, were mated, it might be the case that their possible offspring would range in intellect from 0 to 20, or it might be that they could not go below 8 or above 12.

If the number of germinal determiners of intellect is increased to five or six, the task of telling the constitution of the germs produced by any individual of known original intellectual capacity is enormously increased; and the research needed to guide the best possible breeding of man is very, very much more laborious. Moreover, instead of hoping to bring man to the best possible status (subject to the appearance of new desirable mutations) by a few brilliant rules for marriage, we must then select indirectly and gradually by parental achievement rather than directly by known germinal constitution, just as animal and plant breeders had to do in all cases until recently, and just as they still have to do in many cases. Only after an elaborate system of information concerning family histories for many generations is at hand, can we prophesy surely and control with perfect economy the breeding for a characteristic which depends on the joint contribution of five or six determiners. For it is just as hard to "breed in" a determiner that raises intellect or morality only one per cent. as it is to "breed in" one which raises it a hundred per cent.—provided, of course, the latter determiner exists. And it is thousands of times harder to discover the distribution of a determiner in the human race's germs when it is one of ten that determine the amount of a trait, than when it is one of two.

The germinal determination of intellect, morality, sanity, energy or skill is, so far as I can judge, much more like the second complex state of affairs than the first simple one. Important observations of the inheritance of feeble-mindedness and insanity have been made by Davenport, Goddard and Rosanoff, which they interpret as evidence that original imbecility is due to the absence of a single determiner, and that an originally neurotic, unstable mental organization is explainable almost as simply. It is with regret that I must assure you that these observations are susceptible of a very different interpretation. Much as I should like to believe that these burdens on man's nature are each carried in heredity in a single package, which selective breeding can shuffle off in a generation or so, I can not. A eugenics that assumes that intellect, morality, sanity and energy are so many single niches in the germs which selective breeding can, by simple transfers, permanently fill, is, I fear, doomed to disappointment and reaction. I dare to believe that the time will come when a human being idiotic by germinal defect will be extinct like the dinosaur—a subject for curious fiction and for the paleontology of human nature; but I have no hope that such a change can be made with the ease with which we can change short peas to tall, curly-haired guinea pigs to sleek, or plain blossoms to mottled ones.

There is another fundamental question whose answer is needed for the most economical selective breeding of human nature, a question which time permits me only to mention, not to describe clearly. Stated as a series of questions, it is this: Do the germs which a man produces—his potential halves of offspring—represent a collection peculiar to *him,* or only a collection peculiar to some *line,* or *strain,* or *stock,* or *variety,* of mankind of which he is one exemplar?

Suppose a hundred men and a hundred women to exist, each with identical germinal constitutions, so that, say, in every case one tenth of the germs (or ova) would be of quality 5; one fifth, of quality 6; two fifths, of quality 7; one fifth, of quality 8; and one tenth, of quality 9. Suppose that they mated and had five hundred offspring. Suppose that the best fifty of this second generation married exclusively among themselves; and similarly for the worst fifty. Would the offspring of *these* two groups differ, the children of the best fifty being superior to the children of the worst fifty? Or would this third generation revert absolutely to the condition of the grandparental stock whence they all came; and be alike, regardless of the great difference in their parentage?

Does the selection of a superior man pay because his superiority is, in and of itself, a symptom of probable excellence in his germs; or only because his superiority is a symptom that he is probably of a superior "line" or strain?

That the second answer of each pair may be the true one is a natural, though not, I think, an inevitable, inference from the work of Johanssen, Jennings and others. They have found selective breeding within any one pure line futile, save when some peculiar and rare variations have taken place within it. Their work is of very great importance and forms the best introduction to the general problem of the limits to human racial improvement. I regret that time is lacking to describe these studies of heredity within one "pure line." It is from such that eugenics may hope to learn valuable lessons in economy of effort and exactness of expectation. I have, however, already taken too much of your time with the problems of the exact laws whereby good men have good offspring and whereby breeding for strength, wisdom and virtue may be most effective.

In the few minutes that remain let me sum up what might perhaps have been entitled the A B C of eugenics in the realm of mind.

I have tried to show that, in intellect and character, men differ, by original nature, in some sort of correspondence to the ancestry whence they spring, so that by selection of ancestry the intellect and character of the species may be improved; to show also that injurious by-products of such selective breeding are very easily avoided, if indeed they occur at all; and, finally, to state some of the problems whose answers will inform us of just how the original intellect and character of one man does correspond with that of his ancestors, and so of just the best ways to discover the best strains and to perpetuate them.

I hope to have made it clear that we have much to learn about eugenics, and also that we already know enough to justify us in providing for the original intellect and character of man in the future with a higher, purer source than the muddy streams of the past. If it is our duty to improve the face of the world and human customs and traditions, so that men unborn may live in better conditions, it is doubly our duty to improve the original natures of these men themselves. For there is no surer means of improving the conditions of life.

It is no part of my office to moralize on these facts. But surely it would be a pitiable thing if man should forever make inferior men as a by-product of passion, and deny good men life in mistaken devotion to palliative and remedial philanthropy. Ethics and religion must teach man to want the welfare of the future as well as the relief of the cripple before his eyes; and science must teach man to control his own future nature as well as the animals, plants, and physical forces amongst which he will have to live. It is a noble thing that human reason, bred of a myriad unreasoned happenings, and driven forth into life by whips made æons ago with no thought of man's higher wants, can yet turn back to understand man's birth, survey his journey, chart and steer his future course, and free him from barriers without and defects within. Until the last removable impediment in man's own nature dies childless, human reason will not rest.

23. A Physicist Describes the Mobilization of Scientists in World War I. 1920

This revealing essay might be subtitled, "Nobel Prize Winner Plays Soldier." Written under the spell of the idealism of World War I, this article gives a striking epitome of what the war did to American science. Why was this creative scientist so impressed by a lieutenant colonelcy and a network of important-sounding committees? The names of the leaders of the patriotic efforts of American science constitute an incredibly distinguished roster of productive scientists. Their efforts, detailed here, were heroic. But the difficult questions of the appropriateness of the tasks to which they were assigned and their actual, specific accomplishments must be raised. So, too, must the question of how in America the war effort affected the long-range program of science to push back the frontiers of knowledge. Yet the leaders of American science in the National Research Council found it exhilerating in a way that could not have been predicted and got the Council established on a permanent basis. Again, why?

ROBERT A. MILLIKAN (1868–1953) was born in Illinois and graduated from Oberlin in 1891. He took his Ph.D. at Columbia University and spent 1895–1896 at Berlin and Göttingen. From 1896 to 1921 he was on the faculty of the University of Chicago. After that he was at the California Institute of Technology where as director of the Norman Bridge Laboratory of Physics and chairman of the executive council he in effect ran the institution. He was the best known of the new physicists and often spoke for the profession. In 1923 he was awarded the Nobel Prize.

For further information see L. A. Dubridge and P. S. Epstein, "Robert A. Millikan," National Academy of Sciences, *Biographical Memoirs*, 33 (1959), 241–287; Robert A. Millikan, *The Autobiography of Robert A. Millikan* (New York: Prentice-Hall, 1950); and A. Hunter Dupree, *Science in the Federal Government, A History of Policies and Activities to 1940* (Cambridge, Mass.: Harvard University Press, 1957), chap. XVIII.

CONTRIBUTIONS OF PHYSICAL SCIENCE*

Robert A. Millikan

From the days of Alexander and Caesar, if not from periods even more remote, the engineer has been a vital adjunct of a successful army; for war machines have always had to be built and operated, bridges thrown across rivers, roads rendered passable, new terrain surveyed and new fortifications designed and constructed. These and their like have been from the earliest times the standardized operations of the Engineer Corps of every army. But there is another and a quite distinct rôle which the physical sciences played in the great war. For never in the history of warfare up to the year 1914 had the whole scientific brains of any nation been systematically mobilized for the express purpose of finding immediately new ways of applying the accumulated scientific knowledge of the world to the ends of war.

It is not my purpose in this chapter to deal with the standardized operations of the technical corps of the army and navy during the great war. For this I have no competence. I shall endeavor rather to pass in rapid review the most significant of the newer developments which were due in large measure to the organized activities of scientists who, until the great war, had no association with things military. Many of these scientists, like the writer, became connected during the war either as officers or as civilian employees with the military departments of the Government. But whatever our official connection with the military service, we were all associated in our scientific activities through the National Research Council,[1] which acted in the United States as the great clearing house of scientific information, and as a coördinating and stimulating agency for scientific research and development work in aid of the war.

So far as developments in the physical sciences are concerned this coördinating and stimulating work was done through three main agencies, namely,

* From Robert M. Yerkes, ed., *The New World of Science, Its Development during the War* (New York: The Century Co., 1920), pp. 33–48. Copyright renewed 1947 by Robert M. Yerkes. Reprinted with the kind permission of Mrs. Roberta Yerkes Blanshard.

[1] The National Research Council was set up in 1916 under the auspices of the National Academy of Sciences to do what the National Academy was ineffective in doing, helping the U.S. defense (or preparedness, as it was called then) effort. The National Research Council was able to utilize nonmembers of the Academy, both scientists and engineers and inventors, and to take money from the government for its work. Although established essentially as a government body, the NRC was at first supported by foundation grants.

first, the executive committee of the Division of Physical Sciences of the Research Council, second, the Research Information Service, and third, the weekly conference of the Physics and Engineering Divisions of the Council.

The National Research Council, being itself a voluntary association for research purposes of the scientific agencies of the country, civilian and governmental, industrial and academic, it was to be expected that the Executive Committee of its Division of Physical Sciences would embrace representatives of important scientific and technical agencies. Its membership was as follows: Prof. J. S. Ames, representing the National Advisory Committee for Aeronautics, Dr. L. A. Bauer of the Department of Terrestrial Magnetism of the Carnegie Institution of Washington, Dr. A. L. Day of the Geophysical Laboratory, Major A. L. Leuschner of the Chemical Warfare Service, Dr. C. F. Marvin, Chief of the Weather Bureau, Lt. Col. R. A. Millikan, representing the Signal Corps and the Anti-submarine Board of the Navy, Major F. R. Moulton of the Bureau of Ordnance of the Army, Major C. E. Menedenhall [*sic*] of the Bureau of Aircraft Production, Dr. E. F. Nichols of the Bureau of Ordnance of the Navy, Dr. H. N. Russell, associated with both the Engineer Corps and the Bureau of Aircraft Production, Dr. W. C. Sabine of the Advisory Committee for Aeronautics and the Bureau of Aircraft Production, Dr. Frank Schlesinger of the Bureau of Aircraft Production, General George O. Squier, Chief of the Signal Corps, Dr. S. W. Stratton, Head of the Bureau of Standards and Dr. R. S. Woodward, Head of the Carnegie Institution of Washington.

This committee held stated meetings for the formulation of policies, the initiation of new projects, and for the detailed discussion of the seventy odd major research undertakings which had been initiated in large part at least by the Division and which its members were either directing or closely following. The opportunity both to initiate problems and to follow those initiated elsewhere, particularly abroad, came about chiefly through the most successful functioning of the second agency mentioned above, the Research Information Service.

This service had its inception in the Spring of 1917 when certain British scientists in the British ministry of munitions addressed a letter to General Geo. O. Squier suggesting the development of a liaison between British and American scientists. This letter was referred by General Squier to the Chairman of the Division of Physical Sciences of the National Research Council who laid the matter before the Military Committee of the Council, which committee embraced the heads of the technical bureaus of the navy and army, namely, Admirals Benson, Griffin, Taylor and Earle, and Generals Squier, Black, Crozier and Gorgas, in addition to the heads of civilian technical bureaus like Doctors Marvin and Stratton of the Bureau of Mines and the Bureau of Standards. This body discussed the proposal at some length and concluded that an even more comprehensive plan for bringing about coöperation and preventing duplication was needed. It accordingly appointed a committee consisting of Dr. Walcott, Mr. Howard Coffin, Dr. Stratton and Mr. Millikan to formulate recommendations. The committee formulated a plan which was approved by the Military Committee and then by the Secretaries of War and of the Navy and finally by the President, who appropriated $150,000 from his war emergency fund for carrying the plan into effect.

This plan provided for the establishment of four new offices, one in Washington, one in London, one in Paris and one in Rome. The office in Washington was headed by a group of three men: the chief of the Army Intelligence Service, the chief of the Navy Intelligence Service, and the chairman of the National Research Council; the group in London, by the naval attaché, (Admiral Sims himself) chosen by the National Research Council. The function of the scientific attaché in England, who was Dr. H. A. Bumstead, was to keep in touch with all research activity in that country and to send back almost daily reports to our office in Washington. Similarly, all reports of work done on this side were sent by uncensored mail or by cable to the offices of the scientific attachés in London, Paris and Rome, and distributed from there to the research groups in Europe. The navy coöperated heartily with this plan from the start, and Admiral Sims aided it in every possible way. As for the army, at the request of the General Staff, the Secretary of War issued orders to all army officers who were sent on scientific and technical missions to make duplicate reports, one to the officer who sent them and the other to the office of the scientific attaché, so that there might be a central agency through which an interconnection might be had between all kinds of new developments. The actual functioning of the Research Information Service had most to do with developments in the Physical Sciences.

Furthermore, through the authority conferred by the Military Committee, there was held in Washington at the offices of the National Research Council a weekly conference of the Division of Physical Sciences and of Engineering, which reviewed all the reports from abroad each week and put the workers on this side into the closest touch with the developments on the other side. The whole plan was an admirable illustration of the possibilities of international coöperation in research. In the submarine field, for example, all anti-submarine work in England, France and Italy which was reported by cable and by uncensored mail immediately to the office of the Research Council in Washington, was taken each Saturday night to New London and presented in digested form to the group of scientists which was working there continuously on submarine problems. Similar arrangements were made with the airplane research groups, sound-ranging groups, etc., so that in the Research Information Service we had the first demonstration in history of the possibilities of international cooperation in research on a huge scale, a sort of coöperation which made it possible for any development, or any idea which originated in any of the chief civilized countries of the world to go at once, very frequently by cable, to all the other countries and to be applied there as soon as possible, or to stimulate carefully selected groups of competent technical men in these countries to further developments. The extraordinary rapidity with which scientific developments were made in the war was unquestionably due first, to the forming of a considerable number of highly competent research groups, and second, to the establishment of effective channels for the coöperation between these groups.

So much for the machinery by which the work in the Physical Sciences was stimulated and coördinated. As for the problems themselves it is only possible to sketch briefly the history of a few of the most important. Of them all the submarine problem stood out from the beginning of the war as of paramount importance. Effective attack upon it in this country started with the visit of the scientific mission which was sent to the United States in May, 1917, with definite official instructions from the French, British and Italian governments to hold back nothing, but to lay all the facts and plans of the Allies relating to scientific developments in aid of the war before properly accredited scientific men in the United States. The National Research Council, which acted as the host of this mission in the United States (for the mission had been sent here in return for a similar mission organized and sent abroad by the National Research Council in March, 1917) with authority conferred upon it by the War and Navy Departments, called a conference in Washington of some of the best scientific brains in the United States and for a period of a full week this conference met and discussed in detail the progress thus far made and the plans projected in the fields of submarine detection, of location of guns, airplanes and mines by sound, of ordnance, of signaling and of aviation instruments and accessories.

As a result of these conferences there were organized through the coöperative effort of the National Research Council and several of the bureaus of the army and navy, a considerable number of groups of scientific men, each of which was charged with the development of some particular field. For example, Professor Trowbridge, of Princeton, and Professor Lyman, of Harvard, were selected and placed in charge of the development in this country of the sound-ranging service. They and the group of scientific men whom they associated with them were first given commissions in the Signal Corps, and with Signal Corps authority and funds started development work in sound-ranging at Princeton University and at the Bureau of Standards. This whole group was later transferred to the authority of the Engineer Corps, but its directing personnel remained in the main unchanged and it did extraordinary work in the whole of the fighting of the summer of 1918, locating hundreds of guns by computing the center of the sound wave from observations made on the times of arrival of the wave at from three to seven suitably placed stations. This method had never been used in any preceding war and it proved extraordinarily accurate, a gun being located five miles away with an error of less than fifty feet.

Again it is not an over-statement to say that the most effective part of the anti-submarine work done in the United States grew directly out of that conference, and it grew out of it in this way. As Lord Northcliffe continually reiterated on his trip to the United States in the spring of 1917, the submarine problem[2] was at that time *the* problem of the war, for while Europe might fight with little to eat, it could not fight without iron and oil and other supplies which this country alone could furnish, and in the spring of 1917 civilization trembled in the balance, because the submarine was seriously threatening to destroy all possibilities of transportation from this country to Europe. The English scientists therefore, in particular, came to this country directed by their government to lay before the American scientists every element of the foreign anti-submarine program, whether already accomplished or merely projected, and in the conference under consideration a large part of the discussion centered around the submarine problem, which, as Sir Ernest Rutherford repeatedly pointed out, was a problem of physics pure and simple. It was not even a problem of engineering at that time, although every physical problem, in general, sooner or later becomes one for the engineer, when the physicist has gone far enough along with his work. Hence, since the number of physicists was quite limited, the number of men who had any large capacity for handling the problem of anti-submarine experimentation was small. These men were found mostly in university laboratories or in a very few industrial laboratories which employed physicists, and we unquestionably had gathered a very representative group of them together in the fifty men assembled in the conference at Washington. The success or failure of our anti-submarine campaign, and with it the success or failure of the war, so far as we were concerned, seemed to depend upon selecting and putting upon this job a few men of suitable training and capacity.

At the close of that conference a small committee was appointed to select ten men to give up their work and to go to New London to work there night and day in the development of anti-submarine devices. The men chosen were Merritt of Cornell, Mason of Wisconsin, H. A. Wilson of Rice Institute, Pierce and Bridgman of Harvard, Bumstead, Nichols and Zeleny of Yale, and Michelson of Chicago, although Professor Michelson was almost immediately taken off for other work of much urgency and Chicago was represented in a fashion by the writer who was there a portion of each week. This group worked under the authorization of the Secretary of the Navy and with the heartiest of coöperation from the Navy Department, although it was at first financed by private funds obtained by the National Research Council. In the course of a few months, however, when it had demonstrated its effectiveness it was taken over by the Navy, which spent more than one million dollars on the experimental work at that place. This station with its chief scientific personnel not largely changed became the center of our anti-submarine activity, and with other stations, one at Nahant, Mass., embracing chiefly the

[2] The submarine problem consisted of the fact that German submarines were destroying Allied shipping at so rapid a rate that vital supplies could not reach Europe—and especially Britain—in order to sustain the Allies in the war.

physicists of the General Electric Company, the Western Electric Company and the Submarine Signaling Company, one in New York presided over by Dr. Pupin, of Columbia, and one in San Pedro, Calif., which, like the New York station, was organized under the Research Council, made remarkable progress in the rapid development of anti-submarine devices—devices which exerted a notable influence upon the reduction of submarine depredations, and made it possible even by the fall of 1917, to predict that the submarine menace could be eliminated.

Unquestionably the most effective device developed in America, and one which played a real rôle in the elimination of that menace, was one which had the following origin. The French had already developed an apparatus consisting of a sort of great sound lens which brought the incoming pulses together in the same phase at the center of the lens near the bottom of the hull. This was presented and discussed at length in the conference. A full official report of the device was sent by the French government to the Anti-submarine Board of the Navy, and at a meeting of that board the writer requested to be allowed to take this report to the group of scientists at New London for the sake of a thorough analysis of it, for he felt confident, and so stated at the time, that through such an analysis we would obtain variants of the device which would be an improvement upon it. This procedure was followed and for two days ten men assembled at a hotel in New London and studied that report, drawing up four or five different variants for this device to develop and try out. The most successful and effective detector which actually got into use in the war was one of these variants of the original French device suggested and largely developed by Mason. It consisted of a row of from thirty to sixty sound receivers strung along in two rows one on either side of the keel of the ship, well forward; the sound pulses coming in to all of the receivers on one side were arranged to travel in tubes of just such length as to cause them all to unite in the same phase at the mouth of a tube leading to one ear of the observer, while all the sound pulses received by the other row are brought together in a similar way at the other ear. By now using the binaural sense to equate exactly the sound paths to the two ears, it is possible to locate the direction of the source of sound to within one or two degrees. This instrument could pick up submarines from one to ten miles away depending upon their speed and the weather conditions. A variant of the multiple receiver device, using microphones and electrical compensators to equate phases in place of ordinary sound receivers and sound compensators, was even more effective. Many of our submarines and destroyers which went across during the summer of 1918 were equipped with the acoustical form of this device, and now the electrical form is being still further developed for peace use, rather than for war, for it is possible through it to eliminate the chief terror of the sea, namely, collision in fog. And, when it is remembered that the preventing of a single disaster like the sinking of the *Titantic* [*sic*] or of the *Empress of Ireland* more than pays, without any reference to the value of human lives, for all the time and money spent by England, France and the United States combined in developing detecting devices, it will be seen how shortsighted a thing it is for any country to fail to find in some way the funds necessary for carrying on research and development work in underwater detection. For decades and for centuries we have allowed ships to go down year by year needlessly, simply because we have not realized the possibilities of prevention through properly organized scientific research in this field.

Another device capable of detecting a lurking submarine half a mile or more away by the use of a beam of sound waves of very high frequency was perfected too late to be of use, but it represents a war development of extraordinary interest. The credit for it is due primarily to Dr. Langevin of Paris, though the New York and San Pedro groups of American physicists did excellent work in the same direction following Langevin's lead. Other antisubmarine devices in considerable number were developed and effectively used, but these two are in most respects the most notable.

But it has not merely been in sound-ranging and in submarine detection that the war has demonstrated the capabilities of science. Every single phase of our war activities has told the same story. Turn, for example, to the development of new scientific devices for use with aircraft. How was that handled? The Science and Research Division of the Signal Corps, organized through the coöperation of the Signal Corps and the National Research Council, and later transferred to the Bureau of Aircraft Production, had a group of as many as fifty highly trained men, physicists and engineers, who were working in Washington and in the experimental station at Langley Field, twelve hours a day, seven days a week, on aviation problems—one group on improvements in accurate bomb dropping, another on improvements in airplane photography, another on the mapping of the highways of the upper air in aid of aviation, another upon balloon problems, such as the development of non-inflammable balloons, another on aviation instruments, compasses, speed meters, etc., and producing the best there are in the world, and finally a group on new sensitizing dyes for long wave-length photography, etc. Let me select for special comment the most important physical principles which have just now for the first time found large and effective application in war. I shall classify these under six heads.

The first two of these are (1) the principle of binaural audition and (2) the principle of sound-ranging (locating the position of a gun by plotting the sound wave emanating from it). These two share the honor of having proved themselves the most useful and effective of the new applications of physics to the purposes of war. The second was responsible for the location and destruction of thousands of enemy guns, while the first was responsible for the location and destruction of submarines, airplanes and mines.

The binaural principle itself was unknown even to most physicists before the war, though it is used by all of us when we turn our heads until we think we are looking in the direction from which a sound comes. The accuracy with which this can be done in the absence of disturbing reflections is surprising. When the observer has set his head so that the sound pulses from the source strike the two ears at exactly the same time he has the sense that the source lies in the median plane between the two ears. If the sound pulses strike one ear first, the observer has the sense that the source is on the side of the ear which is struck first. This sense is not due in any appreciable degree to intensity differences produced by shadow effects of the head. It has to do practically entirely with phase differences. The principle is beautifully illustrated by inserting into each ear one end of a piece of rubber tubing four or five feet long and scratching or tapping on the wall of the tubing, first at a point slightly closer to one ear than the other and then moving the tapping object slowly through the mid-point to a position nearer the second ear. The sound of the scratching or tapping will then appear to the observer to be in the ear which is nearest to it and then to move around the head to the other ear as the mid-point is crossed. With the unaided ear one can locate direction in this way to within five or ten degrees. The simplest way to increase the *sensibility* of the method to faint sounds is to increase the size of the ears by providing them with trumpet-like extensions. To increase the *accuracy* of location one stretches out the receiving ends until the distance between them is say, five or six feet instead of five or six inches, as it is in the case of the unaided ear. It is then only necessary to turn the whole receiving system through about one-twelfth the former angle to obtain the same phase difference. The angular accuracy of setting is thus increased twelve fold.

Two methods of applying the principle were used in the war. The one consisted in rotating the whole receiving system, one side of which was connected with a rubber tube to one ear, the other side in the same way to the other ear, until the observer had the sensation of feeling the sound pass from one ear to the other. At this instant he knew that the source was directly ahead of the line connecting the two receivers, or else directly behind this line, the distinction between the two positions being obtainable from the relation between the direction of the motion of the head and the direction in which the sound seemed to pass from one ear to the other. The second method, the one used with the submarine detector discussed above, consisted in keeping the receiving system fixed in space and changing the length of the sound path from each receiver to the ear by means of a so-called rotating commutator until the sound seemed to be passing from one ear to the other. The reading of the dial on the compensator then gave the direction of the source.

This principle proved so effective in locating enemy mining and tunneling operations that according to official despatches received by the Research Information Service both sides gave up such operations practically entirely a year before the close of the war. It was equally effective in anti-submarine warfare, a very simple form of binaural detector having been put out in large numbers by the General Electric Company, in addition to the more elaborate and more effective devices heretofore considered. The principle was less effective in its application to anti-aircraft work though even here it served a very useful purpose.

The third physical principle which was of immense use in the war was the principle of amplification. This extraordinary application of scientific investigations of the past two decades in the field of electron discharges had been reduced to practice in the telephone industry in 1914 when transcontinental wire telephony became for the first time possible through the development of the De Forest audion into a telephone repeater and amplifier—an advance which not only extended enormously the possibilities of communication, but saved at once millions of dollars even in the construction of short telephone lines. With six stage amplifiers of this electronic sort the energy of speech has been multiplied without distortion as much as ten thousand billion fold. Small wonder then that by 1915 enormously amplified wave forms produced by speech had been impressed on the ether from the Arlington Towers[3] with such energy as to be picked up and distinctly understood in Paris and Honolulu. But in spite of the success already attained in this field by the physicists of the telephone company, when the United States entered the war the principle of amplification had not been successfully applied either to inter-communication by wireless phone between ships (for example, submarine chasers) or between airplanes, and one of the most pressing problems which General Squier put up in April, 1917, to the Division of Physical Sciences of the Research Council was the problem of wireless communication between planes. This was solved by the mid-summer of 1917 by the group of physicists of the Western Electric Company to whom it was referred and, on Sunday following Thanksgiving 1917, for the first time in history, airplanes in flight were directed in official tests at the Wright field in Dayton, Ohio, in intricate maneuvers, from the ground or by the commander in the leading airplane, and reports and directions were given and received in clear speech. For wire and wireless telephone receiving, sending and amplifying on sea and land three-quarters of a million vacuum tubes were built by the Western Electric Company alone for the purposes of the war, and half as many more by the General Electric Company, so that the amplifying principle was of scarcely less importance in the successful conclusion of the war than were the principles of binaural location and sound-ranging.

The fourth tremendously important and altogether new application of the principles of physics to warfare was made in the field of airplane photog-

[3] This is an interesting example of the persistence of the idea of the ether as a medium through which light and radio waves, among other things, were propagated. The idea of the ether had presumably been disproved some years before by Millikan's superior at Chicago (mentioned earlier in the essay), A. A. Michelson. Arlington Towers were located at the Navy's experimental station in Arlington, Virginia, where use of one of the early high-vacuum electron tubes was pioneered in the field of radio transmission.

raphy. In this field as in those of submarine detection and sound-ranging, though not in that of amplification, we followed the developments of the British and the French, though contributing important elements ourselves. The war could scarcely have been fought at all without the airplane photographer who was the very eyes of the army. American developments in this field were organized by the Science and Research Division of the Signal Corps which in the summer of 1917 assembled a group of physicists and photographic experts under the direction of Dr. H. E. Ives. This group in closest coöperation with the Eastman Kodak Company of Rochester and the Burke and James Company of Chicago developed what are probably the finest airplane cameras in existence. In addition it developed color filters for detecting camouflage and increasing visibility of such value that forty thousand of them were used in the army and navy. It produced new dyes for use in the production of pan-chromatic plates designed to be used for the penetration of haze in airplane photography and made other advances in this important art which bid fair to revolutionize the whole process of surveying, since an airplane photograph taken in a few seconds can give information which it used to take months to acquire by laborious triangulation methods.

The fifth great new application of physics to warfare lay in the developments in meteorology and in the principles of ballooning. The realization of the possibility of non-inflammable helium balloons and the actual production of small propaganda balloons which dropped their loads a thousand miles from the starting point are among the most spectacular and interesting scientific developments of the war, but neither of them played any actual part in achieving the victory. Of untold importance, however, was the careful though unspectacular work of the meteorological section of the Science and Research Division of the Signal Corps which by thousands of pilot balloon flights accumulated the data that not only aided the flyer in his work at the front, but made possible the so-called ballistic wind corrections upon which the effectiveness of both the artillery and the sound-ranging services largely depended. When it is remembered that the biggest element in the effectiveness of a modern army is its artillery and that the effectiveness of the artillery is dependent entirely upon these wind corrections it will be seen how incalculably valuable the work of the trained physicists and mathematicians proved to be to the practical problems of the great war.

The sixth and last of the new applications of physics to the purposes of the war had to do with the principle of signaling by visible light rays, by infra-red rays, by ultra-violet rays and by super-sound rays. In all of these fields there were developments of great interest and of much importance for the future, though none of them contributed largely to the victory of the Allies. In bombardments all the wire and wireless methods of communication often failed and light signals of some sort were the only reliance. Special signaling lamps were developed by the Science and Research Division of the Signal Corps and ordered in considerable numbers. A notable system of secret signaling with infra-red rays was developed by Theodore Case of Auburn, N. Y., and successfully used in keeping convoys together at night when lights could not be used. The possibility of having secret ultra-violet methods of guiding aviators at night back to their landing fields was demonstrated by R. W. Wood. As already indicated super-sound signaling under water was successfully accomplished by Dr. Langevin and applied experimentally in submarine detection.

Outside the lines of the foregoing classification there were some developments in Physics which deserve mention. Thus a leak proof gasolene tank for airplanes, developed by Dr. Gordon S. Fulcher in collaboration with the Miller Rubber Company of Akron, Ohio, which could be shot through by scores of bullets without leaking a drop of gasolene or catching fire even when the bullets were incendiary, had at the close of the war been ordered placed on all American combat planes. It promised to do away with the chief terror of the American flyer, namely, coming down in flames. An airplane compass and a speedmeter developed by Major Mendenhall and Lieut. Williamson, in coöperation with the General Electric Company were used on all American planes. Dr. Duff, Captain Webster, Captain Sieg and Captain Brown increased notably the accuracy in bombing, a matter of the greatest importance since doubling the accuracy in dropping bombs is more than equivalent to doubling the production of bombing planes. Under the stimulus of the war Dr. Coolidge developed a new and improved x-ray tube for use in field hospitals. Dr. E. F. Nichols developed a new type of mine, which was used in mining operations in the North Sea. Prof. A. A. Michelson developed a new and improved range finder, which was accepted by the Navy Department. Prof. Raymond Dodge developed a new piece of physical apparatus for the selection and training of gunners. This instrument was adopted and used both by the American and foreign navies. Optical glass was produced in large quantities for the first time in the United States under the guidance of a committee of the Physical Science Division of the Research Council, consisting of Drs. A. L. Day, S. W. Stratton and R. A. Millikan.

This is but an incomplete sketch of what look now like the most important developments in Physics which were stimulated by the war. Scores of other problems were undertaken the results of which may in the end be as useful both for the purposes of war and for those of peace as any of those herein set forth.

24. A Protestant Biologist Defends Darwin and Science from a New Attack. 1924

Shortly after this article appeared, the country was titillated by the Scopes trial, in which William Jennings Bryan assisted the state of Tennessee in the prosecution of a young science teacher for teaching evolution in the high school. Although the real issue in the trial was church-state relations and freedom of teaching, most people tended to remember it as a confrontation between fundamentalism (a religious movement that had been sweeping parts of the country since about 1915) and science. On another level the trial represented a contest between those who accepted the modern world and those who did not. Most middle class and urban Americans felt that the defense lawyers "made a monkey out of Bryan," and the authority of science was more firmly established in the public mind than ever. In this article a pious scientist dissects and identifies clearly the issues between Bryan and science. The extent to which these issues had changed since Darwin and Gray first wrote is instructive.

EDWARD L. RICE (1871–1960) was a well known zoologist who taught at Ohio Wesleyan from 1898 to 1945, serving as acting president of the University in 1938–1939. He was graduated from Wesleyan University in 1892 and then went to Berlin and Munich to study zoology. He received his Ph.D. from Munich and taught briefly at Wesleyan and Allegheny College before going to Ohio Wesleyan.

Further literature on this subject can be found in a contemporary book, Maynard Shipley, *The War on Modern Science: A Short History of the Fundamentalist Attack on Evolution and Modernism* (New York: A. A. Knopf, 1927); and Lawrence W. Levine, *Defender of the Faith: William Jennings Bryan: The Last Decade, 1915–1925* (New York: Oxford University Press, 1965).

DARWIN AND BRYAN—A Study in Method*

Edward L. Rice

A few days ago a friend, with whom I was discussing the subject of this address, turned to me with the quick remark, "What I can not understand is why you, a scientist, should pay any attention to the attitude

* *Science,* 61 (1925), 243–250. Reprinted with the kind permission of the American Association for the Advancement of Science.

toward evolution of Mr. Bryan or any other layman." Others may be raising the same question, and an *apologia pro argumento meo* may be in order.

If I were only a scientist, I think I should pay no attention, beyond a smile, to writings like those of Mr. Bryan on evolution. Scientifically, it is of little moment whether Mr. Bryan or any other individual does or does not believe in evolution or in any other scientific theory.

But I am not merely a scientist; in common with the majority of the members of Section F,[1] I am a teacher. As teachers, we may well be jealous of that freedom of investigation and freedom of teaching through which the intellectual progress of the past has been won and through which the intellectual progress of the future must come. There must . . . be limits to this freedom—liberty must not become license; but undue restriction can lead only to mental stagnation. Mr. Bryan's proposition to delegate to state legislature or church council the determination of the orthodoxy of scientific theory savors of the Middle Ages rather than of twentieth century America. And Mr. Bryan wields an influence not to be ignored. Tremendous moral earnestness and extraordinary oratorical power make a combination potent for right, but equally potent for error if misdirected—in no case to be disregarded.

To most of us the matter may have no personal bearing; to others the crisis is immediate. The particularly picturesque attack in the Kentucky legislature was lost by a single vote.[2] Some states have already passed more or less extensive restrictive laws; the question is now pending in other states; and Mr. Bryan promises that the campaign is to be carried into every state legislature.

In some church colleges the crisis is also acute. Permit me to quote, without names, a letter received by the president of my own institution within the current college year:

> Do your professors present the facts of evolution to students in lectures? Do they use text-books which have the theory of evolution in them?
>
> If so, do you regard this policy injurious to the esteem in which the Bible is held by students? Are students more or less Christian on account of such tuition?
>
> In what sense, if any, could evolution and the Bible conflict?
>
> Ludicrous as these questions appear in your environment, I hope you will render me the service which the replies to these questions will bring. We are having down south here a rather heated fight among the different branches of the ________________ church. The fight is very hot in ____________ [name of state], and especially at ________________ University.

As educators it behooves us to take notice of the trend of events and not to sit in smug security.

[1] This was Rice's address as vice-president of the American Association for the Advancement of Science and chairman of its Section F—Zoology in 1924.

[2] That is, the attack on evolution by means of enacting a law forbidding the teaching of such ideas.

But, in common with many members of this section, I am not simply a scientist and a teacher but also a Christian. I recognize and respect the various shades of belief and unbelief represented in this company. The scientific spirit, which recognizes the fallibility of all belief, should exclude the spirit of dogmatism and intolerance toward honest differences in belief. You will not all agree with me, but I hope you will respect my position when I say that it is precisely as a Christian that I most resent the attitude of Mr. Bryan.

Mr. Bryan's scientific belief and religious belief are matters personal to Mr. Bryan; but when Mr. Bryan uses his moral earnestness and his oratorical genius to proclaim to the world that belief in evolution precludes belief in God or, at least, is seriously hostile to religious belief, he becomes fundamentally dangerous, not to science, but to religion. Evolution is the universal belief of science to-day; and modern youth in America is essentially scientific. Confronted with Mr. Bryan's alternative, some young men will give up science; this is unfortunate, but relatively unimportant. Others (more, I believe) will feel themselves compelled to give up religion; this I regard as an inestimable loss to them and to the Christian church. In this day of the world's desperate need of religion, I can not look with equanimity upon any movement which tends to split the forces of the church rather than to bring them into harmony, or upon any attempt to read essentially religious men out of the church because of non-essential differences in scientific or theological belief.

For these reasons I have decided, albeit with some misgivings, to attempt to discuss one phase of this age-old conflict which has been fanned into new flame by the oratory of Mr. Bryan. I shall not attempt a comprehensive defense of evolution nor a systematic harmonizing of evolution and religion; I ask your attention only to the comparison of the methods used by Mr. Darwin and Mr. Bryan in reaching their conclusions and in expounding their views. For an exhaustive treatment of this one phase of the question I have neither the time nor the requisite philosophical training. I can only hope to bring together some interesting and useful items, many of which are familiar to you and all of which are within easy reach. In the snowball fights of our boyhood the snow was available to all alike; but it was found good military tactics to delegate certain individuals to manufacture snowballs for the use of those on the firing line. Similarly, it is my hope to be able to collect material and to shape some scientific snowballs which I trust others may be be able to use to good purpose.

During the last few years Mr. Bryan has been repeating to the world in most categoric form that the work of Darwin is mere guessing. Now emphatic repetition may have a marked psychological effect alike upon the speaker and the hearers, as suggested by the health formula of Coué and the organized cheering of the stadium and the all-too-familiar "pep rally."[3] "Hypothesis equals guess" has a catchy sound; moreover it has a considerable element of truth, particularly if "to guess" be interpreted according to the second definition of Webster (following New England dialect, but with the added authority of Milton and Dryden) as "to judge or form an opinion of, from reasons that seem preponderating, but are not decisive." In passing, it may be noted that the synonymy of the "guess" with the scientific hypothesis is not new with Mr. Bryan; the same parallel was pointed out by Huxley years ago, with the pungent comment—"The guess of the fool will be folly, while the guess of the wise man will contain wisdom." The hypothesis (or "guess") has its place in Mr. Darwin's work, as in all scientific method.

In his volume entitled "The Method of Darwin," Frank Cramer has given an interesting and illuminating analysis of Darwin's work as a conspicuous illustration of scientific method. The inductive method of science includes the use of hypothesis, but does not stop there—except in the judgment of Mr. Bryan. The initial hypothesis may be the result of a long and laborious collection of individual facts, their careful comparison and the selection of the elements common to the series—an induction in the narrower sense of the word; it may come as a flash of inspiration with few data as a foundation—a happy "guess"; or, lastly, an old but unsupported hypothesis may be adopted and rehabilitated. However derived, the hypothesis is only the hypothesis—perhaps that of wisdom, perhaps of folly.

In the case of evolution Darwin found the rival hypotheses of creation and derivation already in the field; but the evolution hypothesis, as worked out, for example, by his own grandfather or by Lamarck, appealed to him but little at the beginning of his scientific career. In common with his scientific friends, he was a strict creationist. But the *Beagle* voyage brought the young Darwin in contact with a wealth of new facts; and to Darwin a new fact was a new starting point for hypothesis—"I can not resist forming one on every subject."

Evolution was still hypothesis to him but becoming ever more attractive. Instead of accepting it and dogmatizing, he opened the "first note-book" in 1837, which, with its successors, was to shake the thought of the world. And note the direction taken in this investigation—not a hit-or-miss collection of data, but an instinctive dash at the crux of the matter. Animals and plants under domestication are particularly subject to variation; here, if anywhere, might suggestions be expected as to the character and cause of variation in nature. Fifteen months of careful collection of data concerning domesticated animals and plants . . . brought him to a new . . . supplementary hypothesis concerning the method of evolution. Evidently a reasonable explana-

[3] Emile Coué popularized changing one's personality and improving one's health by means of autosuggestion, that is, repeating positive thoughts to one's self over and over ("Every day in every way I am getting better and better").

tion of the method of evolution would make more probable the original hypothesis of evolution; historically it was this hypothesis of natural selection which brought evolution into the forefront of scientific . . . discussion.

Again another man might have stopped and dogmatized—not Darwin. It may fairly be said that the balance of his life was given to the verification of these two hypotheses. In part the verification consisted simply in the collection and correlation of more data similar to those already gathered; but, in greater degree, it included the carrying out of one deduction after another from his theory—not as ends in themselves but for verification of the main thesis. Assuming the truth of derivative origin, what was to be expected in the geological succession of the fossils? Then to the rocks for corroboration or contradiction. On the basis of evolution, what was to be expected in the relations of the faunas of Europe, North America, Africa, Australia, the Galapagos? And, again, to the maps, the museums and the journals of explorers for corroboration or contradiction. One after another the most diverse series of data were found, with singular uniformity, to confirm the main hypothesis; and thus was gradually built up that structure of interlocking hypothesis and verification which convinced Darwin himself, made speedy converts of Huxley, Lyell, Hooker, Gray and others of his intimate friends, threw the thought of the third quarter of the nineteenth century into turmoil and to-day dominates the whole scientific world.

It is to the evolution theory in general, not the theory of natural selection, that I refer as dominating the scientific world. In the nature of the case, the inductive method (the method of everyday life as well as of science) can never arrive at a demonstration; it must always remain a matter of less or greater probability. As the testing of a hypothesis brings one and another line of seemingly unrelated data into harmony, its probability increases to that of a theory; the more numerous and the more diverse the correlated data, the greater the probability of the theory, until, finally, practical certainty is attained. In the almost unanimous judgment of biologists the evolution theory has reached this status.

Of the two subsidiary theories to which the name "Darwinism" or "Darwinian theory" is properly restricted, natural selection is accepted by most biologists, although often with decided restrictions, while sexual selection[4] is relegated by many to the rank of a somewhat doubtful hypothesis rather than theory. Note carefully, however, that the evolution theory stands upon its own evidence, independent of the subsidiary theories. Like scaffolding, useful in the work of construction but unnecessary to the permanent edifice, natural selection and sexual selection might conceivably be torn down without materially affecting the evolution theory. Parenthetically, I may add my personal conviction that the destruction of the theory of

[4] The idea that in mating the more attractive members of a species (or human group) are likely to be more successful in surviving and propagating their particular characteristics than other members of the species.

natural selection is hardly less improbable than that of the general evolution theory.

I have laid emphasis upon these familiar distinctions because they are largely ignored by Mr. Bryan. In his writings he refers indiscriminately to evolution in general, human evolution and Darwinism; in at least one passage he explicitly states that he has "used 'evolution' and 'Darwinism' as synonymous terms." If I understand him correctly, Mr. Bryan's objection is not to Darwinism in its technical meaning, but to evolution in any form as applied to man. Evolution of the lower animals and of plants interest him only "as the acceptance of an unsupported hypothesis as to these would be used to support a similar hypothesis as to man." Human evolution and the evolution of the lower forms rest upon similar evidence; and Mr. Bryan denies one as categorically as the other. I believe I am doing him no injustice in limiting my further discussion to evolution in its general sense, applying alike to man and the lower forms of life, but independent of any theories as to its method.

Mr. Bryan outlines the matter as follows:

> The issue can be presented in two questions: First, is Darwin's hypothesis (evolution applied to man) true or false? Second, if false, is it harmful?

Granting, in common with theistic evolutionists, that God "could make man by the long-drawn-out process called evolution just as easily as he could make him by separate act," Mr. Bryan continues: "The question is narrowed down to one of fact—Did God create man by evolution or by separate act?"

Eliminating Mr. Bryan's restriction to man and his confusion of evolution and Darwinism, the first of his questions becomes—Is evolution true? This question Mr. Bryan answers as follows:

> In order that there may be no misunderstanding as to the position of those who believe as I do, let me say that the evidence is not sufficient to establish evolution as the process employed by the Almighty in either plant life or in animal life below man. I am aware that many scientists deal with evolution as if it were an established fact, but no one is compelled to accept any scientist as an authority except as the facts support him. The world can not be warned away from investigation by a scientific gesture. The scientist should be the last to ask that opinion be accepted as a substitute for fact.

But Mr. Darwin has collected a library of facts, and it is Mr. Bryan who is doing the gesticulation, although I grant that it is hardly a "scientific gesture."

As Mr. Bryan demands facts rather than opinions, let us note briefly the main lines of evidence in favor of the evolution theory. This evidence was summarized by Huxley for the ninth edition of the Encyclopaedia Britannica in a statement which is valid to-day as well as in 1878. The seven categories of Huxley's summary (all of which were mapped out in Darwin's

"Origin of Species") are so familiar to every biologist that they may be mentioned by title only, as the evidence from

(1) Embryology
(2) Homologies
(3) Geographical distribution
(4) Rudimentary organs
(5) Classification
(6) Modification under varying conditions
(7) Geological succession.

Let us see how Mr. Bryan meets these various lines of evidence. The numbering corresponds to that of the preceding list.

(1 to 3) To the first three of these items (embryology, homologies and geographical distribution) I find no reference in Mr. Bryan's writings, unless a mere mention of Darwin's emphasis on the similarity of human and simian embryos may be construed as a reference to the argument from embryology.

(4) To the evidence from rudimentary organs he gives hardly more attention—merely a sarcastic reference to a young collegian, whose faith is shaken when his "attention is called to a point in the ear that is like a point in the ear of the ourang, to canine teeth, to muscles like those by which a horse moves his ears." This is hardly an adequate treatment of the one hundred and fifty and more rudimentary organs found in man alone.

(5) Similarity of structure as between man and the apes Mr. Bryan does admit; in fact he says that "the whole case in favor of evolution is based on physical resemblances." But he evidences no appreciation of the universality of gradations in structure upon which the classification of both animals and plants depends.

(6) Mr. Bryan's discussion of modifications is rather astounding—a flat denial, on Biblical authority, that there can be such modification. The italics of the following quotation are his own:

> Evolution joins issue with the Mosaic account of creation. God's law, as stated in Genesis, is *reproduction according to kind;* evolution implies reproduction *not* according to kind. While the process of change implied in evolution is covered up in endless eons of time it is *change* nevertheless. The Bible does not say that reproduction shall be *nearly* according to kind or *seemingly* according to kind. The statement is positive that it is *according to kind,* and that does not leave any room for the *changes* however gradual or imperceptible that are necessary to support the evolutionary hypothesis.

Such changes as have been actually observed in pigeons and cabbages are calmly ignored by Mr. Bryan; he also appears oblivious of the rather palpable fact that not all the present diverse human races can be exactly like the traditional ancestors demanded by his theory.[5]

[5] For fuller discussion, see Piper, C. V., "Does the Bible teach evolution?" Science, Vol. 56, p. 109, July 28, 1922. [*Footnote in original.*]

(7) Mr. Bryan's attitude on geological succession is set forth in a series of statements, not perfectly clear in all points, but apparently intended as a categoric denial of the presence of connecting forms among the fossils. The following passage is representative:

> Wherever there is found living to-day any species of which an ancestor has been found in the rocks the living descendant is like the fossil ancestor. If this is what the evidence proves, why should we assume the truth of an hypothesis which is contradicted by everything which has been found and supported by nothing?
>
> Darwin insisted that his hypothesis should be accepted even tho the missing links had not been found, and evolutionists still insist that the hypothesis should be accepted even tho the missing links have not yet been found. They boldly demand that we substitute a guess for the Word of God even tho the guess has not been proven—in fact, has been disproven by all the evidence.

Mr. Bryan can see no evidence for evolution in the marvelous wealth of fossil forms, some of them obviously intermediate in character between distinct species, genera or larger groups of to-day, and others forming unbroken gradational series between earlier and later fossil forms or between fossil and recent species. Darwin, in 1859, counted the scarcity (not absence) of connecting forms the greatest objection to his theory, and met it with his characteristic frankness. Twenty-one years later conditions had so changed as to justify Huxley's exclamation:

> If the doctrine of evolution had not existed, palaeontologists must have invented it, so irresistibly is it forced upon the mind by the study of the Tertiary mammalia which have been brought to light since 1859.

And H. F. Osborn expresses himself even more strongly in 1910:

> The complete geologic succession of the vast ancient life of the American continent was destined to demonstrate the evolution law.

This difference in reaction of Mr. Bryan and the paleontologists to the evidence of the fossils is exactly paralleled by the difference in reaction of Mr. Bryan and biologists in general to other lines of evidence for evolution. Mr. Bryan advances no new evidence; the data collected by scientists he ignores or denies. To the biologists the evidence seems conclusive for evolution; to Mr. Bryan it has no significance. In large part, doubtless, this difference is due to Mr. Bryan's simple ignorance of the facts. Ignorance of the details of biology is no disgrace to a lawyer; but a lawyer should be slow to pronounce a judicial decision upon technical evidence which he does not understand.

In larger part, however, Mr. Bryan's hostile attitude is due to the fact that he does not approach the matter with an open mind. In theory he recognizes that the "hypotheses of scientists should be considered with an open mind. Their theories should be carefully examined and their arguments fairly weighed"; practically the whole matter is decided for him in advance without reference to the scientific data. "The Bible," he writes, "not only does not support Darwin's hypothesis but directly and expressly contradicts it." Further, the Bible, according to Mr. Bryan, is "the revealed will of God, and therefore infallible"; and other statements imply very clearly that Bible interpretation must be strictly literal throughout.

Here is no scientific method in induction—hypothesis tested out by deduction and verification. Here is no question of greater or less probability; in such deductive reasoning correct logic must lead to a correct conclusion, *provided,* of course, that the first assumption is correct; evolution must be false, *provided* the first two chapters of Genesis are literal and accurate science. But what is Mr. Bryan's guarantee of the literal infallibility of the Bible? This view has not been universally held by the leaders of religious thought in past centuries; it was not accepted, for example, by Luther or Calvin, by Augustine or Jerome. Going back to New Testament times, it was a theologian, not a scientist, who warned the Corinthian Church that "the letter killeth, the spirit giveth life." And, going still a step further back, to the author of our faith, note how whole sections of the Old Testament code are amended in the brusque and authoritative formula, "It was said to those of old time . . .; but I say unto you . . ." In him the Law and the Prophets were fulfilled; but how unique and unexpected the form of the fulfilment! Jesus Christ was a Modernist, not a Fundamentalist,[6] in the matter of Old Testament criticism; and the Gospels are full of his efforts to overcome the deadly literalism even of his own disciples. Take, for example, this dialogue from the fourth chapter of John: "I have meat to eat that ye know not. . . . Hath any man brought him aught to eat? . . . My meat is to do the will of him that sent me and to accomplish his work." Verily, "the letter killeth; the spirit giveth life." The dogma of a literally inerrant Bible is not Biblical, not Christian. It is not with the Bible, but with Mr. Bryan's interpretation of the Bible that evolution is in conflict.

What are the alternatives? Mr. Bryan says: "The Bible is either the Word of God or merely a man-made book." This method of exclusion is always dangerous. Darwin applied it to the Parallel Roads of Glen Roy to his cost. "My error," he writes, "has been a good lesson to me never again to trust in science to the principle of exclusion."[7] There is very likely to be a third alternative hiding somewhere; in the interpretation of the Bible it is exactly this third alternative which is accepted by the theistic evolutionist to-day, as well as by the great majority of intelligent Bible students.

[6] The modernists in religion, as opposed to the fundamentalists, were willing to accept historical and scholarly criticism, symbolic interpretation, and even reason in understanding the meaning of the Bible.

[7] In 1839 Darwin published a paper about the famous geological formation, the Parallel Roads of Glen Roy (actually rock terraces). He assumed that the action of the ocean had caused the formation, for no other cause known then would have accounted for it. He was later embarrassed to have Agassiz suggest a new type of causation, glaciation, that provided a much more satisfactory explanation.

To these men the Bible is not the "Word of God" in the sense of verbal dictation from God; no more is it "merely a man-made book"; but it is a progressive, evolving revelation of God's will to man, changing with the evolution of the human race. Moreover, it is a text-book in religion, not in science. A cardinal of the time of Galileo described the Bible as teaching "how to go to Heaven, not how the heavens go." Mr. Bryan writes that "it is more important that one should believe in the Rock of Ages than that he should know the age of the rocks"; he might well have added that it is of the "Rock of Ages" that the Bible treats—not of the "age of the rocks" nor their contained fossils. The lesson of the first chapter of Genesis is the creatorship of God, not details of the method. With Genesis, thus interpreted, evolution has no quarrel.

I do not question that Mr. Bryan is perfectly sincere in his belief in the falsity of the evolution theory and its danger to the Christian religion; but is he perfectly sincere in the character of his argumentation? Certainly he is not frank. His method is that of the lawyer striving to win his case rather than that of the earnest seeker for truth. The contrast with Darwin is most striking and not to the advantage of the professed defender of the faith of the Christ who characterized himself as "the truth."

Three short quotations show three phases of Darwin's attitude to truth:

> I believe there exists, and I feel within me, an instinct for truth . . . of something of the same nature as the instinct of virtue.
>
> I have steadily endeavored to keep my mind free so as to give up any hypothesis . . . as soon as facts are shown to be opposed to it.
>
> As I am writing my book [The Origin], I try to take as much pains as possible to give the strongest cases opposed to me.

That he succeeded is seen in the impression made upon those who knew him best. Let Huxley speak for all:

> It has often and justly been remarked that what strikes a candid student of Mr. Darwin's works is not so much his industry, his knowledge, or even the surprising fertility of his inventive genius; but that unswerving truthfulness and honesty which never permit him to hide a weak place or gloss over a difficulty, but lead him, on all occasions, to point out the weak places in his own armour, and even sometimes, it appears to me, to make admissions against himself which are quite unnecessary. A critic who desires to attack Mr. Darwin has only to read his works with a desire to observe, not their merits, but their defects, and he will find, ready to hand, more adverse suggestions than are likely ever to have suggested themselves to his sharpness, without Mr. Darwin's self-denying aid.

In Mr. Bryan's writings, on the other hand, no objections are mentioned, no difficulties suggested; instead we find the ex-cathedra statement, oft repeated, that there are no evidences for evolution, therefore no difficulties with Mr. Bryan's position. It is tempting to multiply quotations; one must suffice:

> Neither Darwin nor his supporters have been able to find a fact in the universe to support their hypothesis.

And, in another connection, I have already traced out how the main lines of evidence upon which the evolution theory is based are, one after the other, simply ignored or categorically denied. Very different from the method of Darwin—rather the method of an earlier critic of whom Darwin himself writes:

> The reviewer gives no new objections, and, being hostile, passes over every single argument in favor of the doctrine. . . . As advocate, he might think himself justified in giving the argument only on one side.

As an advocate, yes; as a scientist, no.

Many such omissions of relevant evidence and some apparent misrepresentations of the evolution theory and its supporters may well be due to Mr. Bryan's unfamiliarity with the facts in the case. A lawyer who does not know law or a doctor who has not studied medicine is a quack and subject to legal control; even a school teacher must be duly certificated. Is there not a moral obligation that a man professing authoritative leadership on evolution should first familiarize himself with the subject?

In some cases it is difficult to believe that Mr. Bryan's omissions are due to ignorance. Mr. Bryan lays the utmost stress upon the verbal accuracy of the Biblical story of creation. But which story? We can hardly believe that his attention has never been called to the fact that there are two such stories in Genesis, the first ending with the third verse of the second chapter. Each has its great moral teaching, different from the other but consistent with it. But, literally interpreted, their mutual inconsistency is no less glaring than the inconsistency of either, thus interpreted, with the observed facts of geological succession. There may be a reference to this contradiction in Mr. Bryan's writings; I have never seen one. Is it conceivable that Charles Darwin would ignore a difficulty of this sort?

Contrast, again, the dogmatic certainty of Mr. Bryan concerning a subject wholly aside from his main professional work with the modesty and caution of Mr. Darwin, whose life was devoted to the study of this problem. A review in a prominent and rather conservative church paper contains these words:

> There is something interesting in the naïve notion which Mr. Bryan has of the contrast between the absolute certainty of his own religious opinions and the merely probable opinions of scientific men. He refers to the fact that Darwin is continually using such words and phrases as "apparently," "probably," "we may well suppose." "The eminent scientist," says Mr. Bryan, "is guessing." Because Darwin and other scientific men, in the truly scientific spirit, recognize their beliefs as only more or less probable, and claim for them no absolute certitude, Mr. Bryan considers that their opinions are of no

consequence at all. He *knows*. For him it is a matter of absolute certainty that there is a God, that every sentence of the Bible is the word of God, and that he himself understands aright every sentence of the Bible. All his religious opinions are utterly above the realm of probability, dwelling in a serene and heavenly atmosphere of absolute certitude.

No less certain is he concerning scientific matters. Let me remind you of a sentence already quoted—"*Let me say* [the italics are mine] that the evidence is not sufficient to establish evolution as the process employed by the Almighty, etc." Darwin gave twenty years to the collection of material for "The Origin of Species" and "thirteen months and ten days' hard labor" to the preparation of the manuscript. Perhaps a similar application to the subject would leave Mr. Bryan, however he might decide the main issue, less sure that he had probed the problem to its utmost depths. Perhaps it would bring him to an appreciation of the meaning of "probably" in scientific argument.

Mr. Bryan's writings are done in a style which can hardly be characterized as calmly scientific. He expresses regret at the "epithets" with which the "liberals" attempt to "terrorize the masses of the church into accepting without proof or even discussion the views of those who put their own authority above the authority of the Bible." But his own constant play on the word "guess," his repeated sarcastic parody of the evolution of eyes from "freckles" and of legs from "warts" are hardly conducive to calm discussion; his aphorism that "cousin ape is as objectionable as grandpa ape," and his statement that evolution gives Christ "an ape for his ancestor on his mother's side at least," are suggestive of the famous speech of Bishop Wilberforce in 1860, perhaps also deserving of a reply like that of Huxley on that historic occasion. Such rhetoric is entertaining, and, in this day of slogans, may be effective with the masses—perhaps also with state legislatures; but it is not science; nor it is the method of Darwin. Sarcasm and ridicule are as conspicuous for their absence from Darwin's writings as for their presence in Bryan's.

In addition to the question of the truth of the evolution theory, Mr. Bryan raises the second question of its harmfulness. By his own formulation —"If false, is it harmful?"—this question becomes relevant only in case evolution is proven false. Although by no means granting the falsity of evolution, I wish to call your attention briefly to two points in Mr. Bryan's argument concerning its supposed harmfulness.

First, he argues, evolution, if it does not crowd God out of his universe, at least pushes him so far away in space or time as to make him negligible. I quote:

> Why should we want to imprison such a God in an impenetrable past? This is a living world; why not a *living* God upon the throne? Why not allow him to work *now*?

Is not this exactly the position of the theistic evolutionist, for whom natural law is merely a human attempt to formulate the method of divine activity, and evolution a human attempt to formulate the method of divine creation?

But not all evolutionists are theists; and Mr. Bryan urges that it is evolution which has made them agnostic or atheistic. As his principal illustration he uses the familiar case of Darwin—a gradual drift from an orthodox belief to a condition of agnosticism, albeit with times, even in his later life, when he felt himself "compelled to look to a First Cause having an intelligent mind in some degree analogous to that of man," and in which he deserved "to be called a theist." But is it so certain that evolution was the sole cause or even the chief cause of Darwin's change of belief? Other elements should certainly be considered.

First among these is the matter of continued ill health. When one considers the mass of scientific work accomplished, in connection with the bodily weakness which reduced the working day to a minimum and necessitated frequent periods of complete rest and sanitarium treatment, can one wonder that, in his own words, his mind should become a "kind of machine for grinding general laws out of large collections of facts," and that there should be a corresponding "atrophy of that part of the brain . . . on which the higher tastes depend?" All are familiar with the pictures of the boy Darwin reading Shakespeare in the old window of the school, of the young traveller carrying Milton's "Paradise Lost" on shore trips in South America when only one volume was possible, and the aged scientist realizing with regret that he could no longer "endure to read a line of poetry" and that Shakespeare had become "so intolerably dull that it nauseated" him. May it not well be that his loss of formal religious faith was a parallel of this atrophy of the esthetic sense, seen also in the partial loss of the love for music, art, and, in lesser degree, for natural scenery. It may well be questioned, however, whether Darwin's scientific caution and questioning attitude did not lead him to an over-emphasis of his religious doubts, particularly in reaction against the dogmatic certainty of many of his critics.

There is another element in this problem to which Bryan has not referred. May not the responsibility for Darwin's loss of religious belief be laid, in part at least, upon the impossible character of the dominant orthodox theology of his day. In the storm of invective which burst upon his head after the publication of the "Origin," is it strange that even a man of Darwin's amazing charity and poise should have turned away from organized religion as well as dogmatic theology?

But neither in 1859 nor in 1924 can the blame for the conflict of evolution and religion be placed wholly on the theologians. There is an *odium scientificum* as well as an *odium theologicum*. In 1859 there were materialistic scientists who seized eagerly upon the evolution theory as a new weapon for attacking Christian faith; among the theologians, on the other hand, were strong men who, from the start, recognized the truth of evolution as an aid to faith. To-day, again, very many leading theologians take issue with Mr. Bryan's position as sharply as can the scientists; and some biologists are hardly less dogmatic in their support of a materialistic philosophy than is Mr. Bryan in his attack upon evolution. It may fairly be questioned whether the materialistic scientist is not as responsible for the present anti-evolution flareup as is Mr. Bryan himself. It is unfortunate that Mr. Bryan could not have directed his campaign against this materialism of individual evolutionists rather than against the evolution theory itself.

I have tried in this address to emphasize the hopeless inadequacy of the method exhibited in Mr. Bryan's attack upon the evolution theory, and the illegitimacy of his claim to popular leadership in such an issue. For his religious earnestness and his devotion to moral reform I have profound respect, although I deeply regret the reopening of the age-old conflict of science and religion under his leadership.

From the present phase of this unhappy conflict, two happy results are, however, already becoming apparent. On the one hand, there is an increasing popular interest in evolution and a more intelligent understanding of its significance. On the other hand, an increasing number of our leading scientists are publicly proclaiming their own theistic philosophy, and emphasizing anew the essential harmony of a progressive scientific belief with real religion. I rejoice in the public utterances of such men as Conklin, Coulter, Millikan and Osborn.[8] May their tribe increase! And may their efforts combine with the increasing popular interest in science toward the bringing in of the day when a more scientific religion and a more religious science shall join in a common welcome to truth, whether revealed in nature, in human life, or in the Bible, and shall present an unbroken front in the struggle for the higher evolution of the human race.

[8] These were prominent men of science who had made, as Rice indicates, public statements supportive of theistic belief. The fact remains, as Rice implies, that many prominent scientists of the period tended to be agnostic if not outright antireligious, however much they were unanimous in upholding moral and ethical conservatism.

25. A Theologian Says Science Fails to Produce Certainty and Well-Being. 1929

For a long time science had been gaining prestige in the United States, and a substantial part of the literate population maintained a surprisingly dogmatic faith in the ability of science not only to raise the standard of living but to provide clear and certain answers to all important questions. With the coming of relativity, indeterminacy, and probability, opponents of science at last detected a weakness in the apparent omniscience of the scientist and rushed in to attack. The scientists themselves were abandoning the certainty that science had once promised. How could it offer a sure guideline if it could not anymore even furnish concrete, certain facts, much less predictions? This article represents some of the stored-up animus toward scientific dogmatism and at least one alternative viewpoint that found support among intellectuals in the 1920s.

JOHN WRIGHT BUCKHAM (1864–1945) was a prominent theologian and author and had been professor of Christian theology at the Pacific School of Religion from 1903 to 1939. He was born in Vermont and graduated from the University of Vermont. He also attended Andover Theological Seminary and in 1904 received a D.D. from the University of Vermont. As a conservative theologian, his view was close to that of another important intellectual movement of the 1920s, the so-called New Humanism.

See Charles William Heywood, "Scientists and Society in the United States, 1900–1940: Changing Concepts of Social Responsibility" (Ph.D. dissertation, University of Pennsylvania, 1954).

THE PASSING OF THE SCIENTIFIC ERA*
John Wright Buckham

In spite of age-long racial experience, the illusion that the particular age in which a given generation lives, with its dominant powers and ideas, is final, is one that is continually arising and as continually being shattered.

Our own age has witnessed a striking instance of what may be called the *delusion of finality* in respect to its major interest, Natural Science. It is now a hundred years, roughly speaking, since Auguste Comte (in 1826) began in Paris the first course of lectures on the "Positive Philosophy"

* *The Century Magazine,* 48 (1929), 432–438. Copyright, 1929, The Century Co. Reprinted by permission of Appleton-Century-Crofts, Educational Division, Meredith Corporation.

in which he set forth the doctrine of the "Three States." According to this well-known formula humanity has passed through two successive stages, the theological and the metaphysical, and has entered the third, positive or scientific, stage. Comte himself outgrew his assumption of the finality of positivism, and attempted to supply its lack by founding a new religion, but the idea which he had fathered persisted. The conception of the finality and sufficiency of Natural Science, fostered in one form or another by Herbert Spencer, Huxley, Tyndall and many other men of breadth—rather than depth—of vision, obtained a fixed hold in the nineteenth century and continued—in spite of misgivings—until the World War. The shock of that great upheaval left no dogma undisturbed, and compelled Science, as well as philosophy and religion, to reconsider its *principia.*[1]

Lest there be confusion as to what is meant by the term Science in this article, let me first define it as follows: Science is the organization of a body of experience in a given field, gathered by induction, formulated by hypothesis and deduction, tested by experiment, and progressively related to other new fields of experience and to new conditions. This, of course, covers the phenomena in every field of experience including, for example, religion; but in common usage, and as here employed, "Science" is confined to the examination and organization of sense phenomena—that is, Natural Science.

Within the last ten years a feeling has been growing that Natural Science is not only less certain as to its content but less complete in its scope and far less assuredly beneficent in all of its effects than had been, somewhat unscientifically, assumed. A number of events have conspired to deepen this doubt, until now—although Science is as active and successful as ever—there are many indications that it is imperceptibly but surely losing the strength of its hold, as the spirit of the age restlessly seeks after something to satisfy cravings which Science is unable to meet. A new mental attitude is forming. A new era opens.

> "The cloudy sails are set; the earth-ship swings
> Along the sea of [thought] to grander things."

All this is not mere accident but is in accord with an eternal *law* (logos) which gives meaning to change.

It has been a great and beneficent era in many respects—this of the dominion of Science—but it is due to pass, "lest one good [era] should corrupt the world." Its passing is likely to prove a wholesome and prophetic disillusionment and fraught with future good.

By the passing of the Scientific Era is not meant the passing of Science itself. Science has brought inestimable gain to humanity in knowledge, in method, in human, as well as in material advancement. To challenge this would be the height of folly. Science will go on with its conquests and its

[1] First principles.

beneficences; but its glamour, its dominion over other interests, its inclination toward self-sufficiency are already disappearing.

The causes of the growing disillusionment with respect to Science may be described as, first, a sense of its inability to accomplish what was expected of it and, second, a growing consciousness—springing up within as well as without the domain of Science—of its inner limitations and inadequacy.

The incompetence of Science to meet certain deep-lying human needs has shown itself in at least three directions: first, its failure to throw any real light upon the meaning of human existence and of the universe; second, its incompetence to furnish a satisfactory type of education; and third, its inability to foster moral and spiritual principles sufficient to sustain individual and social integrity and good-will.

The failure of Science to throw light on the ultimate problems of man and of the universe has only slowly and as yet partially become manifest. Its successes have been so signal and so repeated that it has seemed again and again to be on the point of unveiling, now the secret of life, now the nature of matter—or of its increasingly refined and subtle substitutes—and now the inner structure and working of the mind itself. But these hopes have been awakened only to be frustrated. Not only have old difficulties obstinately refused solution, but new and formidable ones have arisen through the agency of Science itself. "Events" are even more puzzling than substances. The atom as "a center from which radiations travel" is as ethereal as *Ariel.*[2] Electricity is a kind of will-o'-the-wisp, and the mind is proving the most elusive of Undines.[3] As the universe has expanded, both as regards its magnitudes and its minitudes, its mystery has increased proportionately until Science is almost pathetically powerless before the Unknown. Science has, in fact, only magnified the causes for the wonder in which, according to Plato, philosophy begins and the awe in which, according to Otto, religion takes its rise. It is characteristic of Science that it can find no meaning in mystery which to religion is so full of significance. To religion mystery means, in the words of Professor Hocking, "I know not, but it is known"; to Science mystery means only impotence and frustration—the end of the trail.

Evolution, which for a time seemed so full of promise, both theoretically and practically, has itself become a problem. It has assumed as many forms as Proteus.[4] There has been an evolution of evolution and the end is not yet. Moreover evolution has disclosed a perverse tendency to work both ways. Devolution has forced its way forward to dispute the field with evolution. The "Escalator Evolution," as Marshall Dawson has termed it, of the nineteenth century has broken down. The escalator when it operates seems to be as ready to carry its passengers downward as upward. This is unfortunate, but must be reckoned with.

[2] The spirit of the air.

[3] Water spirits.

[4] A sea god in classical mythology who had the power of assuming different shapes and forms.

As evolution has assumed the guise of *pan-evolution* and threatened to engulf the soul of man in an Impersonal Nature, the common people, ignorant but not indifferent, have shown a disposition to rise against it and in some sections of our semi-religious country have undertaken to legislate evolution out of their schools and universities. All this is extremely foolish but it raises the question whether the more ardent and clamant protagonists of Evolution have not overpassed their bounds and exceeded what a far-sighted American philosopher long since termed "the limits of evolution." The fear has arisen of evolution becoming, if not "procuress to the lords of hell" at least the insidious, though unintentional, foe of morals and religion, public and private.

What is needed is a re-appraisal of evolution—such as Science of itself alone is unable to make—in the light of the wider law of development, of which it is but the root and prophecy. This alone will reveal the relation of evolution to other cosmic and spiritual laws and will disclose not only its scope and significance but also its necessary limits.

Another at least partial failure of Science lies in the field of education, especially higher education. Science is implicated in this failure since it is the major factor in contemporary education.

It may seem an injustice to charge education with even partial failure, in view of the illiteracy it has banished, the vast numbers whom it is training and the efficiency of method which it has achieved; but any one who could be wholly satisfied with the methods and results of present-day education would thereby prove himself but imperfectly educated. Granted that great progress in educational technique has been made, especially in dealing with various grades of mentality, surely no true educator can be content with the present concentration upon vocational, commercial and technical subjects to the neglect of that knowledge of the Best which Matthew Arnold designated as culture.

There can be no doubt whatever that the study of Natural Science has contributed a certain spirit and method to education which is invaluable and permanent. Thoroughness, accuracy, patience in induction, testing by experiment, readiness to accept new forms of knowledge—these resultants of Science can never henceforth be dispensed with in any adequate system of education. (These virtues, however, characterize the courses *in* Science rather than the courses *about* Science to adopt a distinction used by Professor Shorey.) Nevertheless the mind that has been trained simply, or predominantly, in Science is an unconsciously meager and ill-furnished mind. The range of its interests is mainly technical and specialized. To look into a mind of this type is like looking into a laboratory. It is excellent as a workshop, but there are no pictures on the walls, no books, no flowers, no fireplace, no windows opening on garden, field and sky. What are the resources of such a mind, its points of contact with human-kind? What will happen to it in the years that fail to bring "the philosophic mind" and comprehensive sympathies? Surely this is an insufficient education. Education should release, refine and enlarge to the utmost the powers of imagination, creation, reflection, sympathy, catholicity. True education produces not a mechanized mind that is efficient and nothing more but a *personality*. Its ideal is not mere efficiency but true culture—not a specialized or standardized mind, but a free, harmonious and evaluating mind.

This leads to the third and chief failure of Science: *its inability to supply moral principle and motive*. This was revealed with startling clearness in the recent world cataclysm. Instead of blocking the war, as had been predicted, Science fed its fires with new methods of propaganda and multiplied its destructive agencies a hundredfold, devising deadly explosives and poisons that threatened the extinction of civilization itself. That was not, to be sure, the fault of Science itself, but of those who put it to such deadly uses. Yet the fact that this was possible and that scientists lent themselves to the frightful enterprise places the corporate body of Science beside the Church in undeniable moral default. This has given Science a sinister aspect and possibilities that can never be washed entirely from its once untarnished escutcheon.

Not only so, but Science possesses *within itself* no motive power to lift mankind to a happier and nobler estate. True, its achievements have brought untold gain to human life and happiness and many of its representatives have been among the noblest benefactors of mankind—not a few among the martyrs of the race. Yet Science itself has merely furnished the field and opportunity of this devotion. The motives which inspired its devotees have been moral, humanitarian and spiritual. Not a few scientists have supposed that in devoting themselves to the welfare of humanity they were actuated by Science itself. This is not so. They have found opportunity in Science, not motive. Without those motives which actuate men as men, that is, as moral agents, Science may be used for malevolent as efficiently as for beneficent ends.

Biology has given birth to eugenics, but a eugenics without moral quality or spiritual temper is as perilous as it is obnoxious to those more sensitive instincts which have made marriage, in some instances at least, a sacrament. Anthropology, with its disclosure of a common anthropoid ancestry, is informing, as well as humbling, but it gives no impetus to racial brotherhood. Are we to love each other for the sake of those efficient simian ancestors of ours, "Irascible Strong" and "Tricksy Cunning," who roamed the forests of eld? We may not disdain them, for some of their traits are still ours, but it is not incumbent upon us to hang their portraits in our art galleries and schoolrooms and churches.

A psychology which professes to be scientific has fathered that singularly dehumanizing monstrosity, behaviorism, and has led us into the bat-infested cave, as it has been well termed, of Freudianism.[5] From both behaviorism

[5] These were two types of psychology popularized in the United States in the 1920s. It was not uncommon at this time to try to discredit the entire scientific enterprise by holding up garbled versions of psychological theory as the end result of a scientific approach to life and society.

and psychoanalysis we may learn much, but should we proudly teach them to our students as the *final word* concerning the human mind? It is the assumption of finality that curses such semi-sciences. As for economics and sociology, they are available for wise and fruitful use only when supplied with adequate principles and motives.

The failure of Science to throw light on the ultimate problems of existence, to furnish an adequate education, and to nourish and sustain the human spirit, are not, however, the only, nor yet the most convincing, evidences of the passing of the Scientific Era. There is another and more weighty evidence still. I refer to the growing consciousness of the insufficiency of an isolated and rapidly formulated Natural Science and of the paucity of its resources to meet the problems which [it] itself raises. This sense of insufficiency would not be so significant if it were confined to the critics of Science and the representatives of interests which feel themselves suppressed or slighted. The noteworthy fact is, that doubt of the sufficiency of Science to meet its problems alone is arising *among scientists themselves.*

The advent of Relativity was much more than a disturbing trembler within the domain of physics. It shook the whole structure of Science. The importance of *the observing mind* suddenly thrust itself with startling force upon the attention of the scientific world. The effect has been disconcerting. The observing and calculating mind, by reference to which all objects are relative—why should *it* be so crucial? Why should the observer—mere speck as he is in the vast domain of Nature—loom so large?

> "Thou art the unanswered question;
> Couldst see thy proper eye,
> Alway it asketh, asketh;
> And each answer is a lie.
> So, take thy quest through nature;
> It through thousand natures ply,
> Ask on, thou clothed eternity;
> Time is the false reply."[6]

"We find that space and time themselves, and all objects that fall within them," wrote the late Lord Haldane, "including the self, when so regarded, are there, present, past and future, only in relation to the self that holds them together and in unison. . . . The self reaches over all objects."

Here is a truth that cannot be refuted and yet one that the scientist has never yet adequately recognized. At last the scales are falling from his mental vision. Witness the eye of the scientist turned from *without* to *within!* What astonishment, what perplexity are his! What problems confront him, which he must meet before he can again look out confidently upon Nature!

[6] From Ralph Waldo Emerson, "The Sphynx."

When, at length, after examining thoroughly his own mind, he looks once more upon Nature all is changed; for *Mind* now is there, his own mind, the mind of his fellows, and perchance an All-inclusive Mind. Here is transformation, readjustment, reorientation. Will the scientist make it? Can he make it? The answer is that, through some of his leading representatives, he *is* making it. For, at last, he is coming to see that he cannot neglect or evade philosophy if he would keep Science itself. For this reason the scientist is himself becoming a philosopher—and no mean one either.

One of the most important recent events in the history of thought is the belated but firmly adopted resolve on the part of outstanding scientists to examine the concepts of Science. Space, time, cause, effect, force, induction, hypothesis—these familiar categories which have so long been in use in Science have always been taken for granted, little questioned as to their origin or meaning or validity. Of late, however, the spirit of Socrates seems to have come to haunt laboratory and class-room, asking such penetrative questions as: What do you mean by *Science, Space, Time, Measure, Cause, Motion* and the rest? Is it not high time to inquire into these concepts which you use so freely? An unexamined Science is no better than an unexamined life.

Scientists are giving belated attention to these Socratic queries, thus facing tasks which, though growing out of Science, are essentially philosophical. For Science leads up to philosophy, as does every other examination of experience. In this way there has already arisen a group of *scientist-philosophers* who are commanding world-wide attention—represented by Professors Whitehead, Eddington and Hobson, Dr. Lloyd Morgan and others —who are reëxamining the concepts of Science and reaching extremely important results.

The eminent Cambridge physicist, Professor Eddington, for example, has recently described the external world as "a symposium of the presentations to individuals in all sorts of circumstances." "It is a very relevant fact," continues Professor Eddington, "that physics is now in course of abandoning all claim to a type of knowledge which it formerly asserted without hesitation. Moreover these considerations indicate the limits to the sphere of exact science."

Yet this movement is not so recent as might be supposed. Mach, for example, distinguished three periods in scientific thought, the first experimental, the second deductive and the third theoretical. As Needham states: "Scientific descriptions, mechanistic descriptions, according to Mach, are 'quite fictitious, though still valuable modes of describing phenomena, and to place the laws of physics actually in external nature is to hypostatize the abstractions of purely human origin.' "

Physics is leading in the more comprehensive view of Science which is emerging , but physics is not alone in awaking to the mental factors involved in its subject-matter. From the point of view of biology, for example, comes a very significant word from Joseph Needham, Bio-chemist of Cambridge University, who after an able defense of Biological Mechanism against Neo-vitalism, shows how far Mechanism is from explaining the subjective, *mental* phenomena associated with life. Thus:

"Naturalism gained ground by immense strides during the first seventy-five years of the nineteenth century, but came to a climax about then and has been undergoing severe criticism ever since. . . . There is a profoundly subjective factor in science—quite unrealized by men such as Huxley and Tyndall. The scientific man plays an active part in the selection of the facts before him, and his selection of those facts is determined by the construction of his mind."

Chemistry, too, is passing beyond the test-tube and the retort. "The strength of Science," declares Gilbert N. Lewis, Professor of Chemistry in the University of California (in the Yale University Silliman Lectures of 1926) "lies in its naïveté." Again, in his "The Anatomy of Science," he says: "Science is like life itself; if we could foresee all the obstacles that lie in our path we would not attack even the first, but would settle down to self-centered contemplation. The average scientist unequipped with the powerful lenses of philosophy, is a nearsighted creature, and cheerfully attacks each difficulty in the hope that it may prove to be the last. He is not given to minute analysis of his own methods. Indeed, if he should become too self-conscious he might lose his power, like the famous centipede who, after too profound analysis of his own method of locomotion, found he could no longer walk."

Admirable as this naïveté of Science has been in productive results, it can no longer pass unchallenged, as it has done in the past. The hour has arrived for Science to become self-conscious and, like the hypothetical self-conscious centipede, to halt on its way to consider its method of locomotion and also the direction which it is taking.

Not only have many progressive scientists come to recognize the need of philosophy in achieving a synthetic view-point, but of religion as well. Eminent scientists, like J. Arthur Thomson, Robert A. Millikan, Michael Pupin, J. Y. Simpson[7] and others have not shunned the religious implications of Science, indeed have emphasized them; so that to-day we have a very significant and influential testimony on the part of leading scientists as to the essential place and value of religion. It has remained for a geologist, Kirtley F. Mather, among others, to perceive that, "in the last analysis Science and religion both rest on faith."

All this is quite generally taken to be—indeed scientists themselves take it to be—evidence of the advance of Science. It is an advance of Science, but *an advance across its own borders into the large and free fields of philosophy and toward the wide horizons of religion.* In other words we have here the

[7] These men represented a latter-day version of the religiously orthodox scientist; most leading scientists of this period tended not to give public support to religion.

best evidence of the passing, not of Science itself, but of its undue domination and its independence of other—may we not say even greater—human interests.

26. A Physicist Embraces Uncertainty and Finds a New Vision of Science. 1929

In writing this article Harvard physicist PERCY W. BRIDGMAN (1882–1961) had two purposes: to explain to the general reader the nature and significance of the work of the preceding two or three decades in physics, and to describe a new basis that had developed for a faith in science and in the scientific way of looking at the world and the universe. By 1929 many traditional concepts in both science and popular thinking had given way to an entirely new way of talking about nature. Many thinkers—very probably improperly—saw in other social and intellectual currents of the time analogous and parallel developments, many of which will probably suggest themselves to the reader.

Bridgman was born in Cambridge, Massachusetts. He went to Harvard and spent his whole life there. In 1908 he received his Ph.D. in physics from the University and joined the staff immediately. By 1926 he had been appointed to a named professorship. Bridgman was known both for his philosophical writing (he was an advocate of operationism) and for his experimentation. He received many honors and in 1946 won the Nobel Prize for his work on properties of matter under high pressure.

For more on Bridgman consult Edwin C. Kemble and Francis Birch, "Percy Williams Bridgman, April 21, 1882–August 20, 1961," National Academy of Sciences, *Biographical Memoirs,* 41 (1970), 23–67. William M. Malisoff discusses the issues Bridgman raises, in "Physics: The Decline of Mechanism," *Philosophy of Science,* 7 (1940), 400–414.

THE NEW VISION OF SCIENCE*

P. W. Bridgman

The attitude which the man in the street unconsciously adopts toward science is capricious and varied. At one moment he scorns the scientist for

* *Harper's Magazine,* 158 (1929), 443–451. Copyright © 1929, by *Harper's Magazine,* Inc. Reprinted from the March, 1929, issue of *Harper's Magazine* by permission of the author.

a highbrow, at another anathematizes him for blasphemously undermining his religion; but at the mention of a name like Edison he falls into a coma of veneration. When he stops to think, he does recognize, however, that the whole atmosphere of the world in which he lives is tinged by science, as is shown most immediately and strikingly by our modern conveniences and material resources. A little deeper thinking shows him that the influence of science goes much farther and colors the entire mental outlook of modern civilized man on the world about him. Perhaps one of the most telling evidences of this is his growing freedom from superstitution. Freedom from superstition is the result of the conviction that the world is not governed by caprice, but that it is a world of order and can be understood by man if he will only try hard enough and be clever enough. This conviction that the world is understandable is, doubtless, the most important single gift of science to civilization. The widespread acceptance of this view can be dated to the discovery by Newton of the universal sway of the law of gravitation; and for this reason Newton may be justly regarded as the most important single contributor to modern life.

The point of view for which Newton is responsible is well exemplified by the remark often made that every particle of matter in the universe attracts to some extent every other particle, even though the attraction is almost inconceivably minute. There is thus presented to the mind a sublime picture of the interrelatedness of all things; all things are subject to law, and the universe is in this respect a unit. As a corollary to this conviction about the structure of the universe, an equally important conviction as to man's place in the universe has been growing up; man feels more and more that he is in a congenial universe, that he is part and parcel of everything around him, that the same laws that make things outside him go also make him go, and that, therefore, he can, by taking sufficient pains, understand these laws. These two theses so closely related—that the world is a world of order and that man can find the guiding motif of this order—have come to be tacit cardinal articles of faith of the man of science, and from him have diffused through the entire social structure, so that now some such conviction essentially colors the thinking of every educated person. It is to be emphasized that the justification for this conviction is entirely in experience; it is true that, as man has grown older and acquired more extensive acquaintance with nature and pondered more deeply, he has been increasingly successful in reducing the world about him to order and understandability. It has been most natural to generalize this experience into the conviction that this sort of thing will always be possible, and to believe that as we delve constantly deeper we shall always be able to give a rational account of what we find, although very probably the difficulties will become continually greater.

The thesis of this article is that the age of Newton is now coming to a close, and that recent scientific discoveries have in store an even greater revolution in our entire outlook than the revolution effected by the discovery of universal gravitation by Newton. The revolution that now confronts us arises from the recent discovery of new facts, the only interpretation of which is that our conviction that nature is understandable and subject to law arose from the narrowness of our horizons, and that if we sufficiently extend our range we shall find that nature is intrinsically and in its elements neither understandable nor subject to law.

The task of the rest of this article is twofold. In the first place I shall try to give some suggestion of the nature of the physical evidence and of the reasoning that has forced the physicist to the conclusion that nature is constituted in this way. This task is by no means easy; for not only is it impossible to indicate more than very partially the physical evidence, but it is often necessary to compress into a few sentences steps in the reasoning that can be completely justified only by long and difficult mathematical or logical analysis. The second part of the task is to envisage a few of the far-reaching consequences on the whole outlook of mankind of the acceptance of the view that this is actually the structure of nature. This aspect of the situation can be appreciated without a detailed grasp of the preliminary analysis.

II

The new experimental facts are in the realm of quantum phenomena. Comparatively little has been written for popular consumption about this new realm which has opened in the last fifteen years. The man in the street has been much more interested in relativity, which to him has seemed extremely interesting and revolutionary. Occasionally, however, there has filtered down to him the news that nearly all the theoretical physicists are occupied with a new order of phenomena which they find very much more exciting and revolutionary than any in the realm of relativity. For after all is said and done, the practical effects of relativity, measured in dollars and cents or in centimeters and grams, are exceedingly small, and require specially designed experiments executed by men of the highest skill to show their existence at all. The phenomena with which quantum theory deals, on the other hand, are of the greatest practical importance and involve the simplest aspects of everyday life. For example, before the advent of quantum theory no one could explain why a tea kettle of water boiling on the stove should not give out enough light in virtue of its temperature to be visible in the dark; the accepted theories of optics demanded that it should be visible, but every burned child knew that it was not.

One reason that the man in the street has not sensed this new domain is that it is much more difficult to explain than relativity; this is partly due to the nature of the subject, and partly also to the fact that the physicist himself does not understand the subject as well. I shall not in this article rush in where the angels have not ventured, but it is, nevertheless, necessary to try to give a glimmering of an idea of what it is all about.

Although all the phenomena of ordinary life are really quantum phenomena, they do not begin to stand out unequivocally in their quantum aspect and admit of no other interpretation until we have penetrated very far down into the realm of small things and have arrived at the atoms and electrons themselves. It must not be pretended that the nature of the quantum phenomena met in this realm of small things is by any means completely understood; but a suggestive characterization of the general situation is that atomicity or discontinuity is an even more pervading characteristic of the structure of the universe than had been previously supposed. In fact the name, "quantum," was suggested by the atomicity.

We were a long time in convincing ourselves of the atomic structure of ordinary matter; although this was guessed by the poets as early as the beginning of the Christian era, it was not generally accepted as proved, even by physicists, until the beginning of this century. The next step was the discovery of the atomic structure of electricity; there are indivisible units of positive and negative electricity, and the atoms of matter are constructed of atoms of electricity. This situation was not even guessed until about 1890; the proof and acceptance of the doctrine have taken place within the memory of the majority of the readers of this article. Finally comes the discovery that, not only is matter doubly atomic in its structure, but that there is an atomicity in the way in which one piece of matter acts on another. This is perhaps best understood in the case of optical phenomena. It used to be thought that light was infinitely subdivisible—that I could, for example, receive at pleasure on the film of my camera either the full intensity of the sun's radiation, or, by interposing a sufficiently small stop, that I could cut the intensity of the light down to anything this side of nothing at all. This is now known not to be true; but the light which we receive from the sun is atomic in structure, like an almost inconceivably fine rain composed of indivisible individual drops, rather than like the continuous flood of infinitely subdivisible radiation that we had supposed. If I close the stop of my camera too much I may receive nothing at all on the film, or I may receive a single one of the drops in the rain of radiation, but there is no step between one drop and nothing. The recognition that radiation has this property means that in some respects we have come back very close to Newton's ideas about light.

The proof that this is the structure of light can be given in many ways. Perhaps the most illuminating for our purpose is that discovered by Arthur Compton, for which he received the Nobel prize. Compton's discovery consisted in finding that the drops of radiation behave in certain ways like the material drops of ordinary rain; they have energy and mass and momentum, which means that when they collide with matter they behave in some respects very much as ordinary bodies do. The laws which govern the interaction or collision of ordinary bodies are known to any graduate of a high-school course in physics; he could calculate what would happen after two billiard balls had collided provided we would tell him exactly how each of the balls was moving before the collision, and what were the elastic properties of the materials of which the balls are composed. In making the calculation he would use, among other things, the two fundamental principles of the conservation of energy and the conservation of momentum. Now Compton showed that what happens when a drop, or better a bullet, of radiation collides with an electron is also governed by the same two fundamental principles. The proof consisted in showing that the way in which the electron rebounds is connected with the way in which the bullet rebounds by equations deduced from these principles; this is one of the features which makes Compton's discovery of such a fundamental importance.

But Compton's experiment contains another feature, and it is this which seems destined to revolutionize the thinking of civilization. Go back to the billiard-ball analogy: An expert billiard player can, by proper manipulation of the cue ball, make the two balls rebound from the collision as he wishes; this involves the ability to predict how the balls will move after collision from their behavior before collision. We should expect by analogy to be able to do the same thing for a collision between a bullet of radiation and an electron, but the fact is that it never has been done and, if our present theories are correct, in the nature of things never can be done. It is true that, if someone will tell me how the electron bounces away, I can tell, on the basis of the equations given by Compton's theory, how the bullet of radiation bounces away, or conversely; but no one has ever been able to tell how both will bounce away. Billiards, played with balls like this, even by a player of infinite skill, would degenerate into a game of pure chance.

This unpredictable feature has been seized and incorporated as one of the corner stones in the new theory of quantum mechanics, which has so stirred the world of physicists in the last three years. It has received implicit formulation in the "Principle of Uncertainty" of Heisenberg, a principle which I believe is fraught with the possibility of greater change in mental outlook than was ever packed into an equal number of words. The exact formulation of the principle, which is very brief, is framed in too technical language to reproduce here, but I shall try to give the spirit of the principle. The essence of it is that there are certain inherent limitations to the accuracy with which a physical situation can be described. Of course we have always recognized that all our physical measurements are necessarily subject to error; but it has always been thought that, if we took pains enough and were sufficiently clever, no bounds could be set to the accuracy which we might some day achieve. Heisenberg's principle states, on the other hand, that the ultimately possible accuracy of our measurements is limited in a curious and unsuspected way. There is no limit to the accuracy with which we can describe (or measure) any one quality in a physical situation, but if we elect to measure one thing accurately we pay a price in our inability to measure some other thing accurately. Specifically, in Compton's experiment, the principle states that we can measure the position of the electron as accurately as we choose, but in so doing we must sacrifice by a compensating amount the possibility of accurately measuring its velocity. In particular, if we measure with perfect accuracy the position of the electron, we have thereby denied ourselves the possibility of making any measurement at all of its velocity.

The meaning of the fact that it is impossible to measure exactly both the position and velocity of the electron may be paradoxically stated to be that an electron cannot have both position and velocity. The justification of this is to be found in the logical analysis of the meaning of our physical concepts which has been stimulated by the relativity theory of Einstein. On careful examination the physicist finds that in the sense in which he uses language no meaning at all can be attached to a physical concept which cannot ultimately be described in terms of some sort of measurement. A body has position only in so far as its position can be measured; if its position cannot in principle be measured, the concept of position applied to the body is meaningless, or in other words, a position of the body does not exist. Hence if both the position and velocity of the electron cannot in principle be measured, the electron cannot have both position and velocity; position and velocity as expressions of properties which an electron can simultaneously have are meaningless. To carry the paradox one step farther, by choosing whether I shall measure the position or velocity of the electron I thereby determine whether the electron has position or velocity. The physical properties of the electron are not absolutely inherent in it, but involve also the choice of the observer.

Return to the analogy of the billiard ball. If we ask our high-school physicist what he must be told before he can predict how the billiard balls will rebound after collision, he will say that, unless he is told both how fast the balls are traveling when they collide, and also what their relative positions are at the moment of collision he can do very little. But this is exactly the sort of thing that the Heisenberg principle says no one can ever tell; so that our high-school computer would never be able to predict how a bullet of radiation and an electron behave after collision, and no more could we. This means that in general when we get down to fine-scale phenomena the detailed results of interaction between the individual elements of which our physical world are composed are essentially unpredictable.

This principle has been built into a theory, and the theory has been checked in many ways against experiment, and always with complete success. One of the consequences of which the man in the street has heard a good deal is that an electron has some of the properties of waves, as shown so strikingly in the experiments of Davisson and Germer. Of course no one can say that some day a fact may not be discovered contrary to the principle, but up to the present there is no evidence of it; and it is certain that something very much like this principle, if not this principle exactly, covers an enormously wide range of phenomena. In fact the principle probably governs every known type of action between different parts of our physical universe.

One reason that this principle has not been formulated before is that the error which it tells us is inherent in all measurement is so small that only recently have methods become accurate enough to detect it. The error is unimportant, and indeed immeasurably small when we are dealing with the things of ordinary life. The extreme minuteness of the effect can be illustrated again with the billiard balls. Suppose that at the instant of collision the position of the balls is known with an uncertainty no greater than the diameter of a single atom, a precision very much higher than has ever been attained. Then the principle says that it is impossible to measure the velocity of the balls without a related uncertainty; but on figuring it out we find that this uncertainty is so small that after the lapse of one hundred thousand years, assuming a billiard table large enough for the balls to continue rolling for one hundred thousand years, the additional uncertainty in the position of the balls arising from the uncertainty in the velocity would again be only the diameter of a single atom. The error becomes important only when we are concerned with the ultimately small constituents of things, such as the action between one atom and another or between an atom of radiation and an electron.

III

It is easy to see why the discovery that nature is constituted in this way, and in particular is essentially unpredictable, has been so enormously upsetting. For the ability to predict a happening is tied up with our ideas of cause and effect. When we say that the future is causally determined by the present we mean that if we are given a complete description of the present the future is completely determined, or in other words, the future is the effect of the present, which is the cause. This causal relation is a bilateral relation; given the cause, the effect is determined, or given the effect, the cause may be deduced. But this means, in the particular case that we have been considering of collision between a bullet of radiation and an electron, that the causal connection does not exist, for if it did the way in which the electron rebounds after the collision would be determined, that is, it could be predicted, in terms of what happens before the collision. Conversely, it is of course impossible to reconstruct from the way in which the electron and the radiation rebound the way in which they were moving before collision. Hence the rebound of the electron is not causally connected with what goes before.

The same situation confronts the physicist everywhere; whenever he penetrates to the atomic or electronic level in his analysis, he finds things acting in a way for which he can assign no cause, for which he never can assign a cause, and for which the concept of cause has no meaning, if Heisenberg's principle is right. This means nothing more nor less than that the law of cause and effect must be given up. The precise reason that the law of cause and effect fails can be paradoxically stated; it is not that the future is not determined in terms of a complete description of the present, but that in the nature of things the present cannot be completely described.

The failure of the law of cause and effect has been exploited by a number of German physicists, who have emphasized the conclusion that we are thus driven to recognize that the universe is governed by pure chance; this conclusion does not, I believe, mean quite what appears on the surface, but in any event we need not trouble ourselves with the further implications of this statement, in spite of their evident interest.

One may be sure that a principle as revolutionary in its implications as this, which demands the sacrifice of what had become the cardinal article of faith of the physicist, has not been accepted easily, but there has been a great deal of pondering and searching of fundamentals.

The result of all this pondering has been to discover in the principle an inevitableness, which when once understood, is so convincing that we have already almost ceased to kick against the pricks. This inevitableness is rooted in the structure of knowledge. It is a commonplace that we can never know anything about anything without getting into some sort of connection with it, either direct or indirect. We, or someone else, must smell the object, or taste it, or touch it, or hear it, or see it, or it must affect some other object which can affect our senses either directly or indirectly, before we can know anything about it, even its existence. This means that no knowledge of any physical property or of even mere existence is possible without interaction; in fact these terms have no meaning apart from interaction. Formerly, if this aspect of the situation was thought of at all, it would have been dismissed as merely of academic interest, of no pertinence at all, and the justification of this would have beeen found in the supposed possibility of making the inevitable interaction as small as we pleased. The defender of the old point of view might have flippantly remarked that a cat may look at a king, by which he would have meant that the act of observation has no effect on the object. But even in the old days a captious critic might have objected to this easy self-satisfaction by pointing out that light exerts a pressure, so that light cannot pass from the king to the cat without the exercise of a certain amount of mechanical repulsion between them. This remark of the captious critic now ceases to be merely academic because of the discovery that light itself is atomic in structure, so that at least one bullet of radiation must pass if any light at all passes, and the king cannot be observed at all without the exertion of that minimum amount of mechanical repulsion which corresponds to a single bullet.

This evidently alters the entire situation. The mere act of giving meaning through observation to any physical property of a thing involves a certain minimum amount of interaction. Now if there are definite characteristics associated with the minimum interaction, it is conceivable that no observation of anything whatever can be made without entraining certain universal consequences, and this turns out to be the case. Let us return again to the useful billiard-ball analogy. What must our high-school calculator know in order completely to calculate the behavior of the balls after collision? Evidently, if he is to give a complete description of the motion, that is, give in addition to direction and velocity of motion the exact time at which the balls are in any particular location, he must know how long the collision lasts. This means that the act of collision itself must be analyzed. This analysis is actually possible, and in fact rapid-moving pictures have been taken, showing in detail how the balls are deformed during their contact together.

Returning now to the collision between a bullet of radiation and an electron, in order to determine completely the behavior after collision we must similarly analyze the details of the process of collision. In particular, if we want to predict where the electron is after collision we must analyze the collision sufficiently to be able to say how fast the electron is moving at each instant of the collision. But how shall this analysis be made? If the analysis means anything, it must involve the possibility of observation; and observation involves interaction; and interaction cannot be reduced below a minimum. But the collision, or interaction, between the electron and radiation that we are analyzing is itself the minimum interaction. It is obvious that we cannot discover fine details with an instrument as coarse as the thing that we are trying to analyze, so that the necessary analysis of the minimum interaction can never be made, and hence has no meaning, because of our fundamental dictum that things which cannot in principle be measured have no meaning. Therefore, the act of collision cannot be analyzed, the electron and radiation during collision have no measurable properties, and the ordinary concepts, which depend on these properties, do not apply during collision, and have no meaning. In particular, the ordinary concept of velocity does not apply to the act of collision, and we are prepared to expect something curious as the result of the collision. In fact, the detailed working out of the theory shows that the meaninglessness of velocity during the act of collision carries with it the consequence that the electron emerges from the collision with a certain nebulosity or indefiniteness in properties such as position, which according to the old point of view depend on the velocity, and it is precisely this nebulosity which is described in Heisenberg's principle.

The infinitesimal world thus takes on a completely new aspect, and it will doubtless be a long while before the average human mind finds a way of dealing satisfactorily with a situation so foreign to ordinary experience. Almost the first necessity is a renunciation of our present verbal habits and of their implications. It is extraordinarily difficult to deal with this new situation with our present forms of expression, and the exposition of this paper is no exception. The temptation is almost irresistible to say and to think that the electron *really* has *both* position and velocity, only the trouble is that our methods of measurement are subject to some limitation which prevents us from measuring both simultaneously. An attitude like this is justified by all the experience of the past, because we have always been able hitherto to continue to refine our methods of measurement after we had apparently reached the end. But here we are confronted by a situation which in principle contains something entirely novel, and the old expectations are no longer valid. The new situation cannot be adequately dealt with until long-continued familiarity with the new facts produces in our subconsciousness as instinctive a grasp as that which we now have of the familiar relations of everyday experience.

IV

The implications of this discovery are evidently most far-reaching. Let us first consider the scientific implications and, in particular, the implications for physics. The physicist is here brought to the end of his domain. The record of physics up to the present has been one of continued expansion, ever penetrating deeper and deeper, and always finding structure on a finer and finer scale beyond previous achievement. Several times in the past even eminent physicists have permitted themselves the complacent announcement that we were in sight of the end, and that the explanation of all things was in our hands. But such predictions have always been set at naught by the discovery of finer details, until the average physicist feels an instinctive horror of the folly of prediction. But here is a situation new and unthought of. We have reached the point where knowledge must stop because of the nature of knowledge itself: beyond this point meaning ceases.

It may seem that we are getting back pretty close to the good Bishop Berkeley, but I think that actually nothing could be wider of the mark. We are not saying that nothing exists where there is no consciousness to perceive it; we are saying that existence has meaning only when there is interaction with other existence, but direct contact with consciousness need not come until the end of a long chain. The logician will have no trouble in showing that this description of the situation is internally self-contradictory and does not make sense; but I believe that, nevertheless, the sympathetic reader will be able to see what the situation is, and will perhaps subscribe to the opinion that to describe it the development of a new language is necessary.

The physicist thus finds himself in a world from which the bottom has dropped clean out; as he penetrates deeper and deeper it eludes him and fades away by the highly unsportsmanlike device of just becoming meaningless. No refinement of measurement will avail to carry him beyond the portals of this shadowy domain which he cannot even mention without logical inconsistency. A bound is thus forever set to the curiosity of the physicist. What is more, the mere existence of this bound means that he must give up his most cherished convictions and faith. The world is not a world of reason, understandable by the intellect of man, but as we penetrate ever deeper, the very law of cause and effect, which we had thought to be a formula to which we could force God Himself to subscribe, ceases to have meaning. The world is not intrinsically reasonable or understandable; it acquires these properties in ever-increasing degree as we ascend from the realm of the very little to the realm of everyday things; here we may eventually hope for an understanding sufficiently good for all practical purposes, but no more.

The thesis that this is the structure of the world was not reached by armchair meditation, but it is the interpretation of direct experiment. Now all experiment is subject to error, and no one can say that some day new experimental facts may not be found incompatible with our present interpretation; all we can say is that at present we have no glimmering of such a situation. But whether or not the present interpretation will survive, a vision has come to the physicist in this experience which he will never forget; the possibility that the world may fade away, elude him, and become meaningless because of the nature of knowledge itself, has never been envisaged before, at least by the physicist, and this possibility must forever keep him humble.

When this view of the structure of nature has once been accepted by physicists after a sufficiently searching experimental probe, it is evident that there will be a complete revolution in the aspect of all the other physical sciences. The mental outlook will change; the mere feeling that boundaries are set to man's inquiry will produce a subtle change of attitude no less comprehensive in its effects than the feeling, engendered by Newton's conquest of celestial mechanics, that the universe was a universe of order accessible to the mind of man. The immediate effect on scientific inquiry will be to divert effort away from the more obviously physical fields back to the fields of greater complication, which have been passed over by the physicist in his progress toward the ultimately little, especially the field of biology.

Another important result of the realization of the structure of the world is that the scientist will see that his program is finite. The scientist is perhaps only a passing phase in the evolution of man; after unguessable years it is not impossible that his work will be done, and the problems of mankind will become for each individual the problem of best ordering his own life. Or it may be that the program of the scientist, although finite, will turn out to need more time than the life of the world itself.

But doubtless by far the most important effect of this revolution will not be on the scientist, but on the man in the street. The immediate effect will be to let loose a veritable intellectual spree of licentious and debauched thinking. This will come from the refusal to take at its true value the statement that it is meaningless to penetrate much deeper than the electron, and will have the thesis that there *is really* a domain beyond, only that man with his present limitations is not fitted to enter this domain. The temptation to deal with the situation in this way is one that not many who have not been trained in careful methods of thinking will be able to resist—one reason is in the structure of language. Thought has a predisposition to certain tendencies merely because of the necessity of expressing itself in words. This has already been brought out sufficiently by the discussion above; we have seen how difficult it is to express in words the fact that the universe fades away from us by becoming meaningless without the implication that there really is something beyond the verge of meaning.

The man in the street will, therefore, twist the statement that the scientist has come to the end of meaning into the statement that the scientist has penetrated as far as he can with the tools at his command, and that there is something beyond the ken of the scientist. This imagined beyond, which the scientist has proved he cannot penetrate, will become the playground of the imagination of every mystic and dreamer. The existence of such a domain will be made the basis of an orgy of rationalizing. It will be made the substance of the soul; the spirits of the dead will populate it; God will lurk in its shadows; the principle of vital processes will have its seat here; and it will be the medium of telepathic communication. One group will find in the failure of the physical law of cause and effect the solution of the age-long problem of the freedom of the will; and on the other hand the atheist will find the justification of his contention that chance rules the universe.

Doubtless generations will be needed to adjust our thinking so that it will spontaneously and freely conform to our knowledge of the actual structure of the world. It is probable that new methods of education will have to be painfully developed and applied to very young children in order to inculcate the instinctive and successful use of habits of thought so contrary to those which have been naturally acquired in meeting the limited situations of everyday life. This does not mean at all that the new methods of thought will be less well adapted than those we now have to meet the situations of everyday life, but on the contrary, since thought will conform to reality, understanding and conquest of the world about us will proceed at an accelerated pace. I venture to think that there will also eventually be a favorable effect on man's character; the mean man will react with pessimism, but a certain courageous nobility is needed to look a situation like this in the face. And in the end, when man has fully partaken of the fruit of the tree of knowledge, there will be this difference between the first Eden and the last, that man will not become as a god, but will remain forever humble.

27. A Biologist Reviews the Contributions of Thomas Hunt Morgan to Genetics. 1937

This was a Sigma Xi (honorary science fraternity) lecture to honor Thomas Hunt Morgan (1866–1945), a pioneer American geneticist, on his seventieth birthday. FERNANDUS PAYNE (born 1881), the lecturer, reviews evolutionary theory since Darwin so that he can put Morgan's work into context. American scientists gained preeminence in genetics far out of proportion to their world standing in other fields of science. The most famous American geneticist was Morgan, and Payne indicates two major contributions that Morgan made in his laboratory demonstrations of the actual workings of evolutionary processes in the rapidly multiplying fruit fly, *Drosophila.* Only later, when biochemistry threw new light on the mechanisms of inheritance, did the direction genetics—a really American science—was to take become clear.

Except for the years 1907–1909, when he took his Ph.D. at Columbia, Payne spent his entire life at Indiana University, not far from his birthplace at Shelbyville. From 1927 to 1948 he was chairman of the department of zoology, and for several of those years (and at the time of this address) he served as dean of the graduate school. He was well known in the fields of cytology and genetics, and he held important posts in government science agencies.

See Everett Mendelsohn, "Science in America: The Twentieth Century," in Arthur M. Schlesinger, Jr., and Morton White, eds., *Paths of American Thought* (Boston: Houghton Mifflin, 1963), 433–434; Bernard Jaffe, *Men of Science in America* (2d ed., New York: Simon and Schuster, 1958), chap. 16; A. H. Sturtevant, *A History of Genetics* (New York: Harper and Row, 1965); Elof Axel Carlson, *The Gene, A Critical History* (Philadelphia: Saunders, 1966); and Garland E. Allen, "Thomas Hunt Morgan and the Problem of Natural Selection," *Journal of the History of Biology,* 1 (1968), 113–139.

GENETICS AND EVOLUTION*

Fernandus Payne

I have chosen to speak to you briefly on the subject of "Genetics and Evolution." Evolution is without doubt the central biological problem to

* *Sigma Xi Quarterly,* 25 (1937), 131–140. Slightly abridged. Reprinted with the kind permission of the *American Scientist.*

which all others are subsidiary and so it is proper that we ask ourselves whether the recent advances in genetics . . . have contributed to a better understanding of the methods or theories of the ways of evolution.

Before proceeding, however, I wish to make clear the distinction between evolution and the method of evolution. By evolution I refer to the theory, or perhaps we might even call it a fact, that living organisms have not always been as they are and that they have come to be what they are, through a series of changes, from pre-existing forms. By methods of evolution I refer to the ways by which these changes have taken place, the causal factors involved, the directing forces. For the moment, then, we are concerned with methods of evolution. Evolution we shall take for granted.

In discussing these problems it is well to have in mind what any theory of evolution must explain. There exists several hundred thousand species of plants and animals on the earth. The concept of just what a species is, is not sharply defined, but for our purposes we may think of a species as a group of individuals having for the most part similar characteristics, living in the same geographic region and freely interbreeding. Our concept should also include variability among the individuals, genetic as well as somatic. Observations of living forms show that species are not static. They are ever changing, some perhaps more rapidly than others. The geologic record emphasizes this fact by demonstrating that species of the past were different from those of today. What is the cause of these changes? Are they random? What directs species in the way they go? If it is chance why do we not find many other species, for certainly all possibilities have not been exhausted. Why are there genera, families, orders, and phyla? Why has evolution, in instances like the horse's foot and the elephant's tusk, gone in definite directions? Some species have arisen, flourished for a time, and then disappeared. Why? Why do the principal phyla of animals as we know them today appear in that early geologic formation known as the Cambrian without any clearly established relationship? Why have organs, such as the eyes of the blind fishes, degenerated after having been functional in earlier phylogenetic history and why have the sensory touch buds become more highly developed? Why do not closely related species occupy the same areas? Why are some animals protectively colored, and why do some change their coats with the seasons? What is the explanation of mimicry? Of secondary sexual characters? Why are some animals and plants restricted to cold regions, some to warm, some to dry, and some to moist regions? Why do some animals feed on plants, others on other animals? Why are some forms parasitic, others free living? Perhaps they are adapted to the conditions in which they live, but characteristics are not necessarily adaptive. In fact, some seem to have actually been injurious, if we may judge from such extinct forms as the dinosaurs. At least these highly specialized forms could not continue to exist under changed environmental conditions, even though these changes came about gradually. Even man continues to live on, in spite of wars, diseases, and the handicap of many physical defects and the presence of many vestigial organs. One might continue to ask many more questions for theories of evolution to answer, but perhaps these are sufficient to show you that the methods (note the plural) of evolution may be many, and that the operating causal factors may likewise be many.

Basic to all our thinking of evolutionary problems are two fundamental postulates which all take for granted. There must be change or variation, and secondly these changes or variations must be inherited. You see immediately, then, the intimate relations between genetics and evolution because genetics concerns itself primarily with variations and their inheritance.

What I propose to do is to look in summary fashion at the most popular theories of evolution in the light of our newer knowledge of genetics.

There have been two theories based almost wholly on the effects of the environment. The earlier one developed by Buffon, St. Hilaire and others thought of the environment as acting directly upon living organisms. As little was known of germ cells, the effect was upon the individual, molding it, so to speak, in one direction or another.

In the Lamarckian theory the effect of the environment is indirect. It causes the individual to do things which it formerly had not done. Use and disuse bring about changes which are passed on to the offspring, and thus acquired characters are said to be inherited. Both these theories are concerned primarily with the origin of variations.

The Darwinian theory of Natural Selection was based on several assumptions, as well as the observed facts that animals and plants vary and that man has brought about radical and diverse changes in domestic animals and plants by means of artificial selection and crossing. It was assumed that most of the variations occurring in nature are inherited. It was also assumed that, on the average, parents produce more than two offspring so that the population increases in a geometric ratio. It was further assumed that such a rapid increase in the population would give rise to considerable competition in the struggle for existence, to get food or to keep from being eaten. It was likewise assumed that some variations would occur which would be of advantage in the struggle for existence and that individuals possessing these favorable variations would more likely survive and reproduce offspring similar to themselves. The fittest would survive and hence would be adapted to the conditions under which they lived. In addition it was assumed that variations would continue, in the following generations, to occur in the same direction, that is, they would be cumulative. Subsidiary hypotheses were added to explain special cases such as secondary sexual characters and vestigial organs. Darwin even embraced Lamarckism in his later years to help explain the origin of variations.

The mutation theory as developed by DeVries was based upon observations and experiments with the evening primrose, Oenothera lamarckiana. I should like to emphasize the fact that for the first time in the history of evolutionary thought the experimental method was effectively used. It is true that Mendel and other hybridizers had used the experimental method but they had not correlated their work with the existing theories of evolution. It is also true that Darwin used the method but only to a limited extent.

Oenothera lamarckiana is probably an American species introduced into Europe. DeVries found plants growing in the fields about Hilversum and discovered among these wild forms a number of variations, some of which were markedly different in several respects from the parental type. When brought into the garden and selfed,[1] these variants bred true. The parent species, lamarckiana, when bred in the garden gave rise to more variations like those discovered in the wild. Since the variations appeared suddenly and bred true, DeVries called them mutations, and since the differences between mutants and parental forms were so great, he thought of the mutants as elementary species. He had discovered a method of species formation or evolution by means of large sudden heritable variations. DeVries called upon the Darwinian theory to supplement his theory of evolution by means of mutations, as the mutation theory alone concerned itself primarily with the origin of variations as did the Lamarckian theory.

Recently a new theory of evolution has emerged. I say emerged because it is called emergent evolution. It is not actually so new but it seems to have had a recent eruption, judging by the amount of discussion it has evoked. By some . . . it is hailed as a deliverer from bondage. What is this theory of emergence? I admit I am somewhat dense and so may get lost in the jungle of words with which it is enmeshed. If, however, I interpret correctly the adherents of the theory, an individual is not merely the sum of its parts. They illustrate it by saying that two separate gases, H and O, combine to produce water, which has properties entirely different from either gas. Water has emerged. Countless other examples of a similar sort could be given. New characteristics of animals and plants may emerge from genic changes and new combinations of genes, and so the door is opened for unlimited possibilities for evolution.

Paleontology has revealed many instances of evolution in a given direction. Notable examples already mentioned are the changes from a four-toed horse to a horse with a single toe and the progressive enlargement of the tusk of elephants. Many other examples could be given. In fact, if we look only at the end products, it would seem that all complex organs must have developed step by step in the direction they have gone. Eimer, who first developed this idea of evolution in a straight line, thought of the directive force or forces as within the organism. Others suggested that the environment might be the causal factor, while a third possibility is that only those variations which go in a given direction are preserved.

[1] That is, inbred, pollinated with pollen from the same plant.

A factor in evolution, if not a complete explanation of species formation, is isolation. When we speak of isolation we usually think of geographic isolation, where barriers of one sort or another keep groups of individuals from freely intermingling. In the past it had been assumed that new variations had but little or no chance of survival because of the swamping effects of inbreeding if the variation appeared in a large population. If, however, a variation occurred in a small group of isolated individuals, its chance of survival would be far greater. . . . It can be illustrated best by a study of island faunas and floras. Volcanic islands must, of course, be populated by forms which migrate from the mainland. Darwin in his studies had noticed that the fauna and flora of the Galapagos Islands were similar to those of the mainland of South America, but that many differences existed. He also found that while the different islands of the group were very similar with respect to their faunas and floras, each island had forms peculiar to it. It would seem that isolation such as mentioned, while not encouraging variation, does possibly play a part in the preservation of variations and hence in this way, play a part in the evolution of species. Barriers may be of other sorts such as climatic, or physiologic differences, which prevent interbreeding.

The theories which I have outlined all too briefly do not exhaust the list, but they are the ones of major interest to us and time does not permit a consideration of others. You may have noted that no one of them gives complete and satisfactory answers to all our questions.

I have said that we are to look at these theories of evolution in the light of the recent discoveries in the field of genetics. I have also said that genetics is one aspect of the larger problem of evolution because without variation and inheritance evolution cannot take place. What are these developments in genetics which might influence our thinking about evolutionary problems? There are two aspects to these developments, one cellular, the other experimental, the cellular giving the physical basis for the interpretation of the experimental. With respect to the cellular basis we know:

[1] That an egg cell of maternal origin unites with a sperm cell of paternal origin and that from the fertilized egg a new individual arises by cell division, growth, and differentiation. Thus the germ cells are the only connecting links between parents and offspring.
[2] That each germ cell, maternal and paternal, brings in a series of chromosomes, so the fertilized egg has two series or double the number of the mature germ cell.
[3] That in cell division each chromosome divides so that every cell of the body has the double number of chromosomes.
[4] That in germ cell formation chromosomes pair, like with like, maternal with paternal, and then separate in the cell division which follows, each mature germ cell thus receiving the half number of chromosomes.

[5] That when the maternal and paternal elements of the pairs of chromosomes separate, the distribution is a random one.

From the experimental side a considerable body of knowledge has accumulated and certain laws or principles have been derived. The following are pertinent to our further discussions:

Characters are represented in the germ cells by means of elementary particles, genes. In the individual these genes are in pairs. Sometimes the two genes are alike, in which case the individual is said to be homozygous. Sometimes they are unlike as in hybrids, in which case the individual is said to be heterozygous. In either case when the germ cells are formed the two genes of a given pair separate. This separation of the genes in germ cell formation is the law of segregation. The cellular explanation is that one gene of the pair lies in one member of a pair of homologous chromosomes and the other gene lies in the other chromosome of the same homologous pair.

In his original experiments Mendel discovered that when he crossed parents differing with respect to two pairs of characters, each pair behaved independently of the other. This is known as the law of independent assortment and is explained on the basis that one pair of genes lies in one pair of chromosomes and the second pair of genes lies in a second pair of chromosomes. The law is not of universal application as shown by more recent work.

One of the early discoveries of Professor Morgan and his students was that genes do not always assort independently; that sometimes the characters which go into a given cross together come out together. They called this linkage and the explanation is that the linked genes lie in the same chromosome. In the male Drosophila[2] linkage is complete, but in the female linkage is only partial, the characters exhibiting recombinations in a certain percentage of cases. In the formation of these recombinations, crossing-over is said to take place and the assumption is that pieces of the homologous chromosomes actually change places. Genes which lie in the same chromosome show linkage in crossing over. Those which lie in different chromosomes show independent assortment.

One of the most interesting developments of the genetic studies in Drosophila is that the number of groups of linked genes is four, and that there are four pairs of homologous chromosomes. In corn there are ten pairs of chromosomes and ten groups of linked genes.

Studies of crossing-over have given evidence that genes are arranged in linear order within chromosomes, those lying near,[3] crossing-over less, those lying farther apart, crossing-over more.

Other genetic facts pertinent to our evolutionary thinking are that in

[2] Fruit fly, and more particularly Drosophila melanogaster, or common vinegar fly, utilized with sensational success by Morgan to experiment with inheritance. Its hardy nature, prolific breeding, and short life cycle make the creature a superb laboratory subject.

[3] That is, near to each other.

Drosophila and other forms where extensive studies have been made, many mutations have occurred and are still occurring. Most of these mutations are genic changes, but some are due to chromosomal rearrangements, or since the genes lie in the chromosomes, to genic rearrangements. It would seem that these mutations are random, that they are as apt to occur in one direction as in another, but certain mutations have occurred again and again, while others have occurred only once, and other possible mutations have never occurred.

Most of the mutations which have appeared are recessives, that is, when crossed to the wild-type, the wild-type character is dominant and so the hybrid looks like a wild-type.

While in general the effects of a gene are fairly definite, precise—due perhaps to the uniform conditions under which they exert their influence—still the environment may modify the effects of a gene. A good example is the red primrose which, when reared at 30° to 35°C. and in moist shady places, has white flowers. The same plants reared at 15° to 20°C. have red flowers. In other cases the effects of the gene can be suppressed entirely. Abnormal abdomen in Drosophila, when bred in moist cultures, is a very marked defect, but when reared on dry food all flies are normal. Even though the character is suppressed for several generations by continuously breeding on dry cultures, the gene is in no way changed, for the moment moist cultures are provided, the abnormality reappears as before.

That the gene is stable is demonstrated by the fact that in hybrids genes separate sharply without influencing each other. In fact, a gene for white coat color may be kept in a black rabbit for generations without effect, for when two white genes again come together, a white coat is the result.

The popular concept is that a single character is dependent upon a single gene. Such a concept is far from the truth, for a character may be the resultant of a given combination of genes. The best known example perhaps is the cross between rose and pea comb in the domestic fowl. The hybrid is walnut, a comb unknown before this cross was made. In Drosophila orange eye color is due to two genes, one, salmon, the other, salmon modifier. But the probability is that every character is due to the combined action of a number of genes. . . .

While we usually think of a single gene as affecting only a given character, the fact is that a gene may affect several characters. The seed capsules of the common jimpson [*sic*] weed have spines. In Blakeslee's cultures a mutant with no spines occurred. These two forms not only differ with respect to spines, but also with respect to thickness of stems, length of internodes, and habits of growth.

The causes of genic changes have not been discovered, but mutations have been produced by means of heat and X and radium rays . . . These effects are direct effects on the germ cells or on the genes. While there is much speculation as to the cause of genic changes, the factual knowledge is so meager that such speculations have little value.

Another experiment, which may or may not be thought of as falling within the field of genetics, but which has had considerable influence on recent discussions of evolution theories, is that of Johanssen on Pure Lines. Darwin made use of artificial selection in building his theory of Natural Selection. It was well known in Darwin's time and is known now, that if, in bisexual forms, extreme variants with respect to a given character are selected and inbred, the mean for that character may be shifted in the direction of selection. There is a limit, however, as we now know, to such effects, and usually selection must be continued or the mean for the character reverts to what it was originally. In other words, change may be produced but it is not permanent.

Johanssen repeated this type of an experiment with beans, a form in which self-fertilization occurs. By using a single bean as the parent, in order to start with homozygous genes, and weight as the character, he discovered that he might shift the mean in the first generation, but after that, continued selection for lighter or heavier beans, was ineffective. The descendants of a single bean, he spoke of as a Pure Line. How, then, can evolution take place in Pure Lines? By means of mutation, Johanssen said.

Let us return now to our theories of evolution outlined earlier. . . .

[Early writers] thought that the environment acting directly was the causal factor in evolution. Since, under experimental conditions, extreme heat and X-rays, by acting directly on the genes, may cause mutation, there is no reason why these same factors and others as well, might not account for some evolutionary changes if not for all. Even if we did grant, however, that environmental factors could account for all mutations—a supposition not supported by genetic evidence—there would still remain the question of the forces moulding evolution in the directions it has taken. Of course, the environment must play a large part, even here, but it is difficult to conceive just how the environment could determine the direction of the development of a complex organ like the vertebrate eye or ear.

The evidence from genetics gives little encouragement to the Lamarckians. Mutations appear suddenly. They are due either to genic changes or genic rearrangements and are in no sense due to somatic effects of the environment transferred to the gene. Genetics gives no evidence for any type of variation, other than mutation, as effective in evolution. It is true, of course, that the environment may cause mutations by acting directly on the genes, as just stated, but this is not variation in the Lamarckian sense. In spite, however, of the negative evidence of genetics, the theory still has its many followers. It seems so simple, so satisfying, where other theories fail. While many experiments have been made purporting to demonstrate the transmission of acquired characters, I think it may be truthfully said, even without a re-examination of the evidence, that not one of them will withstand repetition under the eyes of a critical investigator. . . .

In the original development of the theory of Natural Selection by Darwin, he made but little effort to explain the causes of variation. He saw variation everywhere in nature, and while he knew some variations were inherited and some not, he more or less took it for granted that most variations were inherited. He knew that occasional sports or mutations occurred, but he thought they were too infrequent to be effective in evolution, and he also thought that such large sudden variations, or sports as they were then called, would tend to throw things out of balance, or cause a lack of coordination of parts. In his later years Darwin accepted the Lamarckian principle to explain the origin of variations. The variations Darwin saw and upon which he built his theory of National Selection were, for the most part, the small fluctuating variations which we now know to be due to the environment, and which we now know are not inherited. With respect, then, to the most crucial part of any theory of evolution—that is, variations and their origin—the theory of Natural Selection failed. It was built upon an insecure foundation. Considering, however, the weak foundation, the superstructure has withstood attack remarkably well, due largely to the fact that the error included the truth. Perhaps Darwin's error can be excused in part on the ground that the experimental method has not yet come into general use, and there is no way of distinguishing between fluctuating variations due to the environment and mutations except by the breeding test. It is true that genetics has not answered the question of the origin of mutations, but it has said emphatically that mutations are the only variations which can be used in the building of any theory of evolution, and that they are the result either of genic changes or genic rearrangements.

Even though Darwin's error was a gross one, it must be noted that the variations upon which he built his theory included mutations, as we define that word today—size does not enter into our definition—and so Natural Selection may still be, and probably is, a factor in evolution by operating upon mutations. In fact, Wright, in his recent discussions of evolution, is very positive on this question. He says, "The conclusion seems warranted that the enormous recent additions to knowledge of heredity have merely strengthened the general conception of the evolutionary process reached by Darwin in his exhaustive analysis of the data available seventy years ago."

Since Darwin thought of variations as random and varying about a statistical mean, the question arises whether mutations so vary. Are they as apt to occur in one direction as in another? Perhaps our data are too few for a final answer to such a question, but they are indicative. First, mutation frequencies are different for different genes. Some genes in Drosophila have mutated many times, others not at all. These mutations are usually in the same direction, but in some instances like the series of multiple allelomorphs[4] of eye color, they are in different directions. But even here there would seem to be limitations, for the colors are variations of red, shading down by degrees to white. As Shull has pointed out, blue and green have not appeared. In this

[4] An alternative character for which a gene is the vehicle.

instance, at any rate, it would seem that variations in eye color can occur only within certain limits—limits determined by the structure of the gene itself and the physiology of the organism in which it operates. Even though mutations are not random, in the sense that they are as apt to go in one direction as another, they may still furnish the variations on which Natural Selection acts. It merely means that the possibilities for evolution in diverse directions will be limited.

Another aspect to the theory of Natural Selection is the assumption that variations are cumulative. That they have been cumulative in some instances, at least, no one can doubt; but the evidence from genetics—inadequate we admit—is negative.

Much discussion has hinged about the question of the chance of survival of a new variation in a large population. Prior to our present knowledge of genetics we talk about the swamping effects of crossing. The idea seemed to be that if a new variant mated to a normal or wild-type individual the new variation would be diluted; that if this hybrid mated to another wild-type individual there would be still further dilution; and that if this outcrossing continued long enough the new variation would finally disappear. On this basis a variation appearing only in one individual or in only a few individuals would have very little or no chance of survival. In the light of genetics this question is now an entirely different one. Mutations are due to genic changes or genic rearrangements. Genes and their effects are in no way diluted by continued outcrossing. They persist, even though recessive and in the heterozygous state, as long as the individuals possessing them persist, and they will be passed on to future generations as long as these individuals reproduce. . . .

Isolation—not merely geographic isolation—is considered by some, particularly the taxonomists, as an important factor in evolution, and I must admit their evidence is convincing. The thought is that in restricted groups of individuals, variations will have a better chance of survival than in larger populations, due to closer inbreeding and to an absence of crossing back to the main population. Hence, evolution is encouraged.

Genetics has demonstrated that there is much heterozygosity existing in populations. Hence anything which would bring about inbreeding, or anything which would prevent free intercrossing in the population, would tend to segregate already existing genes. In other words, isolated groups of individuals may become different from the general population without the occurrence of new mutations. Of course, if new mutations did occur there would be a better chance of survival than in the [general] population.

It should be kept in mind, however, that in a restricted population there will be fewer mutations than in the general population for the very obvious reason that there will be fewer opportunities. It should also be kept in mind that the mutations of our cultures are mainly losses and that close inbreeding soon produces homozygosity with its deleterious effects. Wright, in his studies on population size and evolution, concludes that small and large populations are not as conducive as those intermediate in size—small populations for the reasons mentioned, large populations because there is so little opportunity for the production of homozygous individuals.

DeVries' mutation theory, as previously stated, was based upon the appearance of large heritable variations or mutations. The changes were so large that the mutants were described as elementary species. Species formation was thus a simple process—a sudden change, the cause of which was unknown.

Genetics has shown very clearly that size does not enter into our definition of a mutation; that most mutations are really small and hence species do not suddenly, by a single jump, come into existence. The result is that the mutation theory is now merely a statement with respect to the kind of variation which can be effective in evolution. Other factors have had to come to the rescue.

Orthogenesis or straight line evolution[5] is very obvious in numerous instances, even though we do not know its cause. Genetics, however, gives no evidence for orthogenesis. The only orthogenetic series, if the end products alone are studied, is the series of eye colors from red to white—changes of a single gene. But these changes did not occur in an orthogenetic series. White first rose from red, and all the others except eosin have arisen from red.

It is obvious that my remarks have been somewhat superficial and incomplete, but nothing else was attempted. I have tried to point out to you the contributions of genetics to evolutionary thought because Professor Morgan has played such a large part in these genetic discoveries.

[4] A series of changes in a progression of forms that appears to have some end purpose to it.

Part 4

THE MIDDLE OF THE TWENTIETH CENTURY: BIG SCIENCE

One of the challenges of the years of the Great Depression lay outside the United States, beyond the wreckage of the domestic economic system. The external threat was the anti-intellectual activity of Nazi Germany and the political absolutism of other totalitarian regimes. The dictators threatened intellectual honesty and even freedom of research in the laboratory. Scientists who believed that public political involvement was inimical to the progress of scientific research began to doubt their own beliefs. Archibald MacLeish, America's unofficial poet laureate, enjoined the artists, scholars, and scientists to leave their ivory towers and laboratories. Those who did not he characterized as "the irresponsibles." Such an unwelcome demand that scientists involve themselves in positive public policy underlined the problems elucidated by physicist W. James Lyons in his gloomy article about the world in which scientists found themselves in the late 1930s.

Despite depression and dictatorship, despite the continued attacks on science, scientific research went on throughout that decade as before, with perhaps some stringencies in local situations. University and foundation funding, for example, provided the cyclotrons and other instruments that put Americans such as Ernest O. Lawrence of Berkeley into the forefront of investigation into radioactivity and atomic phenomena. Although the increasingly large demands for money for big-instrument research drove scientists to make a virtue of the necessity of cooperation, as astronomer Otto Struve pointed out, they had not forgotten the lesson of the effectiveness of team research learned during World War I.

The World War II Manhattan project (to build an atomic bomb), which encompassed cooperative and team research on a stupendous scale, stands as the landmark ending independent effort and individualism as characteristic of American science. Yet accompanying the team approach, and, as Vincent du Vigneaud shows, anticipating it during the decades before the war, was still another characteristic of American science: the tendency of traditional disciplinary distinctions to disappear. Who can say where the DNA-RNA revolution was inaugurated? Chemistry and biology shared the field, and Wendell Stanley's crucial research (about which du Vigneaud spoke) grew out of medical bacteriology and agricultural research. As a result of the decline in scientists' disciplinary identity in the Manhattan and other projects, in the Big Science of the mid-twentieth century, interdisciplinary and cooperative investigations seemed to hold the most promise of yielding dramatic progress.

The depression era gave way to the Big Science era under the impact of four influences. The first was the as yet unexplained rebirth of optimism at the end of the 1930s when the general public tended to look to an expected "World of Tomorrow," a clean, utopian society in which everyone would be healthy and machines would solve all problems. Science fiction, in either the "Buck Rogers" comic strip or a more sophisticated version, seemed to furnish concrete images for this expected new society. In later years, when products predicted for the "World of Tomorrow," such as jet planes and television, became commonplace, the science fiction writers' vision of the future appeared justified. Scientists themselves talked in the late 1930s and early 1940s about "science as the great discoverer of new worlds for man to conquer, explore, and eventually transform into worthy habitations for the future" (a quote from a 1941 book appropriately entitled, *New Worlds in Science*). F. R. Moulton, permanent secretary of the American Association for the Advancement of Science, reported that the dominant note of the 1939 annual meetings of the Association (held after the fighting of World War II had begun) was one "of serene confidence in nature and man. . . . Although politicians are despairing of the future, scientists have no serious misgivings."

A second major impact came from the arrival of émigré scientists driven out of their homelands by dictatorship and war. As late as 1963, for example, 163 out of 670 members of the honorific National Academy of Sciences were foreign born. Many of the émigrés, of whom Enrico Fermi and Albert Einstein can serve as examples, were extremely eminent when they arrived, and such figures of course made an impact. But the larger number of new arrivals, such as Leo Szilard, were already outside of the scientific establishment of Europe, both politically and intellectually. When they arrived and began working within American traditions, they found the New World environment intensely stimulating and the laboratory milieu in the United States liberating. So much excitement did they generate that for a time American science seemed to be world science itself. In the 1950s and 1960s, however, Americans were competitive rather than dominant in most fields of endeavor. The most remarkable change stimulated if not caused by the newcomers, was the fact that Americans were doing outstanding work not only in applied and experimental science but even in theoretical science. It had taken a century and a special kind of immigration, but the dreams of mid-nineteenth century Americans that their country would contribute in the most fundamental way to new scientific thinking were finally fulfilled. Between 1939 and 1969, 69 American scientists, native and émigré, won or shared in Nobel prizes.

World War II constituted the third of the great transforming forces that produced Big Science. The impact of the war was complex but not altogether fortuitous. At the beginning, engineer Vannevar Bush, then president of the Carnegie Institution of Washington, took the initiative in bringing scientists into the service of the government in order to improve the defense effort (as it was then called). By 1940 he had his efforts legitimated by the creation of the National Defense Research Committee. A year later, the name and function of the NDRC was changed, and the Office of Scientific Research and Development (now including engineering) came into being. Other scientific research and development units within the government, especially within the War and Navy Departments, came to play major roles in the utilization of scientific resources for war purposes.

The decisions of Bush and his associates as to how science would be mobilized was of the greatest importance because the wartime institutions provided the pattern within which was implemented the new relationship between science and the federal government that developed after the war. Basically, their decision was to leave the scientists in their civilian status, working in the university, research institute, or industrial laboratories where they already were or might easily be attracted. Thus the enormous amount of government money that was funneled into scientific work went not into new government agencies as much as into contract work in existing laboratories around the country (about 70 percent of the research and development funds in the later years of the war went into contracts with civilian institutions). As with defense contracts of all kinds, the primary recipients of federal funds were the largest units; in the case of science, places such as the Massachusetts Institute of Technology and the laboratories of the great manufacturing corporations.

Since present institutions always seem to be superior to discarded alternatives, so in the postwar world the pattern of letting panels or committees allocate government funds to nongovernmental agencies and persons appeared to be the best possible way of sluicing money into scientific endeavor. (The system had been well known in the 1930s from its use in the National Advisory Committee for Aeronautics and the National Cancer Institute.) This model was therefore used after the war for most military research and development and appeared most conspicuously in the procedure of the National Science Foundation of the 1950s and 1960s. By flooding the universities, especially, with money, the federal government more than ever made it virtually impossible for science to be centered outside of the combined research and teaching unit of the graduate school. Most independent research institutions, such as the Mellon Institute of Pittsburgh, sought university affiliation, and the Rockefeller Institute, the most elite and most like the European model, in 1955 transformed itself into Rockefeller University. That the teaching-research model might not be the best one is indicated by the major exception in the pattern of government support of science, the National Institutes of Health. After World War II, the National Institutes of Health to a large extent fell heir to the health research of the armed forces and were built up additionally by health-conscious congressmen. The conspicuous success of this governmental research institute raises questions as to whether Bush's original plan was necessarily the best one.

World War II had one further type of impact on American science: federal money brought an incredible prosperity to science, a prosperity based in large part on the idea that one could simply by spending money accelerate inevitable scientific progress. Few people (although there were such as Frank B. Jewett) dissented from this opinion, and the atomic bomb project and then developments such as the computer (widely utilized in business as well as research) confirmed in the minds of scientists and public alike the idea that they could buy science and progress. Such views ultimately complicated what became known as the politics of American science.

At the end of the war the fourth of the great forces that brought forth Big Science appeared suddenly in the form of the nuclear explosions that obliterated most of two Japanese cities. The public reacted by viewing the scientist as a sort of magician and science as a means of attaining an almost indecent and superhuman omnipotence. Nothing could have confirmed better the belief that the world of tomorrow was within reach if not already here.

The impact of the bomb on the scientists was far more fundamental, and on an entirely different level. A significant minority of them, often led by the founders of *The Bulletin of the Atomic Scientists,* felt that the development and utilization of nuclear fission called for scientists to take responsibility for their creations in a way never seriously envisaged before, even in the face of Nazism. As the dangers of fallout became manifest—and political—the life scientists also became directly as well as ideologically concerned, and without willing it, all American scientists found themselves deeply involved in moral, social, and political questions.

The forces making for Big Science, however, were to a large extent brought into play because of another war-born crisis. During the war, the United States tended to put all scientists into defense work. "We gambled on a short war," wrote an alarmed scientist-administrator, Arthur H. Compton. "Science professors and students alike left the universities." As a result, not only had basic research come to a halt but young scientists, if they existed at all, were generally engaged in weapons-related research. It would be, concluded Compton, "at least six years before a normal supply of young professionals is available." The government was implicitly committed to helping close these gaps in both research and personnel.

One of the major events of the immediate postwar years was the establishment of the National Science Foundation in 1950 to channel money into scientific investigation and education. The designers of NSF hoped that pure rather than applied research would thrive off National Science Foundation funds. Before and even after the establishment of NSF, the Atomic Energy Commission and the Office of Naval Research had been especially active in keeping up a high level of federal subsidy of research laboratories. A substantial amount of pure research was actually carried out under grants from these and other agencies officially interested only in applied science. Even the limited amounts of money available from the National Science Foundation came as a welcome relief to scientists who had been diverting development program funds for the purpose of pure science as their predecessors in the geological surveys had done generations earlier. Because of this and other types of "bootlegging" of funds, official figures as to the exact number of billions of dollars put into scientific research by the federal government will probably never be known, except insofar as one can measure the results of the expenditures. In 1938, 28,000 names appeared in *American Men of Science;* in 1965, 150,000 merited inclusion. In truth, in the affluent years of the 1950s and 1960s, no one looked too closely at government investment in the growth of science.

The field of nuclear physics (along with a number of fields closely allied and derivative) was for many years the recipient of the most uninhibited spending of the federal government. The various health fields likewise enjoyed for decades an unbelievable congressional largesse, supplemented in a powerful way by a whole new generation of foundations the grants of which, although covering all of science, were most important in research related to disease. A rather unexpected development of investigations into disease-causing viruses led, as noted above, into the birth of a new glamour area, molecular biology, with implications for humankind far beyond any but the wildest of imaginations. It appeared that the discovery of the ultimate in mind-body relations might be within sight and that the dreams of the earlier eugenicists were capable of realization with a precision and immediate effectiveness never really comprehended by nonfiction writers. By 1970 Max Delbrück, the biophysicist, could declare flatly in his Nobel Prize address, "We may say in plain words, 'This riddle of life has been solved.' " The men who unlocked these new secrets of nature were for some years extremely backward about making even cautious public statements about the implications of the work. George W. Beadle, for instance, who won a Nobel Prize for his research in the area, could as late as 1959 discuss molecular biology and still in the same essay talk in terms of classical eugenics when discussing the balance of genetic elements in the population. And not everyone, as geneticist T. M. Sonneborn made clear before long, was optimistic about what the biochemists and molecular biologists had been discussing privately among themselves. But just when the popularizers of science were letting the general reader know about the implications of the DNA-RNA revolution, the climate of opinion in the United States changed.

For about twenty years after World War II, science enjoyed a romance with the public. Surprisingly little dissent was heard as the federal government put incredible amounts of money into scientific endeavors of one kind or another. Large corporations used the size of their research programs as gauges of company prestige, and in various parts of the country local spokesmen boasted of the economic affluence as well as cultural prosperity that grew out of the role of the area as a research center. The public's image of science was appropriately as confused as that of the legislators and stockholders who allocated funds for science: No one was able to distinguish between research and development. Scientists, engineers, and even technicians seemed to be indistinguishable from one another, and differences between pure and applied research were not elucidated by trying to separate "research" from "development." Who could say, for example, exactly how one ought to classify various space projects, from mere rocketry to retrieving geological specimens from the moon?

In addition to the television-automobile-jet plane benefits of science, national pride and fear motivated segments of the public who supported spending for scientific work. The Cold War with the Soviet Union inspired an American policy of keeping ahead of any country in scientific weaponry, from nuclear bombs to germ warfare. Particularly after the shock of the first Sputnik, orbited by the USSR in 1957, the spigots of money could easily be turned on by the magic words, "How are we going to beat the Russians?" And, indeed, federal science expenditures jumped sharply after Sputnik.

It was inevitable that the honeymoon science was enjoying with the public—and at public expense—would end. Someone figured in 1959 (when there were perhaps a million scientists and engineers in America, 85 thousand holding doctorates) that if the rate of growth of recent years was projected into the future, within a century everyone in the United States, including children, would be a scientist or engineer, and if the rate of growth of expenditures on research and development continued, within 40 years the sum would exceed the total national income of the year 2000. By the mid-1960s disenchantment had clearly set in. It could be measured not only by the leveling off of federal expenditures but by an increasing public hostility to some of the results of postwar science and technology.

The atomic bomb had resurrected the literary image of the mad, destructive scientist, especially as, after a time, the incineration of Hiroshima and Nagasaki appeared to have been unnecessary. As atomic explosions continued in the guise of "scientific tests," public hostility intensified still more after it became clear that responsible officials, including scientists, were lying about the extent and danger of the aftereffects. Science was blamed for major and to many people alarming alterations in the physical environment, because scientists had produced or condoned the pesticides, gases, rays, and chemicals of pollution. When it was revealed in the 'sixties that even the symbol of American civilization and "science," the new automobile, was insufficiently equipped with scientifically designed safety features, a revulsion of opinion occurred, not only against the irresponsible use of technology but against science. Utilizing the occasion of the mismanagement of a magnificently conceived geological probe of the crust of the earth (the Mohole project) and the financial pressures of the Vietnam War, Congress did indeed level off government expenditure on science. Tight university and foundation budgets tended to reinforce the impression that the great growing period of subsidy of American science had ended in the mid-1960s. As a *New York Times* editorial writer suggested, a new 200-Bev accelerator involved in a civil rights dispute was anyway a "scientific luxury," inappropriate for the times. The inner city was presenting more pressing problems than the moon, the atom, and the virus.

Despite the tightened budgets, American science still had tremendous momentum, substantial leadership, and two significant assets: real power and continuing vision. Both require some comment. For the solution of social problems, science could offer information and a method and approach, even if they had to be used more often than before in social science research that would grow on funds diverted from the "hard" sciences. And there was always the dream of the use of material innovation and application of science to bypass, destroy, or obviate rather than solve social problems. Perhaps, as physicist Alvin Weinberg pointed out, such "technological fixes" were not chimerical but hardheaded. Perhaps this conception had always provided the real social rationale for American science.

By the 1960s American science was no longer innocent—if it ever had been—of the corruptions of power. Within science itself there ruled an unofficial but effective "Establishment," by legend concentrated in the exclusive Cosmos Club in Washington. The group in power was descended directly from the concatenation brought together by Bush at the beginning of World War II and perhaps even from the National Research Council of World War I. To outsiders, the Establishment appeared to have a monopoly on government and foundation funds, which were used to make the rich laboratories richer; to beneficiaries of the Establishment, its powers were used to sustain quality work and honor high standards. Dissidents and mavericks within American science existed relatively comfortably, as long as certain basic rules honoring the scientific community were not broken. Considering the size and diversity of scientific enterprise in the United States, it is remarkable how effective the Establishment was in shaping and setting the tone and image of science during the second third of the twentieth century.

Since American science had relatively concentrated control, it was inevitable, given the public's romance with it, that it should appear to have considerable potential for wielding power in American society. After the explosion of the atomic bomb, a number of scientists, most notably Leo Szilard, continued to attempt to realize this political potential. As the amounts of money siphoned into research and development grew, a new type of nonlaboratory, nonscience wielder of power proliferated—the science administrator, who could allegedly mediate between nonpolitical, research-oriented scientists and the public and government. The momentum generated by the atomic scientists to influence public policy directly, however, increased, and by 1964 Donald W. Cox could somewhat optimistically write about *America's New Policy Makers: The Scientists' Rise to Power.*

Scientists in politics were interested in several main topics: peace and control of atomic weapons, and later all chemical and biological warfare; environmental contamination; and, inevitably—and not necessarily for selfish purposes—the increase of support for science. What science might have to offer was both abstract and practical. As in the very beginning, scientific work and education, even pure science, continued to be justified because of their social and material usefulness. But, as Charles H. Townes of maser and laser fame suggested, power and control in the end were relatively ineffective. No one could foresee or control the directions that research might take or what the social impact of that research might be.

More and more frequently as the 1970s began, thoughtful Americans were reminded that earth is a mere spaceship on which the passengers have to engineer their own comfort and survival. Not only did they increasingly realize that the totality of nature on the earth and even beyond is an indivisible unit—a grand ecosystem—but that the world environment, the whole "web of life," was in danger of becoming poisoned by artifacts of human technology and mismanagement. What had started out in the 'sixties as concern about pollution developed into acute anxiety about what man is doing to his planet in general.

The young, especially, raised searching questions about what the scientists were bequeathing to later generations. Many people talked, as their predecessors had in the 1930s, about the misuse of science and about the fact that nevertheless the world needed scientists to solve its problems. A spokesman for the rebellious generation of 1970, for example, while denouncing the abuse of scientific resources asserted that "science has given us an understanding of the evolutionary play in the ecological theatre and has awakened us to a true and challenging comprehension of man and of man's place on this planet." Ecologists Iltis, Loucks, and Andrews suggested that to preserve an environment fit for survival many facets of science, some of them unconventional, would have to be brought into play. Where once much research had been devoted to meeting man's needs, in the 1970s the scientists would have to determine what man's needs might be. The New World environment, which in colonial times had been the occasion and cause of scientific endeavor in America, in the last third of the twentieth century was becoming the central concern of scientists who owned the colonials as forebears.

28. An Astronomer Discusses His Science and the Need for Cooperative Research. 1940

One of the few scientific disciplines in which Americans have often been internationally preeminent is astronomy. In the twentieth century, especially, Americans have had the advantage of superior research facilities in the form of large telescopes. Here an eminent astronomer, OTTO STRUVE (1897–1963), explores some of the effects that this investment of very large sums of money was having on science. Already before World War II the team research aspect of "big science" was present in the embryonic form of "cooperation." Of particular interest is Struve's discussion of problems in astronomy that he considered exciting and, at the same time, his attitude toward theoretical work. No doubt he had in mind at least partly the popularizations of cosmology by the English scientists, Eddington and Jeans, which were enjoying sensational success in America during this period.

Born in Russia, Struve came to the United States and took a Ph.D. in astrophysics in 1923 at the University of Chicago. He was appointed instructor at Chicago's Yerkes Observatory in 1924 and became professor in 1932. In 1946 he was made MacLeish Professor. In 1932 he became director of the McDonald Observatory, and in 1950 director of Leuschner Observatory. His many contributions included work in spectroscopy, interstellar matter, stellar evolution, and cosmology.

For more information see Bernard Jaffe, *Men of Science in America* (2d ed., New York: Simon and Schuster, 1958), chap. 18, and *Current Biography*, 1949, 597–599.

COOPERATION IN ASTRONOMY*

Otto Struve

The McDonald Observatory of the University of Texas was officially dedicated on May 5, 1939. It is equipped with a reflecting telescope 82 inches in aperture, and it is, by virtue of this fine equipment, one of the important centers of astronomical research of the world. The observatory is operated jointly by the University of Chicago, which supplies the staff from its own Yerkes Observatory at Williams Bay, Wisconsin, and provides the larger part of the operating funds, and by the University of Texas, which built the entire

* *The Scientific Monthly*, 50 (1940), 142–147. Reprinted with the kind permission of the American Association for the Advancement of Science.

plant from the bequest of W. J. McDonald, of Paris, Texas, and which, in addition, pays for part of the maintenance.

The plan of collaboration, entered into by the two universities, has been very successful. Without this plan, the 82-inch telescope would not have come into existence; the Yerkes Observatory would not have been able to increase its scientific output and would not have attracted to its staff new astronomers of high international standing; the University of Texas would not now own one of the world's most powerful telescopes, and would not have become the sponsor of some of the most intensive research in astronomy.

Although the new telescope has been in operation for only nine months, some interesting scientific results have been secured with it. Dr. Polydore Swings, visiting professor at the University of Chicago, has already announced the discovery of a large number of "forbidden" radiations, hitherto unknown, in stellar spectra photographed by him last month at the McDonald Observatory. Forbidden radiations can not be observed in the physical laboratory, and without astronomical observation they would have remained unknown. Although these radiations are "forbidden" on the earth, they occur in the vast spaces of the universe where low density of gas, almost limitless space and permanent semi-darkness are conditions required for their origin. It is unnecessary to dwell upon the importance of this knowledge: the physicist knows that to understand the properties of matter he must understand the radiations which matter is capable of producing. He has studied in the laboratory the "permitted" radiations, but his data would be unintelligible without a knowledge of the "forbidden" radiations.

The most important discovery made by Dr. Swings is that of a forbidden line of the element Fe X (iron ten). Under normal conditions the atom of iron possesses twenty-six electrons. When energy in the form of light or in the form of a shock—perhaps a collision with another particle—is applied to an iron atom, it may lose its outermost electron, after which it would be designated as Fe II. If more energy is applied, it may lose two electrons and it then becomes Fe III, and so on. The energies required to produce these changes are great. In the atmosphere of the sun we observe Fe I, with a little admixture of Fe II, and with no Fe III at all. Little imagination is required to realize the tremendous amount of heat and light in the surface layers of the sun. And yet, they are barely sufficient to create and maintain Fe II in a gas whose density is ten thousand times less than that of the air we breathe. In some of the hottest stars Fe III was found last year by Dr. Swings, and was confirmed through a study of photographs made with the 82-inch reflector.

Astrophysicists have established a scale of energies to measure the forces which must be applied to an atom to break off its various electrons. The unit of this scale is the electron-volt that is set, by agreement, equal to the energy which is acquired by an electron when it passes through a potential difference of one volt. To knock off the first electron of an iron atom requires 8 electron-volts; to produce Fe III requires 16 electron-volts. But to produce Fe X requires about 200 electron-volts—an energy which has not hitherto been contemplated in astrophysical studies! This is in itself remarkable, but to the physicist a still more significant point is the fact that the forbidden line is produced by a transition between the sub-levels of the same multiple state. No other forbidden line of this type had ever before been observed, although Dr. Theodore Dunham at Mount Wilson had inferred, theoretically, that such transitions must occur in the gases which fill the spaces between the stars.

Dr. Gerard P. Kuiper, of the Yerkes Observatory staff, is now engaged in the discovery and study of some very strange stars which we designate "white dwarfs." This name gives but an imperfect idea of the remarkable properties of these stars. They are called "dwarfs" because they are small—some are smaller even than the earth. And they are "white" because their temperatures are high—of the order of 10,000° Centigrade at their surfaces. What the name does not indicate, and what is really of most importance, is the fact that in mass they are not dwarfs, for their masses are of the order of that of the sun. If you could squeeze a mass comparable to that of the sun into a volume smaller than that of the earth, you would have a density so enormous that it defies the imagination. Kuiper has computed for one of his stars a mass of one thousand tons per cubic inch. Such matter is not ordinary matter at all—we have nothing like it on the earth or on the sun, and we call it "degenerate matter." The problem is to find, by means of astronomical observations, the properties of this "degenerate matter." We already know that it follows quite different laws of physics than does ordinary terrestrial matter, and we hope that we shall ultimately discover how it reacts to changes in pressure and temperature.

For years astronomers have been interested in the problem of the chemical composition of the stars. The question is an intricate one. The tremendous energies radiated by the stars into space in the form of heat and light are almost certainly produced by the slow conversion of certain kinds of atoms into other forms. We have heard much in recent months of the so-called carbon cycle, of the proton-proton mechanism, and of other processes which may account for the heat and light of the stars. These processes gradually use up the available hydrogen and thereby reduce its abundance. There is ample indirect evidence that the hydrogen contents of the stellar interiors are not all the same. The only way we can directly observe the composition of a star is to study its spectrum.

It might appear to be an easy task to determine the amount of each chemical element in the outer layers of a star from its spectrum. In reality, the problem is very difficult, and little trustworthy information has thus far been obtained. In a few cases we suspect that hydrogen is less abundant than it is in normal stellar atmospheres, but the evidence is limited to some rather unusual objects whose spectra are quite difficult to interpret. At the

McDonald Observatory, Dr. Jesse Greenstein observed the star υ Sagittarii, which had been previously observed at the Yerkes Observatory and elsewhere, but only in the blue and violet regions of the spectrum. The new work, carried out in ultra-violet light, was facilitated by the high reflecting power of aluminum with which the 82-inch mirror is coated and by the excellent transparency for ultra-violet light of the quartz prisms of the spectrograph.

The work on υ Sagittarii shows that it is a star whose atmosphere contains an abnormally small amount of hydrogen. From the point of view of chemical composition, it is by far the most interesting object in the sky, and it is very remarkable in another respect: In normal stars of its class, which have surface temperatures of about 10,000° Centigrade, the atoms of hydrogen and the free electrons produce a certain amount of haziness in their atmospheres which prevents us from seeing into their deeper layers. In υ Sagittarii, however, the hydrogen content is low, the metals contribute a relatively greater share to the free electrons, and the opacity, or haziness, of the atmosphere is, therefore, not the same as in a normal star. The theory of the formation of stellar spectra depends entirely upon this opacity. There has previously been little opportunity to secure observational evidence in support of this conclusion, but υ Sagittarii provides the ideal material for testing the theory.

Since its installation on Mount Locke the 82-inch reflector has been in constant demand and has provided a continuous flow of scientific material. Dr. A. Unsöld has collaborated with me in studying the ultra-violet spectra of B and A stars; Unsöld has also determined the curve of growth of Arcturus; Roach has cleared up some of the mysteries of P Cygni; and Page has observed the spectra of planetary nebulae. A program of radial velocities of faint B stars is being carried out by Popper and Seyfert.

All this work owes its inception to the cooperative arrangement between the University of Texas and the University of Chicago. Astronomers have always been eager to cooperate. They have carried out large international projects, such as the complete mapping of the sky with photographic telescopes distributed all over the world, and the determination of accurate star positions with meridian circles in a dozen or more observatories. They have organized international bureaus for the distribution of astronomical news, such as discoveries of comets and novae. But universities have already been willing to pool their resources for the maintenance of a large observatory. The Chicago-Texas arrangement, and the somewhat similar arrangement between Ohio State University and Ohio Wesleyan University in maintaining the Perkins Observatory, are a new demonstration that satisfactory results can be obtained by collaboration.

The greater number of existing astronomical observatories in the United States were built and equipped from funds donated by private persons. Most of these institutions were opened when instrumental equipment was not as highly developed and not nearly as costly as it is to-day. When the plans for the Yerkes Observatory were first considered in 1892, the secretary of the university recalled that, not many years before, the old University of Chicago had owned the largest astronomical instrument then in existence—a telescope (now at the Dearborn Observatory) having an objective 18½ inches in diameter. Since that time telescopes had been made with objectives having diameters of 20, 23, 24, 25, 26, 27, 28, 30 and 36 inches. The Yerkes 40-inch refractor was completed in 1897; but a few years later the 60-inch reflector at Mount Wilson and, finally, the 100-inch Hooker reflector at Mount Wilson became, in turn, the world's largest telescopes.

Large apertures mean great light-gathering power. Not all astronomical investigations require great light-gathering power, but many do. An astronomer who has at his disposal a 15-inch telescope can carry on certain types of research, but he can not participate in some of the most interesting and fruitful investigations now in progress at the larger observatories. Doubtless many able astronomers have experienced a deep feeling of disappointment when limitations of instrumental facilities have prevented them from carrying on the type of research which they considered most useful and valuable. Yet, restrictions of university budgets and the rapid decrease in the frequency of large private donations leave little hope for the improvement of small and inadequately equipped observatories. During the period of 31 years between the construction of the Dearborn 18½-inch telescope and the Yerkes 40-inch telescope, the priority in aperture passed through ten different institutions. Since the completion of the Mount Wilson 60-inch reflector in 1908 it has remained with the same institution.

There can be no doubt that the growing disparity between the facilities for research available at various observatories raises a serious problem for the future development of astronomy. Fifty years ago practical work in astronomy was rather uniformly distributed among a large number of observatories in Europe and America. As time goes on we see a growing tendency to reduce observational activities at the smaller institutions. The existing equipment is usually insufficient and funds are not available for a complete modernization. Moreover, at some of the eastern and middle-western institutions the climate is not good enough for the efficient utilization of a large telescope. The organization and maintenance of a separate observing station in the southern hemisphere, or even in our own Southwest, is beyond the resources of the average university.

Hence, there has been a general tendency to substitute theoretical studies for observational work. Fortunately, in the United States this process has only started. But in Europe we have seen the gradual decline of observational work and the rise of theoretical institutions. The Kapteyn Institute at Groningen, Holland, the theoretical astrophysical institute at Oslo, Norway, the substitution of a theoretical department for what was once the University Observatory at Kiel, Germany, and, above all, the rise of theoretical astrophysics in England during and after the World War are striking examples.

I believe we must guard against this tendency in the United States. After all, the success of theoretical study depends essentially upon the supply of observational results. These now come almost entirely from a few large American observatories. There is, of course, no danger that Mount Wilson, Lick, Harvard and a few other observatories will not continue their observational activities on their present scale. But is it wise to restrict all observational work to a small number of institutions? Is this not likely to produce a cleavage between theoretical astronomers and observers which will result in much confusion and unsatisfactory progress? Is it not also rather disquieting when we contemplate the number of young astronomers who are being trained for future careers at institutions which are unable to carry on modern observational activities? The present trend toward pure theory is not a natural process, but is one which is forced upon astronomy by restricting circumstances.

I fear that unless something is done toward equalizing the research opportunities of all astronomers there will be a gradual deterioration of many observatories which, in the past, have been able to carry on investigations of a quality comparable to that of the largest institutions. Half-measures are expensive and are not satisfactory. The difficulty which is normally experienced in a present-day observatory of limited means is not a lack of problems which can be attacked and solved with the available equipment. Even the most modest observatory can do *some* useful work. Important contributions to science are often made with small equipment. Ross's Milky Way Atlas was made with a 5-inch lens, Merrill's catalogue of emission stars is based largely upon observations obtained with a 10-inch telescope and the Harvard spectroscopic surveys were made with instruments of relatively small aperture. But it is significant that all these investigations were made at large observatories, in spite of the fact that such instrumental equipment could have been easily available at a small observatory.

The difficulty is, of course, that the small observatory is compelled to search for something that it is able to do, instead of doing what is scientifically important and interesting. An observatory which has only a 12-inch visual refractor can do useful work on double stars and variable stars, but the limitation is quite likely to destroy initiative even in this restricted field. When a scientist is unable to do what he considers necessary and is constrained to do what his telescope allows him to undertake, he not only loses interest but often loses his contact with modern developments; his department deteriorates; his students suffer from the narrowness of the institution's interests. In the meantime the observatory continues to draw heavily from the university's general budget. With normal deterioration the instruments and buildings are more than likely to require increased appropriations as time goes on, but the scientific output becomes, if not less in volume, certainly less in value.

Fortunately, it is not necessary for this process to continue. A powerful modern telescope provides a wealth of material. Some of it can be utilized immediately by the observer. But more often than not the photographic plates contain valuable material which is not immediately used and which is often not even appraised by the observer. It is certain that by cooperation we can do our telescopic work more efficiently than we have done it in the past, and supply a greater number of astronomers with material.

If several observatories would pool their resources they would be able to construct a series of instruments designed for special purposes. The whole equipment would constitute a first-class observatory, although separately each instrument might not be of great value. For example, the McDonald Observatory has a large reflector which is suitable for work with slit spectrographs. But it has no powerful camera for objective-prism work and may never be able to purchase such an instrument. It would be advantageous for us to collaborate with some other institution which may be contemplating the construction of a new instrument. Without collaboration each observatory builds what appears to be now the most generally useful instrument—a parabolic reflector—even though there is already a considerable amount of duplication in reflectors of moderate size.

The resources of the McDonald and the Yerkes Observatories are sufficient for successful work in many phases of astrophysics. They are not sufficient for the development of new fields of research, such as solar physics, objective-prism spectroscopy, the study of cosmic rays or of cosmic radio disturbances. Moreover, the 82-inch telescope is not particularly suitable for certain types of astrophysical research and for which more efficient types of instruments have been designed. In order to enlarge the scope of the work of the McDonald Observatory it would be reasonable to invite the collaboration of other institutions.

Let us suppose that a plan of collaboration could be worked out which would be satisfactory to all participating institutions. We should then be able to organize jointly an observing station in the Texas mountains, where the McDonald Observatory is located, which would be much more powerful than the present McDonald Observatory alone. The participating institutions would all profit from the fine climate, which yields nearly 300 clear or partly clear nights each year; from the excellent seeing which averages much better than in the Middle West; from the exquisite transparency of the air at an altitude of almost seven thousand feet; and last, but not least, from the latitude of N 30° which permits the observation of a large part of the southern sky. The joint enterprise would benefit from the participation of many competent astronomers whose services no one observatory could possibly afford. Duplication of effort, no less than of telescopes, would be avoided. Many special types of research which require a large amount of preliminary laboratory work—for example, radio-metric measurements, photoelectric photometry and various other applications of electrical methods—could be prepared at the participating institutions and could be divided among them. Altogether, it would seem to be conservative to say that the plan would increase the observational facilities which are now available to many astronomers and would, thereby, make their present connections more attractive. It is my strong conviction that astronomers and university administrators should seriously consider such a project.

The plan would, of course, be expensive. But it would cost much less than equipment for separate new observatories. It would leave the present organizations intact and would secure for each participant a new outlet for research. The local observatories could continue their present functions and could, in addition, serve as laboratories for the measurement and discussion of material obtained at the observing station. The astronomers would from time to time travel to the observing station in order to gather material. They could be available for teaching at all other times and could conduct most of their scientific work in their offices at home. The constant pressure for new telescopes on the various campuses would be materially relieved. Measuring machines, photometers, etc., that would still be needed by each institution, would be relatively inexpensive.

Although the suggested plan for wider cooperation among astronomers presents some difficulties, they are probably no more serious than those that have already been overcome in carrying out earlier national and international undertakings. But, however serious they may be, they should be resolutely met, both because the scientific results to be obtained promise to be very important and because in putting such a plan into effect astronomers will set another notable example of cooperation.

29. A Conservative Scientist Finds Himself in a World Unfriendly to Science. 1940

The title of this essay suggests the concern of the author. Considering the past approbation and support that had been won by American scientists, they confronted with considerable dismay contrary, hostile sentiments arising from such an array of different sources as is depicted here. This selection is particularly interesting in that the author represents the more conservative scientists who came with great reluctance to think that the scientific intellectual might have to leave his laboratory for a time in order to continue in his calling as scientist.

WILLIAM JAMES LYONS was born in 1904 in Duluth, Minnesota. He taught high school for a time in Minnesota and then returned to finish his education at the University of Chicago. In 1935 he took a Ph.D. in physics at Saint Louis University and then joined the Western Cartridge Company as a physicist. At the time he wrote this article he was teaching physics at Loyola University in New Orleans. Later he worked for the Department of Agriculture, the Firestone Tire and Rubber Company, and the Textile Research Institute. Most of his contributions have been in the area of the properties of synthetic textiles and other materials.

See Charles William Heywood, "Scientists and Society in the United States, 1900–1940: Changing Concepts of Social Responsibility" (Ph.D. dissertation, University of Pennsylvania, 1954).

SCIENCE IN AN UNFRIENDLY WORLD*

W. James Lyons

That there are prevalent in the world to-day certain trends of thought and action unfavorable to the future prosperity of science needs no extensive proof. The sentiments range from a general coolness toward technological advance to an active antipathy toward a free science. Each challenge, separately, has been recognized and identified with more or less adequacy, but a more comprehensive view of them as parts of what is in effect a more general movement is lacking. While it may be said that one of the challenges to science merely implies "friendly competition" of a sort, or that another is only a vague rumbling, they deserve a brief consideration by the student of affairs, as well as by the scientist. Because they are not isolated in time, but confront science together, their significance is enhanced. It is not my intention, however, to "uncover with alarm" an immediate danger or one of appalling magnitude.

These current dangers to science are clearly distinguishable from each other. The first, specifically, is the attack on the dominant positions which science, and the *scientific method* have come to achieve, during the past seventy years, in the national philosophy of education. The most prominent and energetic leader of this thrust is, of course, Robert Maynard Hutchins,[1] of Chicago, whose ideas have been developed in numerous publications and addresses. The second adverse trend, less well-defined, is a growing coolness toward science, and suspicion of it, in the public mind, more

* *The Scientific Monthly,* 50 (1940), 258–263. Reprinted with the kind permission of the American Association for the Advancement of Science.

[1] Robert M. Hutchins was the leader of the neo-Thomist or traditionalist school of thinking. These intellectuals believed that all necessary ideas were already available in the great intellectual documents of the Western world and that giving more attention to the Greeks and their successors was more substantial and beneficial than pursuing new ways of gaining wisdom.

markedly at the low-income level. The subtle nature of this threat precludes its having any leaders. Aside from some political talk about declaring a legal holiday for invention, the attitude is yet largely negative. What well may be taken as the representative attitude in this respect has been presented sympathetically by Norman Foerster.[2] Finally, science is confronted with the demoralizing programs of the totalitarian governments. The problem for science here is not the cruelties and injustices which have been inflicted on scientific workers in those countries on account of race or religious creed, reprehensible though such activities are. But a problem and a threat to free science does exist in the system or regimentation and delimitation, of thought and inquiry, in vogue in the dictatorships.

These three unfavorable currents, in common, reflect a lessening in the prestige which science has enjoyed for decades. They reflect, it would seem, a conviction spreading vertically and horizontally that science in one or another of its aspects is not as worthwhile to mankind as it had seemed to be. Whether, further than that, any causal or logical relationships between them exists, I am not prepared to say. It does not appear, however, that the existence of such relationship is pertinent to the present discussion.

Bound up with the question of whether scientists are meeting these challenges satisfactorily is the one whether some of the criticisms or attitudes are justified. What has become the traditional mode of pleading the case for science is to point with pride to past achievements and to predict a future of more abundant creature comforts and less pain. A latter-day expansion of this theme has been an invitation to the world to put itself in the hands of science for the great cure of all spiritual, economic and social ills. Perhaps this proposal is in the nature of a counter-offensive. But the whole argument is becoming a little worn and empty. It was effective thirty years ago, but is not pertinent to the dominant interests of the troubled world of to-day. Some of its implications have been rather badly riddled by observers of the modern scene.

But what can and ought the apologists of science say or do in its behalf?

I

It is true that the analysis and criticism of Dr. Hutchins and his associates are focussed on the theory and pattern of higher education which the American system had evolved by, say, 1930. Nevertheless, science is greatly implicated, since the dominant theme in that theory was the *scientific method*. In the reaction against the indoctrination which characterized earlier practice in higher education, objectivity and the empirical basis of knowledge were now emphasized. Less and less, with the passage of time, was there

[2] Norman Foerster was a leading spokesman of the New Humanism movement in literature. The New Humanists emphasized traditional forms, standards, and values as opposed to both naturalism (what is) and romanticism (what one wishes).

a willingness to teach anything that could not be rigorously proved. Norms for evaluation and appreciation became themselves subjects for dissection and objective analysis. The emphasis turned to equipping the student with a fund of specialized knowledge, and the special techniques by which more facts could be uncovered.

This concept of education the new critics have branded as anti-intellectual. It is claimed that in the modern teaching of science, reason has been suppressed, in deference to the accumulation of an ever-expanding body of facts. In consequence, then, of the influence which science, as well as some other factors, has had on education, we are told that the modern college graduate has lost the art of thinking and knowing. The factual deposits which have been left by his collegiate sojourn lack integration in any broad sense. He is without general principles and standards according to which he may recognize, understand and evaluate the large variety of problems with which he will be confronted in later life.

This case against science, materialism and the elective system in higher education has not been unconvincing. More than one college has hastened to revise its curriculum and, in re-announcing its aims, pay tribute to the *disciplined mind*—a term which a few years ago connoted the extreme of intellectual atrophy. It is the theory of the New Program, says Stringfellow Barr, that "In addition to being able to read, write, and reckon more skillfully and fruitfully than the American college graduate of to-day commonly does, in addition to being able and willing to face present issues, he should be able to recognize those eternal problems which his ancestors faced before him and which recur for every generation of human beings . . . . Experience had taught that the best statements of those problems are the 'classics' of human thought." The variance between the point of view and mode of thought here disclosed, and the modernistic and iconoclastic, scientific attitude is readily apparent.

Not unlike the advocates of science in education, the new humanists appeal to the democratic traditions of American society. Only a system of liberal education, embracing the humanities to an important degree, it is indicated, can free men's minds, and safeguard their freedom of thought and speech.

None of the new curricula designed for a more humane education appear to have completely eliminated mathematics and the sciences. With varying prominence, these branches of knowledge are given places in the courses of study. Their *purity* is the norm for admittance. But the sciences are under surveillance, if not suspicion. No longer, under the new plans, are they to be the pacemakers or patterns for the literature, philosophy and history courses. There is a feeling that science as a whole, after sixty or seventy years, has forsaken culture, even civilization.

Scientists and technologists, particularly those associated with education, would do well to recognize that much of the rationale of the new humanist movement is sound. It appears that only by first doing that are they likely to get an attentive audience.

It is fairly certain that there has always been a demand, of varying vigor, for a type of education which looks to the development of the whole individual. The old *classical* education of a hundred years ago attempted to meet this demand by reducing all graduates to one accepted type, the conservative, respectable, widely but vaguely informed gentleman. All facets of the personality were polished, albeit none very brilliantly. But while every educational theory has embraced the idea of well-rounded development, as an ideal at least, the modern, scientifically-inspired higher education has, in practice, avoided the responsibility of itself pursuing that objective. Under the urge to advance and spread impersonal, scientific knowledge, the colleges and universities, their faculties and curricula, have lost sight of the individual personality. Thus, while higher education at its best has been producing rather well-equipped scholars and research workers, in so far as it has contributed, the graduates are without any well-formulated objective in life, are unconscious of their moral and social responsibilities, have no confirmed esthetic or cultural appreciations. Even the functions which the colleges have been performing well lose much of their attraction when it is realized that but few of their graduates are associated with scholarship or research in later life.

The scientist-educator, rather than continuing to proclaim the adequacy of scientific knowledge as the framework of higher education, will render a greater service to society, and to the cause of science ultimately, if he will cooperate in the design of curricula which more effectively meet existing needs. Such curricula will aim to realize, within the bounds of each individual's capacity, men of enlarged vision, sensibilities and intellect. To accomplish this will require a re-orientation of science education; some "scientistic" dogmas will have to be shed, and some concessions made to the non-scientific branches. There is need for a renewed recognition of the limitations of science, particularly in the fields of human relations and conduct. Here the point of view which continually stresses standards of good citizenship and moral responsibility, is more profitable to the average student than is the indifferent, factual and analytical approach, which aims to be *scientific*. Many elementary courses employing this approach never get far beyond describing the approach itself, and conclude by leveling all standards and principles. The student gains nothing positive.

All too often have advocates of science discounted and disapproved whole sections of the cultural deposit of the race, largely because they were accumulated by non-scientific methods. In many of these fields science is still groping for an effective approach, and has thus far established only general, abstract principles. Whatever motive is associated with this immature sort of science can not be its own. Realism would dictate that here science can not yet offer a workable substitute for the structure which men have built up from experience and expediency, by trial-and-error, hammer-and-tongs methods.

While educators and scientists may wisely make concessions to the non-scientific branches, nothing indicates that all claims for modern science should be relinquished. It must be recognized that the physical and biological sciences have made contributions of facts and generalizations about nature, with which, for tangibility and value, the unhampered speculation and magic of earlier ages have nothing to compare. Educators may well take precautions against the current movement leading back to the old type of classical training. This type of education, adapted to an aristocratic society, and accepting ignorant bliss for philosophic calm, has tended to train narrow-minded and snobbish young men and women.

II

The historians of science are responsible, it would appear, for the unpopularity of science among those most acutely affected by the depression. In their clamor to enhance the scientific tradition, and hoard for science all credit for the remarkable and unprecedented material advances which studded the century and a quarter preceding 1930, these historians have been more enthusiastic than accurate. The steamboat, the locomotive and railroad, the power mill, the telegraph, etc., were all hailed as triumphs of the scientific method. Passing over the inaccurate and illusory nature of this interpretation of technical and commercial progress, the fact remains that it achieved wide currency. In the more intellectual circles, technology and the development of new commercial practices were credited with facilitating the adaptation of science. But, as the doctrine filtered on down through the schools and the press to become a part of the prevailing thought-pattern, science emerged as the most prominent force responsible for making this modern world so startlingly different from all preceding ages. Thus when, for many people, the modern world, in spite of all its resources, began to slip from its role of "best of all imaginable worlds," science came in for a proportionate share of blame.

Had a more accurate picture of the part science has played been presented, science would not now be the object of so much suspicion and resentment.

It is true that the modern world is a product of science, but it is also a product of the commercial and industrial expansion which seems to have had its inception in the city-states of northern Italy about the fifteenth century. Only by distorting the meaning of science to be synonymous with all types of initiative and daring having an intellectual content, can this expansion be regarded as a phase of scientific activity. This economic expansion and the rise of scientific inquiry are more properly to be interpreted as separate aspects of the more general cultural renaissance.

It is true that, in technology, industry and science have cooperated to a large extent. But for *pure* science, facts and laws have their gradations of importance regardless of whether or not they have immediate application. The criteria here are the fundamental nature of the facts, and the generality of the laws, as they bear on a fuller understanding of nature. Admittedly, the scientist, in mapping out his programs of research, is not always oblivious of the ultimate utility of his possible findings. But his immediate, and primary objective is to uncover more elements of behavior in nature, and to correlate these elements under a general principle or theory.

The inventor or technologist, on the other hand, is not concerned with facts, old or new, which do not contribute directly to the perfection of the device or method upon which he is working. It may be said that, in general, it is only the technique of experimentation that the inventor has in common with the scientist. Many early inventors not merely ignored, but actually defied, the accepted scientific doctrines of their time, and evidently with success.

Still another agency, distinct from either science or invention, has contributed to the material comfort and convenience of modern life. It is the spirit of initiative and adventure without which, though we still had our science, the world would be very different. It is this spirit which has thrown great bridges across rivers, railroads across continents, airways and cables across oceans, given us by mass production . . . automobiles, electric refrigerators, radio receivers and the host of other accessories of present-day life. It is this same spirit which leads to competition and friction between groups of people, within borders and across borders.

Thus, a more comprehensive view discloses that science was not the main spring in the development of our modern material facilities, but rather, one of several contributing factors. The role of science was passive; it made available facts and laws which men of action in time saw fit to exploit advantageously. Condemnation of science for the present social and economic derangement of the world is as unwarranted as was the undivided credit it received in earlier, more prosperous decades.

III

What attitude scientists are to adopt toward those political régimes which are unfavorable, if not outright antagonistic, to scientific progress, already appears to have become a matter for debate. At the outset of any discussion of this type it should be emphasized that scientific groups have no business taking up the torch for any political or religious "ideologies," however attractive some of these, or abhorrent others, may be to the individual scientist personally.

Those of us adhering to the liberal tradition should like to identify pure science with democracy on the theory that, in the realm of ideas, science stands, as does democracy in the social domain, against authoritarianism. We note that in both modern science and ideal democracy, the statement or conclusion of no man is beyond criticism or scrutiny; that in science a great principle is never irrevocably established, but is accepted only in so far as it is in accord with a wide array of experimental facts, just as in a republic, a government and its laws exist only so long as they are in accord with the will of the popular majority. But, however attractive this analogy may appear, as with all analogies, it can not be pushed too far, to be confused with an identity. That the positions of science and democracy are merely analogous, and that the two not always have been even allies and sole mutual supporters, is indicated by the disinterested prospect of history. We have only to recall the brilliant researches of the French mathematical physicists of the eighteenth century, whose work was supported by the aristocracy of a decaying despotism. Again, no one can say that pure science did not thrive in imperial Germany. On the other hand, we recall that the first truly popular, democratic movement in Europe found "no use for scientists," and forthwith beheaded the father of modern chemistry.[3]

Science can not endorse this or that ideology, or social system simply because, as has been often pointed out, science can not choose between objectives or ultimate values. Any such an alignment is artificial. It does not appear, however, that scientists must remain oblivious of current social and political developments having implications in the activities of science.

Science has a job to do: to enlarge man's knowledge of nature and of the world in which he lives. The materials of science are facts. Its method is not to stop at arbitrarily established boundaries, if beyond may lie truths to be uncovered and examined. And because this method is of the essence of modern scientific inquiry, scientists have the right to insist that the integrity of the method be preserved. The public proscription by scientific organizations of those agencies or régimes which endeavor, by coercion, if not by law, to circumscribe the activities of science, to supplant objective data with propaganda, or to circulate flimsy, immature opinion as scientific doctrine, can not be declared inappropriate. Thus, the recent indictment by a group of American anthropologists of the National Socialist doctrine of "Aryan" supremacy, as having no factual basis, was a commendable move.[4]

In general, toward recent ideological developments the attitudes of scientists thus far appear to be divided. Some scientific workers are in-

[3] This refers to the execution of Lavoisier during the French Revolution. Lavoisier was sentenced not for his laboratory work but because he happened to be one of the hated tax farmers. In this whole passage Lyons's grasp of history is something less than ideal, and one has to read his words for their significance in terms of 1940 and not take his amateurish historical generalizations too seriously.

[4] The Nazis in Germany professed to believe that they embodied the virtues of the Aryan "race," virtues which were apparently inherited and which gave their possessors the right to enslave and exterminate "non-Aryan" peoples.

clined to remain aloof and complacent, concerning themselves with the solution of their immediate technical problems. Their view, evidently, is that the infringement on free inquiry by political and religious ideologies is not something "new under the sun," and that it has not in the past proved fatal. With that view goes the conviction that the very accomplishments of science make its acceptance and support inevitable. The spirit of this group, it would appear, dominates the scientific societies and accounts for the extensive silence of the latter in the face of current developments.

Scientists of another temper, alarmed at the prospect of a worldwide censorship of research and inquiry, and being unable to register their apprehension and disapproval as a scientific body, have aligned themselves with pro-liberal organizations whose objectivity is questionable. Neither one of these courses is highly effective in preserving a progressive, untrammelled science, free from potentially discrediting alliances.

While the sense of the complacent scientist may be justified, I propose that he re-examine his premises, bearing in mind that earlier successful cultures, becoming assured and complacent, have been supplanted. And not always has the new order been superior; one civilization at least, that of Rome, disappeared in the darkest barbarism. It is not advocated that scientists organize to entrench modern science as an established system, simply for the sake of self-perpetuation. Indeed, if it is conceivable that some method of inquiry other than the one we now know as the *scientific,* could more readily give man an accurate picture of his world, scientists should be the first to welcome the method. The real problem is the preservation of a proven method of inquiry against the suppression of all significant inquiry.

30. Two Nuclear Scientists Try to Influence Who Controls Atomic Energy. 1946

After the explosion of the atomic bomb, the scientists who had worked on it had to face an entirely new dimension of the old question of the responsibility of scientists for the way society used their brainchildren. Following their complete failure to influence policy with regard to the use of the atomic bombs in 1945, the scientists in the Manhattan project organized themselves to try to do something about the awesome force that they had released. *The Bulletin of the Atomic Scientists* was in the beginning a little newsletter. In the first issue of the *Bulletin* that indicated that it was to be more than an organ for the Chicago group appeared this editorial article. The point in question was the attempt of the military—who had obstructed the scientists, primarily with security regulations, all through the war and even afterwards—to gain effective control over the national monopoly of atomic energy. What the scientists had in mind as they advocated their case vigorously in Congress is made clear in this editorial.

The editors of the *Bulletin* at this time were H. H. GOLDSMITH (1907–1949) and EUGENE RABINOWITCH (born 1901). Goldsmith was born in Austria and educated at the City College of New York and Columbia University. He was a nuclear physicist and, beginning in 1943, information officer for the Manhattan project in Chicago and later at Brookhaven National Laboratory. Rabinowitch was born in Russia and took a Ph.D. in chemistry at Berlin in 1926. He was a member of the staff of the Kaiser Wilhelm Institute, the University of Göttingen, and the University of London. In 1938 he came to the United States and taught at the Massachusetts Institute of Technology (with time out to work on the bomb). Since 1947 he has been research professor at the University of Illinois. His field is biophysics.

See Richard Hewlett and Oscar E. Anderson, Jr., *A History of the United States Atomic Energy Commission, The New World, 1939–1946* (University Park, Pennsylvania: Pennsylvania State University Press, 1962), I; Morton Grodzins and Eugene Rabinowitch, eds., *The Atomic Age, Scientists in National and World Affairs, Articles from the* Bulletin of the Atomic Scientists *1945–1962* (New York: Basic Books, 1963); and Alice Kimball Smith, *A Peril and a Hope. The Scientists' Movement in America, 1945–1947* (Chicago: University of Chicago Press, 1965).

MILITARY OR CIVILIAN CONTROL OF ATOMIC ENERGY?*
[H. H. Goldsmith and E. Rabinowitch]

The controversy "should atomic energy in America be under military or civilian control" has been brought to a showdown in the Senate Special Committee on Atomic Energy.

On the extreme militaristic side, there is Senator Eugene D. Millikin of Colorado who believes that even General Groves[1] favors too much civilian influence. But in justice to General Groves, his scheme of an Atomic Energy Commission containing four military among its nine members with an active officer as administrator, can be relied upon to bring about complete military control.

* *Bulletin of the Atomic Scientists*, 1 (March 15, 1946), pp. first and last. Reprinted by permission of *Science and Public Affairs, the Bulletin of the Atomic Scientists*. Copyright © 1946 by the Educational Foundation for Nuclear Science.

[1] General Leslie R. Groves, self-confident commander of the Manhattan District, that is, the atomic bomb project. He found scientists more interested in results than in regulations and discipline. He emerged as a leading advocate of military control of nuclear energy.

The argument in favor of leaving research and development of atomic energy in the hands of the military goes as follows: "The atomic bomb is the most powerful weapon in existence. The applications of atomic energy to peacetime industrial processes can be postponed without endangering our national economy. With the world political situation as it is, we must give first priority to the military problem, and this can best be achieved by leaving the control of atomic energy in the hands of the military. This will ensure the maximum production of atomic weapons in America, prevent leakage of our atomic bomb secrets to other nations, and thus delay the atomic rearmament of our potential enemies."

This argument is fallacious. In the first place, it is not true that the world political situation is independent of our policy on atomic energy; in the second place, it is not true that the best way to ensure our continuous superiority in atomic armaments is to put the military in full control of research on atomic energy.

For the present state of mutual suspicion and fear between the wartime Allies is to a considerable degree the result of our failure to provide a clear lead in dealing with atomic energy. The international repercussions of the establishment of a permanent military control over atomic energy in America may well be disastrous for the cause of peace. Some politicians may view this step merely as a warning to the Russians, as an additional show of toughness in dealings with the Soviet Union. However, putting a law on the statute book will be taken by the world, not as a tactical step, but as a part of a long-range policy. International tensions come and go; but policies which have crystallized out in the form of laws or treaties remain. Permanent military control of atomic energy in America will signify to the world that America is basing its long-range policies on the assumption that a new war is inevitable, and this will help to **make** it inevitable.

While making the prevention of a new war more difficult, military control will not assure our continued advantage over the rest of the world in atomic armaments (and in scientific war potential in general). The Army will inevitably put emphasis not on creative research, but on building the largest number of atomic bombs of the existing type and protecting the secrecy of the present processes by strict security regulations, and other police methods.

If we expect a war this year or next year, this unrealistic attempt to perpetuate a momentary advantage will be of little avail, but will do no particular harm. But beyond this short term, this policy bodes ill for our security (as far as security can at all be obtained by maximum striking power). It is already depriving the atomic bomb laboratories of their most valuable brainpower. The really good scientists feel that their effort will be wasted if they are forced to work under the conditions of secrecy and compartmentalization, without the benefit of free exchange of ideas. Research laboratories which are run in this way will soon lapse into routine work. The importance of secrets that can be safeguarded by military supervision of science is widely exaggerated in the public mind. As repeatedly pointed out by informed scientists, and confirmed by engineering experts who have participated in the building of our atomic bomb plants, the delusion that we are safely protected by a secret stock of atomic bombs and an esoteric knowledge, which we share with nobody, may easily become a "scientific Maginot line."

Secrecy regulations, which military mentality is likely to force upon fundamental scientific research, will cause a paralysis of scientific progress. This paralysis will spread from governmental atomic bomb laboratories to all laboratories working in the nuclear field. The necessity may arise of establishing secret courses in our universities, leaving the majority of students in ignorance of basic facts of their science. With nuclear physics as an opening wedge, the same disintegration may permeate the fields of bacteriology, medicine, and other sciences all of which may be used in the next total war.

Since public opinion does not distinguish between science on the one hand, and technology (in which "secret processes" are common), on the other hand, it is likely to consider the opposition of scientists to military rule and compartmentalization only as a selfish fight for a comfortable way of life, or as the defense of certain liberal ideals, which have to be scuttled in the face of the "hard facts of life." What may be scuttled in the process, is America's leadership in scientific and technological developments, decisive for its future position in the world.

But, it may be argued, the Senate Committee did not follow General Groves and Senator Millikin. By a vote of 10:1, it has accepted on March 12 the "compromise" proposal of Senator Vandenberg:

"There shall be a Military Liaison Board appointed by the President composed of representatives of the Departments of War and Navy, in such number as the President may determine. The commission shall advise and consult with the board on all atomic energy matters which the board deems to relate to the common defense and security. The board shall have full opportunity to acquaint itself with all matters before the commission.

"The board shall have authority to make written recommendations to the commission from time to time as it may deem appropriate. If the board at any time concludes that any action or proposed action of the commission, or failure to act by the commission, is inimical to the common defense and security, the board may appeal such actions or proposed actions of the commission to the President, whose decisions shall be final."

Senator Vandenberg claims that this proposal leaves all power with the civilian Commission, and merely allows the military to interfere whenever it feels that the interests of national defense are threatened by action (or inaction) of the Commission. When President Truman, on March 13 confirmed his support for civilian control as envisaged in the McMahon bill, but added that "the military had an important part to play and should be consulted"—Senator Vandenberg said that that describes his proposal "exactly." Why then are the proponents of civilian control still alarmed? Why did Senator Millik[i]n vote for Vandenberg's proposal and Senator McMahon against it, and not the other way around?

The answer is that Vandenberg's proposal leaves it up to the military board to determine what is of concern to "national defense and security." This gives it the right—or rather, the duty—to try to impose **its** concept of security upon as wide an area of fundamental research as possible. In other words, the military advisory board is certain to attempt a continuation of the Manhattan District policy of secrecy and compartmentalization.

The reason why a military board cannot but try to carve out the largest possible chunk out of the living body of science, is because they don't understand it, and therefore have to "play safe." How can they know what field of nuclear research (or any other branch of natural science) is "important for national defense?" After all, it was not the military who first guessed the explosive potentialities of atomic fission in 1939!

Nations are accustomed to looking to their military men whenever their security is endangered. The revolutionary fact of the present situation is that military have ceased to be experts on security. In fact, they can offer no security. Some of them—like General Arnold—freely acknowledge it. But the spokesmen of the Army and Navy do not feel like coming to the Congress and saying: "You have entrusted us with guarding the security of the American nation. We have to return the mandate, because we cannot fulfil it. There is no security for a nation in the world of atomic armaments." Instead, they proclaim themselves guardians of secrets in which fictitious security is supposed to reside. This is the dangerous delusion which the scientists are fighting when they oppose the apparently innocent right of the military to "advise and consult" with the Atomic Energy Commission.

The testimony of the Federation of Atomic Scientists before the McMahon Committee, presented on January 28, clearly stated that the scientists "do not exclude efficient liaison between the Atomic Energy Commission and the military. Provision to make this liaison mandatory will not be opposed by them." Neither do the scientists object to the Army undertaking research and development of atomic ordnance. But they do oppose the extension of Army control into the field of fundamental science under the pretext of guarding "secrets vital for national security." It was Secretary of War Patterson himself who stated before the McMahon Committee that the military are not competent to draw the line between basic research and its military applications, and that this is one of the policy decisions which must be left entirely in the hands of the Civilian Commission.

31. A Biochemist Recounts the Discovery of the Chemical Nature of Viruses. 1946

This speech, on the occasion of the presentation of an award to Wendell M. Stanley (born 1904), explains how he started out as an organic chemist interested in synthesizing organic compounds. He then turned his attention to the field of bacteriology and the study of viruses. It was he who proved that a virus is not a conventional living organism but a definite crystal composed of molecules—molecules with a complicated atomic structure, but molecules nonetheless. The fact that a virus seems to have the ability to carry out biological activity meant that new questions were raised about the nature of life, what is, and what is not, living. When Stanley and others went on to investigate the chemical nature of the virus crystal molecule, they discovered the presence of ribonucleic acid and opened up the exciting era of "RNA" and "DNA" in which the mechanism of heredity and the chemistry of life itself came increasingly under the investigation—and possibly control—of the scientist. It was this new molecular biology that was the heir to the new genetics. In Stanley's career we see how in the period from the 1930s to the 1950s chemistry, bacteriology, and genetics all came together in the new biochemistry.

VINCENT DU VIGNEAUD (born 1901) is a biochemist who was born in Chicago and received his Ph.D. in 1927 from the University of Rochester. He taught at the University of Pennsylvania and George Washington University and then in 1938 was made head of the department of biochemistry at Cornell University Medical College. In 1955 he was awarded the Nobel Prize.

Details and background can be found in George W. Corner, *A History of the Rockefeller Institute, 1901–1953, Origins and Growth* (New York: The Rockefeller Institute Press, 1964); John Cairns, Gunter S. Stent, and James D. Watson, eds., *Phage and the Origins of Molecular Biology* (Cold Spring Harbor: Cold Spring Harbor Symposia, 1966); *Current Biography,* 1956, 160–162; and *Chemical and Engineering News,* 23 (1945), 620–622.

SCIENTIFIC CONTRIBUTIONS OF THE MEDALIST*

Vincent du Vigneaud

* Reprinted from *Chemical and Engineering News,* Vol. 24, March 25, 1946, pp. 752–755.

On the occasion of the presentation of the Gold Medal of the American Institute of the City of New York some five years ago to Dr. Stanley, Thomas M. Rivers quoted Pasteur. I should like to quote again this statement of Louis Pasteur, made in 1888.

> Two contrary laws seem to be wrestling with each other nowadays; the one, a law of blood and death, ever imagining new means of destruction and forcing nations to be constantly ready for the battlefield—the other a law of peace, work, and health, ever evolving new means of delivering man from the scourges which beset him. The one seeks violent conquests, the other the relief of humanity. The latter places one human life above any victory; while the former would sacrifice hundreds and thousands of lives. . . .

Such thoughts as these expressed by Pasteur some 60 years ago, and so applicable to our present-day world, cannot help but pass through our minds as we pause to pay tribute tonight to a man who has followed in Pasteur's footsteps; who, as a chemist, likewise turned to the study of, as Thomas Rivers has so neatly put it, "the infinitely small in biology". As one reflects on the virus diseases, such as infantile paralysis, certain types of encephalitis, yellow fever, influenza, measles, and chicken pox, one realizes that advancement in the understanding of the viruses is indeed a step in "evolving new means of delivering man from the scourges which beset him".

As we all know, our medalist made, some ten years ago, one of the truly outstanding contributions of this generation, a discovery that changed the course of development of an entire field of study, and marshalled in a new era of virus research. I refer, of course, to his discovery that a virus, tobacco mosaic virus, could be isolated in crystalline form and that this high molecular weight protein, a definite, tangible, physical entity, could be correlated with virus activity. We are gathered here tonight, however, not to pay homage to this man for this signal discovery, for which he has received recognition and honors throughout the land. We are gathered instead to honor him, through the Nichols medal, for the elegant symphony of research which he has composed during the past decade, and for which his original discovery served as the prelude. In the medalist tonight, we have a man who not only made a great discovery, but a man who knew what to do about it.

The medalist is not only recognized for his outstanding research but for creating the leading laboratory in the world on the chemistry of viruses. From this laboratory at the Rockefeller Institute at Princeton has emanated during the past 14 years some 150 contributions of the highest quality to this subject by him and his coworkers. In addition, men have come forth from his laboratory with inspiration, well-disciplined thinking, and excellent techniques to carry on the work he so ably instigated.

It is a real pleasure for me to discuss the scientific accomplishments of the medalist, particularly because I have followed his scientific work from the time he was a postgraduate assistant to Roger Adams at the University of Illinois in 1929, working on chaulmoogric acid,[1] on through the years to his recent achievements on the influenza virus. It has been my rare privilege to follow intimately the transition of the medalist from a young organic chemist to a biochemist of note and rich achievement, and to observe the experiences and influences during his earlier training which were reflected in his later work.

STUDY AT ILLINOIS

I can, perhaps, best bring out the different phases of the medalist's work, their importance and interrelationship, by dividing the work into three periods, the period at Illinois, an interim period consisting of work carried out at the University of Munich and at the Rockefeller Institute in New York, and the main endeavor which covers the work on viruses conducted at the Rockefeller Institute in Princeton during the past 14 years. The medalist's interest in the application of chemistry to problems in medical research undoubtedly stems from the work which he carried out with Professor Roger Adams at the University of Illinois. While working there as an organic chemist, he synthesized a great many fatty acids similar to hydnocarpic and chaulmoogric acids, the active constituents of oils which have been found to be of value in the treatment of leprosy. While at Illinois, he carried out much of the bacteriological testing of the synthetic fatty acids and, in addition, conducted studies on physical properties of the various synthetic compounds in an effort to relate the biological activity of these compounds to their chemical and physical properties. In this work, you will note his contact with the bacteriological realm and the development of his interest in the relationship of biological activity to chemical and physical properties which undoubtedly later played a role in his major work.

Another aspect of his work at Illinois dealt with the stereoisomerism of diphenyl compounds,[2] a field which was being rapidly developed at that time by Roger Adams. In his work with Professor Adams, Dr. Stanley showed that through the use of x-ray data on interatomic distances, it should be possible to calculate in advance which compounds should be resolvable[3] and which should not. He showed how well the theoretical calculations applied to the compounds which had already been resolved and then predicted the structures of several compounds, as yet unprepared, which should be capable of resolution. On the basis of the theoretical calculations, it was also predicted that it should be possible to prepare a resolvable disubstituted diphenyl derivative provided the two α,α' groupings were of sufficient size. This problem of a resolvable disubstituted diphenyl intrigued him greatly,

[1] A complex organic acid; see below in this selection.

[2] The atomic and molecular structure of an important group of hydrocarbon chemicals.

[3] That is, capable of being broken up into relatively smaller units making up the molecule.

possibly because his attempts to secure such a compound at Illinois resulted in failure, but he did not accept failure.

MUNICH PERIOD

From Illinois, Stanley went abroad as a National Research Council Fellow to study at the University of Munich with Professor Wieland with whom he had elected to work, undoubtedly because Wieland, an organic chemist, was interested in the application of organic chemistry to natural products.

During that year at Munich he conducted work on the isolation of sterols[4] from yeast, but he still thought of the unfinished problem at Illinois. He found that one of the chemists at the University of Munich had prepared a compound, which, on the basis of the earlier calculations, should be convertible into a resolvable disubstituted diphenyl. While Professor Wieland was making a trip to the United States, Stanley had the opportunity of testing this prediction. He obtained some of this material and set about to prepare the desired compound and to resolve it into its stereoisomers. The outcome was successful and he thus obtained the first resolvable α,α'-disubstituted diphenyl.

ACHIEVEMENT AT ROCKEFELLER

Upon his return from Germany Stanley accepted a position at the Rockefeller Institute in New York to work with Dr. Osterhout, who had for years been interested in the permeability of the cell walls of the large unicellular plant Valonia.[5] Dr. Osterhout wanted a model semipermeable membrane which would not only provide for the transportation and accumulation of sodium and potassium ions, but would permit the preferential accumulation of potassium ions, and across which electrical currents could be measured. Dr. Stanley translated this problem into finding a nonaqueous material in which potassium salts would have a greater solubility than sodium salts and preferably one in which these salts were capable of conversion into a less soluble form. He therefore spent many a day examining the International Critical Tables in an effort to locate such a material. From his survey, he predicted that the properties of guaiacol[6] were such that it should serve as an acceptable material for the semipermeable membrane. This material was incorporated in an artificial model with a slightly alkaline solution containing sodium and potassium ions on one side and a slightly acidic aqueous solution on the other side. It was found that this model permitted not only the accumulation of potassium and sodium ions in the inner aqueous phase, but the potassium accumulated preferentially with respect to sodium. It was then observed by

[4] A particular type of important organic compound.

[5] A kind of algae.

[6] An organic derivative of wood tar, often used pharmaceutically.

Dr. Osterhout that the electrical potential existing across the guaiacol nonaqueous phase was in the same direction and possessed the same order of magnitude as the potential that exists across protoplasmic membrane of the Valonia cell. This model proved useful in subsequent work. Again a problem set before him was solved by forthright, direct thinking and experimentation —an approach characteristic of the medalist.

Thus ended the first two periods of Dr. Stanley's research activities. He had demonstrated his ability as an organic chemist, he had acquired considerable command of the disciplines of bacteriology and physical chemistry. He had also become acquainted with microchemical methods and techniques. In this work he showed his resourcefulness in utilizing tools of the physicist in application to an organic chemical problem, and had demonstrated interest and insight into the geometry of organic structures. He had also shown his ability to orient himself quickly in a new field and had demonstrated that curious knack of "getting results". Perhaps most important of all, he had shown no inclination to be afraid of the complexities of the living cell.

WORK WITH VIRUSES BEGUN

When the board of directors of the Rockefeller Institute decided to establish a laboratory of plant pathology in Princeton in 1931 in conjunction with the then existing Department of Animal Pathology, Dr. Stanley was invited to go to this new laboratory to conduct chemical work on viruses. Because of the tremendous advances that have been made in the field of viruses during the past number of years, it is extremely difficult to visualize now the general state of mind that existed at that time. This new class of infectious agents had been discovered in 1892 and most of the work on them had been conducted by pathologists. Viruses were generally regarded as microorganisms because they possess the ability to grow or multiply and to change or mutate when within the living cells of certain susceptible hosts. When Dr. Stanley went to Princeton in 1932, he was, of course, well aware of this general feeling regarding the nature of viruses. However, he also knew that, with the possible exception of the large elementary bodies of vaccinia[7] virus, physical entities carrying virus activity had not been isolated and demonstrated as such. He knew, too, that in the case of tobacco mosaic virus there was a sufficient amount of the active agent in the extracts of diseased tissues so that such extracts could be diluted over a million times and still be capable of infecting normal plants. He reasoned that extracts capable of such high dilution must contain a physical entity responsible for the biological activity. Whether, as has been said in fun, this entity was an organism, an "organule", a "molecism", or a molecule, he thought that it should be possible to concentrate and purify the entity and eventually learn something of its true nature.

[7] Cowpox.

The effect of certain enzymes upon the infectivity of extracts containing tobacco mosaic virus had already been studied. It had been reported that pepsin had no effect, but that trypsin[8] caused a loss of virus activity, and it was inferred that the virus might be protein in nature. Dr. Stanley in initiating his studies reinvestigated the effect of trypsin on tobacco mosaic virus and proved that the loss of virus activity was not due to the proteolytic[9] activity of the trypsin, but was due to an inhibitory effect, and therefore the inference that it was a protein could not be substantiated on the basis of the action of trypsin. It was necessary therefore to conduct experiments to determine its general nature before it would be possible to engage upon an intelligent attack on isolation. He continued with studies on the effect of various enzymes and demonstrated that the enzyme pepsin, contrary to earlier reports, actually inactivated tobacco mosaic virus only under conditions favorable for proteolytic activity. As a result of his studies with pepsin, in 1934 he felt he was justified in believing that tobacco mosaic virus was a protein or at least very closely associated with a protein which could be hydrolyzed[10] by pepsin. At about this time at the institute at Princeton, Dr. Northrop and his collaborators were assembling proof that showed beyond reasonable doubt that certain enzymic activities were the specific properties of certain crystalline proteins which they had isolated. It was therefore natural for Dr. Stanley to employ in his work with tobacco mosaic virus the methods which Dr. Northrop had used so successfully with the enzyme proteins. This was done, and Dr. Stanley succeeded, towards the end of 1934, in isolating, from the juice of Turkish tobacco plants infected with tobacco mosaic virus, a crystalline material which appeared to have the properties of the virus. He worked with this material for about six months and eventually published the results of his preliminary experiments in June 1935. The conclusion that tobacco mosaic virus was a crystallizable protein was received with general skepticism. Despite the fact that his work on the isolation of the virus was soon confirmed in laboratories all over the world, the long-held idea that viruses must be organisms seemed to prevail. A great many individuals clung to the idea that solutions of the purified protein contained a microorganism as an impurity. Although the many arguments that were raised against the idea of a virus being a crystalline protein were unconvincing, Dr. Stanley felt that the burden of proof rested on his shoulders. He therefore took up the challenge and set about to obtain experimental data concerning the relationship of virus activity to the specific protein. Through the next decade he brought to bear experimental attack, ingenious and varied, with results that forced a skeptical world to accept this crystalline protein as the virus entity. Appre-

[8] Pepsin and trypsin are enzymes active in breaking down substances to be digested and absorbed.

[9] Having the capacity to break down proteins chemically.

[10] A chemical reaction involving the addition of hydrogen and hydroxyl and obtaining two or more new compounds.

ciation that other viruses were proteins soon developed, although this did not mean that all viruses were proteins, a point which had been misunderstood by some, for there is evidence that the viruses range in order of complexity.

In meeting the challenge, our medalist showed first that the same characteristic virus protein could be obtained from a variety of mosaic diseased plants, some of which were so distantly related to Turkish tobacco that their normal constituents failed to react serologically with each other. His next important series of contributions was the demonstration that different strains of tobacco mosaic virus could be isolated. They were similar to the tobacco mosaic virus protein yet possessed properties which were distinctive and characteristic. He even mixed purified preparations of two different strains and then separated them by chemical means and showed that the two distinctive virus activities had also been separated, each remaining associated with its characteristic protein. He then demonstrated in a series of investigations that the virus protein could be inactivated in different ways with retention of certain of its characteristic properties, and that in certain instances it was possible to reactivate the material.

In addition to this work a study of the composition of the virus material was being conducted in Stanley's laboratory. While this work was going on, Pirie, in England, drew attention to the fact that the virus protein contained phosphorus and was probably a nucleoprotein.[11] Stanley's early analyses had not indicated the presence of phosphorus, but phosphorus was found in his later analyses. The isolation of nucleic acid from the crystalline virus protein was reported by Pirie in December 1936 and by Stanley a few days later. The first analytically pure nucleic acid from tobacco mosaic virus was not described until 1939. It was in that year that Dr. Loring, a former student of mine who was working with Dr. Stanley, demonstrated that about 90% of the phosphorus in tobacco mosaic virus preparations could be isolated in the form of a ribonucleic acid.

The work on viruses was further extended in Stanley's laboratory. In rapid succession, the viruses of tobacco ring spot, latent mosaic of potato, tomato bushy stunt, cucumber mosaic, alfalfa mosaic, tobacco necrosis, and jaundice of silk worms were isolated in pure form and found to consist of distinctive nucleoproteins. Studies were made on these various purified virus preparations and in every case the virus activity was found to be associated with a characteristic nucleoprotein of high molecular weight.

While the isolation of these nucleoproteins was being accomplished, extensive studies relating tobacco mosaic virus activity to the nucleoprotein were carried out. Throughout these investigations the virus activity was found to be a specific property of the nucleoprotein. Extensive experiments on the physiochemical and optical properties of the virus nucleoprotein were conducted in Dr. Stanley's laboratory by Dr. Lauffer. By means of deter-

[11] A protein linked or associated with large nucleic acid molecules.

minations of viscosity, sedimentation, diffusion, and stream double refraction, data were obtained which permitted the conclusion that the particles of tobacco mosaic virus consisted of rods about 15 mμ in diameter and 280 mμ in length with a molecular weight of about 40,000,000. These studies and conclusions were made before the advent of the electron microscope. The conclusion regarding the size and shape of the virus nucleoprotein was received in some quarters with considerable skepticism because it was, of course, based on sheer physicochemical theory. When the electron microscope became available, pictures of the virus were made. It was found that the virus particles actually consisted of rods about 15mμ in diameter and 280 mμ in length. This was, of course, a very pleasing result to Dr. Stanley and his collaborators, for it showed that their extensive use of physicochemical theory was completely justified.

At this stage Dr. Stanley started to devote attention to the basic nature of virus activity. He undertook a twofold attack on this problem. He set out, first, to study the chemical structure of tobacco mosaic virus and of its distinctive strains in an effort to demonstrate the nature of the structural differences between strains. Secondly, he set out to prepare various chemical derivatives of tobacco mosaic virus in an effort to learn whether or not it would be possible to alter the structure with retention of biological activity.

In the first phase of this work the amino acid composition of tobacco mosaic virus was investigated. With these data as a foundation, similar studies were started on various strains of tobacco mosaic virus. It was found that tobacco mosaic and aucuba viruses contained different amounts of tyrosine, tryptophane, and phenylalanine, from the ribgrass strain. The cucumber viruses also contained quite different proportions of these amino acids. These preliminary results of amino acid determinations were of the greatest importance, for they provided confirmation of the view that virus strains actually differed in chemical composition and also provided the first information concerning the nature of some of the chemical differences that exist between the strains of a virus and, hence, of the changes in chemical structure that must accompany the mutation of a virus. Similar variations in the concentrations of other amino acids were encountered, and in the case of sulfur there appeared to be a qualitative difference amongst the strains. These results demonstrate conclusively that the changes which accompany the mutation of a virus with the production of a new virus strain, and hence a new disease-producing agent, involve deep-seated and fundamental alterations in the structure of the virus particles.

However, it appeared possible that some virus strains might result from structural changes of lesser magnitude, which, conceivably, it might be possible to duplicate in the laboratory. The virus had been demonstrated to possess certain functional groups, such as carboxyl, amino, sulfhydryl, phenolic, and indolic groups, which in lesser molecules are susceptible to different kinds of chemical reactions. It seemed possible that definite structural changes might be produced by means of chemical reactions involving one or more of these groups. The effect on the biological activities of such changes could then be determined and most importantly, the nature of the progeny of any derivative possessing virus activity could be determined.

If the infecting molecules of a given derivative serve as exact models for reproduction, one would expect to reisolate the derivative from plants so infected. The accomplishment of such a result would, of course, correspond to the mutation of a virus in vitro.[12] If, on the other hand, the inoculation of a given derivative resulted in the production of normal virus, it might be concluded that the structural changes were reversed within the cells of the host or that the portion of the molecule involved in the structural changes was unimportant and played a subordinate role in the reactions of virus reproduction. In the first studies of this nature, which Dr. Stanley conducted with Dr. Anson, it was found that the sulfhydryl groups of tobacco mosaic virus could be oxidized with iodine without changing the specific virus activity, but that the inoculation of this oxidized virus to normal Turkish tobacco plants resulted in the production of ordinary virus. Later, Dr. Stanley and Dr. Miller, another former student of mine, treated tobacco mosaic with ketene and phenylisocyanate, and thus prepared acetyl and phenylureido derivatives of tobacco mosaic virus. It was soon found that the disease caused by the derivatives was indistinguishable from the ordinary tobacco mosaic disease and that the virus isolated from such plants was indistinguishable from ordinary tobacco mosaic virus. It was concluded therefore that the derivatives were either converted into ordinary virus within the cells of the host or that the infecting molecules may not necessarily serve as exact patterns for reproduction. The only indication that was obtained in this work of an effect on virus activity came when the derivatives were tested on various hosts. It was found that some of the derivatives which possessed their full biological activity when tested on one host possessed only about one-sixth this activity when tested on another host. These results made Stanley suspect that a property of the virus, which perhaps can best be described as virulence, may remain consistent for one host but be modified with respect to another host upon formation of a chemical derivative. The fact that the specific activity of the derivative is unchanged with respect to one host is proof that the result is not due to the inactivation of a portion of the preparation.

These results provide the first example of the modification of virus by means of known and reproducible changes in the structure of the virus. They lend encouragement to the belief that eventually chemists will be able to make what may be called heritable structural changes in viruses. The accomplishment of this objective would, of course, have a far-reaching significance, for it would indicate that man might be able to utilize viruses for useful

[12] Outside of a living body, as in a test tube or other artificial environment.

purposes. He might, for example, eliminate the disease-producing viruses by supplanting them with innocuous viruses manufactured in the laboratory by chemical means from pathogenic[13] viruses. It is even possible that chemical modification might result in the production of viruslike entities having other special beneficial results on higher organisms.

We have been privileged to witness, in the span of our own scientific life, a revolution in virus research. It is just, meet, and proper that we should gather together tonight to honor the man who was responsible for it, and who has continued to pioneer and lead the way toward a fuller understanding of these infectious agents.

32. A Scientific Leader Assesses the Uncertain Future of Postwar American Science. 1947

FRANK B. JEWETT (1879–1949), a physicist, engineer, and executive, was a conservative who sensed the way that American science was developing in the years immediately after World War II. He recognized the momentum given by the war effort to large-scale, organized, interdisciplinary research, and he was aware of widespread enthusiasm for exciting areas of investigation opening up in the 1940s. But he was conscious also of unfavorable tendencies engendered by the role of science in the war and the period just after. He saw the middle and late 1940s as a time of irresolution, and he had qualms not only about what science could do to society but about what society and war-scarred science itself could do to the development of scientific endeavor in America. He in effect drew up a balance sheet of favorable and unfavorable factors affecting the future of research. Testing his prescience by later events shows him to have had impressive insight into his own times.

Jewett was born in Pasadena and was graduated from Throop Institute (later the California Institute of Technology). After taking a Ph.D. in physics at the University of Chicago, he taught briefly at Massachusetts Institute of Technology and then in 1904 joined the American Telephone and Telegraph Company as an engineer. He worked his way up to the vice-presidency of the Western Electric Company, and in 1925 he became vice-president of American Telephone and Telegraph and president of the Bell Telephone Laboratories. He was in charge of research at the Bell

[13] That is, causing disease.

Laboratories until 1940, and in that period he emphasized the importance of basic research in industrial development. He was elected a member of the National Academy of Sciences in 1918 and served as president from 1939 to 1947.

To learn more about Jewett and the context within which he was speaking, see James L. Penick, Jr., *et al.*, eds., *The Politics of American Science, 1939 to the Present* (Chicago: Rand McNally & Company, 1965), and Oliver E. Buckley, "Frank Baldwin Jewett," National Academy of Sciences, *Biographical Memoirs,* 27 (1952), 238–264.

THE FUTURE OF SCIENTIFIC RESEARCH IN THE POSTWAR WORLD*

Frank B. Jewett

Because of the vastness of the field of science, anyone is necessarily limited to a small sector in what he can say with any degree of assurance about the specific future of scientific research. The easy thing to do is to confine his discussion to the fields he knows most about, make the best assumptions he can about future conditions, and then proceed to speculate.

Rather than do this I have chosen in what follows to examine as best I can what seem to me certain broad underlying factors in social affairs, which are destined to influence profoundly the future of all scientific research in the postwar years. Some are directly connected with matters of science. Some, however, are only indirectly connected with it, but because they are concerned with human affairs on the big stage they will dominate all lesser actions on it. In a way they are analogous to the long slow massive ocean swells which roll in on the shore and dwarf the surface waves that local gales produce.

It is of course trite to say that prognostication and prophecy in any field of human endeavor is always a highly dangerous pastime. Except possibly as to a few rather obvious things this is particularly true in the domain of science. Dangerous as it is in normal times to make bold statements about the future of science, it is doubly or trebly dangerous at a time when the world is emerging from a gigantic war in which science and scientists as never before played a dominant role and when it is embarking on an era of what men everywhere hope will be a long period of peace. At the moment we are in the turmoil of a transition period. We know and are dazzled by the accomplishments of applied science mobilized for a single purpose—war. A few of us know how these accomplishments were attained; the price that was paid by science for them; the futility of hoping to duplicate many of

* From George A. Baitsell, ed., *Science in Progress, Fifth Series* (New Haven: Yale University Press, 1947), pp. 3–23.

them even in small measure in a world governed by a peace economy; the fact that in fundamental science particularly, and to a large extent in applied science also, we have lost irrevocably the better part of a generation of creative research men and the better part of a generation of creative additions to our stock pile of fundamental knowledge.

These are but samples. There are scores of other matters that scientists know about or sense in which the intrusion of war has played havoc with the orderly progress of science and presented us with gigantic tasks which must be undertaken with a depleted supply of trained man power and in an atmosphere of popular misconception of what science really did do and what it can do under peacetime conditions. Supplementing this, of course, is the fact that in addition to the normal elements of rancor, distrust, hatred, and the like which are the aftermath of all wars, there is a great universal fear resulting from that most spectacular of all applied science accomplishments—the atomic bomb. It is a type of fear which men have never before experienced and which no other weapon of destruction man has yet devised can create. It is a fear that is completely destructive of morale and as such produces an effect on men's thinking which makes it difficult if not impossible to forecast how human beings will act in their mass appraisal of what should be done in the years ahead.

Just a word more about this abject fear. The results of aerial block buster bombing during the war have shown that although cities can be reduced to rubble, tens of thousands of people killed, and the morale of the nation seriously affected, it is a long slow process to reduce a nation to the point of capitulation by it. This is because men know that in such bombing only a fraction of the people are killed in each operation and each knows that he may be among the lucky survivors. In consequence, his morale, though shaken, is destroyed slowly.

When, however, he knows that *all* or nearly all life within a great lethal area will perish in a single blast and that the area of destruction may be anywhere, any time, the case is quite different and his fear becomes an abject fear. All the evidence I have seen indicates that as a rsult of Hiroshima and Nagasaki all morale and all normal social restraints were in a fair way to vanish in Japan. Had speedy capitulation not followed, the entire population might easily have been turned into a frantic mob where anything could have happened had one or two more bombs been dropped.

For us, fortunately, this devastating fear is still only that of an intellectual process, not one pointed up by a vivid national experience—Japan is a long way off and was a detested enemy. It is a type of fear which, nevertheless, has permeated our entire population and which consciously or unconsciously is certain to affect many of our actions for a long time to come. It is the major of the many dislocations in our prewar thinking and functioning which the war has produced and which bedevil our attempts to prognosticate the future in order to plan for an orderly advance.

EFFECT OF ATOM BOMB

This apparent digression from my topic has been made for a very pertinent reason. Whatever the ultimate results may be, the atom bomb has produced a revolutionary change of major proportions in human affairs. Since it and all the phenomena involved in it are so clearly and uniquely the results of fundamental scientific research and its technological developments, it is inevitable that the spotlight of attention should have been focused on science and its future. It is inevitable also that the multitude of lesser, or at least less spectacular, achievements of science during the war should be given a degree of prominence they might not otherwise have attained. Similarly, scientific men are being turned to and in some instances are giving advice on matters concerning which they are not well qualified to speak. Clearly then science, scientific research, and the men of science have, for the time being at least, been put in a front row position in human affairs. It is an extremely dangerous position for us to be in and one that will require a maximum of wisdom to administer wisely.

There are other factors of uncertainty also, which, though only indirectly connected with science in the war, do spring from the nature and outcome of the conflict and make it hazardous to be too certain in prophesying the future of scientific research. While we all hope that the years ahead will be those of peace we cannot yet be sure. For the time being the world is trying to play both sides of the street of the future. On the one side the nations have joined in giving a solemn pledge of faith in a United Nations, aimed to insure lasting abolition of war as an instrument of national policy. At the same time we, in common with all other major powers, are going ahead on a reduced scale with the development of new and yet more powerful instruments of destruction—all based on science. Until time resolves these doubts as to the future, and we have assurance of which side of the street the world is destined to walk on permanently, science and scientific research will remain in a quagmire of uncertainty.

If the final answer is that of a long period of peace then scientific research can be oriented powerfully in that direction, and the old freedoms of intercourse and publication which have made science great can be restored. Some degree of prophecy can then be made safely. If, however, the decision is that the world is to walk on the dark side of the street, the freedoms which mean so much to scientists and more to the progress of science cannot be restored. Technology in some sectors may flourish for a time, but fundamental science generally and applied science in many useful fields will wither or die. The irksome regimented restraints of active warfare will have to be largely retained, and science will continue to be the servant of political government.

Partly because of this present uncertainty in which we are involved and partly because of the achievements of applied science when mobilized by government for war, we are today debating whether, to what extent, and in what manner political government should support and control science in the future. If we as a nation come to the conclusion that the years ahead are but an armistice, then government must play a dominant role in the field of science. If not, then it is highly questionable whether in the long run it is in the interest of the nation to alter radically the methods of the past which have made this nation preëminent in the fields of fundamental and applied science.

SUPPORT OF SCIENCE BY GOVERNMENT

All of the proposals for federal support of science now being considered involve some measure of direct or indirect political control of science. Some, such as S. 1850,[1] involve strong controls which are far reaching, uncertain, and dangerous. In some directions the immediate advantages which would accrue from federal monetary support are obvious. The things that are not so obvious are the restraints on free scientific research which the dispensing of such money would inevitably place in the hands of political government and in an unknown, essentially uncontrolled bureaucracy. It is a question on which scientific men can and do differ widely. Before a final decision is reached, it is a matter that should be thoroughly debated. At the moment, some of the forces at play in the debate are based on assumptions which may or may not be valid. Principal among these is the tacit assumption that it is clearly in the interest of society greatly to stimulate the progress of science in all sectors far beyond that which would occur if things were left to the normal balances of interest in a complex society. It is an assumption and urge participated in by men of science and laymen alike. For the former, i.e., the scientists, it is natural and understandable. It is their field of interest and they see such a multitude of interesting things to do that they are avid to speed up the exploratory machinery. For the latter, i.e., the laymen, however, the urge is to a large extent motivated by a feeling that more "gadgets" of the kind science and technology can produce will *ipso facto* lead to a better life. In some sectors, notably those which radiate from the roots of the biological sciences, this may be true. In other sectors the assumption is at least open to serious question.

To anyone who has reviewed objectively the progress of the past 50 or 60 years, during which the flood of utilitarian things based on science has increased enormously and at such an accelerated rate, a question must arise as to whether there is not some limit to the speed with which society can wisely assimilate radically new things of science en masse into the social structure.

[1] The so-called Kilgore Bill, which would have established a federal National Science Foundation not unlike the one ultimately set up except that it would have, under the Kilgore Bill, embodied controls and patent protection for the government opposed by men such as Jewett and Vannevar Bush.

Experience has shown that it is relatively easy for men and women of limited skills to learn to operate the most complex things of science. It is not nearly so easy for the society to learn how to control that operation wisely for the common good. Many if not most of the major social problems which now bedevil us are concerned with matters which are connected with the technology of the last two or three decades. They are evidence that we have not yet solved satisfactorily the problems of incorporating the science and technology of the past into the community living of the present. A discouraging element is the fact that in some directions our avidity to avail ourselves of certain obvious material advantages has caused us unconsciously to foster social forces and schisms which have divided us into groups which have little real understanding of each other. As such they tend to be suspicious if not actively antagonistic one to another. While not new in human society, it is, I think, a phenomenon which has increased greatly with the rapid growth of applied science.

What I have just stated should not be construed as meaning that I would advocate even a partial holiday in scientific research—fundamental or applied. I have no such thought, and even if I had it would be a futile one since nothing can stop research so long as there is an unknown to be probed and so long as men continue to be curious about that unknown.

My sole [*sic*] reasons for discussing the question as I have are two. First, because I think there is a real question on how far it is wise to go in artificially stimulating science and scientific research beyond the normal rate it would have in a balanced society. Second, because the question has, I think, a direct bearing on the attitude of the scientist of the future both toward his work and toward the society of which he is a part.

As to the first of these, there seems to be little if any serious thought given, in the current welter of discussion about putting more money and the power of government into stimulation of science, as to the results in other sectors of society. Modern society is too complicated a thing to warrant overstimulation of any sector at the expense of the rest. As to the second, we have already gone a long way from the more or less current philosophy of my early manhood.

It was then not uncommon for the fundamental scientist to consider desire for the application of his methods or results to be on a lower plane than the desire to use them solely to increase man's knowledge of nature. Or for the technologist to consider his task done when he had produced a new thing and to consider the social effects of its use none of his concern. In passing, I might remark that in both cases these feelings of detachment were rarely contested by those next in the chain from the unknown to the utilitarian—rather they were fostered. The industrial scientist looked upon the college professor as a curious person who was afraid of life outside his ivory tower and the businessman considered the engineer or industrial research man as a superior technician, but one little capable of understanding the adaptation of his creations to the complexities of society. If this war has done nothing else, it has stirred us all up to a better appraisal of each other and of the part we can and must play in the complex structure we call society.

SCIENCE AT THE BEGINNING OF THE WAR

To make any reasonable appraisal of the future of scientific research in the postwar years, we must have some picture of where we were at the beginning of the war; of the changes which the war brought about; of the present situation; and of the kinds of scientific research we are talking about.

In the field of fundamental science we had, at the beginning of the war, gone a long way since the beginning of the century. We had rather completely broken down the segmental barriers which, even in my early days, still separated the named sciences. We realized that they were all interrelated and that advances in one sector were likely to have direct or indirect bearing on those in another, even when for convenience the names we used to designate the sectors seemed to indicate complete separation. The number of brilliant men and women doing creative work in every field of fundamental science and in imparting their knowledge and the knowledge of their methods to a younger generation was great and growing. The number of eager young people seeking this knowledge, either with a view to carrying on in fundamental science or of transplanting it to one of the fields of biological or physical science technology, was very great.

The power and validity of the so-called scientific method had stood the test of time not only in the fields of fundamental science but in those of the technologies as well. It had demonstrated that while genius would produce superior results in science as elsewhere, the method itself was a lever which, when understood, could be effective in the hands of many less than genius caliber men.

The barriers which once existed between men of science interested primarily in fundamental research and those interested primarily in applied science had largely disappeared. This was evidenced by the membership of the learned societies and by the constant interchange of activity between the two sectors. Men had learned that except for the size of the tools sometimes employed, the elements of cost and the time as factors determining the direction of inquiry and the objectives of applied science, there was no essential difference between the kind of men or paraphernalia required in fundamental or applied research. A properly trained man could be operating as a fundamental scientist one day and as an applied scientist the next if occasion demanded. The war gave added proof if any were needed.

At the beginning of the war an enormous number of powerful research tools had been developed for use in every field of inquiry. The stockpile of established fundamental knowledge was large and growing rapidly. It was, of course, more bountiful in some sectors than in others, but nowhere was there any evidence that the limit of the unknown was anywhere yet in sight.

It was and is still true that every bit of proven new knowledge is a potential springboard to further acquisitions. This stockpile of fundamental knowledge was the raw material of industrial research. So extensive and effective had this latter become that in some sectors it was treading close on the heels of the producer of new fundamental knowledge. At times its needs forced it to undertake fundamental research although this was not the proper function of industrial research.

The picture then at the beginning of the war was that of an army of research scientists distributed in two sectors; of a stream of young men being trained and flowing out into productive life and of a rapidly increasing store of fundamental knowledge. As might be expected, the extent of this knowledge was not uniform over the whole field of science, so that the contour of the stockpile was irregular. Outside this centrol core of basic knowledge and surrounding it was a vast structure representing the work of the industrial research men. The contour of this structure was much more irregular than that of the core. This was partly because utilitarian applications were in many cases limited by limitations in basic knowledge on which were superimposed the limitations of economic or social demands.

SCIENCE DURING THE WAR

Now the war comes. What were its temporary and lasting effects on the existing science structure? Even before active warfare began and while the storm clouds were gathering, industrial research men began to direct their energies to implements or instrumentalities of war and scientific men in universities began to forsake fundamental research and for the duration became industrial research men. It wasn't long before practically all fundamental research in the physical sciences ceased, as did the advanced training of men. Soon also there were no young men taking that broad preparatory training which is requisite to advanced training. The biological sciences were a bit slower in feeling the full effects of the war drain. Ere long, however, they were in the same situation as the physicial sciences and for the time being all American science had gone industrial and to war. The streams of new knowledge and those of an oncoming generation of new research men dried up and ceased to flow. The reservoir of accumulated fundamental knowledge stopped enlarging; fortunately its contents were of such a kind that they did not decrease to any marked degree through the evaporation of extensive use.

The result of all this was that we here in America created a vast industrial research and development organization, the like of which had never before been seen. Largely it was the work of the OSRD and its three main operating agencies, the NDRC, CMR,[2] and Office of Field Services. Though OSRD was the largest single unit in the overall structure and the dominating

[2] That is, Office of Scientific Research and Development, National Defense Research Committee, Committee on Medical Research.

one in that it covered the whole domain of science, it did not represent our major effort. The total effort of the greatly enlarged scientific departments of industry, the army, the navy, and the agencies mobilized under the National Academy of Sciences and the National Research Council, and that of the Manhattan District project of the atomic bomb, greatly exceeded the amount of work and money administered by OSRD. Measured by the standards of normal industrial research, the total effort was inefficient. Measured in terms of the standards of war, where time and success in achieving an objective are paramount and where cost measured either in terms of money or effort wasted on abortive work is of no concern, it was highly efficient. It was so efficient, in fact, that even a very few of the more outstanding successes would have more than justified all the cost even had there been no value whatever in the host of lesser achievements.

In no significant degree was this vast undertaking a fundamental science research organization, however. It was purely and simply a great industrial research and development effort with a single main objective—war. It was dependent for its raw material on the existing stockpile of accumulated knowledge and its additions to that pile were insignificant and to a large extent incidental.

In a way it is unfortunate that the term "research" ever became part of the designation of the work, since it has led to a widespread misconception of what was done and an equally widespread misconception in the lay mind as to what can be done in peacetime by following the pattern of the war organization. These misconceptions are not surprising since we ourselves have used the term "research" loosely and the public, dazzled by the obvious accomplishments of the technological undertaking, is justified in attributing much of them to the kind of reasearch it is thinking about.

Although the methods and tools employed were those of fundamental science research, the objectives were solely those of industry. Further, while during the war scientists were content to put up with the restraints imposed by war and by government authority, the conditions imposed by these restraints were the very antithesis of those in which fundamental science grows and flourishes and in which creative men work happily and productively.

No matter what plausible arguments are advanced, fundamental science cannot flourish in peacetime under the regimentation of a wartime setup or under bureaucratic control. Fundamental science can be aided—it cannot be directed. Its fruits are those of the free mind and no one is wise enough to know what another man's brain cells may produce if afforded opportunity to function freely.

Even in normal industrial research there is very little "direction" of effort of the kind we commonly associate with that term. About all an Industrial Research Director does or can do is to see to it that his team is headed in the right general direction. If he could do more he could dispense with most of the team. His main job is to provide the wherewithal for their work and a satisfactory environment and atmosphere in which it can be carried on. He more frequently than not approves support of ideas in which his men are interested but of which he himself is skeptical. He rarely withholds support if it can be given. He does these things because he knows that if he is wrong and his men right, the result is desirable. *Per contra,* if he is right and the men wrong, it is far better to have the facts prove the case than to have the idea dismissed arbitrarily by fiat.

With these things in mind let us take account of the situation in which scientific research finds itself at the beginning of a new era of peacetime operation.

1. The corps of highly trained research men in fundamental and applied science is substantially what it was at the beginning of the war. Its average age is four or five years greater than it was and its creative potentiality is probably less.

2. There is an almost complete absence of fully trained men in the next lower age group. To some extent this bracket is filled by men who were partially trained when they went into war work, who have gained experience offsetting their loss, and who will continue in the productive field. Many, particularly the younger men, will elect to resume their formal training and will not become active producers for some time.

3. There is a dearth of men who would normally have completed their undergraduate training and been candidates for advanced training. A large number of them will wish to resume their training. Others, for one reason or another, will forego it and take up work in some other sector. Many of them who have received highly specialized training will elect to continue as technicians in the field of this training. In rare cases they will have the inherent capacity to overcome the things they have been deprived of. Frequently, however, they will have a false notion of their ability to do the really worth-while work they could have done had war not intervened, and aspire to more than they can accomplish.

4. The stockpile of fundamental knowledge is not much greater than it was four or five years ago and its contour is about the same.

5. The stockpile of applied science adaptations is enormously expanded in some sectors and not at all in others. The contour is, therefore, one of higher peaks and lower valleys than it once was and a great problem of industrial research is, therefore, to bring technology to a more balanced state in relation to social needs.

6. The number, diversity, and, in many cases, the power of research tools, have been greatly increased, particularly in those sectors of application which were of most interest to the war effort.

In some sectors, such, for example, as nuclear physics and some sectors of biological science, the war needs have developed tools, methods, or materials which, although not earlier unknown, were rare, expensive, or difficult to obtain but which have been made common, cheap, or relatively easy of access.

7. The exigencies of war have in many cases established more clearly the interdependence of different sectors of science on one another for their mutual advancement and have developed the beginnings of a better association and teamwork.

On the whole, the biological sciences and their technologies have probably suffered less than the physical sciences by the incursion of war and have derived more of permanent benefit from it. This is particularly true of medicine, public health, and nutrition. One principal reason for this lies in the fact that while the main war problems of the physical sciences were concerned with implements of destruction—a field essentially alien to their main peacetime objectives—those of the biological sciences were concerned with the saving of life, the prevention of disease, and its speedy cure. In other words, the war problems of the biological sciences were not different in kind from the peace problems. Added to this the clinical facilities which war afforded the biological technologies were unique and not obtainable in peacetime—huge numbers of men living under a regimen of controlled conditions. In contrast to this, the physical science technologies, although they produced prodigies of achievement by way of rapid mass production of radically new things, did so to a large extent by adaptation of established techniques. In many cases these adaptations were possible only because cost was a minor factor. In such cases the experience gained gives promise of little help in the solution of postwar problems.

RESEARCH PROBLEMS

Let us turn now more specifically to the research problems immediately ahead in both the fundamental and industrial fields. We have seen, among other things, that in both fields there is much to be done to restore us to the position we would have been in except for the war; that we must embark with a depleted crew of trained men and the prospect of its remaining depleted for a considerable period; and with, for the moment, a great public interest in science and an urge from the public to go faster even than we were going before.

Of all the difficulties which now beset scientific research, dearth of adequately trained man power is far and away the most important. Further, it is not something which can be speedily overcome by the expenditure of money, even if that were fully available. In this one lack will be found the main headaches and disappointments of the next few years. They will be many, particularly the disappointments. In both the institutions of learning, which are our main source of fundamental science knowledge and of trained men, and in industry, great plans have been made for expansion. Many of them are doomed to partial or complete failure because they cannot be staffed adequately. In many cases partial failure will be masked and unrecognized because the semblance of success will exist, but not the substance. Men of mediocre ability will be going through the motions of genius and producing an end product which, because of their mediocrity, is itself mediocre. There is a long-run danger in such a situation because frustration of expectation may easily be ascribed to the deficiences of scientific research itself rather than to its true cause.

Nothing which the war developed, so far as I know, cast any doubt in any field on the validity of the scientific method of controlled experimentation. Whatever its direction, scientific research will, therefore, proceed in the future along the established lines of the past. It will be implemented with more efficient tools than it had before the war and by a closer association of work in different fields.

INDUSTRIAL RESEARCH

Industrial research early took a leaf from the book of industrial experience and learned that organized teamwork could vastly strengthen and expedite research. Fundamental science research has been slower to adopt this method generally. This is partly due to the nature of such research and partly because of the way recognition of achievement has been accorded in the past. Credit for results in fundamental science has been largely a jealously guarded personal thing rather than something to be shared by a team. As knowledge has increased and the distance between the scattered fields of inquiry has narrowed, the beneficial results to be obtained from team attack on a complicated problem have become increasingly apparent. More and more we are seeing fundamental research undertaken by some sort of group attack. There is evidence that in the postwar years this tendency will increase to the advantage of science.

Of the two fields, that of industrial research seems likely to be less hampered in the years ahead by the adverse factors mentioned earlier than fundamental science. This is partly because its problems are more specific, and are based on a firmer foundation; also, because of teamwork, attack can frequently be undertaken simultaneously from several directions. The industrial laboratory—particularly one already well established—is also frequently in a good position to attract just the right men for its particular problems. By this I do not mean merely its ability to pay higher salaries, but more, its ability to provide facilities and the attraction of association with a considerable group of men having a common general interest. In addition the industrial laboratory, being an integral part of a technological undertaking, has access to a reservoir of engineers who, while usually not schooled in research work, are nevertheless so trained that they can relieve the research man of much of the simpler work of experimentation and in addition aid him greatly in the creation of new research tools.

The smaller and particularly the newer industrial laboratories will have a harder time than the larger, longer established ones in the years of man-power shortage. They will, however, have one advantage over the larger laboratories in their quest for top-flight men. Being small, with relatively few research problems and in many cases with an urgent need for a few A-1 men, they will frequently be in position to offer salaries with which neither the universities nor the larger industries can compete. The larger an institution, industrial or educational, becomes, the more it is forced to conform to standards of treatment for personnel and the less is its ability to handle such things as monetary compensation on a purely individual basis. Fairness and the necessity of maintaining group morale requires reasonable conformity to standards in the regular compensation of men of comparable age, experience, and worth. Like the university, the large industrial laboratory must therefore in the last analysis frequently depend on something other than money to hold top-flight men against the lure of high salaries. In the older of the large industrial laboratories the great diversity of problems has brought together a group of men who for variety of interest are not unlike those gathered in a university.

In the field of fundamental research the most obvious thing is the vast variety and number of problems, solution of which seems certain to produce valuable new knowledge. There appears to be no end to them. The limitation to their solution seems nowhere to be inadequacy of method or of tools, but solely one of limited trained creative man power. For the most part money does not seem to be a limiting factor despite the fact that some of the problems, especially in the field of the physical sciences, do require the use of large and costly equipment. Such equipment is usually not so unique to an isolated investigation that it is inapplicable to others, however. Past experience has been that even in cases of this kind, when a convincing case of need and reasonable prospect of success can be established, the requisite funds are forthcoming: witness Mt. Palomar and the great cyclotrons. For the most part, however, the really valuable new fundamental knowledge will probably still be produced with relatively inexpensive tools.

In trying to look ahead in the sector of fundamental science research some things seem quite clear. Scientific research, like most human activities, is given to fads. While the fad lasts (and it usually lasts until the problem is pretty well exhausted or until some independent venturesome soul has discovered a new and more exciting lead) substantially the whole interest in that particular branch of science is devoted to it, including that of the oncoming generation of research men. At the moment there are several of these fields of focal interest to most of which the technological accomplishments of the war effort have given added impetus. In physics, for example, it is nuclear phenomena. In the biological sciences it is the astounding possibilities which present themselves in the further study of such things as the vitamins, penicillin, streptomycin, antimalarials, viruses, and the like; in the possibilities which inhere in the further development of biochemistry and biophysics and in the employment of the things produced by nuclear physics as aids to biological research. In the various branches of science there is a host of these current fads, all intensely interesting and with intriguing rewards which beckon imaginative men onward. They are a lure for the applied science research man as well. Whatever the more distant unknown future of science may have in store for men, I think it is reasonably safe, therefore, to say that if fundamental science is afforded the freedom it needs, it will for the next few years be mainly concerned with developing the things which now hold the center of interest.

Earlier I mentioned the growth of coördinated attack on complicated research problems by teams of men skilled in special sectors of science. One striking result of war research, which for the time being turned all scientific men into industrial research workers, was to spotlight the place of mathematics, as never before, as a powerful factor in research work. All research men, both fundamental and applied, have always had some training in mathematics and all have made some use of it. For the most part, however, this use has been rather incidental and primitive. It is only in recent years that research men outside certain sectors such as astronomy have come slowly to realize how powerfully mathematics in the hands of skilled mathematicians can aid in the solution of complicated problems. To a large extent this gradual realization has evidenced itself mainly in the applied science field—many sectors of fundamental science seem still to be unaware of its possibilities.

In part this is the result of inadequate understanding of mathematics by physical or biological science research men and in part it is due to a failure of mathematicians themselves to become interested in furthering practical application of their science—the most exact and powerful of all sciences. The exigencies of war research have brought this deficiency vividly to view and I think it safe to say that research in the years ahead will see a rapidly expanding employment of applied mathematics. I suspect that impetus will be given also to the training of able men in this field analogous to the impetus of 30 or 40 years ago in the fields of the physical sciences.

One striking result of the war necessities in this field has been that through collaboration of the physical scientists and the mathematicians a host of powerful mathematical tools have been and are being developed to aid the mathematicians. Aids of this sort seem to be imperative in the solution of many problems which are arising in numerous fields as a result of physical science expanding beyond the old limits and into fields of complex nonlinear phenomena.

RESEARCH INSTITUTES

A matter of interest in connection with future research is the question of where it is most likely to be done and the changes, if any, in this sector which the war years have brought about. So far as industrial research is concerned, the answer seems fairly clear. The larger industrial units will continue to maintain expanded research organizations. This is both because such organizations can be sufficiently comprehensive to cover most if not all of the science sectors involved, and because in industrial research many of the problems are so intimately connected with other parts of the business that separation of their solution into an outside agency involves delays and inefficiency. Even in these essentially self-contained units, however, special problems will arise from time to time which necessitate recourse to university laboratories or institutions like the Mellon or Battelle Institutes.[3]

For smaller units forming part of an industry, there will, I think, be an increase in the number of association laboratories which concern themselves with basic problems of common interest to the industry as a whole —the specialized application-development features being left to the individual members.

I should expect to see a considerable expansion in the number and size of laboratories along the lines of Mellon or Battelle and regional laboratories supported by all the adjacent industries and serving all impartially.

In specialized fields there will doubtless be growth in the number and size of laboratories equipped to undertake solution of problems on some sort of fee basis.

To a limited extent there will doubtless continue to be laboratories for industrial research connected with institutions of learning. There is clearly some limit, however, in the extent to which educational institutions can enter this field without creating serious internal problems of administration or raising questions as to the status of the institution under its charter.

In the field of fundamental science research the picture is not quite so clear. There has been much talk that in the future a great deal of the work that has been done in educational institutions will be taken over by institutes devoted exclusively to research. Doubtless there will be an increase in the number of such laboratories and doubtless also they will make noteworthy contributions to knowledge. I have a strong feeling, however, that the great bulk of the contributions will continue to come from educational institutions where research and teaching are combined.

We must continue to look to the universities and colleges for our supply of trained men—no other institutions are equipped for this task. There is a great stimulus to both teaching and research to have the two combined. Any purely research organization has a large tendency to become narrowly specialized since it lacks the necessity of reviewing its work continuously from the standpoint of imparting its results to other than already highly trained specialists. The maintenance of our graduate schools at a high level of efficiency is, therefore, a matter of the utmost importance if we as a nation are

[3] Mellon and Battelle Institutes were independent, nonprofit research institutes supported at least in part by gifts and endowment.

to maintain leadership in science and derive maximum benefit from what may be in science.

In looking ahead at scientific research in the immediate future there is one heritage of the war with which we will have to cope that is rather distressing. War is always a corroding thing. At its worst it is debasing and degrading to the participants. Fortunately those so affected are a minority of all who are engaged. For all, however, and scientific men are no exception, it destroys those standards of thrift, frugality, and efficiency of effort which are necessary in a productive civil economy. We have just been through a long experience where waste, squandered money, and lack of efficiency have come to be taken as a matter of course. In the field of science this has bred in many young men (and many older ones too) a feeling that nothing worth while can be accomplished without elaborate facilities, a wealth of assistance, and a lack of desire to plan economically which will continue to plague us in our endeavor to get back on an even keel. We will have demands for more and more elaborate equipment, for more assistance on the plea that work can't be done otherwise, and, in the field of education, more pleas for scholarship aid from those who before the war would not have expected it.

In the present discussion I have dealt mainly with factors which seem to me likely to have a controlling effect on scientific research in the postwar years. Except in a general way I have not attempted to prognosticate the future in specific sectors of science. There are so many promising leads that seem worth following up that to do so adequately would be a long job even if I were competent to do it. Actually, of course, the extent of exploration or lack of exploration of any field will be determined by general conditions; by the personal interest and curiosity of research men; by the number of such men that we produce and the facilities they have to work with.

At the moment and for some time to come the principal limiting factor in doing all that we might wish will be the number of able young research men and women we can train. This is a problem primarily for the universities and technical schools. Doubtless we can increase somewhat the number of really able research men by stimulating through scholarships and monetary aids a larger number of men to embark on research as a career than would normally choose it otherwise. The increase due to such a process will be less than the number we hope for, however. This is because many who are not really qualified through inherent capacity or deep-seated interest will embark in science because of the lures or because for the moment science is in the spotlight of public interest.

It will take all the wisdom which science and society possess to steer a safe course through the uncertain years just ahead.

33. A Technical Editor Details the Efforts of the Scientific Establishment to Discredit a Writer. 1963

For a number of years after 1950 the American scientific community was embarrassed by the writings of Immanuel Velikovsky. Velikovsky was a physician and psychoanalyst who took up residence in the United States in 1940. His background included experience in promoting and editing scholarly publications, and he was well educated in both literature and science. During the 1940s he developed interpretations of historical records that suggested that many suppositions of the scientists of his day were erroneous. A number of these men utilized their considerable influence in what they considered to be the interests of science. To this day American scientists, with much good reason, do not take Velikovsky's work seriously, even though he has been shown to have made some astonishing predictions. The following selection, then, should be read not as evidence for or against the validity of Velikovsky's contentions (the discussion of which would take at least one book by itself) but as a study in the operations of the scientific Establishment. The haunting question here is, what should respectable scientists do about thinkers whom they believe to be crackpots? Do scientists as a group have a responsibility to maintain the public image of science? How is the public to decide whether a scientific innovator is a crackpot or not? Should a scientist rebuffed by his colleagues turn to the public for support? How much error or unproved assertion is tolerable in a scientist's work before he shows himself to be unsound? The motives of the defenders of Velikovsky also need close examination. One obvious question in this connection is why this article appeared in a journal the clientele of which consisted primarily of behavioral scientists.

RALPH E. JUERGENS was born in 1924 in Lakewood, Ohio. He was graduated in civil engineering from Case Institute of Technology in 1949 and then worked as a structural engineer. From 1957 to 1961 he was a general contractor and then joined *Construction Methods Magazine,* where he has been successively assistant, associate, and contributing editor.

A much fuller discussion of this affair is in Alfred de Grazia, Ralph E. Juergens, and Livio C. Stecchini, *The Velikovsky Affair, Scientism vs. Science* (New York: University Books, 1966). The general setting of the problem is taken up in Warren O. Hagstrom, *The Scientific Community* (New York: Basic Books, 1965).

MINDS IN CHAOS: A Recital of the Velikovsky Story*
Ralph E. Juergens

Thirteen years ago the appearance of Immanuel Velikovsky's *Worlds in Collision* precipitated an academic storm. Prominent American scientists, roused to indignation even before the book was published, greeted it with a remarkable demonstration of ill will that included a partially successful attempt to suppress the work by imposing a boycott on its first publisher's textbooks. The reading public witnessed the unique spectacle of a scientific debate staged not in the semi-privacy of scientific meetings and journals, but in the popular press, with scientists—in rare accord—on one side and lay champions of free speech on the other. With the might of authority all on one side of the issue, the debate was resolved in a predictable manner; Velikovsky and his book were discredited in the public eye.

From the start there was more to the controversy than the simple question of a dissenting scholar's right to be published and read; the atmosphere generated by scientific consternation was charged with a peculiar emotion that *Newsweek* termed "a highly unacademic fury." Even if Velikovsky's book were, as one astronomer put it, the "most amazing example of a shattering of accepted concepts on record," the violence of the reaction against it seemed all out of proportion to the book's importance if, as most critics insisted, the work was spurious and entirely devoid of merit. Many non-scientist observers concluded that Velikovsky's work was not run-of-the-mill heresy, but a thesis that presented a genuine threat to the very ego of science. It seemed that *Worlds in Collision* was being attacked with a fervor "reserved only for books that lay bare new fundamentals." Caught up in this fervor, more than one scientist-reviewer of Velikovsky's book adopted tactics even more surprising than the overt and covert deeds of the would-be suppressors.

Before attempting to trace the course of the Velikovsky affair, we might first recall the unsettling message of the book that initiated that strange chain of events. In Britain, where *Worlds in Collision* was also rejected by almost all scientists, but with a lesser show of emotion, Sir Harold Spencer Jones, the late Royal Astronomer, summarized its thesis this way:

> The central theme of *Worlds in Collision* is that, according to Dr. Velikovsky, between the fifteenth and eighth centuries B.C. the earth experienced a series of violent catastrophes of global extent. Parts of its surface were heated to such a degree that they became molten and great streams of lava welled out; the sea boiled and evaporated; . . . mountain ranges collapsed, while others were thrown up; continents were raised causing great floods; showers of hot stones fell; electrical disturbances of great violence caused much havoc; hurricanes swept the earth; a pall of darkness shrouded it, to be followed by a deluge of fire. This picture of a period of intense turmoil within the period

* "Minds in Chaos: A Recital of the Velikovsky Story" by Ralph E. Juergens, is reprinted from the *American Behavioral Scientist* (September 1967) pp. 4–17, by permission of the publisher, Sage Publications, Inc.

> of recorded history is supported by a wealth of quotations from the Old Testament, from the Hindu Vedas, from Roman and Greek mythology, and from the myths, traditions and folklore of many races and peoples. . . .
>
> These catastrophic events in the earth's history are attributed by Dr. Velikovsky to a series of awe-inspiring cosmic cataclysms. In the solar system we see the several planets moving round the sun in the same direction in orbits which are approximately circular and which lie nearly in the same plane. Dr. Velikovsky asserts that this was not always so, but that in past times their orbits intersected; collisions between major planets occurred, which brought about the birth of comets. He states that in the time of Moses, about the fifteenth century B.C., one of these comets nearly collided with the earth, which twice passed through its tail. [The earth experienced] the disrupting effect of the comet's gravitational pull, . . . intense heating and enormous tides . . . incessant electric discharges . . . and the pollution of the atmosphere by the gases in the tail. . . . Dr. Velikovsky attributes . . . oil deposits in the earth to the precipitation, in the form of a sticky liquid (naphtha), of some of the carbon and hydrogen gases in the tail of the comet, while the manna upon which the Israelites fed is similarly accounted for as carbo-hydrates from the same source.
>
> This comet is supposed to have collided with Mars . . . and, as the result of the collision, to have lost its tail and to have become transformed into the planet Venus. . . .
>
> Further catastrophes . . . ensued . . . Mars was shifted nearer to the earth so that in the year 687 B.C. . . . Mars nearly collided with the earth.
>
> These various encounters are supposed to have been responsible for repeated changes in the earth's orbit, in the inclination of its axis, and in the lengths of the day, the seasons and the year. The earth on one occasion is supposed to have turned completely over, so that the sun rose in the west and set in the east. Dr. Velikovsky argues that between the fifteenth and eighth centuries B.C. the length of the year was 360 days and that it suddenly increased to 365¼ days in 687 B.C. The orbit of the moon and the length of the month were also changed. . . .

In short, Velikovsky's research among the ancient records of man—records ranging from unequivocal statements in written documents, through remembrances expressed in myth and legend, to mute archaeological evidence in the form of obsolete calendars and sundials—and his examination of geological and paleontological reports from all parts of the globe led him to conclude that modern man's snug little world, set in a framework of celestial harmony and imperceptible evolution, is but an illusion. Velikovsky's reappraisal of world history ravages established doctrine in disciplines from astronomy to psychology: universal gravitation of masses is not the only force governing celestial motions—electromagnetic forces must also play important roles; enigmatic breaks in the geological record denote, not interminable ages of languorous erosion and deposition gently terminated by cyclic submergence and emergence of land masses, but sudden, violent derangements of the earth's surface; the remarkably rapid annihilation of whole species and genera of animals and the equally remarkable, almost simultaneous proliferation of species in other generic groups bespeak overwhelming catastrophe and wholesale mutation among survivors; the mechanism of evolution is not competition between typical and chance-mutant offspring of common parents, but divergent mutation of whole populations simultaneously exposed to unaccustomed radiation, chemical pollution of the atmosphere, and global electromagnetic disturbances; ancient cities and fortresses were not brought low individually by local warfare and earthquakes, but were destroyed simultaneously and repeatedly in worldwide catastrophes; calamities described in clear-cut terms in surviving records of the past—records almost universally interpreted allegorically by late-classical as well as modern scholars—were common traumatic experiences for all races of mankind, and as such have been purged from conscious memory. . . .

[In 1940] reflecting upon events in the life of Moses, Velikovsky began to speculate: Was there a natural catastrophe at the time of the Exodus of the Israelites from Egypt? Could the plagues of Egypt, the hurricane, the parting of the waters, and the smoke, fire, and rumblings of Mt. Sinai described in the Bible have been real and sequential aspects of a single titanic cataclysm of natural forces? If the Exodus took place during—or because of—an upheaval, perhaps some record of the same events has survived among the many documents of ancient Egypt; if so, might not such a record be a clue to the proper place of the Exodus in Egyptian history?

After weeks of search Velikovsky came upon the story he sought. A papyrus bearing a lamentation by one Ipuwer had been preserved in the library of the University of Leiden, Holland, since 1828. Translation of the document by A. H. Gardiner in 1909 had disclosed an account of plague and destruction closely paralleling the Biblical narrative, but the similarities escaped Gardiner's attention. Ipuwer bewailed the collapse of the state and social order during what seemed to be a calamity of natural forces. Mention of Asiatic invaders (Hyksos) made it appear that the sage Ipuwer had witnessed the downfall of the Middle Kingdom (Middle Bronze Age) in Egypt.

For nearly 2000 years scholars have conjectured and debated about the proper place of the Exodus in Egyptian history. But the end of Middle Kingdom, which is conventionally assigned to the eighteenth century B.C., had never been considered; it seemed much too early according to Hebrew chronology. All efforts have been directed toward finding a likely niche in New Kingdom history. Velikovsky, however, felt confident that his method of correlation was valid; he resolved to establish the coevality of the Exodus and the Hyksos invasion as a working hypothesis and pursue the inquiry through subsequent centuries. He discovered so much apparent substantiation for the novel synchronization that he was soon compelled to face up to its inherent dilemma: either Hebrew history is too short by more than five centuries—an inconceivable premise—or Egyptian chronolgy, a proud joint achievement of modern historians, archaeologists, and astronomers, and the standard scale against which all Near Eastern histories are calibrated, is too long by an equal number of centuries. The latter alternative seemed just as inconceivable; all the excess centuries would have to be found and eliminated from post-Middle Kingdom history, that portion of Egyptian history considered by all scholars to be unalterably reconstructed and fixed in time. But soon Velikovsky found the apparent explanation for the discrepancy: certain Egyptian dynasties appear twice in conventionally accepted schemes—first, their stories appear as they have been pieced together from the monuments and other relics of Egypt; then in history gleaned from Greek historians, the same characters and events are given secondary and independent places in the time table. "Many figures . . . are 'ghosts' or 'halves' and 'doubles.' Events are often duplicates; many battles are shadows; many speeches are echoes; many treaties are copies."

In the fall of 1940 Velikovsky traced events similar to those described in the Pentateuch and the Book of Joshua in the literature of ancient Mexico. This confirmed his growing suspicion that the great natural catastrophes that visited the Near East had been global in scale. Immediately he expanded his research to embrace records of all races. The next five or six years he spent developing parallel themes—reconstructions of ancient political history and recent cosmic history—and as month followed month the intimate details of a new concept of the world emerged. Two manuscripts were the product of his labors: *Ages in Chaos* traced Near Eastern history from —1500 to —300; *Worlds in Collision* documented the evidence and sequence of catastrophes on earth and in the solar system.

The late Robert H. Pfeiffer, then Chairman of the Department of Semitic Languages and Curator of the Semitic Museum at Harvard University, read an early draft of *Ages in Chaos* in 1942 and conceded that the revolutionary version of history might well be correct. He felt the work should receive a fair trial and objective investigation. He also read subsequent drafts of the manuscript and made efforts to help find a publisher for it. To one prospective publisher he wrote: "I regard this work—provocative as it is—of fundamental importance whether its conclusions are accepted by competent scholars or whether it forces them to a far-reaching and searching reconstruction of the accepted chronology." Notwithstanding Pfeiffer's endorsement, eight publishers returned the manuscript.

Before seeking a publisher for *Worlds in Collision,* Velikovsky tried to enlist the help of scientists in arranging for certain experiments that would constitute crucial tests for his thesis, which was essentially three-fold: (1) There were global catastrophes in historical times; (2) these catastrophes were caused by extraterrestrial agents; and (3) these agents, in the most recent of the catastrophes, can be identified as the planets Venus and Mars, Venus playing the dominant role. All three postulates would be largely substantiated if it could be shown that, contrary to all conventional expectations, Venus (1) is still hot—evidence of recent birth, (2) is enveloped in hydrocarbon clouds—remnants of a hydrocarbonaceous comet tail, and (3) has anomalous rotational motion—evidence suggesting that it suffered unusual perturbations before settling in its orbit as a planet. The first two of these points were selected by Velikovsky in 1946 as the most crucial tests for his entire work. . . .

He was confident of ultimate vindication for his conclusion that Venus is hot despite the fact that the outer regions of its envelope were known to have a temperature of −25°C. Even as recently as 1959 astronomers believed that because of the great reflecting power of its clouds the ground temperature on Venus could differ little from that on earth. Venus orbits closer to the sun, but more solar radiation is reflected away from Venus than from the earth. Nevertheless, Velikovsky argued that the seeming contradiction in evidence long available—apparent slow rotation, yet nearly identical temperatures on shadowed and sunlit surfaces of the envelope of Venus—is illusory because the planet is young; it is hot and radiates heat from day and night hemispheres alike. [Fifteen years later, in 1961, radio astronomers announced that radiation from Venus indicated that its surface must have a temperature of 600°F. And in February 1963, after analyzing data from Mariner II, scientists raised this temperature estimate by another 200 degrees. No convincing explanation has yet been advanced to square this evidence with orthodox cosmologies.]

Velikovsky thought his second deduction about Venus—hydrocarbon dust and gases must be present in its atmosphere and envelope—might be investigated spectroscopically. To this end in April 1946 he approached Prof. Harlow Shapley, then director of Harvard College Observatory. Without going into detail, Velikovsky explained that he had developed a hypothesis about recent changes in the order of the solar system and that his conclusions might be checked in part by spectral studies of Venus. Shapley pointed out that sudden changes in the planetary order would be inconsistent with gravitational theory; nevertheless, he agreed to consider performing such experiments if another scholar of known reputation would first read and then recommend Velikovsky's work. At Velikovsky's behest, Prof. Horace M. Kallen, co-founder of the New School of Social Research and at that time dean of its graduate faculty—a scholar already familiar with the work—wrote Shapley to urge that he conduct the search for hydrocarbons on Venus if at all possible. But to Kallen's plea, Shapley, who had refused to read the manuscript, replied that he wasn't interested in Velikovsky's "sensational claims" because they violate the laws of mechanics; "if Dr. Velikovsky is right, the rest of us are crazy." Nevertheless, Shapley recommended that Velikovsky contact either Walter S. Adams, director of Mt. Wilson Observatory, or Rupert Wildt at McCormick Observatory.

In the summer of 1946 Velikovsky directed identical inquiries to both Wildt and Adams, stating that he had a cosmological theory implying that "Venus is rich with petroleum gases and hydrocarbon dust." So strong were these implications that he believed the presence or absence of these materials in the atmosphere and envelope of Venus would constitute crucial support or refutation for his thesis, and therefore he wished to know if the spectrum of Venus might be interpreted in this sense. Wildt replied that the absorption spectrum of Venus shows no evidence of hydrocarbons. Adams pointed out that the absorption bands of most petroleum molecules are in the far infra-red, below the range of photographic detection, and that hydrocarbons known to absorb in the detectable range are not apparent in the spectrum of Venus.

All this notwithstanding, Velikovsky elected to defer once more to his historical evidence; he left in his manuscript and later in the published book the statement that a positive demonstration that petroleum-like hydrocarbons are or are not present in the envelope of Venus would be a decisive check on his work. [On the basis of an apparent ability to condense and polymerize into heavy molecules at a temperature near 200°F in the atmosphere, the clouds of Venus must consist of heavy hydrocarbons and more complex organic compounds; thus concluded Mariner II experimenter Lewis D. Kaplan in February 1963.]

At the end of July 1946 the late John J. O'Neill, science editor of the New York *Herald Tribune,* agreed to read Velikovsky's manuscript. O'Neill was immediately impressed, and he devoted his column for August 14 to the work. In his opinion, "Dr. Velikovsky's work presents a stupendous panorama of terrestrial and human histories which will stand as a challenge to scientists to frame a realistic picture of the cosmos."

Between June and October 1946 Velikovsky submitted his manuscript to one publisher after another, but the consensus was that the heavily annotated text was too scholarly for the book trade. Eventually, however, the trail led to the Macmillan Company, where trade-books editor James Putnam saw possibilities in the book. In May of 1947 an optional contract was signed, and then, after another year in which various outside readers—among them, O'Neill and Gordon Atwater, then Curator of Hayden Planetarium and Chairman of the Department of Astronomy of the American Museum of Natural History—examined the manuscript and recommended publication, a final contract was drawn and signed.

By March 1949 word of the book Macmillan was preparing for publication had spread among people in the trade. Frederick L. Allen, editor-in-chief of *Harper's Magazine,* sought authorization to present a two-article synopsis of *Worlds in Collision* and had Eric Larrabee, then an editor on the *Harper's* staff, prepare a tentative condensation from galley proofs. Allen wished to submit this for approval, but Velikovsky did not respond to the proposal for more than six months. In the fall, however, after more urging, he agreed to see Larrabee to discuss a one-article presentation of his theme; Larrabee then rewrote his piece completely.

Larrabee's article, "The Day the Sun Stood Still," appeared in *Harper's* for January 1950. The issue sold out within a few days, and so great was the demand from readers that a number of dailies both here and abroad reprinted Larrabee's text in full.

In February 1950 *Reader's Digest* featured a popularization of Velikovsky's findings prepared by the late Fulton Oursler, who emphasized their corroboration for Old Testament history. . . .

Early in February 1950, when *Worlds in Collision* was about to go to press, Putnam called on Velikovsky to show him two letters Macmillan had received from Harlow Shapley. In the first, dated January 18, Shapley expressed gratification over a rumor that Velikovsky's book was not going to appear, and astonishment that Macmillan had even considered a venture into the "Black Arts." In his second letter, written on January 25 after Putnam had answered the first, discounting the alleged rumor and assuring him that the book would appear on schedule, Shapley, who had still not seen the manuscript, remarked: "It will be interesting a year from now to hear from you as to whether or not the reputation of the Macmillan Co. is damaged by the publication of 'Worlds in Collision'." At the very least, release of the book would "cut off" all relations between Shapley and Macmillan. He also announced that, at his request, one of his colleagues who was also a classicist was preparing a "commentary" on Larrabee's article. He concluded with an expression of his hope that Macmillan had thoroughly investigated Velikovsky's background; however, "it is quite possible that only this 'Worlds in Collision' episode is intellectually fraudulent."

The second letter apparently struck close to home for Macmillan president George Brett, for he personally answered Shapley to thank him for "waving the red flag." Brett promised to submit the book to three impartial censors and to abide by the majority verdict of the three.

Apparently the majority again voted thumbs up; the book was published on schedule. The identities of the last-minute censors were never officially revealed, but one of them, Prof. C. W. van der Merwe, Chairman of the Department of Physics at New York University, later disclosed to John O'Neill that he had been enlisted by Macmillan and had been one of the two who voted in favor of publication.

Meanwhile, the February 25, 1950, issue of *Science News Letter,* a publication then headed by Harlow Shapley, printed denunciation of Velikovsky's ideas by five authorities in as many fields: Nelson Glueck, archaeologist; Carl Kraeling, orientalist; Henry Field, anthropologist; David Delo, geologist; and Shapley himself speaking for astronomers. This medley of protest came forth just as *Worlds in Collision* went to press—none of the critics had seen the work.

On March 14, the commentary on Larrabee's article by Shapley's colleague, astronomer Cecilia Payne-Gaposchkin, appeared in *The Reporter.* (An earlier draft of the article had been mimeographed and circulated widely by direct mail to scientists, science editors, and publishers.) Stringing phrases from three sentences appearing on as many pages of Larrabee's article into a sentence of her own, Gaposchkin set it in quotation marks and introduced it as "Dr. Velikovsky's astronomical assertions." The gist of her thoroughly abusive article was that electromagnetic phenomena are of no importance in space, and in a purely mechanical solar system the events of *Worlds in Collision* are impossible. The March 25 issue of *Science News Letter,* in a "Retort to Velikovsky," who had as yet not been heard from, cited Gaposchkin's critique as recommended reading for all scientists—"a detailed scientific answer to Dr. Velikovsky."

On April 11 *The Reporter* reproduced letters to the editor from Larrabee and Gaposchkin. Larrabee challenged the propriety of her attack on a book she had not yet seen, and Gaposchkin acknowledged that her review had been based on popularized preview articles only; she remarked that she had since read the book (published April 3, 1950) and found it to be "better written . . . but just as wrong."

The last few weeks before *Worlds in Collision* made its appearance were spent in strategic maneuvering by the leaders of the resistance forces. The late Otto Struve, then director of Yerkes Observatory at the University of Chicago and an ex-president of the American Astronomical Society, penned letters to both John O'Neill and Gordon Atwater, requesting them to abandon their earlier positions with respect to *Worlds in Collision.* Atwater, unaware that he was facing an inquisition, replied that he believed Velikovsky's work had great merit, and though he did not accept all its conclusions in detail, he was preparing a favorable review of the book for *This Week* magazine. He was planning—indeed had already publicly announced—a planetarium program to depict the events of *Worlds in Collision.* O'Neill composed a heated reply, but then destroyed it. He let it be known that his earlier appraisal of the book had not since been altered in any way.

Atwater's planetarium program was scuttled immediately. During the last week of March he was summarily fired from both his positions with the museum—as Curator of Hayden Planetarium and Chairman of the Department of Astronomy—and requested to vacate his office immediately. Thus, when his review in *This Week* appeared on April 2, an article in which he pleaded for open-mindedness in dealing with the new theory, the credentials printed alongside Atwater's name were already invalid. Last-minute attempts to influence *This Week* not to publish this cover story failed when the editor sought and followed O'Neill's advice. . . .

O'Neill's prepared review for the *Herald Tribune* had been scheduled to appear on April 2. But instead of O'Neill's article readers of that Sunday's issue found a review written by Struve. No concrete arguments were presented by Struve to justify his rejection of the book; "It is not a book of science and it cannot be dealt with in scientific terms." He went on: "It was necessary for readers to wait until a recent issue of the 'Reporter' to learn, through Mrs. Cecilia Payne-Gaposchkin . . . that the observations of Venus extend back five hundred years before the Exodus, thus refuting the absurd theory of a comet that turned into a planet." Velikovsky, however, had specified no date for the eruption of Venus from Jupiter, except that it had occurred some time before the Exodus. And, as Velikovsky pointed out in his book, the Babylonian tablets (Venus Tablets of Ammizaduga) cited by Gaposchkin to support her claim ascribe such erratic motions to Venus that translators and commentators have been baffled by them ever since they were discovered in the ruins of Nineveh in the last century; he also pointed out that even if the apparitions and periods of Venus recorded on the tablets date from early in the second millennium, which is disputed among scholars, they prove only that Venus already then moved erratically and quite unlike a planet.

Reviewing *Worlds in Collision* in the *New York Times Book Review,* also on April 2, the late chief science editor of the *Times,* Waldemar Kaempffert, followed Gaposchkin into the same territory and falsely accused Velikovsky of suppressing the Venus Tablets of Ammizaduga. Kaempffert seemingly had not read the book very carefully before condemning it, for not only did Velikovsky describe the tablets and quote the complete texts of observations from five successive years out of twenty-one, but he discussed opinions written by various orientalists and astronomers who had studied the tablets (Rawlinson, Smith, Langdon, Fotheringham, Schiaparelli, Kugler, Hommel).

In the next few months, "a surprising number of the country's reputable astronomers descended from their telescopes to denounce *Worlds in Collision,"* to quote the *Harvard Crimson* of September 25, 1950. Newspapers around the country were barraged with abusive reviews contributed by big-name scientists; some of these writings were syndicated to ensure better coverage.

Ignoring Velikovsky's alternate explanation that, perhaps in the grip of an alien magnetic field, a "tilting of the [earth's] axis could produce the visual effect of a retrogressing or arrested sun," Frank K. Edmondson, director of Goethe Link Observatory, University of Indiana, wrote: "Velikovsky is not bothered by the elementary fact that if the earth were stopped, inertia would cause Joshua and his companions to fly off into space with a speed of nine hundred miles an hour." This argument, first formulated by Gaposchkin, is at best disingenuous, for the all-important time factor—the rate of deceleration—is completely ignored.

Paul Herget, Director of the Observatory, University of Cincinnati, derided the ideas expressed in *Worlds in Collision,* but advanced no specific counterarguments on scientific grounds. Nevertheless, he concluded that all the book's basic contentions were "dynamically impossible." Frank S. Hogg, director of David Dunlop Observatory, University of Toronto, and Oregon astronomer J. Hugh Pruett both reiterated the erroneous Gaposchkin-Struve notion that observations of Venus made before the time of the Exodus refute

Velikovsky's theme. California physicist H. P. Robertson chose the easy path of invective: "This incredible book . . . this jejune essay . . . [is] too ludicrous to merit serious rebuttal." . . .

On May 25, 1950, when sales of his book were at their peak, Velikovsky was summoned to Brett's office and told that professors in certain large universities were refusing to see Macmillan salesmen, and letters demanding cessation of publication were arriving from a number of scientists. Brett beseeched Velikovsky to save him from disaster by approving an arrangement that had been tentatively worked out with Doubleday & Company, which had no textbook department. Doubleday, with Velikovsky's consent, would take over all rights to *Worlds in Collision.* As evidence of the pressure being brought to bear, Brett showed Velikovsky a letter from Michigan astronomer Dean B. McLaughlin, who insisted Velikovsky's book was nothing but lies. On the same page McLaughlin averred he had not read and never would read the book.

While Velikovsky pondered his next move—whether to approve the transfer of rights to Doubleday, or to make an independent search for a new publisher—his scientist-critics apparently began to see their problem in a more serious perspective. Inability to dismiss the events of *Worlds in Collision,* gleaned from a multitude of sources, suggested that a substantial assault upon his method and sources was in order.

The June 1950 issue of *Popular Astronomy* carried another attack on Velikovsky by Cecelia Payne-Gaposchkin. Her words were prefaced by a few lines from the magazine editor (who explained, "We are giving greater prominence to this analysis of 'Worlds in Collision' than is usually accorded to book reviews . . . for two reasons. 1. This book has been brought to the attention of a large reading public by having been mentioned favorably in several popular magazines. 2. The analysis here given is by a recognized authority in the field of astronomy, the science with which the book comes into closest contact, or sharpest conflict." Gaposchkin's "analysis" was divided into two parts, first place being devoted to "the Literary Sources." By the simple ruse of ignoring both contextual material and corroborative references, she purported to show that Velikovsky had misrepresented his source. Her "Scientific Arguments" included restatements of undemonstrable dogmatisms and a highly sarcastic synopsis of Velikovsky's thesis.

Prof. Otto Neugebauer of Brown University, a specialist in Babylonian and Greek astronomy, in an article for *Isis* that was mailed far and wide in reprint form, accused Velikovsky of willfully tailoring quoted source material. To support this charge, Neugebauer specified that Velikovsky had substituted the figure 33°14′ for the correct value, 3°14′, in a quotation from the work of another scholar. When Velikovsky protested in a letter to the late George Sarton, then editor of *Isis,* that the figure given in his book was correct and the 33°14′ was in fact Neugebauer's own insertion, not his, Neugebauer dismissed the incident as a "simple misprint of no concern" that did not invalidate his appraisal of Velikovsky's methods. And the reprint was circulated by an interested group long after its errors had been pointed out.

The fundamental position of Neugebauer is that the voluminous Babylonian astronomical texts from before the seventh century B.C., all of which are inconsistent with celestial motions as we know them, were composed in full disregard of actual observations; Velikovsky regards these records as representing true observations of the heavens before the last catastrophe. . . .

After receiving assurances from Doubleday that it was immune to pressure from textbook writers and buyers, Velikovsky approved the transfer of rights on June 8, 1950. On June 11, columnist Leonard Lyons spread the news, and on June 18 the New York *Times* noted: "The greatest bombshell dropped on Publishers' Row in many a year exploded the other day. . . . Dr. Velikovsky himself would not comment on the changeover. But a publishing official admitted, privately, that a flood of protests from educators and others had hit the company hard in its vulnerable underbelly—the textbook division. Following some stormy sessions by the board of directors, Macmillan reluctantly succumbed, surrendered its rights to the biggest money-maker on its list."

Leonard Lyons reported that the suppression was engineered by Harlow Shapley. When queried, however, Shapley told *Newsweek*, "I didn't make any threats and I don't know anyone who did." The late George Sokolsky also discussed the case in his column, and shortly afterward received a letter from Paul Herget, who was apparently disappointed that all the credit was going to Shapley. Herget wrote, and Sokolsky quoted: "I am one of those who participated in this campaign against Macmillan. . . . I do not believe that [Shapley] was in any sense the leader. . . . I was a very vigorous participant myself. . . ." Dean McLaughlin wrote to Fulton Oursler: "*Worlds in Collision* has just changed hands. . . . I am frank to state that this change was the result of pressure that scientists and scholars brought to bear on the Macmillan Company." . . .

Dumping its offensive best seller, however, was but the first step in the re-establishment of Macmillan's reputation. There remained matters of purgatorial sacrifice and public recantation.

James Putnam, a 25-year veteran with Macmillan, had been entrusted with making the arrangements to contract for and publish Velikovsky's manuscript. His judgment in urging that Macmillan accept *Worlds in Collision* had been confirmed in spectacular fashion when the book became a best seller. Nevertheless, the negotiations to transfer publishing rights to Doubleday were carried on without his knowledge, and as soon as the transfer had been consummated, Putnam's good friend, editor-in-chief H. S. Latham, was delegated to inform him that his services were being terminated immediately. [In January 1963 Latham expressed in a letter to Velikovsky the great regret he still feels for Macmillan's capitulation.]

At the annual meeting of the American Association for the Advancement of Science held in Cleveland in December 1950, a Mr. Charles Skelley, representing the Macmillan Company, addressed the members of a committee specially appointed to study means for evaluating new theories before publication. He pointed out that, as a contribution to the advancement of science, his firm had "voluntarily transferred" its rights to a "book that the panel regarded as unsound. . . ." His remarks were duly recorded and reported by panel chairman Warren Guthrie. Harvard geologist Kirtley Mather was the main spokesman before the panel, discussing possible methods of censorship. . . .

Laurence Lafleur, then associate professor of philosophy at Florida State University, brought a new argument to bear against Velikovsky in the November 1951 issue of *Scientific Monthly:* ". . . the odds favor the assumption that anyone proposing a revolutionary doctrine is a crank rather than a scientist." Lafleur itemized seven criteria for spotting a crank. Examples:

> *Test 6.* Velikovsky's theory is in no single instance capable of mathematical accuracy. Its predictions, if capable of any, would certainly be so vague as to be scientifically unverifiable.
>
> *Test 7.* Velikovsky does show a disposition to accept minority opinions, to quote the opinions of individuals opposed to current views, and even to quote such opinions when they have been discredited to the point that they are no longer held even as minority views. For example, we may cite the notion that the earth's axis has changed considerably.

So Lafleur concluded that Velikovsky qualified as a crank "perhaps by every one" of these tests. But having established this "we must still deal with the feeling, first, that scientists should have attemped to refute Velikovsky's position, as a service both to him and to the public. . . ." Thus the professor acknowledged that much of earlier criticism—thousands of words printed in the span of more than a year and a half—was denunciation rather than refutation. But in his own attempt to perform the recommended "service," Lafleur, even with the aid of astrophysical theorems contrived for the occasion, fared no better than the scientists. On the assumption that an electroscope would detect it, he denied that the earth carries an electric charge. (No scientist corrected, in print, this mistaken notion or any other wrong statement by any critic during the entire *Worlds in Collision* controversy.) Laffeur also claimed that an approach between two celestial bodies close enough to bring their magnetic fields into conflict must inevitably bring about collision, evaporation, and amalgamation of the bodies. . . .

Earth in Upheaval appeared in November 1955. Velikovsky examined the century-old principle of Lyellian uniformity by comparing its tenets with anomalous finds from all quarters of the globe: frozen muck in Alaska that consists almost entirely of myriads of torn and broken animals and trees; whole islands in the Arctic Sea whose soil is packed full of unfossilized bones of mammoths, rhinoceroses, and horses; unglaciated polar lands and glaciated tropical countries; coral and coal deposits near the poles; bones of animals from tundra, prairie, and tropical rainforest intimately associated in jumbled heaps and interred in common graves; the startling youth of the world's great mountain chains; shifted poles; reversed magnetic polarities; sudden changes in sea level all around the world; rifts on land and under the seas.

Then Velikovsky took up the question of evolution, arguing that Darwin had rejected catastrophism in favor of Lyell's uniformity because the catastrophists of his day would not acknowledge the antiquity of the earth. But in reality catastrophes suggest the only plausible mechanisms for the phenomenon of evolution by mutation. Thus Darwin's contribution to the theory of evolution, which dates from Greek times, consisted only in the as-yet undemonstrated hypothesis that competition can give rise to new species. In the controversy that followed the publication of *The Origin of Species,* the issue revolved around whether or not evolution was a natural phenomenon, and it was resolved quite properly in the affirmative. But what was obscured in the uproar, argued Velikovsky, was the inadequacy of Darwin's hypothesis; "If natural selection . . . is not the mechanism of the origin of species, Darwin's contribution is reduced to very little—only to the role of natural selection in weeding out the unfit." Velikovsky proposed in *Earth in Upheaval* that evolution is a cataclysmic process; ". . . the principle that *can* cause the origin of species exists in nature. The irony lies in the circumstance that Darwin saw in catastrophism the chief adversary of his theory." . . .

Favorable discussion of *Earth in Upheaval* may have [created] some pressure to discuss it in other scientific media. In March 1956 *Scientific American* presented a review by Harrison Brown. His words, however, were devoted to an apology for the misbehavior of scientists who had suppressed *Worlds in Collision* and to a restatement of his own earlier position with respect to that book. In a seven-column article, Brown dismissed *Earth in Upheaval* without challenging one of its points. He dealt with the new book in a single paragraph, then reverted to the old controversy. But he again refrained from producing any of the arguments against *World*[*s*] *in Collision* which he had claimed would fill thirty pages. [In 1963, Brown declared in a letter to one of Velikovsky's Canadian readers that his review of *Earth in Upheaval* had been directed against the "abominable behavior of scientists and publishers."]

In December 1956, when the International Geophysical Year was in the planning stage, Velikovsky submitted a proposal to the planning committee through the offices of Prof. H. H. Hess of Princeton University: ". . . It is accepted that the terrestrial magnetic field . . . decreases with the distance from the ground; yet the possibility should not be discounted that the magnetic field above the ionosphere is stronger than at the earth's surface." Also, "an investigation as to whether the unexplained lunar librations, or rocking movements, in latitude and longitude coincide with the revolutions of the terrestrial magnetic poles around the geographical poles" might well be included in the program. Hess was notified by E. O. Hulburt of the committee that should the first proposition be proven right by experiments already planned, the second might be investigated later. [As it turned out, the most important single discovery of the IGY was that the earth is surrounded by the Van Allen belts of charged particles trapped in the far reaching geomagnetic field. . . .]

About the time Mariner II approached Venus, late in 1962, Princeton physicist V. Bargmann and Columbia astronomer Lloyd Motz wrote a joint letter to the editor of *Science* to call attention to Velikovsky's priority in predicting three seemingly unrelated facts about the solar system—the earth's far-reaching magnetosphere, radio noise from Jupiter, and the extremely high temperature of Venus—which have been among the most important and surprising discoveries in recent years. They urged that Velikovsky's thesis be objectively reexamined by science.

Also at that time it was announced that ground-based radiometric observations at the U.S. Naval Research Laboratory in Washington and at Goldstone Tracking Station in California had shown Venus to have a slow retrograde rotation, a characteristic that puts it in a unique position among the planets.

Feeling vindicated by these developments and encouraged by the publication of the Bargmann-Motz letter in *Science,* Velikovsky sought to publish a paper showing that the points brought out in that letter were but a few among many other ideas set forth in his books that have already been supported by independent research. The attempt was in vain; Philip Abelson, the editor of *Science,* returned Velikovsky's paper without reading it and published instead a facetious letter from a Poul Anderson, who claimed that "the accidental presence of one or two good apples does not redeem a spoiled barrelful."

Mariner II, when its findings were revealed, confirmed Velikovsky's expectations, showing the surface temperature of Venus to be at least 800°F. and the planet's 15-mile-thick envelope to be composed, not of carbon-dioxide or water as previously supposed, but of heavy molecules of hydrocarbons and perhaps more complicated organic compounds as well.

Retrograde rotation, organic molecules in the envelope, and extreme heat on Venus find no convincing explanation, though they have already caused much deliberation; yet in *Worlds in Collision* two of the three phenomena were claimed as crucial tests for the thesis that Venus is a youthful planet with a short and violent history, and the third (anomalous rotation) supports the same conclusions. . . .

Seldom in the history of science have so many diverse anticipations—the natural fallout from a single central idea—been so quickly substantiated by independent investigation. One after another of Velikovsky's "wild hypotheses" have achieved empirical support, but not until December 1962, in the Bargmann-Motz letter to *Science,* was his name ever linked in the pages of scientific journals with any of these "surprising" discoveries and never yet by the discoverers themselves. A platitude repeated on various occasions, has it that any one who makes as many predictions as Velikovsky is bound to be right now and then. But he has yet to be shown wrong about any of his suggestions. Prof. H. H. Hess, who is now Chairman of the Space Board of the National Academy of Science, recently wrote to Velikovsky: "Some of these predictions were said to be impossible when you made them; all of them were predicted long before proof that they were correct came to hand. Conversely, I do not know of any specific prediction you made that has since proven to be false."

This record would appear to justify a long, careful look at *Worlds in Collision* by the guild that not only refused to look before condemning it in the past, but actively campaigned to defame its author.

34. A Geneticist Explores the Possibilities for Controlling Life. 1963

Here is the statement of a social and self-styled scientific conservative in the face of the radical implications of the discoveries of biochemistry—the wonderful world of DNA and RNA—which apparently embody the secret of life, of inheritance, and of the formation, growth, and functioning of cells. Some of the potentialities of this new science can be inferred from the caution and cautions of the author concerning the neo-eugenics and reductionism of biochemists. Whether the evidence cited here is representative of developments in biology, whether genetics has really been transformed, is still not clear. This selection was the dedicatory address at the opening of the new McCort-Ward Biology Building, The Catholic University of America, in 1962.

T. M. SONNEBORN was born in Baltimore in 1905 and took both his bachelor's and doctor's degrees at Johns Hopkins. His field has been zoology, and he has done distinguished work in genetics. He has been a member of the department of zoology at Indiana University since 1931 and Distinguished Service Professor since 1953. Sonneborn has received many honors and in 1961 was elected president of the American Institute of Biological Sciences.

When the issues involved in this selection were better understood later in the 1960s, many excellent discussions appeared. Perhaps the most relevant is to be found in John D. Roslansky, ed., *Genetics and the Future of Man* (New York: Appleton-Century-Crofts, 1966).

IMPLICATIONS OF THE NEW GENETICS FOR BIOLOGY AND MAN*

T. M. Sonneborn

During the last decade, genetics has been spectacularly fruitful, developing with ever-increasing speed and ever more challenging prospects. Each month, almost each day, advances are announced. Questions are being posed—and answered—that seemed, until recently, beyond the scope of foreseeable experimental analysis. And no end is in sight!

Intoxicated with their successes, many practitioners of the new revolutionary genetics exhibit unbounded assurance that no secret of living nature and no obstacle to controlling it are beyond their powers to expose or overcome. This leads some foremost biologists to startling views about the future of biology and of man. Biology, they contend, is undergoing a fundamental change which will soon make much of it—especially much of its methodology—obsolete. Further, they warn that the new genetics may provide the means of remaking man—and soon. They urge that no time be lost, for mankind may well be confronted with powers potentially greater than those of atomic energy; that they may be used for good or ill; that it behooves man to become aware of this new knowledge and its power and to hasten to acquire the wisdom needed to use it well, lest we are once again caught unprepared.

To close our eyes and ears to these visions and warnings might well be perilous. On the other hand, if there is another side to the story, we should hear it, too, before making judgments. For some time I have wanted to think through these problems. In arriving at my judgments which, I confess, turned out to be different from what I at first supposed they would be, I have had to take into account both the past and the present, both the classical and the new genetics: their essential natures, findings, implications, limitations, and prospects. I shall try to lead you over the path I have traversed.

Let us first look upon the past, upon classical genetics. Its great findings are now common knowledge. They include the concept of the gene as the indivisible unit, the atom, of heredity; the rules of genetic transmission from parents to offspring; the organization of genes into linked groups; the linear architecture of each array of linked genes; the numerically assignable position or locus of each gene in the linear group; and the assignment of each linkage group to a chromosome.

These discoveries have two major implications for biology in general. First, by their complete generality—their universal applicability to all organisms from microbes to man—they conferred upon the whole of biology a magnificent unity matched by very few other generalizations. Second, and perhaps even more important for the issues that now divide biologists, they reveal the power and the fruitfulness of purely biological methods.

* *AIBS Bulletin*, 13 (1963), 22–26. Reprinted with the kind permission of AIBS Publications and the author.

The essential nature of classical genetics was almost completely developed in the work of the genius who created it, the monk, Gregor Mendel. In the hands of his followers, the method he employed—without essential modification or addition—led to every feature of the classical theory of the gene with one exception: the location of the genes in chromosomes. The method consisted of making inferences—completely reliable inferences—about the existence, organization, and behavior of unseen objects, genes, from very simple purely biological observations. One had only to note the relative frequencies with which alternative characteristics and combinations of characteristics occur among individuals in successive generations. Nothing could be simpler or more purely biological, yet it was tremendously revealing. In its simplicity, indirectness, and power, the methodology is comparable to that by which classical physicists and chemists were led to the concepts of molecules and atoms.

We may go even further. Some of the most important advances of the new revolutionary genetics are based solely on the Mendelian method of inference from observations on hereditary traits and their combinations. In this way, for example, Benzer has been able to infer that the gene is divisible into cistrons, recons, and mutons, the genic subunits of function, recombination, and mutation.

Let us not lose sight of the main point. Simple, purely biological methods, without the essential aid of chemistry or physics, of biochemistry or biophysics, led to the exposure of some of the deepest secrets of living nature. Modern researchers and students often forget—or never knew—that they could not have guessed what questions to ask or what problems to attack at the biochemical and molecular levels, had not a sound and profound theoretical structure first been established by the elegant and powerful methods of pure biology.

Classical genetics has great implications for man. Man has the same genic system as other organisms, and it operates in the same way, with impact upon every feature of his development, structure, functioning, and behavior. Why, then, could one not apply to man the same genetic principles that were applied with such success to the production of desired strains of domesticated animals and plants? The question was quickly considered. Early in this century a vigorous, enthusiastic, and highly vocal eugenics movement developed. But it bore within itself the seeds of its own destruction. It was based upon grossly insufficient knowledge and much error in detail, quite aside from all moral, ethical, religious, or political considerations.

Nevertheless, knowledge of human genetics along classical lines does provide the means by which we could decisively affect the hereditary make-up of mankind many generations hence. To some serious, critical, and excellently informed thinkers, this prospect has the tremendous appeal of a grander social consciousness directed not just to one's living fellows, but to the welfare of remote descendants.

The purely genetic, as well as the ethical, moral, religious, social, and political questions involved have been eloquently, imaginatively, and repeatedly dealt with by eminent geneticists. They are well aware of the present genetical and nongenetical limitations of a eugenics program. Among them are our still relatively great ignorance of human genetics, the considerable admixture of hidden recessive lethal, detrimental, and other unfavorable genes in all of us, the slow progress to be expected at best, and the undesirability of forcing the program upon anyone. Nevertheless, they hold persuasively that slow progress is better than none; that more rapid progress could be made as knowledge increases, especially knowledge that might reveal the presence of masked recessives; and that diffusion of knowledge and cultivation of long-range social consciousness would foster voluntary individual eugenic actions. Although the situation is genetically complex and loaded with nongenetical overtones, eugenic possibilities for the future of man are implicit in classical genetics.

The revolutionary and triumphant new genetics is often characterized as being distinctively biochemical and molecular and as differing thereby from classical genetics. In my opinion, this characterization is wrong for two reasons. First, this aspect of the new genetics is not in itself new. To some extent, genetics has been biochemical and molecular almost from the start. Nearly 60 years ago, Garrod pointed out the existence of gene-controlled biochemical traits in man. There has never been a time since then when comparable studies and theoretical constructs were not current. Second, some of the most spectacular triumphs of the new genetics—such as Benzer's revelations of the number and array of subunits in a gene—were achieved by classical genetic methodology.

The main point of difference is more fundamental. The essential nature of the new genetics is its deliberate refusal to concern itself with secondary complications and its hard concentration on a small number of the simplest possible systems, with the conscious objective of attacking the most basic problems and of putting everything else aside until they are solved.

Classical genetics worked with peas and corn, with flies and mice. It worked mainly with hereditary traits that appeared after long and complicated developmental processes. They were mostly far removed from the direct action of the gene itself—so far, that it is marvelous how much their study did, in fact, reveal about the gene. The new genetics began by turning away from such organisms and such traits. It abandoned the idea that, with them, one could work backwards step by step from the final trait to the gene and its primary action. It has stripped away the complexities introduced by multicellularity and interactions among cells, tissues, organs, and organ systems. It has concentrated attention on organisms with relatively simple organization: on Fungi, Algae, and Protozoa to some extent, but mainly on still simpler unicellular or subcellular organisms—Bacteria and Viruses, and, more recently, to the isolated and separately cultivated *cells* of higher organisms, including man.

The new genetics went even further in the directions of simplification and concentration. First, it accepted the common assumption that visible morphological and anatomical traits are preceded developmentally by, and are outcomes of, biochemical traits. The objective of simplication therefore demanded shifting attention from the visible to the underlying biochemical traits, and from the visible chromosome to its underlying chemical structure. Second, it assumed that in principle the problems of the chromosome, gene, and gene action are essentially the same for all organisms. Regardless of whether this assumption is 100% true, work proceeded in the faith that the deepest insights would come by concentrating efforts on a relatively simple type of chromosome and a particularly favorable gene rather than by the comparative method of diffusing attention among many. Whole schools of researchers therefore confined their studies to the chemical constitution of the genetically important component of a relatively simple chromosome or to the composition or action of a single gene—such as the gene controlling the enzyme β-galactosidase, or to the gene for the enzyme tryptophane synthetase in the colon bacterium, or to the gene for host specificity in a virus that attacks the colon bacterium.

This remarkable concentration on one or a few relatively simple favorable materials constitutes the essential nature of the new genetics. It has paid off handsomely. Among its most important accomplishments are the discoveries of the chemical composition and structure of the genetic material of the chromosome and gene and the molecular events in genic replication, mutation, and action. These accomplishments are pertinent to our theme. I shall state them only in the most general terms.

The genetic material of the chromosome is DNA, deoxyribonucleic acid, an aperiodic linear polymer[1] consisting of but four main kinds of repeating units whose chemical composition and structure are known. A sequence of more than 100 to less than 10,000 of these units carries out a simple direct function; such a sequence constitutes the classical gene, or the modern cistron. Different genes differ in the order in which the four kinds of units occur along the length of the gene, not in the kinds of units. As has been pointed out, the four kinds of units are like an alphabet with four letters, the genes being words composed of hundreds to thousands of these letters, each word having a definite sequence of letters.

The molecular events in genic replication are, in essence, astonishingly simple. Each unit in the DNA polymer is composed of two subunits, but there are altogether only four main subunits in the four units. Symbolizing the four subunits as A, C, G, and T, the four units are A-T, T-A, C-G, and G-C. Thus

[1] That is, a large molecule made up of repeated units of other, small molecules, in a linear series rather than in a structure with branching and linked units.

A is always paired with T and C with G. Hence, longitudinal splitting of the polymer between the two subunits of each unit yields reciprocal half-polymers, each of which can regenerate the missing half by following the pairing rule. That is the essence of the disclosure of the deepest secret of biology, the nature of genic replication.

On the molecular level, genic mutation consists in the loss or gain of one or more units of its DNA, in the inversion of a segment of the polymer consisting of one or more units, or in the change of one or more units from one to another of the four kinds, e.g., change of an A-T unit into T-A, C-G, or G-C. Since a gene consists of a length of DNA including hundreds or thousands of units, there are also that many *different* sites at which a mutational event can occur within a single gene. About 400 different mutational sites have been found and mapped in the most fully studied gene.

The molecular events in genic action are still under investigation and current conclusions may well have to be modified in the light of further discoveries. However, there is much evidence in support of the following statement. One of the half-polymers in a gene serves as a template on which a so-called *messenger* is formed; this is a ribonucleic acid, RNA, polymer with a complementary sequence of subunits. The messenger peels off, passes into the cytoplasm, and become[s] associated with ribosome.[2] There, beginning at one end of the messenger, the first three subunits become associated with a complementary set of three in a small RNA molecule (t-RNA) which bears elsewhere a particular amino acid. This process continues down the length of the messenger, each successive triplet of subunits in the messenger combining with a t-RNA molecule bearing a complementary triplet of subunits and an amino acid. As the amino acids are brought in, they link up by peptide linkages and peel off as a polypeptide chain. The action of a gene is thus the making of a polypeptide. Each gene makes a particular polypeptide defined by its linear sequence of amino acids. This is accomplished by a three-step linear point for point correspondence: each sequence of three subunits in one strand of the genic DNA corresponds to a sequence of three subunits in the m-RNA; each sequence of three in the latter corresponds to a sequence of three in a t-RNA; and each of the latter corresponds to one amino acid. These correspondences constitute the so-called genetic code. As there are 20 kinds of amino acids in proteins, there must be at least 20 corresponding different kinds of t-RNA molecules, each with a different triplet of subunits in the part of the molecule that combines with m-RNA. Likewise, there must be at least 20 different meaningful sets of three subunits in m-RNA and in the genic DNA. Deciphering of the code has already made great progress. The composition (but not the order) of the triplets in DNA, m-RNA, and t-RNA that correspond to each of the 20 amino acids has been inferred. Moreover, a beginning has been made in discovering exactly which subunits

[2] Within the cell but outside of the nucleus a small, dense particle consisting of ribonucleic acid and protein and believed to be associated with all growth and reproduction of the cell.

are present at particular sites in a few genes. The methods are in hand for a complete description of the chemical composition and structure of a gene, its messenger, and the various t-RNAs. Colossal though the task is, in principle it can now be done.

The most basic questions of genetics and biology thus have been answered within a decade. In the process of doing so, there has been a marriage between biochemistry and genetics from which has issued a new generation of investigators well acquainted with both parental disciplines. They work in the faith that every problem of biology will sooner or later be solved in molecular terms by them or their descendants. They expect to proceed by the same concentrated and simplified mass effort that has already been shown to be so rewarding, going first to the next more complicated set of problems and then on and on to successively greater levels of complexity until they achieve complete molecularization of the whole of biology.

This is, I believe, what most of the brave, new molecular geneticists would tell you is the chief significance of the new genetics for the whole of biology. They say: "The old biology is sterile, dying, or dead. It is a waste of time. Only the molecular approach is fruitful. And the only sensible way is to proceed from the bottom up, tackling successively increasing orders of complexity as each underlying level has been reduced to molecular terms."

These molecular biologists have shown what they can do. Their views are not lightly to be dismissed. Yet, there are other biologists who take a diametrically opposed view. All molecular genetics and molecular biology is held by them to be a gross and fallacious oversimplification. They maintain that the essence of biology is complexity and that the simple schemes of molecular genetics are not biology at all but chemistry. Biologists of this sort maintain that *true* biology—like taxonomy, ecology, and evolution—will never be reduced to molecular terms. They point out that the idea of *Reductionism*—i.e., reduction of biology to physics and chemistry—is more than a century old and, in principle, no closer to achievement than it ever was.

What does our review imply about this conflict of views? The great success story of the new genetics in its approach from the bottom up, after a long prior history of failure by classical genetics to penetrate fully to the molecular level by starting at the other end, certainly forces us at least to say: "Well, let's see how much further this new approach can go." Twenty years ago, the enormous successes of today would have been judged impossible or tremendously remote; the problems of the gene, mutation, and gene action were then considered to be typically biological and complex. The successful solution of these problems in molecular terms came and came fast. Further progression upwards may also come.

Nevertheless, admitting all this is not the same as prophesying the sterility, doom, and demise of the old-fashioned, purely biological approach to biological problems. Our review of classical genetics showed how that biological approach yielded the knowledge without which the new molecular geneticists would have been unable to see the problems or to know how to start attacking them. In the light of these facts, there can be no doubt that *both* the biological and biochemical approaches have been indispensable.

The question before us is whether the same interrelation will repeat itself. I believe it will. The biological approach at each level will have the task of finding the most suitable materials, of discovering the phenomena to be attacked at the molecular level, and of inferring by indirection—in so far as it can—the nature of the mechanisms and the problems. Then, and only then, can the molecular biologists proceed to a deeper attack.

Actually, the biologist has already, to some extent, done this for the next domains to be attacked at the molecular level—the domains of the regulation of gene activity and of cellular differentiation. And much still remains to do in this domain. For example, in the past few years my coworkers and I have attempted to obtain a decisive answer, by purely biological methods, to a molecular question: whether certain hereditary, morphological differentiations of the cell are solely determined by properties of the products of gene action. These biological methods have permitted us to demonstrate that another factor is involved: namely, the *pre-existing* structure of the cell. Where the products of genic activity go and how they become integrated into the structural and functional features of the cell are decisively determined by the pre-existing structure and organization of the cell, particularly of its outer thin layer or cortex. These visible cortical differentiations and their experimentally produced variations are perpetuated in the absence of genic or nuclear differences in kind or in activity. To use the current cliché, a new dimension—pre-existing cell structure—has been added to the picture of cellular heredity and differentiation, one which would not be discovered by working exclusively with *in vitro*[3] systems or with a molecular approach.

History thus appears to be repeating itself and probably will go on doing so. In the long run, it seems that the progress of biology would be cut off and the stream of advances of molecular biology would dry up unless we assure the lively continuation of both the biological and the biochemical approaches. I hope this heartens the many bright youngsters whose real joy lies in research at the biological level, but who feel terrible pressures to forsake it in order to survive academically in this molecular age. There is no need to hearten the biochemically minded. But I hope my portrayal of the past and present and my judgment of the future of biology will give at least some of them not just tolerance of "old-fashioned" biological methodology but well-warranted appreciation of the indispensable role it has played, is playing, and will probably have to continue to play in order to provide them with the materials, problems, and inferences they need to keep going.

So much for the implications of the new genetics for the future of biology. What are its implications for the future of man? As I hinted at the start, there

[3] That is, outside a living body, as in a test tube or other artificial environment.

is division of judgment on this topic among biologists. Some believe that recent advances have brought us close to the day when man could use his knowledge of molecular genetics to remake man's hereditary constitution according to his wishes. Others do not agree that such knowledge is just around the corner or is likely to come as far ahead as we can now foresee. There are two ways of approaching the question. One is general and argues from the well-attested universality of genetic phenomena that whatever we know about fundamentals and can do with bacteria and viruses, we shall soon know and be able to do with man. The other approach is specific and argues that we are now directly and rapidly acquiring the needed molecular and genetic information about man himself.

There are only three known methods of making new hereditary constitutions. One is to crossbreed diverse types and select among the progeny the combinations of hereditary traits that we wish to establish. That is the method of classical genetics and I have already referred to its possibilities. In any case, it is not this but the other two methods, those made possible by the new genetics, which are under dispute. One of these is the method of directed mutations. The new genetics has gone far beyond the old, purely random induction of mutations. It employs chemicals that act specifically on the component units of DNA in known, suspected, or discoverable ways. And each of several chemicals acts on different subunits and in different ways. Such action is clearly directive. Will it then become possible—as it certainly is not yet—to produce exactly the changes one wishes in the genes of an organism or in any one gene? The great difficulty in precise control here is that there are so few different kinds of subunits in all genes—just four main kinds. Each of the presently known directive chemical mutagens acts on one or two of these units. And roughly ¼ to ¾ of the thousand-odd subunits in each cistron or gene are exactly the same. So the degree of directiveness is not very great. But neither is it as weak as I have portrayed. The same chemical agent does not appear to be effective or equally effective in mutating all of the subunits of the same kind. This might be due to what is characterized as a neighbor-effect—some neighboring groups of subunits making a given kind of subunit more or less vulnerable to change by the mutagenic agent. Nevertheless, all chemical mutagens now known can mutate any one of very many sites in the same gene, probably in every gene.

Equally important is the fact that any one particular mutation occurs in very few treated cells. The process is inefficient. This is a very long way from being able to control the making of desired genetic constitutions in each treated cell. Possibly by building mutagens into molecules of much larger size, which fit only a very long stretch of a particular DNA molecule, one might produce specific action, at the desired site, on one and only one kind of gene. Thus far, nothing remotely approaching this has been accomplished or, so far as I am aware, either tried or designed. There is no immediate or clearly foreseeable way to control efficiently the hereditary constitution of even bacteria or viruses by directed mutations.

The third and last of the known methods of changing the hereditary constitution of bacteria is by a sort of infection with the gene one wants to incorporate into the hereditary constitution. (This is not known in the case of viruses.) This can be done in two ways. One is to isolate the DNA from a bacterium containing the desired gene and then exposing to this the bacteria in which the change is desired. Some of the exposed bacteria take up the gene and some of their descendants transmit this in place of the corresponding gene which their ancestors had. This is known as bacterial transformation. It was, in fact, this phenomenon which proved that genes are DNA.

The same result can be achieved by letting a virus attack a cell, destroy it, and then infect another cell but not destroy it. In the process, the virus sometimes carried with it one (or a few) genes of the first victim bacterium and introduces it (or them) into the second recipient bacterium, where it may replace the corresponding gene of the recipient. This is known as transduction.

Transformation is always random. *Any* gene of the bacterium that provides transforming DNA may be acquired by the bacteria exposed to it. It is also incompletely efficient. Only a fraction of the exposed cells and their descendants acquire any particular gene. Many attempts have been made to fractionate the total DNA as a step towards isolating single genes for transformation, with little success. Similar limitations usually attach to the use of viruses in the transduction method. However, not all viruses transduce genes randomly. One well-studied virus transmits only genes from a small, localized part of the bacterial gene string. It is conceivable, but not very likely, one might find a battery of carrier viruses, each of which was a specific carrier for a different one of the many genes in the normal set. If that should ever be achieved, the present low efficiency of transfer would still remain to be overcome.

The prevalent failure to achieve specificity in bacterial transformation conceivably could be overcome. Bautz and Hall recently isolated the RNA messenger of one viral gene, the one that Benzer's group has studied so exhaustively. This magnificent achievement suggests that their method might be reversed in such a way as to permit isolation of the gene itself. That would, you may think, mean easy sailing to generalize the result. However, this gene happens to be nonessential for the virus, under certain conditions, and the methodology is built upon that fact. Unfortunately, perhaps, almost every gene is indispensable for the cell that carries it. Therefore, without modification, this lovely method cannot be generalized. Some other way must be sought, if single known viral or bacterial (or other) genes are generally to be made available as transformants.

In sum, present knowledge of bacteria and viruses still gives no secure basis for concluding that directive control of their genetic constitutions is in prospect. And, if it were possible to conclude otherwise, we would still be left with the important fact that the sorts of controls now imaginable are inefficient. This means that of a large number of treated cells, few would respond to the control methods. This is no problem to viral and bacterial geneticists, who routinely use highly selective methods and ruthlessly throw away the million failures and keep the one success. But such wastage is out of the question with man. I, therefore, cannot yet visualize for the foreseeable future any important directive control of the hereditary constitution of man based upon carryover of our new knowledge of viruses and bacteria. Of course, some will say: "But 10 or 20 years ago you could not have foreseen what we now know." True enough. We can't foresee what the next decade or two will bring forth in the control of heredity in viruses and bacteria. That's just my point! The knowledge is not in hand yet! Even if we have confidence that such control eventually will come, there is no present basis for estimating whether it would come tomorrow, a century hence, or in the even more remote future.

What then about the second approach, that the new genetics of man himself poses the possibility of soon controlling his hereditary constitution? Let's see what we now know or are in process of discovering. As I mentioned at the start, the cells of man are now being handled and studied like bacteria, or more correctly, like Protozoa. Moreover, some of the main, new biochemical, genetic knowledge is on substances found in man, such as hemoglobin; and human chromosome study is proving highly feasible, popular, and fruitful. Human genetics is indeed at the focus of attention as never before.

Among the new discoveries are evidences that the DNA-RNA-protein story of microorganisms is essentially the same in man. This gives further foundation to the expectation of being able to extrapolate in principle from microbes to man. However, there are thus far, to my knowledge, no successes in extending transduction to mammals and very few claims for success in extending transformation. It would be premature to judge the validity or potential generalizability of these claims. One great difficulty in judging such claims is the present lack of reliable methods for distinguishing among several known kinds of changes in heredity when the cells under study, such as human cells in culture, cannot be analysed by crossbreeding. To be sure, there is some slight evidence that cell and nuclear fusions occur rarely and this raises the hope that a satisfactory substitute for ordinary mating may occur in human cell cultures.

It is also well known that human cells in culture occasionally show losses of one or more chromosomes. If nuclei of different cells fused and then loss of a chromosome occurred occasionally among descendants, the techniques of genetic analysis for sexless fungi, brilliantly developed by Pontecervo and his school, would be available for the study of the genetics of sexless human cells. Unfortunately, up to the present, in spite of intensive search, this sort of work hasn't made substantial progress.

In fact, the genetic analysis of mammalian cell cultures has, as yet, hardly gotten off the ground. There are several reasons. First, unlike bacteria and viruses, human cells are diploid—they contain two genes of each kind. So, dominance and recessiveness complicate the picture. Consider just the first step: selecting mutations. If mutation occurs once in a hundred thousand genes—a reasonably high rate—then both genes of a pair would mutate in the same cell only once in 10 billion cells. And, since nearly all simple gene mutations from normal are recessive, usually both genes of the kind would have to mutate in order to detect the mutant cell. That would be pushing it a bit even for viruses or bacteria; for the enormously larger human cells, it is on the verge of impracticality. Although there have been a number of claims for finding and selecting genic mutations in human cell culture, alternative explanations of the observations are difficult to exclude at present. For, what seems to be a genic mutation may, in reality, be a very different phenomenon, discovered many years ago in our work on Protozoa and subsequently found by others in bacteria. A given set of genes may not always act the same. Certain external conditions activate or repress the functioning of certain genes, and, once done, cells with the active and inactive genes may reproduce side by side, true to their different phenotypes. In the absence of breeding analysis, such hereditary changes behave like gene mutations; but breeding analysis shows them to be merely persistent differences in gene *action*. Nearly all, if not all, of the persistent hereditary changes reported for human cell cultures could be of this kind. They cannot be accepted as gene mutations—especially if they occur with high frequencies—so long as this alternative is not excluded.

Finally, there is the associated fact that in cell culture most of the known genic actions that specifically characterized the phenotype in the intact body cease. These genes just don't act in the isolated cell cultures. Only a few, such as those determining galactosemia and acatalasia, have proved workable in cell culture.

There is, thus, nothing yet in our knowledge of the new human genetics that provides a basis for seeing an early control of human hereditary constitutions. Not only are the methods devised for bacteria and viruses and the results obtained with them far from providing such a basis, but we are almost completely ignorant of the genetics of human cell cultures, and we find in these cultures several nasty, special, and as yet unsolved difficulties. There is still a long way to go. Tomorrow or 10 or 20 years hence there *may* be breakthroughs, but they cannot at present be foreseen.

I, therefore, feel forced to conclude that there is no immediate prospect of radical new powers of controlling human heredity and evolution. This may well come eventually, but it does not seem to be just around the corner. There are still vast areas of ignorance to be converted into precise knowledge before such human engineering will be possible. For the present and for the foreseeable future, sound application of genetic knowledge to the improvement of man—if it is to be done at all—will necessarily have to be restricted to the slower and less spectacular, but sound, methods based upon classical genetics.

35. A Science Writer Inquires into the Place of Scientists in Public Life. 1964

As the federal government more and more provided support for science, and as its social consequences became increasingly momentous after World War II, scientists began taking part in public affairs—both as specialists and as citizens—far more than had previously been the case. Some observers credited a scientists' lobby for the ratification of the treaty banning atmospheric nuclear testing in 1963. Not all scientists, of course, approved of leaving the laboratory, even briefly, to speak out on public issues, unless, perhaps, they spoke in a strictly technical advisory role. This selection summarizes the opinions of many Americans of the early 1960s concerning the problem of scientists and engineers who have both expert knowledge and their own private opinions.

DONALD W. COX (born 1921) is a science writer, lecturer, and educator who believes that the scientists should exercise as much power in society as possible. He attended Montclair State College and received his Ed.D. from Columbia University. From 1952 to 1962 he was associated with various space projects and with the National Aeronautics and Space Administration. Since 1965 he has been with the Philadelphia Public Schools. He is the author of several books, including *The Space Race.*

For further information see Robert Gilpin and Christopher Wright, eds., *Scientists and National Policy-Making* (New York: Columbia University Press, 1964); H. L. Nieburg, *In the Name of Science* (Chicago: Quadrangle Books, 1966); and Daniel S. Greenberg, *The Politics of Pure Science* (New York: World Publishing Company, 1967).

CAN SCIENTISTS BOTH ADVISE AND MAKE POLICY, TOO? Should Scientists Run for Political Office?*

Donald W. Cox

In September, 1959, Senator Kenneth Keating (who more recently has been President Kennedy's chief critic of our Cuban policy) found himself involved in a debate with Dr. Pascual Jordan, professor of theoretical physics at the University of Hamburg. The New York lawmaker felt that in the Western world scientists should be on tap as counselors and advisers to

* From *America's New Policy Makers: The Scientists' Rise to Power* by Donald W. Cox.

Government; but they should not try to be politicians, since their training does not equip or qualify them for the job.

After praising the scientists for educating the general public and succeeding as informants of mankind, Keating then took them to task for dabbling in politics. "When they nourish a campaign for banning or controlling atomic weapons and for stopping atomic tests," he said, "or rush into a crusade to compel the U.S. Administration to reach some peaceful sort of misunderstanding with the Soviets before it is too late, or campaign for world government as the sanest solution" without enough apparent realization of the difficulties, "they have often contributed not to the enlightenment, but to the confusion of free men everywhere." He ridiculed the way Szilard and other atomic scientists had tried to oppose the further use of nuclear weapons.

Scientists fail as politicians—according to Keating—because they do not "gauge the political, that is, the human factors of the situation" and they "overestimate the influence of reason in human affairs." He also thought that the scientists "tend to consider the elimination of future war as a problem of logic, rather than of passions, power, and truly conflicting human aims and interests." Politics is not as superficial and easy as most scientists feel, because there is no sure test, no certainty, and no objective criterion, according to the New York Senator. He further believed that "once a scientist is convinced of an idea, he often becomes incapable of changing his mind even when faced with new facts." For this reason, Keating asked the scientists to exercise self-restraint by making a clear-cut division between science and politics, by confining their political activity to providing information, and by not trying to become policy makers.

On the other side of the argument, Dr. Jordan felt that scientists *should* participate in politics. Western scientists concerned with threats to the democratic way cannot divorce their scientific efforts from political efforts to insure democracy's continued existence, he declared. Since scientists feel that scientific progress depends on an atmosphere of intellectual freedom, they cannot remain aloof from politics, because the democratic way requires participation of *all* segments of the population. If we disqualify one group, Dr. Jordan said, then we can keep others out, too; so the Government would then be run only by political experts, and democracy would cease to exist.

Jordan warned his listeners that we must not expect scientists to become "political miracle men." Given such status, scientists would approach great world problems with the conviction that they had to introduce entirely new methods into politics. Jordan recommended instead that scientists should give greater support to the political institutions of democracy. But he avoided any discussion of *how* scientists were to cooperate and carry out their new civic duties.

Jordan forgot to point out that, since the typical scientist is *more* than just another John Doe, his advisory pronouncements on controversial questions carry a weight *far* greater than those of an ordinary man. Scientists' use of public forum to air personal prejudices without giving sufficient study to the problem has caused many of our political leaders to be wary of their pronouncements. Many observers feel this breach can be mended only if scientists realize that, by acting responsibly, they will earn more respect in political circles.

The Keating-Jordan debate disclosed the gulf between the politicians and the scientists. Jordan and his fellow scientists' preocupation with "the scientific method," which, they feel, only they are capable of understanding, is the cause of much of the conflict between them and the rest of society. The scientists' attitude was well expressed by a Nobel Prize winner, Dr. Albert [Szent-Györgyi] of the Marine Biological Laboratory in Woods Hole, Massachusetts, in an article in the October 20, 1962, issue of the *Saturday Review*. He concluded: "The only way out is to apply to political questions the same humble, honest, objective approach that has characterized the development of science. Yet there is not at present one scientist in Congress. Scientists are consulted, but only on technical questions, not on basic issues of policy. Unfortunately, the electorate will not vote for scientific-minded people until it can think scientifically itself, and in this it has a long way to go."

Tragically, when the scientist does apply his brains to a pressing human problem, he usually comes up with a "scientific" solution—that is, mechanistic, tidy, and oblivious to human passions, prejudices, traditions, fear, greed . . . and political realities.

The recent emergence on the American scene of the technological monster known as "Big Science" is the most obvious result of increased Government doting on science for advice and policy help. "Big Science" is a title coined by Dr. Alvin Weinberg, director of the Oak Ridge National Laboratory, to connote the large science organizations—such as the atomic bomb laboratories and the massive, complex industrial and governmental research facilities—that have sprung up in recent years. It is a label for the *groups* of scientists who are called upon today to do the jobs that *individuals* like Edison and Steinmetz performed alone in the past. Some scientists, among them Weinberg, believe that Big Science is ruining science. He analyzed the problem recently in the authoritative journal, *Science,* with the following astute observation:

> "Since Big Science needs great public support it thrives on publicity. The inevitable result is the injection of a journalistic flavor into Big Science which is fundamentally in conflict with the scientific method. Issues of scientific or technical merit tend to get argued in the popular, not the scientific press, or in the Congressional committee room rather than in the technical society lecture hall. . . .
>
> "One sees evidence of the scientists' spending money instead of thought. This is one of the most insidious effects of large-scale support of science. . . . The line between spending money and spending thought is blurring.

"Finally, the huge growth of Big Science has greatly increased the number of scientific administrators. Where large sums of money are being spent there must be many administrators who see to it that the money is spent wisely. Just as it is easier to spend money than to spend thought, so it is easier to tell other scientists *how* and *what* to do than to do it oneself. The big scientific community tends to acquire more and more bosses . . ."

Dean Don Price of Harvard has coined a similar-sounding phrase—the Scientific Establishment—to describe Big Science. He noted recently, "The plain fact is that science has become the major establishment in the American political system; the only set of institutions for which tax funds are appropriated almost on faith." He has elevated this establishment to that of a "new priesthood allied with military power."

These scientific administrators that are multiplying faster than any other type in Washington these days are policy makers in their own right. They are not predominantly advisers, although they are still called upon from time to time to render advice to their nonscientific political superiors. Because their growing policy-making activities were seen as a threat to their departmental chieftains' authority, a warning to the scientists was handed down by a high White House official in late 1962.

Just before he triggered a crisis in the Canadian Parliament by trying to force nuclear weapons on our northern neighbor without first consulting either them or the State Department, McGeorge Bundy, President Kennedy's special assistant for national security matters, told the scientists that they should not do what he himself was to do so shortly thereafter. He pleaded with several thousand scientists on December 27, 1962, to respect their calling. "This means that he [the scientist] should carefully limit the occasions on which he speaks ex cathedra so that he is not placed in danger of losing his reputation," Bundy said.

He went on to point out how dangerous it was for scientists to take unbending positions on such acute problems as the diffusion of atomic know-how among countries such as China, and the chances of accidental explosions touching off a nuclear war. "These are very great problems indeed," said Bundy, former dean of Harvard, "and they have engaged the prayerful attention of our national leaders for years. It is dangerous for the scientist to adopt an either-or attitude on these complex problems." He then lambasted Sir Charles P. Snow for predicting two years earlier that the Chinese would have the atomic bomb within five years. This nonscientist went on record suggesting that the British scientist's flat assertion on this atomic problem was just "not so" because it "touched on the political realm," a sacred territory into which he should not have trespassed.

Yet Bundy saw no contradiction between this advice and his own offhand policy making less than two months later, when he spoke publicly about American "gifts" of nuclear armed missiles to Canada.

Sir Charles Snow had previously pointed out in one of his Godkin Lectures at Harvard the *hazards* of a head of state's getting his scientific advice principally from *one* individual, especially on matters not open for public review and debate. "Both the policy maker and the science adviser have a responsibility to minimize the danger of biased or highly personalized advice on secret matters, especially in areas of science and technology that can be fateful to the security of a nation," the controversial British scientist declared.

The "objective" science expert will inevitably be called upon to make political judgments, along with scientific ones, in advising on confidential matters of state. Robert Gilpin, in his book *American Scientists and Nuclear Weapons Policy* (Princeton, 1962), has divided this confused middle ground of scientific advice in Government into the realm of *science,* or WHAT IS, and the realm of *policy,* or WHAT IS TO BE DONE. Most scientists advising on nuclear weapons have to combine scientific and political judgment. Therefore, there is no longer any "pure" scientific advice. Although the typical scientist tries to remain coldly objective at all times, it is difficult for him to free himself from his nontechnical assumptions and his moral values. "The values of a scientist are an integral part of his research," wrote Gilpin. "These values affect the problems he selects for study, the facts which he believes are relevant and the implications or hypotheses he draws from the facts."

Gilpin concluded that "both scientists and political leaders have acted as if it were possible to make a clear delineation between the political and the technical realms. This simplistic view of the scientist's role as adviser has created expectations which the scientist cannot fulfill. He is expected by political leadership, fellow scientists, and his own conscience to render only objective technical advice. As a consequence . . . scientists have been assigned many apparently "technical" tasks whose performance has required a political skill far beyond their competence; scientists and political leaders have failed to realize the nature of the nontechnical assumptions underlying the scientists' advice; and scientists have charged one another with intellectual dishonesty when they have disagreed strongly with advice which has been given." . . .

The one flaw in Gilpin's thesis is his statement that "American scientists, at least in the area of national policy towards nuclear weapons, have become full partners with the politicians, administrators, and military officers in the formulation of policy." He failed to point out *who* picks the scientists called upon to participate in policy making and *when* they are listened to by Government administrators. We need some way of judging their advice earlier in the game, so that we do not have to wait for their memoirs to discover the reasons for their stands.

Dr. Killian,[1] on the other hand, believes that in certain kinds of technical questions scientists will respond with objectivity and integrity. "In these

[1] James R. Killian, in 1957 appointed special assistant to the President and chairman of the President's Scientific Advisory Committee (PSAC). He had been president of MIT.

situations," says Killian, "the adviser must state the facts, including their differences, and list the alternative interpretations of which the policy maker must be aware. As science gets more and more involved in controversial policy decisions, both the policy maker and the body politic must understand that scientific method and analysis do not always yield a single, incontrovertible answer. There are, of course, many questions to which the scientist can give positive answers, the correctness of which can be demonstrated, but he sometimes has to resort to 'dusty' answers when the policy maker understandably is 'hot for certainty.' "

Killian feels that checks and balances to insure the best scientific advice can be attained through proper organization. He feels that the direct-access door to the President, which his Science Advisory Committee had at all times, opens the way to a route the committee can use to bring in alternative opinions and independent positions whenever its members disagree with the Special Assistant for Science and Technology. As a board of consultants to the special assistant, the PSAC can also give the President the benefit of varied points of view, and, Killian says, "if he is wise, he can use it to test his judgment and insure against his own prejudices" becoming dominant. "This mechanism has worked well under the last two Presidents, since both the Science Adviser and PSAC realize that their effectiveness rests on their ability to maintain the confidence of both the President and the scientific community."

Dr. James Warwick of the High Altitude Observatory in Boulder, Colorado, goes beyond Killian. He does not believe that there is "any distinction in a real sense between scientific and political advice. Scientists don't consider that the world of science stops at the skin-level of the two billion human beings on the surface of the globe; and, equally, political officials do not consider that their domain fails to extend to the atmosphere and beyond. The question . . . is not *what* we discuss but *how* we discuss it."

"The problem," he points out, "is that over a vast range of human affairs, we do not, nor are we in the measurable future likely to have, nor perhaps is it even desirable that we ever have, predictability. But, in order for the biological organisms on this planet to survive in this scientific world, decisions must be taken that are essentially nonscientifically based just in order to get on with the affairs of the world."

He feels there is a great danger today that these essentially political decisions will be turned around and justified to the world on a scientific basis. "The dividing line between science and politics seems to be more and more crossed," he says. Warwick admits that a politician might underestimate his own powers of assimilation of the scientific facts about the real world around him. Since the real world is so important to the politician, Warwick hopes that the close relations already established between senior scientists and legislative and executive department leaders will continue to grow. "One could even hope," he adds, "that this kind of arrangement would become normal at the state and local legislative level."

Not all scientists share Warwick's views on the issue of advice *vs.* policy. Many take the more conservative view expressed by Dr. Harold Brown, director of defense research and engineering, who has stressed that, while it is important for scientists and engineers to give technical advice to those responsible for decisions, "technical advice is only one factor in the decision."

"Military advice," he wrote, "necessary to given information on the importance of new weapons that might come from additional testing, and political advice is necessary on the impact on international affairs of a decision to do nuclear testing. The advice on each of these factors—technical, military and political—must come from experts in the field, and the scientists and engineers have the responsibility for only *one* of them."

Brown conceded, however, that "a scientist or engineer in government, being an informed person, may have opinions on matters beyond his technical sphere of responsibility." He saw no objection to the scientist expressing such opinions on nontechnical matters so long as the views are clearly identified as opinions of an informed nonexpert. But these opinions often spill over into the gray area between advice and policy. Dr. Jerome Wiesner, the President's assistant for science and technology, has pointed out that the scientist, seeking *the* solution to a political problem is usually not satisfied with short-step accommodations, such as those that comprise the bulk of political actions. Wiesner says scientists often impatiently strive to solve our problems with larger measures, despite their high degree of risk.

Another prominent scientist, Dr. Harrison Brown, recently commented on one of the reasons for the impatience shown by so many scientists in Government. "One of the more serious difficulties in the past," he said, "has been that, with few exceptions, department heads have not been research-minded, with the result that research and development have been relegated to the lower regions of the priority list. The advice of top scientists in various departments and agencies has, all too often, gone unheeded. The lack of attention given to scientific-technological problems by political appointees has often resulted in the perpetuation of provincialism in a number of the agencies, as well as in the creation of atmospheres which have seriously impeded proper intellectual and physical growth."

Brown sees some hope of having the National Academy of Sciences-National Research Council play a mediator-catalyst role when scientific-political policy matters reach a temporary stalemate. The NRC is a quasi-governmental organization which, during the last few years, has contributed substantially in a variety of ways to the formulation of Government programs. Being outside the Government, its members can theoretically examine problems in a more detached way than can governmental persons themselves. Brown hopes the Academy-Research Council will become the major source of scientific-technical advice to Congress. In the past, Congressional Committees have often obtained scientific advice in a rather haphazard way, and the persons giving the advice have not always been as competent as one might wish.

The Academy-Research Council structure needs to be "strengthened to the point where government agencies and Congressional committees alike, in need of advice, would turn confidently to the organization for recommendations as to how that advice might best be obtained," Brown says. "Congress is going to be concerned more and more during the years ahead with legislation involving scientific and technical matters. The establishment of a strong Academy-Congressional bond appears to make a great deal of sense, provided the Academy-Research Council strengthens itself to the point where it can handle the increased load in a systematic way."

The most noted exemplar of a country where the national academy of science acts as the official advisory and policy coordinating body to the Government is the U.S.S.R. In the spring of 1961, responsibility for coordinating Russia's total research and development effort was established at the highest level in the Soviet government by the creation of the State Committee for the Coordination of Research and Development, with a deputy chairman of the Council of Ministers as its head. The membership of this central coordinating committee includes the president of the Academy, the minister of higher and secondary specialized education, and representatives of each of the important state committees involved with research and development work. The Academy retains its role of exercising scientific and methodological leadership, but its work is focused primarily on the most important long-run problems of science, and it is not concerned with the myriad Soviet technological institutes.

Scientists are now pre-eminent in the Kremlin hierarchy, with 8 out of 15 members of the Communist Party presidium having a technical background. Over half of the party Secretariat and the Central Committee of the Communist Party are made up of leading Soviet scientists and engineers. In 1962 a new Central Steering Committee of the Council of Ministers of the U.S.S.R. was established to coordinate scientific research work and to strengthen the position of science in the Soviet government. This move was in line with Russia's aim to surpass the U.S. by 1980—not just economically, but in making better use of scientific and technological advances.

We in America have a long way to go if our scientists are ever to achieve a comparable status by holding down an equal percentage of key seats in our Government. But the American people may never wish to achieve such a heavily weighted group of scientist-managers in our pluralistic democratic society.

When Khrushchev sacked the vituperative Valerian Zorin from his post as chief U.N. negotiator during the last week of October, 1962, he picked a top Soviet scientist, Vassily Kuznetsov, as a replacement. First Deputy Foreign Minister Kuznetsov, who speaks flawless English, is by trade an engineer. In 1941, he received the Stalin Prize for inventing a high-quality alloy steel. In diplomatic circles Kuznetsov has made an excellent reputation as a flexible conciliator with a sense of humor—a characteristic that so many diplomats and scientists do not possess.

The typical American scientist-diplomat is not as well prepared as Kuznetsov and his Soviet counterparts. Too many of the Americans are in a hurry to return to their former posts; their more professional Soviet colleagues realize their careers are greatly furthered by their conscientiousness and their success in putting across a point for the U.S.S.R. Soviet scientists are forever coming up with new subjects to advance their Government's cause in international negotiations, while most Americans restrict themselves to the topics on the agenda. It is not surprising that in such an unequal contest, the American is often bested. What we need is more flexibility in our scientific-diplomatic negotiations—a flexibility such as the British have practiced so well over the years.

The confusion and controversial questions still remain unresolved since the days of the dispute over the development of our H-bomb. What does the scientist do when a President or high Government official decides *not* to take the advice of his science advisers? Does he go to Congress and the people with his case, if he feels that the national interest is in danger? Under our Constitutional system there is no simple answer to this dilemma: Should the adviser put loyalty to his country above loyalty to the man he advises?

If a scientist's advice continues to prove unpopular to higher authorities, the risk of criticism and downfall of the adviser might result. High Government scientists can suffer the same fate as Harold Stassen and Lewis Strauss,[2] should they become involved in a controversy from which there is no way out. Oppenheimer's[3] advice against pushing the development of the H-bomb was essentially the reason for his fall from power. As more of our top scientists assume dual policy and advisory posts in the Government, they will face the need to adjust to the dual role of policy maker and adviser in Government.

On March 21, 1963, the Kennedy Administration picked Gerard F. Tape, a 47-year-old physicist and president of the Associated Industries, Inc., to be the new scientist member of the AEC. The appointment of another scientist to the AEC was in line with a policy decision of the administration to bring more scientific influence into the management of the agency. The White House had already broken with past tradition by bringing in a scientist, Dr. Glenn T. Seaborg, to head the AEC.

Capitol Hill did not greet the new scientist domination for the commission with enthusiasm. There was some grumbling among members of the Joint Atomic Energy Committee that scientists were not necessarily best

[2] Lewis Strauss as head of the Atomic Energy Commission behaved in such a way that President Eisenhower was unable to appoint him Secretary of Commerce because of opposition from scientists and others.

[3] J. Robert Oppenheimer, a distinguished physicist who suffered political persecution in the 1950s. There exists an enormous literature on the Oppenheimer case and the circumstances alluded to here by Cox.

qualified to run an agency confronted with management and policy problems transcending strictly technical issues. The administration, however, felt that this experiment had proved to be a success and should be continued. White House officials also believed that a good balance had been struck in the commission with the present membership of two scientists, two lawyers, and one businessman.

Tape happened to be well qualified within the scientific community, since he possessed the somewhat rare combination of a broad background in scientific research as well as considerable administrative experience. He had been deputy director of the AEC's Brookhaven National Laboratory on Long Island (New York) from 1951 to 1961. In 1962, he had been made president of the nonprofit corporation known as the Associated Universities, set up after World War II by nine Eastern universities to manage the Brookhaven Laboratory for the AEC. Associated Universities subsequently took over the operation of the National Radio Astronomy Observatory in West Virginia for the National Science Foundation.

Tape replaced on the AEC Dr. Leland J. Haworth, a 58-year-old physicist, who, in turn, subsequently replaced the retiring 71-year-old Dr. Alan Waterman as director of the NSF. Haworth—a Republican—who had served for two years as a controversial member of the AEC, made his mark as a scientist-administrator of the AEC's Brookhaven National Laboratory from 1948 to 1961. In his two short years as an AEC commissioner, Haworth developed—somewhat to the consternation of his colleagues—into a powerful political figure who combined his scientific prestige and political acumen to fashion a sphere of influence that stretched from the White House into the commission staff. At times there were jealous grumblings from other commissioners that his influence was too great and that his perspective was too narrowly oriented toward science.

Haworth took his oath of office as the new head of the NSF on June 28, 1963, at a difficult juncture in the rapidly expanding, evolving agency. For the Foundation had suddenly found itself not only responsible for basic research, but also with broadened responsibilities—including the support of scientific education. Its current budget had swelled to a high of $589 million, which was quite a jump from the meager $4.5 million budget with which the NSF began its life in 1951.

His most pressing problem was to disentangle the mismanagement of the $60 million Project *Mohole,* an ambitious engineering endeavor to drill a hole through the earth's crust to the Mohorovicic Discontinuity, a geological region generally accepted as the dividing zone between the crust and mantle. It was hoped that Haworth would carve out a more aggressive policy-making role for the NSF, which his predecessor had looked upon as being more of an advisory agency.

In Britain, there is a clear distinction between advising and policy making. The British setup avoids placing excessive responsibility on the expert, in the belief that excessive responsibility often paralyzes thought. Herman Finer, professor of political science at the University of Chicago, recently pointed out that American politicians have created an atmosphere hostile to the nonpartisan expert. "American public administration," said Finer, "has refused to make the distinction between career administrators and scientists. In American government, scientists get to high posts in departmental hierachies as Heads of Bureaus. This affects the scientist's impartiality." This situation does not exist in Britain where there is a protective shield supplied by the British administrative class.

To prevent the life of the scientist in U.S. government service from becoming unbearable, Finer feels that there is a dire need for an "intervening layer" of career servants between the scientists and the politicians, similar to that which exists in Britain, France, and Germany. He recommended that a group of these experts should be recruited from the best minds of the young graduates, who have had a science-oriented liberal education. This group could then take over the jobs of fighting for money in Congress, justifying departmental policy, briefing the political chieftains, and taking up arms in the eternal internal struggle among Government departments. In this way, we could treat our intellectual science advisers with more freedom and respect than we have in the past.

The scientists have come of age politically. While most of them have based their opinions and advice on a combination of scientific-political factors, their political acumen and wisdom are not necessarily more correct than or superior to the acumen and wisdom of nonscientists. Scientists must speak out on the public and private uses of their technological inventions. They have been almost universally silent on these issues. It is not so important that they just speak out or be asked for advice, but that they have *something worth while to say* on controversial subjects. This means an awareness of the true nature of our weapons, as well as a willingness to do something about the problem. They must learn to work with the other professional groups who hold views similar to theirs. But no intellectual grouping of the professions will be able to combat the present industry-military elite until they align themselves with social scientists, labor leaders, teachers, and farmers. Together, they can carry the weight to exert constructive influence in our society and in Congress.

The most crucial area of advice [given] by qualified scientists to the Federal government is at the point where Congress is about to vote on appropriations. Philip Abelson, editor of *Science,* testified on this sensitive subject before the Senate Committee on Aeronautical and Space Sciences in mid-1963. "At that point," Abelson said, "the Congress gets advice that is almost unanimously of a self-interest kind. There has not been a mechanism for some kind of devil's advocate.

"The Bureau of the Budget tries to function in such a capacity," he went on to say, "but the people in the Bureau of the Budget are not sufficiently independent, are not sufficiently informed. I think it would be very useful if, in these areas involving great appropriations, a group of first-class scientists could be asked to point out what are the negative aspects of scientific projects that are glossed over by the advocates [of these projects].

"Some of my scientist friends are great enthusiasts. They feel there is no end to the amount of money that can be usefully appropriated for science. So there is no restraint on them and they will advocate, and advocate, enthusiastically and forcefully.

"Somehow or other, the role of the devil's advocate should not only be established but the devil's advocate should be given a little bit of due. He should be available to the legislative branch and both the majority and minority party should have some say in calling upon that kind of talent."

In line with this scientist-editor's suggestion, several Congressmen, led by Republican Representative Abner W. Sibal of Connecticut, have recommended that each House of Congress retain not just one, but at least three scientists on its payroll to aid in decision making on these vital issues affecting our future. Ultimately, Congressmen will have their own scientific advisers to help them assess the differences between advice and policy, and what is right for the nation.

The scientist's rise in politics could become the most important single development in the changing political patterns wrought by the technological impact of the Space Age. Dr. Killian recently urged his colleagues to run for Congress. He noted that many were already in various state legislatures. Killian suggested that a technologically oriented society must have scientists "in the public arena if it is to deal wisely with all the great policy matters arising out of science and technology."

Sidney Hook, chairman of the philosophy department at New York University and a noted social scientist, has serious reservations concerning Killian's call to his scientific brethren to help save the world from calamity. "To a scientist," said Hook, "the arena of history and political affairs is bewildering, because he soon discovers that telling the truth is one of the best ways of deceiving his opponents."

As a prime example of the political innocence of the natural scientific mind, Hook pointed to the belief held by Nobel physicist Dr. I. I. Rabi concerning possible Russian evasion of any political agreement over a nuclear test ban. In an August, 1960, *Atlantic* article, Rabi expressed his feeling that we shouldn't mention our fear of this probable course of cold war events because this would be an admission that we thought the Russians would be cheating on us. (The Soviets did just that—without signing an agreement—when they resumed nuclear testing in the atmosphere two months later.)

In late 1961, Vannevar Bush admonished his fellow scientists to respect "those individuals who are masters of the art of operating in the confused area of the American political scene." He felt that if our scientists are to have their "full influence in the days to come, many of them will indeed need to learn to practice this difficult art" of politics.

A need for re-education and development of political sophistication has plagued our scientists since the mushroom cloud blossomed over Hiroshima in 1945. The problem, however, is twofold. Our politicians also must re-educate themselves in science to be able to properly frame their questions to the scientists who appear before the various Congressional committees. More importantly, perhaps, politicians must realize that scientists are not infallible.

During 1948—the year preceding the first Soviet atomic explosion—Oppenheimer and his fellow scientists continually assured the administration that the Russians could not achieve an atomic bomb until far in the future. They were wrong. At the same time, Vannevar Bush staked his reputation on the declaration that an intercontinental ballistic missile was "impossible" to build. He, too, was wrong.

As the scientists have not always been correct in their *technical* assessments, so likewise they have also faltered in their *social* and *political* judgments concerning the problems which scientific advances pose. As long as they remained in an advisory role, the scientists usually did not stir up any great controversy. It was only when they meandered into the whirlpool of policy making that they found themselves in troubled waters. AEC Chairman Glenn Seaborg has praised the development of advisory relationships between scientists and Government. "Scientists are asked for advice; scientists offer advice," he said bluntly. "But the permeation of science into the whole fabric of our society requires that science be utilized other than as a reference book —other than a Noah Webster or a Dr. Spock or a Dr. Gesell. It means that science must become a *general* and *participating partner* in government, not merely a limited or advisory partner. Men who *know* science and technology —whether or not they are scientists or engineers—must join in creating our *laws,* in forming our social order and in establishing our national policy."

Seaborg further pointed out that, during our nation's founding and early development, we drafted lawyers into all three branches of Government service. "We still need the lawyer," he said, "but we must extend the draft to a new class. We must conscript science and technology. We can no longer afford to exempt the scientists and the engineers. We must reclassify him. We must convince him that there is as much challenge and excitement in the laboratories of government as in the laboratories of science."

Dr. Wallace Sayre, professor of public administration at Columbia University, coined a precise definition of the problem of technological experts in Government when he wrote: "Scientists influential in the creation, maintenance, and modification of American science policy are scientists in politics. Their spokesmen need not always be occupants of public office or party officials, but, whoever they are, they cannot escape politics and remain leaders of science. They will have to share all the hard knocks of the political process and the difficulties imposed upon those who attempt to shape public policy. They will sooner or later have to recruit allies from organized groups of nonscientists, since they will soon discover that they cannot exert a unilateral dominance in the formulation of science policy.

"As they move into alliances with other similar interest groups on the peace issue, etc., they will discover that a price will have to be paid in the form of a mutually acceptable compromise on policies and priorities." Sayre concluded with this warning: "It is time that the scientists who have been sternly lecturing the nonscientists about their need to understand science, in turn realize that they themselves have an obligation to understand politics."

The scientists' efforts to present their case to the public have too often been marred by their limited interests. Although such groups as the NAS, AAAS, FAS,[4] and other[s] have contributed on many occasions to public enlightenment, by failing to go far enough in taking strong, sound stands on controversial issues, they have not fulfilled their responsibilities to the community at large.

Scientists who wish to influence science policy in a democratic society must enter the political arena along with other lobbying groups. They cannot remain in an ivory tower above the turmoil. No single approach to their problems of plunging into politics will suffice for more than a very small segment of the scientific community. The activities of the various *ad hoc* scientific groups which arose to cope with separate problems like fallout and nuclear testing have usually been unsustained. Through these sporadic social protests, many scientists have made their voices heard for the first time beyond the confines of their laboratories.

In a December, 1961, address to a Phi Beta Kappa assembly, Dr. Harrison Brown said more scientists should become active in partisan party politics. "I wish that many scientists would select the political party which comes closest to their beliefs," he said, "and then work actively within that party. I personally took this step many years ago and have found it to be a most rewarding experience.

"It seems to me that if democracy is to thrive, it is important that our system of political parties be made to function as well as possible. This means that thinking people should become active in the political process, over and above the simple activity of voting. I have seen scientists contribute substantially to political programs by providing information which is helpful in the drafting of campaign positions and speeches. Persons who are candidates for office usually are appreciative of this help, as are the workers around them. Also, a two-way educational process takes place which is healthy. The [scientist] learns something about the complexities of the political process."

[4] That is, the National Academy of Sciences, the American Association for the Advancement of Science, and the Federation of Atomic Scientists.

In 1960, Vice-President Nixon was advised to make the main basis of his Republican presidential campaign the assertion that the full development of the United States depended on a national program of scientific and technological exploitation, to which he would devote his full energies if elected. The Republican platform stressed the importance of science to the nation. Unfortunately, the Republicans failed to put any scientists on their Committee on the Impact of Science and Technology. Their advisory group was composed of a medical doctor, two businessmen, a college administrator, two architects, and a radio broadcaster.

Back in 1958, a group of 20 scientists agreed to form an Advisory Committee on Science and Technology within the Advisory Council of the Democratic Party. The committee produced two major documents which were subsequently widely used by Democratic candidates and by party workers as background material.

One of the studies involved the problems of disarmament. The committee came forward with a specific proposal for the creation of a special agency within the executive branch to undertake the research and development that would be necessary if we were ever successfully to approach negotiations in this area.

The Democratic Advisory Committee on Science and Technology made a formal report, *Defense, Disarmament and Survival,* on December 27, 1959. The scientist-authors were not as worried over the decline of American technology as they were over the rapid upsurge of the U.S.S.R. and Red China.

The committee recommended five steps that could be taken in the military area to create a quasi-stable situation until an effective disarmament agreement could be made. These steps were: (1) Reduction of the vulnerability of our own retaliatory system; (2) Reduction of the vulnerability of our defensive system; (3) Improvement of our detection system to reduce the possibility of a false alarm; (4) Introduction of further safeguards against surprise attack; (5) A major effort, with the cooperation of the U.S.S.R., to slow down the rate of spread of nuclear and missile technology.

The advisory scientists also recommended that disarmament be made a major national goal. They suggested the immediate establishment of a National Peace Agency—as an independent organization—within the executive branch, with a starting budget of $1 billion.

In the judgment of these scientists, the first step in the process of eliminating the possibility of a surprise attack would be a bilateral U.S.-U.S.S.R. agreement, setting limits to nuclear testing and establishing controls on which later abolishment of testing could be used. The report stressed that Red China should be involved in these negotiations—even though the U.S. had no official relations with that power.

Defense, Disarmament and Survival, approved by Senators Humphrey, Kennedy, and Symington, Governor Adlai Stevenson, former Secretary of the Air Force Thomas K. Finletter, and former President Harry Truman, became a plank in the 1960 Democratic platform. Defense and disarmament matters were brought up often during the campaign. Following President Kennedy's inauguration, the proposed peace agency legislation was introduced in Congress. After considerable debate and some modification, the United States Arms Control and Disarmament Agency was formally created.

After this first legislative triumph of the young Kennedy Administration, Dr. Harrison Brown concluded: "I would like to stress that had scientists not involved themselves in politics, had this group of men elevated their noses and stated that politics was beneath their dignity—or had they stated simply that they were too busy—we probably would not have a Disarmament Agency today. It is, I believe, a beautiful example of how citizens of good will and knowledge can function effectively and with dignity in the democratic process."

This episode also marked the first example in American politics of a group of scientists working together as a team to achieve a mutually agreed upon nonscientific goal. Previously, scientists had worked jointly only on predominantly scientific projects.

Glenn Seaborg, the 51-year-old AEC chairman, sees scientists as lawmakers "if we can find scientists who are willing to have a try at this mode of life and if we can find the scientists who have the necessary combination of what I call scientific and political capabilities." He says that he already knows several people who would fit his description, although he has politely hesitated to name them outright. He agreed that it "was important to convince them to run for office, because it would be very helpful" to have more people with deep training in science in Congress to serve on such committees as the Committee on Astronautics in the House and Senate and the Joint Congressional Committee on Atomic Energy.

Seaborg cautioned, however, that we should not emulate the Soviets "who have virtually enthroned the engineer and the scientist in the seats of power so that they number a *majority* of the members of the Presidium, the Council of Ministers, and the Party Secretariat. But we must bring the engineer and the scientist across the threshold and into the chambers where our national policy is created—not merely allow them to stand in the corridors where it is discussed. We can no longer afford to insulate the body politic from the contributions, influence, and impact of such a significant segment" of our society. He assumed that the scientists who take the plunge into Government service could acquire a sufficient degree of proficiency for success in the science of politics.

Dr. Herbert York, chancellor of the University of California at San Diego and former chief scientist for the Advanced Research Projects Agency and the Pentagon, agrees wholeheartedly with Seaborg that "a small fraction of the national Congress should consist of scientists and engineers." But he feels that scientists will probably find it hard to "persuade the electorate of any particular constituency that their particular problems, as opposed to the nation's general problems, would be better handled by a scientific representative." Most political leaders, according to York, are likely to believe "that they will best be served directly by someone who is known to be adept at something, which, for want of a better name, may simply be called "politics." They will prefer to have someone representing them who has a background in law or local political affairs."

York is also worried about the length of time it would take an elected scientist senator or representative to become effective. He believes that this period might be so long that the frustrated scientist-legislator might get so out of touch with his original calling that "he would no longer be able to do the same kind of job he would have done had he not served in Congress." York believes that this problem would not be as serious for a scientist serving in an appointed post in the executive branch, since most of the time he would be involved largely with matters of science and technology.

Dr. Harold Urey, Nobel Prize winner for the discovery of heavy water, believes that a scientist or anyone else who has a desire to do so should run for political office, but he feels that scientists "are not likely to be good vote getters. Good scientists have too much the quality of being forthright, where many times in a political situation it is necessary not to say everything that one thinks."

The scientist who chooses to run for office to be a policy maker often finds himself fighting a rather lonely battle. Unfortunately, there is no large politically active group of scientists to which a scientist might go to seek political support. Leo Szilard's rather small Council for Abolishing War, which actively supported three Democratic Senatorial peace candidates in the 1962 election, has been the lone exception. The Federation of American Scientists tries to influence legislation but does not openly support anyone running for public office.

There have been recent rumblings that American scientists should become more active politically—both as officeholders and as lobbyists in the Congressional cloakrooms—because the time left to engineer man's survival on this planet is running out. These outcries have not come from the missile and space scientists and engineers, who for the most part are conservatively rooted to cold war weapons technology, and have not as yet felt free to criticize the hand that feeds them. Rather, it has been the older atomic scientists, like Szilard, Rabi, Seaborg, and Feld, who have taken the lead in urging scientists to increase their civic activity. Rabi, who teaches physics at Columbia, said recently that he would "like to see active political clubs in our great engineering schools where the issues of the day are debated by keen minds accustomed to inquiring why, accustomed to getting down to basic questions, and also oriented toward finding new methods and new solutions to new problems and old. If such activity interfered with the hard grind to which these people are ordinarily subjected," the Nobel laureate continued, "I would be prepared to lengthen the course of study or even to narrow its professional focus somewhat in order to gain general utility of the human material involved."

Rabi went on to say that "we have to dispel the notion that a man of science who devotes part or all of his time to some public office has somehow deserted a sacred cause for a field much less worthy of his talents." Although most contemporary scientists have shunned Government posts, except in wartime, Rabi pointed out that many scientific greats have served their countries long and well. Newton filled the post of Master of the British Mint with great distinction. Benjamin Franklin served his young country in numerous political capacities, including his ambassadorship to France and his participation in the framing of the Constitution.

"We have enough knowledge, when properly applied and integrated, for men to have a much deeper understanding of themselves, and to use the knowledge for transcendant [*sic*] purposes," Rabi said. "Yet the real possessors of this knowledge are in their studies, in their laboratories, seeking more and more information, more knowledge, and living apart from their fellow men while their destiny is being shaped by people who have little realization of the power of the tools at our command."

Rabi has made an important contribution with his suggestion that *all* science students in the colleges become more oriented to the political problems of the day. Whether they eventually decide to leave the laboratory or remain within it for the rest of their professional careers, the scientists of tomorrow would at least have the opportunity for a better grounding than their predecessors in the social, historical, political, and economic impacts of the new technology. In this respect, they might take a cue from the late eighteenth-century country squire of Monticello, Virginia.

Thomas Jefferson, our first and only scientist-President (Herbert Hoover was an engineer), believed the safeguard of freedom was the education of the individual. To Jefferson, science and mathematics were basic in training the mind for rational thinking. He himself was a semiprofessional botanist, paleontologist, archaeologist, astronomer, cartographer, meteorologist, geologist, mineralogist, horticulturist, mathematician, medical doctor, architect, and zoologist. No other President since Jefferson has understood science as well or done so much for it—relatively speaking. He promoted scientific agriculture, foresaw the potentials of steam power, and enthusisatically supported small-pox vaccination. His detailed instructions to Lewis and Clark showed his grasp of the importance of obtaining practical scientific knowledge about the West.

Obviously, no twentieth-century President can ever expect to become another all-around scientific Chief Executive in the Jeffersonian tradition because the immense increase in scientific knowledge since the turn of the eighteenth century has made it physically impossible for any man to become a specialist in, at the most, more than two or three scientific disciplines. But it is important for every politician, whether his eyes are set on the State House, the White House, or Congress, to become better acquainted with the general technological changes in the major sciences if he is not to be an ignorant and ill-informed citizen.

Jefferson left us a noble legacy in this point in a letter that he wrote to DuPont de Nemours in 1816: "Enlighten the people generally, and tyranny and oppressions of body and mind will vanish like evil spirits at the dawn of day." The mission that Jefferson outlined a century and a half ago remains the supreme challenge today.

36. A Physicist Suggests that Scientific Technology Can Solve Many Social Problems. 1966

Not all the talk of the mid-twentieth century about how science could make the world a better place was vague or futuristic. Nor was such improvement of man's condition necessarily always inhibited by the backwardness of modern society, which presumably lagged far behind the technology which it had spawned. ALVIN M. WEINBERG (born 1915), director of Oak Ridge National Laboratory, in his speech accepting the 1966 University of Chicago Alumni Award pointed out that science still had not lost its potential as the source of utopia, or at least a force capable of solving social problems.

Born in Chicago, Weinberg attended the University of Chicago, completing his Ph.D. in physics there in 1939. After two years as research assistant in mathematical biophysics at his alma mater, Weinberg was drawn into the atomic bomb project, first at Chicago and then at Oak Ridge where he became research director and finally, in 1955, director. His major scientific contributions have been in the area of construction and improvement of atomic reactors. As an administrator, he has been interested in peaceful uses of atomic energy.

Facts about his career can be found in *Current Biography,* 1966, 431–433.

CAN TECHNOLOGY REPLACE SOCIAL ENGINEERING?*

Alvin M. Weinberg

During the war, and immediately afterward, the federal government mobilized its scientific and technical resources, such as the Oak Ridge National Laboratory, around great technological problems. Nuclear reactors, nuclear weapons, radar, and space are some of the miraculous new technologies that have been created by this mobilization of federal effort. In the past few years there has been a major change in focus of much of our federal research. Instead of being preoccupied with technology, our government is now mobilizing around problems that are largely social. We are beginning to ask what we can do about world population, the deterioration of our environment, our educational system, our decaying cities, race relations, poverty. President Johnson has dedicated the power of a scientifically oriented federal apparatus to finding solutions for these complex social problems.

Social problems are much more complex than are technological problems, and much harder to identify: How do we know when our cities need renewing, or when our population is too big, or when our modes of transportation have broken down? The problems are, in a way, harder to identify just because their solutions are never clear-cut: How do we know when our cities are renewed, or our air clean enough, or our transportation convenient enough? By contrast the availability of a crisp and beautiful technological solution often helps focus on the problem to which the new technology is the solution. I doubt that we would have been nearly as concerned with an eventual shortage of energy as we now are if we had not had a neat solution —nuclear energy—available to eliminate the shortage.

There is a more basic sense in which social problems are harder than are technological problems. A social problem exists because many people behave, individually, in a socially unacceptable way. To solve a social problem one must induce social change—one must persuade many people to behave differently than they have behaved in the past. One must persuade many people to have fewer babies, or to drive more carefully, or to refrain from disliking Negroes. By contrast, resolution of a technological problem involves many fewer individual decisions. Once President Roosevelt decided to go after atomic energy, it was by comparison a relatively simple task to mobilize the Manhattan Project.

The resolution of social problems by the traditional methods—by motivating or forcing people to behave more rationally—is a frustrating business. People don't behave rationally; it is a long, hard business to persuade individuals to forego immediate personal gain or pleasure, as seen by the indi-

* *Bulletin of the Atomic Scientists,* 22 (1966), 4–8. Reprinted by permission of *Science and Public Affairs, the Bulletin of the Atomic Scientists.*

vidual, in favor of longer-term social gain. And indeed, the aim of social engineering is to invent the social devices—usually legal, but also moral and educational and organizational—that will change each person's motivation and redirect his activities to ways that are more acceptable to the society.

The technologist is appalled by the difficulties faced by the social engineer; to engineer even a small social change by inducing individuals to behave differently is always hard even when the change is rather neutral or even beneficial. For example, some rice eaters in India are reported to prefer starvation to eating the wheat we send them. How much harder it is to change motivations where the individual is insecure and feels threatened if he acts differently, as illustrated by the poor white man's reluctance to accept the Negro as an equal. By contrast, technological engineering is simple; the rocket, the reactor, and the desalination plants are devices that are expensive to develop, to be sure, but their feasibility is relatively easy to assess, and their success relatively easy to achieve once one understands the scientific principles that underlie them.

It is therefore tempting to raise the following question: In view of the simplicity of technological engineering, and the complexity of social engineering, to what extent can social problems be circumvented by reducing them to technological problems? Can we identify Quick Technological Fixes for profound and almost infinitely complicated social problems, "fixes" that are within the grasp of modern technology, and which would either eliminate the original social problem without requiring a change in the individual's social attitudes, or would so alter the problem as to make its resolution more feasible? To paraphrase Ralph Nader,[1] to what extent can technological remedies be found for social problems without first having to remove the causes of the problem? It is in this sense that I ask: "Can technology replace social engineering?"

THE MAJOR TECHNOLOGICAL FIXES OF THE PAST

To explain better what I have in mind I shall describe how two of our most profound social problems—poverty and war—have in some limited degree been solved by the Technological Fix, rather than by the methods of social engineering.

The traditional Marxian view of poverty regarded our economic ills as being primarily a question of maldistribution of goods. The Marxist recipe for elimination of poverty, therefore, was to eliminate profit, in the erroneous belief that it was the loss of this relatively small increment from the worker's paycheck that kept him poverty-stricken. The Marxist dogma is typical of the approach of the social engineer: One tries to convince or coerce many people to forego their short-term profits in what is presumed to be the long-term interest of the society as a whole.

[1] See below in this essay.

The Marxian view seems archaic in this age of mass production and automation, not only to us, but apparently to many East European economists. For the brilliant advances in the technology of energy, of mass production, and of automation have created the affluent society. Technology has expanded our productive capacity so greatly that even though our distribution is still inefficient and unfair by Marxian precepts, there is more than enough to go around. Technology has provided a "fix"—greatly expanded production of goods—which enables our capitalist society to achieve many of the aims of the Marxist social engineer without going through the social revolution Marx viewed as inevitable. Technology has converted the seemingly intractable social problem of widespread poverty into a relatively tractable one.

My second example is war. The traditional Christian position views war as primarily a moral issue: If men become good, and model themselves after the Prince of Peace, they will live in peace. This doctrine is so deeply ingrained in the spirit of all civilized men that it is blasphemy to point out that it has never worked very well—that men have not been good, and that they are not paragons of virtue or even of reasonableness.

Although I realize it is a terribly presumptuous claim, I believe that Edward Teller[2] may have supplied the nearest thing to a Quick Technological Fix to the problem of war. The hydrogen bomb greatly increases the provocation that would lead to large-scale war, not because men's motivations have been changed, nor because men have become more tolerant and understanding, but rather because the appeal to the primitive instinct of self-preservation has been intensified far beyond anything we could have imagined before the H-bomb was invented. To point out these things today, with the United States involved in a shooting war, must sound hollow and unconvincing; yet the desperate and partial peace we have now is far better than a full-fledged exchange of thermonuclear weapons. One can't deny that the Soviet leaders now recognize the force of H-bombs, and that this has surely contributed to the less militant attitude of the USSR. And one can only hope that the Chinese leadership, as it acquires familiarity with H-bombs, will also become less militant. If I were to be asked who has given the world a more effective means of achieving peace—our great religious leaders who urge men to love their neighbors and thus avoid fights, or our weapons technologists who simply present men with no rational alternative to peace—I would vote for the weapons technologist. That the peace we get is at best terribly fragile I cannot deny; yet, as I shall explain, I think technology can help stabilize our imperfect and precarious peace.

THE TECHNOLOGICAL FIXES OF THE FUTURE

Are there other Technological Fixes on the horizon, other technologies that can reduce immensely complicated social questions to a matter of

[2] Physicist Edward Teller, known ambiguously as "father of the hydrogen bomb" because of his political and technical contributions to its development.

"engineering"? Are there new technologies that offer society ways of circumventing social problems and at the same time do not require individuals to renounce short-term advantage for long-term gain?

Probably the most important new Technological Fix is the intra-uterine device for birth control. Before the IUD was invented, birth control demanded the very strong motivation of countless individuals. Even with the pill,[3] the individual's motivation had to be sustained day in and day out; should it flag even temporarily, the strong motivation of the previous month might go for naught. But the IUD, being a one-shot method, greatly reduces the individual motivation required to induce a social change. To be sure, the mother must be sufficiently motivated to accept the IUD in the first place, but, as experience in India already seems to show, it is much easier to persuade the Indian mother to accept the IUD once than it is to persuade her to take a pill every day. The IUD does not completely replace social engineering by technology. Indeed, in some Spanish-American cultures where the husband's manliness is measured by the number of children he has, the IUD attacks only part of the problem. Yet in many other situations, as in India, the IUD so reduces the social component of the problem as to make an impossibly difficult social problem much less hopeless.

Let me turn now to problems which, from the beginning, have had both technical and social components—those concerned with conservation of our resources: our environment, our water, and our raw materials for production of the means of subsistence. The social issue here arises because many people by their individual acts cause shortages and thus create economic, and ultimately social, imbalance. For example, people use water wastefully, or they insist on moving to California because of its climate. And so we have water shortages; or too many people drive cars in Los Angeles with its curious meteorology, and Los Angeles suffocates from smog.

The water resources problem is a particularly good example of a complicated problem with strong social and technological connotations. Our management of water resources in the past has been based largely on the ancient Roman device, the aqueduct. Every water shortage was to be relieved by stealing water from someone else who at the moment didn't need the water or was too poor or too weak to prevent the theft. Southern California would steal from Northern California, New York City from upstate New York, the farmer who could afford a cloud-seeder from the farmer who could not afford a cloud-seeder. The social engineer insists that such expedients have gotten us into serious trouble; we have no water resources policy, we waste water disgracefully, and, perhaps, in denying the ethic of thriftiness in using water, we have generally undermined our moral fiber. The social engineer, therefore, views such technological shenanigans as being shortsighted, if not downright immoral. Instead, he says, we should persuade or force people to

[3] That is, the birth control pill.

use less water, or to stay in the cold middlewest where water is plentiful instead of migrating to California where water is scarce.

The water technologist, on the other hand, views the social engineer's approach as rather impractical. To persuade people to use less water or to get along with expensive water is difficult, time-consuming, and uncertain in the extreme. Moreover, say the technologists, what right does the water resources expert have to insist that people use water less wastefully? Green lawns and clean cars and swimming pools are part of the good life, American style, 1966, and what right do we have to deny this luxury if there is some alternative to cutting down the water we use?

Here we have a sharp confrontation of the two ways of dealing with a complex issue: The social engineering way which asks people to behave more "reasonably," the technologist's way which tries to avoid changing people's habits or motivation. Even though I am a technologist, I have sympathy for the social engineer. I think we must use our water as efficiently as possible, that we ought to improve people's attitudes toward the use of water, and that everything that can be done to rationalize our water policy will be welcome. Yet, as a technologist, I believe I see ways of providing more water more cheaply than the social engineers may concede is possible.

I refer to the possibility of nuclear desalination. The social engineer dismisses the technologist's simpleminded idea of solving a water shortage by transporting more water, primarily because in so doing the water user steals water from someone else—perhaps foreclosing the possibility of ultimately utilizing land now only sparsely settled. But surely water drawn from the sea deprives no one of his share of water. The whole issue is then a technological one: Can fresh water be drawn from the sea cheaply enough to have a major impact on our chronically water-short areas like Southern California, Arizona, and the eastern seaboard?

I believe the answer is yes, although much hard technical work remains to be done. A large program to develop cheap methods of nuclear desalting has been undertaken by the United States, and I have little doubt that within the next ten to twenty years we shall see huge dual-purpose desalting plants springing up on many parched sea coasts of the world. At first these plants will produce water at municipal prices. But I believe, on the basis of research now in progress at Oak Ridge and elsewhere, water from the sea at a cost acceptable for agriculture—less than ten cents per one thousand gallons—is eventually in the cards. In short, for areas close to the sea coasts, technology can provide water without requiring a great and difficult effort to accomplish change in people's attitudes toward the utilization of water.

The Technological Fix for water is based on the availability of extremely cheap energy from very large nuclear reactors. What other social consequences can one foresee flowing from really cheap energy eventually available to every country, regardless of its endowment of conventional resources? While we now see only vaguely the outlines of the possibilities, it does seem likely that from very cheap nuclear energy we shall get hydrogen by electrolysis of water, and thence the all-important ammonia fertilizer necessary to help feed the hungry of the world; we shall reduce metals without requiring coking coal; we shall even power automobiles with electricity, via fuel cells or storage batteries, thus reducing our world's dependence on crude oil, as well as eliminating our air pollution insofar as it is caused by automobile exhaust or by the burning of fossil fuels. In short, the widespread availability of very cheap energy everywhere in the world ought to lead to an energy autarchy in every country of the world, and eventually to an autarchy in the many staples of life that should flow from really cheap energy.

WILL TECHNOLOGY REPLACE SOCIAL ENGINEERING?

I hope these examples suggest how social problems can be circumvented or at least reduced to less formidable proportions by the application of the Technological Fix. The examples I have given do not strike me as being fanciful, nor are they at all exhaustive. I have not touched, for example, upon the extent to which really cheap computers and improved technology of communication can help improve elementary teaching without having first to improve our elementary teachers. Nor have I mentioned Ralph Nader's brilliant observation that a safer car, and even its development and adoption by the automobile industry, is a quicker and probably surer way to reduce traffic deaths than is a campaign to teach people to drive more carefully. Nor have I invoked some really fanciful Technological Fixes: like providing air conditioners, and free electricity to operate them, for every Negro family in Watts[4] on the assumption, suggested by Huntington, that race rioting is correlated with hot, humid weather—or the ultimate Technological Fix, Aldous Huxley's "soma pills"[5] to eliminate human unhappiness without improving human relations in the usual sense.

My examples illustrate both the strength and the weakness of the Technological Fix for social problems. The Technological Fix accepts man's intrinsic shortcomings and circumvents them or capitalizes on them for socially useful ends. The Fix is therefore eminently practical and in the short term relatively effective. One doesn't wait around trying to change people's minds: If people want more water, one gets them more water rather than requiring them to reduce their use of water; if people insist on driving autos while they are drunk, one provides safer autos that prevent injuries even in a severe accident.

But the technological solutions to social problems tend to be incomplete and metastable, to replace one social problem with another. Perhaps the

[4] An area in Los Angeles famous for the destructive riots that occurred there in the summer of 1965.

[5] In the novel, *Brave New World,* soma pills are essentially satisfying narcotic doses without bad effects; they serve to keep the labor force happy.

best example of this instability is the peace imposed upon us by the H-bomb. Evidently the *pax hydrogenium* is metastable in two sen[s]es: In the short term, because the aggressor still enjoys such an advantage; in the long term, because the discrepancy between have and have-not nations must eventually be resolved if we are to have permanent peace. Yet, for these particular shortcomings, technology has something to offer. To the imbalance between offense and defense, technology says let us devise passive defense which redresses the balance. A world with H-bombs and adequate civil defense is less likely to lapse into thermonuclear war than a world with H-bombs alone, at least if one concedes that the danger of thermonuclear war mainly lies in the acts of irresponsible leaders. Anything that deters the irresponsible leader is a force for peace: A technologically sound civil defense would therefore help stabilize the balance of terror.

To the discrepancy between haves and have-nots, technology offers the nuclear energy revolution, with its possibility of autarchy for haves and have-nots alike. How this might work to stabilize our metastable thermonuclear peace is suggested by the possible political effect of the recently proposed Israeli desalting plant: I should think that the Arab states would be much less set upon destroying the Jordan River Project if the Israelis had a desalination plant in reserve that would nullify the effect of such action. In this connection, I think countries like ours can contribute very much. Our country will soon have to decide whether to continue to spend $\$5.5 \times 10^9$ per year for space exploration after our lunar landing. Is it too outrageous to suggest that some of this money be devoted to building huge nuclear desalting complexes in the arid ocean rims of the troubled world? If the plants are powered with breeder reactors, the out-of-pocket costs, once the plants are built, should be low enough to make large-scale agriculture feasible in these areas. I estimate that for $\$4 \times 10^9$ per year we could build enough desalting capacity to feed more than ten million new mouths per year, provided we use agricultural methods that husband water, and we would thereby help stabilize the metastable, bomb-imposed balance of terror.

Yet I am afraid we technologists will not satisfy our social engineers, who tell us that our Technological Fixes do not get to the heart of the problem; they are at best temporary expedients; they create new problems as they solve old ones; to put a technological fix into effect requires a positive social action. Eventually, social engineering, like the Supreme Court decision on desegregation, must be invoked to solve social problems. And of course our social engineers are right: Technology will never replace social engineering. But technology has provided and will continue to provide to the social engineer broader options, making intractable social problems less intractable; perhaps most of all, technology will buy time, the precious commodity that converts violent social revolution into acceptable social evolution.

Our country now recognizes—and is mobilizing to meet—the great social problems that corrupt and disfigure our human existence. It is natural that in this mobilization we should look first to the social engineer. Unfortunately, however, the apparatus most readily available to the government, like the great federal laboratories, is technologically, not socially oriented. I believe we have a great opportunity here for, as I hope I have persuaded the reader, many of our social problems do admit of technological solutions. Our already deployed technological apparatus can contribute to the resolution of social questions. I plead, therefore, first for our government to deploy its laboratories, its hardware contractors, its engineering universities, on social problems. And I plead secondly for understanding and cooperation between technologist and social engineer. Even with all the help he can get from the technologist, the social engineer's problems are never really solved. It is only by cooperation between technologist and social engineer that we can hope to achieve what is the aim of all technologists and social engineers—a better society, and thereby a better life, for all of us who are part of society.

37. The President's Science Adviser Talks to Physicists About Research. 1967

During the mid-1960s the feeling of affluence among American scientists was giving way to apprehensions about the level of the support that they might expect from the federal government. Early in 1967 it was the unenviable job of the President's science adviser, DONALD F. HORNIG (born 1920), to confront the most prosperous segment of American science, the physicists. He told them that they could continue to expect government support but at the same time he warned them that federal largesse was going to have limits to it. In the course of his remarks he surveyed the achievements of American physics and listed the symbols of the investment government was making in physical research. He also voiced some of the doubts of science policymakers concerning what directions scientific research ought to take.

Born in Milwaukee, Hornig was educated at Harvard. After taking his doctorate in physical chemistry there in 1943, he spent a brief period at Woods Hole and then joined the nuclear bomb project. After World War II he taught chemistry at Brown and then at Princeton, where he became Donner Professor in 1959. In 1964 he transferred from Princeton to the White House. After a year as vice president of the Eastman Kodak Company and professor of chemistry at the University of Rochester, Hornig in 1970 became president of Brown University.

See Philip M. Boffey, "The Hornig Years: Did LBJ Neglect His Science Adviser?" *Science,* 163 (1969), 453–458.

ADDRESS BY Dr. Donald F. Hornig, Special Assistant to the President for Science and Technology at the American Physical Society Banquet, Wednesday, April 26, 1967, Sheraton Park Hotel, Washington, D.C.*

Although my physicist friends all prefer to have me labeled a chemist, and my chemist friends prefer to have me labeled a physicist, I claim both as it is convenient for me. In particular, I feel a special affection for this meeting because I gave my first scientific paper at a Washington meeting of the Physical Society, and only three years ago I had the opportunity of talking to you about my initial views of my job and the scientific outlook for the country. At that time I was able to present an outline of our plans for high energy physics, and naturally I am pleased that the prognostication was reasonably correct and that the initial design funds for the 200 Bev accelerator are in the President's budget this year. Since that time a lot has happened and the climate has changed, both in this country and the world. I would like, therefore, to say a few words tonight as to how the science scene, and particularly how the physics scene, looks from the White House.

The last three years have seen growth and significant progress in many areas of physics. Large numbers of important experiments on fundamental particles and their interactions have shed new light on the subnuclear domain. The real breakthrough, however, in our understanding of what constitutes a fundamental particle and of the relation between the particles —of which there are more than I at least understand—still lies ahead. New theoretical principles continue to be introduced and explored. The attempts to bring order to the subnuclear chaos have given us new insights into the fundamental symmetry properties of our universe. But clearly, the solid foundation of new laws and the experiments to unambiguously support them have yet to be constructed. It is an exciting time: within a year the elegant experiments at Brookhaven are challenged by still more detailed experiments at CERN, in Switzerland;[1] results of experiments at Harvard are challenged by newer results at Hamburg. This competition is good and is the stuff out of which progress is made. In these same three years, new progress has been made in our picture of the universe and its formation. And detection of the microwave radiation, which appears to be due to the primordial explosion 10 billion years ago, is surely a landmark. New work on the synthesis of the

* Pages 1–12. Copyright by Donald F. Hornig and reprinted with his permission.

[1] Referring to the Brookhaven National Laboratory of the American Atomic Energy Commission and a cooperative nuclear research facility located in Berne but supported by a number of European governments.

elements and nuclear interactions begins to give us a better look into that original cooking pot. Who knows where all these trails will lead, but surely the future looks exciting.

In solid state physics we have progressed from a basic understanding of superconductivity, that remarkable phenomenon by which electricity flows indefinitely in sufficiently cold matter, to a sophisticated understanding of its details.

Solid state physics continues to open up completely new areas both for science and for practical applications. Experiments in tunneling and especially the quantum interference effects are surely among the major achievements of the past few years.[2] The quasar, that great new energy source on the fringe of our universe, is one of the great mysteries of our cosmology. These objects seem to radiate energies at a rate exceeding any known source of energy or any process we can currently imagine.

Plasma physics, the physics of highly ionized gases, which we hope will one day lead us to peaceful thermonuclear energy sources and to rocket propulsion systems, has seen its ups and downs, or perhaps oscillations would be a better term. The problem of confining a plasma at a temperature of millions of degrees in chambers without material walls is still a critical one in the controlled thermonuclear program, but great progress has been made in understanding the instabilities which currently limit the confinement and make the plasmas penetrate the magnetic walls which hold them.

The properties of metals, which used to rest on a largely empirical foundation, is becoming more and more of a science, and during the past few years full details of some extraordinarily complicated Fermi surfaces have been mapped. The day is in sight when the properties of new metals and alloys will be predictable from purely theoretical considerations.

These three years have seen a great variety of important new physical tools coming into use or being started as facilities for the future. In this time:

- The Stanford Linear Accelerator, the world's highest energy and highest intensity electron accelerator, has been put into use and has met its technical specifications. Most remarkably, it has been completed within the original cost estimates.
- The conversion of the AGS 33 Bev proton accelerator at Brookhaven to much higher intensities has been authorized. The design study has been funded and funds to get on with the job are in this year's budget.
- A 150 inch telescope has been started at the Kitt Peak National Observatory which will for some purposes be the most powerful instrument available.

[2] Tunneling is a quantum mechanics phenomenon in which a weak current of electrons flows between two conductors separated by a thin layer of insulating material; quantum interference effects have to do with the wave properties of particles, also an aspect of quantum mechanics theory.

- Just two weeks ago President Johnson and President Frei of Chile announced that a parallel telescope will be built in the Southern Hemisphere at Cerro Tololo InterAmerican Observatory. For the first time, a really powerful telescope will be available to study the southern skies, most particularly the region of the center of our own galaxy and the Magellanic Clouds, our nearest galaxy.
- A site for the 200 Bev accelerator has been chosen and initial funding requested from the Congress.

These facilities, built and requested, are of course only part of what has been happening. One might mention the new electron accelerator at Cornell; the large new "meson factory" and the new thermonuclear facility, Scylac, at Los Alamos; the materials research centers which are operating so effectively around the country; and the new physics buildings which are springing up in every state in the country.

Nevertheless, in spite of this real progress, there has been a general unease in the physics community at what is sometimes called the "tight money" situation. I think there is little doubt that the growth in funds has not matched the aspirations of physicists for new ventures, or the aspirations of more and more universities to strengthen their efforts in physics. In this sense, the growth of funds is being strained by the very vitality of the field, but the facts do not point to any cut-backs or even severe restraints. Funds for university physics, independent of associated Federal laboratories, have risen from $107 million in the year ending June 1964 to about $143 million in the year which will end this June. Physics research *per se* in universities has grown from $101 million to $132 million and the effort on science development and education in physics from $6 to $17 million. I can assure you of the President's very strong interest in the progress of science and in the progress in training of physicists.

What is true for physics, as it is true for other areas of basic science, is that more funds are being spent, more buildings being built, more instruments being bought, more results being published, and more exciting advances being made, and there is no sign of diminution in the future. The President's budget this year calls for a significant increase in our basic research expenditures along with expanded programs in education, facilities, and scholarships, fellowships and traineeships. The recent pattern, as you know, has been one of steadily rising basic research and education expenditures while funds for development have stayed very close to $9 billion. Nevertheless, it would be wrong not to recognize the very genuine concerns not only in the scientific community but in Congress and among ourselves as to the course of the future, and I would like to say a few words on that score.

At the end of World War II we awoke with a start to the realization that this country was not properly cultivating its scientific base, not only in physics but in other areas like health research. On the other hand, the achievements of physicists in the war, and particularly the emergence of nuclear energy, pointed out to the country in a dramatic way the possibilities and promise for the future inherent in scientific advance. There was a vacuum to be filled and we proceeded to fill it at a breathtaking pace. At times, in some fields, the doubling period was two or three years; overall the doubling period was of the order of 4 to 5 years through much of the two decades following World War II. What has changed now is not that there are restraints to be imposed on science either by the Congress or by the Executive, but that the initial vacuum has largely been filled and a new situation has arisen which requires new thought.

When I say that the vacuum has been filled, I mean that we have built a strong, viable scientific establishment in this country. In a whole variety of fields from particle physics to molecular biology the quality of American science is second to none. The quality of American scientific education is equal to or superior to that existing anywhere in the world, and the number of university centers at which good science is carried out has been enormously expanded. American science-based industries lead the way to producing what is called the technological gap in Europe, and are producing a new industrial revolution which will shape the future of our country in many years. In all these respects the goals set by Bush in his magnificent *Science—The Endless Frontier,* and laid out for us in the Steelman Report, have been achieved.[3]

The country need not be convinced any longer that we need strength in basic research. This is accepted by the Executive, by the Congress, and by the people of the country. We accept as the goal that America must be second to none in most of the significant fields of science. Everything about our actions signifies our intention of moving vigorously forward in seizing new opportunities as they are presented. What is *not* accepted is the notion that every part of science should grow at some automatic and predetermined rate, 15% per year or any other number, as a consequence.

The simple fact is that science and technology, research and development, have changed from being frosting on the cake of defense expenditures, health expenditures, and so on to being a significant national expenditure which must compete with other claimants on national resources. The question is not whether we should have basic research, whether we should have research and development, or even whether it should continue to grow—but rather in what ways and for what purposes it should be expanded. The answer to this question will have to be supplied not by me but by all of us.

What has happened seems plain enough to me. Not so long ago, science was "pure" and could be conducted by people who talked largely to each other; now the country has become convinced of its significance and has

[3] These are two important post-World War II reports on how the United States ought to shape its science policies: Vannevar Bush, *Science—The Endless Frontier. A Report to the President on a Program for Postwar Scientific Research* (Washington: Government Printing Office, 1945), and John R. Steelman, *Science and Public Policy* (Washington: Government Printing Office, 1947).

provided the resources which have enabled it to grow into an important national activity. By any standards, we provide a higher proportion of our very high national income to science than does any other society in the world. But now, instead of languishing in the wings, science is on front stage center; it is in the spotlight and the quality of its performance is reviewed by public critics in the popular press.

The goals of our scientific effort and the nature of our scientific effort are not only being examined within the scientific community but by various organs of my office and, more importantly still, by numerous committees of the Congress. There is every reason why they should do so, just as they do for every other important national activity. The heightened interest in this case undoubtedly arises because it is new and has not been so examined in the past. In short, if support is to continue to grow, it is no longer adequate to arrive at a subtle conviction of the needs within the scientific community or to communicate those needs to me and to the relevant agencies. The scientific community is going to have to learn to articulate its hopes, to describe the opportunities which are before us for practical advance, to express the excitement of the new intellectual thrusts—but to do these in terms which the American people, who are expected to pay the bill, will generally understand and have faith in. There is no alternative.

An excellent start has been made in the Pake Report, *Physics—A Survey and Outlook,* and in the Whitford Report, *Ground-Based Astronomy.* But the dialogue will have to be carried to the newspapers, to the schools, to the public and to the Congress, as well as to the Federal agencies and the Bureau of the Budget. It is not that we have entered a period of restraint—it is that science has matured, and to move ahead we must explain over and over again why and how. I have no doubt that if we all undertake this task and identify the goals and opportunities of our efforts, a vigorous new surge is possible.

Now I would like to say a word about basic research in comparison with applied research and development. The facts are very simple. We are determined that the knowledge and understanding we have gained from science will be put to use to meet the needs of our people and the world as expeditiously as possible. We want and expect to have progress in our defenses, in the health care we provide to our people, in the education we provide to our children, and in the advance of our industrial economy. We want the progress of science to show up in the quality of our cities and in the elimination of the pollution of our water and atmosphere. We want to explore the oceans and the regions of space. To this end the Federal Government supplies research and development funds where the results are technically feasible and economically or socially worthwhile.

But, because we are determined to make use of every bit of available knowledge whose application is feasible, economic and useful, it does not follow in the slightest that this implies a decreased interest in basic research. The two activities are separate and usually done by different groups of people. On the one hand, there are people who feed the pool of knowledge and understanding into which we dip for our practical achievements, and on the other hand there are people who recognize human needs and find new ways to meet them. Both are important, both demand creativity, imagination, enterprise and talent, and both will go forward.

The President has put this very clearly in his recent message to the Congress transmitting the Annual Report of the National Science Foundation. After describing the practical benefits provided by scientific advance, he said:

> We know that we can continue this flow of benefits to mankind only if we have a large and constantly replenished pool of basic knowledge and understanding to draw upon. For the path between basic discovery and its application can be both long and uncertain.
>
> We intend to maintain such a pool with all our talents and resources, so that we can apply it to our needs. Perhaps most important, we intend to maintain this pool of basic knowledge and understanding because of the stimulus it provides to our young minds in the challenge of ideas. Knowledge, as we have learned from our rich experience, is not a laboratory curiosity. It is a critical tool for our national health, our national growth, and the sound education of all of us. The very process of generating knowledge produces the highly trained scientists and engineers that are needed to man our universities, industries and government.

Unhappily, these points have not always been understood by government project officers, and there undoubtedly are unhappy instances of efforts to mix the two and to warp basic research projects in the direction of application—or even to judge basic research projects not by the standards of scientific excellence but by the likelihood of practical advance. This we are trying to change. We are trying to get a clear recognition that even when basic research is supported by a mission oriented agency, its role is to build up the basic reservoir on which applications will rest rather than to define an application supporting the mission in each and every project. . . .

38. Nobel Prize Winner Doubts That Scientific Discovery Can Be Directed. 1968

Still one more of the glamour fields of American science in the 1960s was quantum electronics, known to the public in the form of the maser and laser. One of the primary developers of this area of research here reads an eloquent lesson from what happened. After three centuries of science in America, the basic questions of justification for support of science, control of the direction of its development, and the relations between pure and applied research had changed in only a limited number of aspects. The answers to such fundamental questions apparently would always be marked with enough irony so that any overambitious attempts to plan or control science were doomed to frustration. Society, even a society deeply influenced by scientists, it appeared, would have to be content to furnish a general environment in which science could prosper and not attempt much active manipulation or direction of scientific efforts.

CHARLES H. TOWNES (born 1915) grew up in Greenville, South Carolina, and attended Furman University there. He did his graduate work at Duke and at the California Institute of Technology. His accomplishments in both pure and applied physics are many. He served in a variety of institutions besides the Bell Laboratories and Columbia University mentioned in the article. Since 1967 he has been professor of physics at large at the University of California, Berkeley.

Further information concerning him and the problem he discusses can be found in Dael Wolfle, ed., *Symposium on Basic Research* (Washington: American Association for the Advancement of Science, 1959); Bela A. Lengyel, "Evolution of Masers and Lasers," *American Journal of Physics*, 34, (1966), 903–913; and *Current Biography*, 1963, 423–425).

QUANTUM ELECTRONICS, and Surprise in Development of Technology, The Problem of Research Planning*

Charles H. Townes

The evident importance and the considerable expense of scientific research stimulate frequent efforts to assess its contributions to our society, and to optimize its planning. Such efforts are usually undertaken on the premise that we can and should make decisions about the support of scientific research on the basis of what we foresee as its tangible contributions to the nation. While hard-nosed assessment of the contributions of research is clearly appropriate and worthwhile, I am convinced that devotion to this premise is often self-defeating, as will be illustrated here by the obstinate and sometimes bruising facts of past experience.

If we forget the cultural values of knowledge, and evaluate science only by the touchstone of "practical" results, we may at first seem to have a straightforward guide for planning research. We know well that basic research develops many of the new ideas and new information from which technology is derived. Hence, it is easy to conclude that we need primarily to consider what types of technology are wanted for the future, and sponsor those forms of basic science which will contribute the background of infor-

* *Science*, 159 (1968), 699–703. Copyright 1968 by the American Association for the Advancement of Science. Reprinted with the kind permission of copyright holder and author.

mation needed for them. There is indeed some truth in this reasoning; it applies particularly to those aspects of technology and of science which we now understand reasonably well, and where we are looking primarily for upgrading of our present abilities or the maturation of developments which are now predictable. But our ability to foresee the practical effects of science is too imperfect. For periods of time as long as a decade or more, or for the really new ideas and startling developments which are not now foreseen, the above approach is unhappily limiting and misleading. Furthermore, it gives a very unrealistic view of the environment needed for high-quality scientific research and of the complex interplay between science and technology, which includes the stimulation of basic science by applied science as well as the reverse.

How, in fact, can we plan for the new idea and the startlingly new, but now unrecognized, technology? Certainly we cannot show that a particular line of basic research will lead to new technological developments if we don't yet even know the nature of these developments. Nor is it possible to satisfy a persistent doubter that present basic research, even though it may be uncovering new knowledge and new ideas, will lead to important though unknown developments for human welfare. Perhaps the best way to examine such questions with some objectivity is the historical method, use of experience.

A general conclusion which seems to me to emerge from a historical approach—the examination of a number of research case histories—is that mankind consistently errs in the direction of lack of foresight and imagination. We continually underestimate the power of science and technology in the long term. Eminently knowledgeable planners and scientists, in attempting responsibly to make realistic appraisals of research, and facing what is at the time uncertain or unknown, all too frequently fall short in foresight and imagination. The element of surprise is a consistent ingredient in technological development, and one we have great difficulty in dealing with on any normal planning basis. Let me now proceed to discuss a particular example with which I happen to be well acquainted—quantum electronics. This is done in some detail, because very specific examples rather than generalities are probably necessary to overcome our natural tendency toward complacency.

ORIGIN OF QUANTUM ELECTRONICS

Quantum electronics became a field of physics and of engineering with development of the devices known as the maser and the laser. They use a new type of amplification, the stimulated emission of electromagnetic waves from atoms or molecules. The two devices are of the same general class; in fact, the laser was originally called an optical maser, although the name maser is sometimes restricted to molecular amplification in the radio or microwave range because it was derived as an acronym for *m*icrowave *a*mplification by *s*timulated *e*mission of *r*adiation. The word laser simply means *l*ight *a*mplification by *s*timulated *e*mission of *r*adiation, an application of the same idea to light waves. The parent device involved an amplification technique so radically different that it could not grow out of previous electronics in any orderly way; in fact, its birth in the early 1950's seems to have almost required prior development of the field of basic research known as microwave spectroscopy.[1] How can I justify such a bald statement? Because the idea for maser amplification originated independently in three different laboratories of microwave spectroscopy, and from research rather universally eschewed in applied laboratories. Each of these three origins had a slightly different timing, and differed appreciably in its completeness and practicality. However, all three came from physicists occupied with basic, university-type research on the microwave spectroscopy of gases.

TECHNOLOGY AS A SOURCE OF BASIC SCIENCE

It is almost equally significant that microwave spectroscopy itself grew out of wartime technology. This, as well as a good deal of closely related radio-frequency spectroscopy, originated with physicists who had acquired experience in electronics during World War II. In particular, microwave spectroscopy—the study of the interaction between microwaves and gaseous molecules—came about because microwave oscillators and technology were well enough developed during the war to allow this new branch of physics to be fruitful. Thus, a field of basic research was made possible by technology, and the first work in microwave spectroscopy in this country was largely carried out in industrial laboratories. Four independent groups of scientists in the United States, at the Bell Telephone Laboratories, at Westinghouse, at the RCA Laboratories, and at Columbia University initiated more or less independently the study of gases by means of microwaves immediately after the war, and pursued it with some vigor because of its evident importance to physics. The historical importance of technology to its origin is quite clear when one finds that the only university group of these four had been heavily involved in microwave technology during the war and initiated its work to solve an important radar problem. A little later than these four laboratories, the General Electric Company and several universities began further work in the field.

MIGRATION OF MICROWAVE SPECTROSCOPY TO THE UNIVERSITIES

No doubt in the industrial laboratories there was some hope that the new field of physics would have a worthwhile contact with commercial

[1] That is, observing the effects of microwave electromagnetic radiation upon various materials.

applications. In the case of the Bell Telephone Laboratories, I had myself written a memorandum with some care to convince research management that this could be the case. However, after several years this type of work died out in the four industrial laboratories where it had an early start and moved to the universities entirely. There it attracted a good number of excellent students, as well as experienced professors, because of the insight it afforded into molecular and atomic behavior. Reasons for growth of the field in universities may seem natural enough. Reasons for its decay in industry are equally important, and illustrate rather clearly our dilemma in the planning of research.

Evidently the four large industrial laboratories, although deeply involved with electronics, did not feel at the time that research on the microwave spectroscopy of gases had much importance for their work. I do not know the detailed reasoning of management at Westinghouse and RCA, but after the small teams of research workers which had been quite successful at these laboratories left or lost interest, research in the field was not rebuilt. At the General Electric Company, the research scientist in this field was transferred by management decision to another field considered more pertinent to the company's business. In the case of the Bell Telephone Laboratories, there was a management decision that, while one senior scientist could be appropriately supported, the work was not important enough to the electronics and communications industry to warrant adding a second one. Yet it was out of just this field that 2 or 3 years later a completely new technique of amplification was born which now occupies hundreds of scientists and engineers in the same laboratories. Clearly, misjudgment of its potential was not a simple human fault of any one company or individual; it was a pervasive characteristic of the system.

SOCIOLOGY OF THE MASER INVENTION

Microwave spectroscopy in the universities utilized some of the new electronics techniques of the time, and was able to examine delicately and powerfully the various types of interactions between electromagnetic waves and molecules in ways which were different from those of normal spectroscopy. My own work, by then at Columbia University, flourished in an environment where a considerable amount of related radio frequency spectroscopy was being carried out, and supported by a rather farsighted Armed Services contract. The resulting development of ideas, in close association with electronics, led in 1951 to invention of the maser at Columbia, and shortly after to other proposals for use of stimulated emission for practical amplification—one at the Lebedev Institute in the Soviet Union and another at the University of Maryland. It is worth noting that basic research in the Soviet Union was at that time primarily concentrated in laboratories of the Soviet Academy, some of whose scientists taught in universities, and that this closest equivalent to our university research laboratories was the setting for the invention there.

By 1954, collaboration with J. Gordon and H. Zeiger produced the first successful oscillator with the new amplifying principle. While a few applied scientists were enthusiastic, overall it evoked only very mild industrial interest. I cannot claim that foresight of the academic community concerning the maser was remarkably greater than that of industrial organizations. But what was important was one of the crucial strengths of academic institutions, that an individual professor by and large makes his own decisions as to what is worthwhile and what might work. This, I believe, generally allows a scientific diversity and utilization of individual insights or enthusiasms in the academic world that are difficult to match in more closely planned and ordered industrial organizations. The latter are especially adapted for a concerted attack on a well-recognized goal. But the diverse and novel ideas for strikingly new approaches to problems are more normally current in communities where vigorous basic research flourishes. Coherent amplification by stimulated emission of radiation, and the idea of gradual quantum transitions rather than quantum jumps, for example, were reasonably well-recognized processes in some academic circles. Applied scientists were at the time characteristically surprised by them. Furthermore, even though there are now many varieties of masers, for some reason the two most complete original suggestions for practical maser systems, from Columbia University and from the Lebedev Institute, both involved molecular beams and Stark effects,[2] techniques and ideas which were of some currency in academic circles but scarcely ever considered in industrial laboratories. But certain ideas of electronic engineering were important too, for example, in providing an understanding of regeneration and of the utilization of coherent amplification. It was the mixture of electronics and molecular spectroscopy inherent in the field of microwave spectroscopy which set appropriate conditions for invention of the maser.

The new type of amplification immediately produced an interesting oscillator, but not so immediately a very usable amplifier. My visit with scientific colleagues at the Ecole Normale Supérieure in Paris generated what seemed to me the first clear view of a practical amplifier by the use of paramagnetic solid materials, because there I was associated with other physicists studying paramagnetic materials and became aware of some of their properties which were otherwise unknown to me. A somewhat similar idea grew up independently from Professor Strandberg, a microwave spectroscopist at M.I.T. He passed on an interest to Professor Bloembergen of Harvard, who had been studying paramagnetic properties for some time, and who provided the variant of the maser which is now its most practical form for amplifiers. By this time industrial laboratories had become more alert to the

[2] Molecular beam techniques involve the production of a well-defined beam of molecules, and the Stark effect is produced by an electrical field applied to spectrum lines.

new possibilities, and it was Feher, Scovil, and Seidel at the Bell Telephone Laboratories who first built a workable amplifier with paramagnetic materials. From this point on, the nation's applied laboratories pursued maser amplifiers for the microwave region with vigor and success.

THE LASER

By 1957, I was eager to try to push the new technique on into the shorter wavelength regions, since it was clear that molecules and atoms had the capability of amplifying wavelengths very much shorter than anything previously done by vacuum tubes. I discovered that my friend Arthur Schawlow, then at the Bell Telephone Laboratories, had also been thinking along somewhat similar lines, and so we immediately pooled our thoughts. It was he who initiated our consideration of a Fabry-Perot resonator[3] for selection of modes of the very short electromagnetic waves in the optical region. This very likely had something to do with the fact that Schawlow had first been trained as a spectroscopist and had done his thesis with a Fabry-Perot, another important technique current primarily among university spectroscopists. From this collaboration came the first fully developed ideas for lasers.

The new device was so far out of the normal tradition that its value for applied work was not immediately obvious to everyone. Bell's patent department at first refused to patent our amplifier or oscillator for optical frequencies because, it was explained, optical waves had never been of any importance to communications and hence the invention had little bearing on Bell System interests. But the potentialities were soon sufficiently clear that a number of laboratories in both universities and industry became strongly interested in the optical maser, later called a laser. In particular, management at the Bell Telephone Laboratories not so much later gave it considerable priority. The first actual operating system, the ruby laser, was produced by Maiman at the Hughes Aircraft Company; this was followed shortly by a second type based on an idea of Javan at the Bell Telephone Laboratories, and then a third one made by Sorokin and Stevenson at IBM. Clearly, the nation's powerful industrial laboratories had begun their push to develop the field. Subsequently, quantum electronics has blossomed to its present level of about $200 million of business per year, with an expectation of about $1 billion per year by 1970 or 1971.

The successive ideas for improvement and extension of the new type of amplification to the point which I have described came primarily from the realm of basic research. Some of them were rather new, some of them older ideas which had been current in laboratories of basic research. Their sources were almost exclusively scientists trained in microwave and radio frequency spectroscopy. In fact, all but one of those I have mentioned or alluded to

[3] A type of interferometer, a device used to separate and compare, and therefore measure, beams of light.

above had extensive experience in this field. The demand for such personnel in industrial and governmental laboratories by the early 1960's was, of course, intense.

PRACTICAL USES

What has come out of this development? A total variety of applications too long to list. Since the new technique allows amplification and control of electromagnetic radiation in the infrared, optical, and ultraviolet regions approximately equivalent to what electronics has provided in the radio region, one needs only to think of the utility of light and of electronics to see that a marriage of these two fields would have possible applications in almost any sophisticated technology. I shall give a few examples.

Maser-type amplification comes very close to providing the ideally sensitive amplifier, which can successfully amplify one quantum of radiation. For microwaves, the new amplifier actually provided a sensitivity about one hundred times better than what had previously been available. While by now there are some other types of improved amplifiers, the maser amplifier remains and will likely remain for all time our most sensitive detector of microwaves. Its use is particularly important in allowing efficient transoceanic commercial communications through satellites, scientific measurements of new sensitivity, and in making practical space communications throughout the solar system.

The constancy of atomic properties and the lack of noise fluctuations also makes a maser oscillator the world's most precise clock. A maser based on hydrogen is so constant that if kept going for 300,000 years, its expected error would be only about 1 second.

Since light waves can be amplified by the new techniques, they can provide light of almost indefinitely high intensity. Already lasers produce light many millions of times more intense than what was previously available. Laser beams can be accurately controlled and focused to drill holes in refractory materials such as diamond, to partially evaporate and thus precisely adjust electronic circuit elements, or to do delicate surgery. As a surgical tool, the laser is particularly useful in the performance of operations inside the eye without any external incision.

The laser allows our most accurate measurement of distance. In the laboratory, it has detected changes of distance as small as 1/100,000 the diameter of an atom. The coherence of laser light allows interferometric measurements to a precision of a fraction of the wave-length of light up to distances of many miles. This is already being used for detection of earthquake phenomena, and for very precise machining. The directivity of laser beams makes them convenient tools for civil engineering; they have been introduced for the boring of tunnels, the dredging of channels, and the grading of roads.

In photography, the new intensities of light available have allowed much higher-speed photography than was previously possible. But still more spectacular is use of the laser as the basis for a new type of photography called holography. Laser light projected through a photographic film with holographic techniques, gives a real three-dimensional image with a wealth of detail and a remarkable depth of focus.

Other uses of laser beams include radar, guidance for the blind, information processing, and information storage and retrieval. In the future there may be wireless power transmission, large-screen color television, and cheaper communications.

THE RESEARCH PLANNER'S PROBLEM AND THE DRIVE FOR PRACTICALITY

Consider now the problem of a research planner setting out 20 years ago to develop any one of these technological improvements—a more sensitive amplifier, a more accurate clock, new drilling techniques, a new surgical instrument for the eye, more accurate measurement of distance, three-dimensional photography, and so on. Would he have had the wit or courage to initiate for any of these purposes an extensive basic study of the interaction between microwaves and molecules? The answer is clearly No. For a more sensitive amplifier he would have gone to the amplifier experts who, after considerable effort, might have doubled the sensitivity of amplifiers rather than multiplied it by a hundred. For a more accurate clock, he probably would have hired those experienced in the field of timing; for more intense light, he would have sought out and supported a completely different set of scientists or engineers who could hardly have hoped to have achieved an increase in intensity by the factor of a million or more given by the laser. For more accurate measurements or for better photography, he would have tried other improvements of known techniques and very likely have achieved moderate success, but no breakthrough by orders of magnitude. It was the drive for new information and understanding, and the atmosphere of basic research which seems clearly to have been needed for the real payoff.

There is at least a superficial similarity between the search for new technology and the pursuit of happiness, each of which is sometimes best approached by indirection. We know some straightforward, but limited, ways to achieve happiness. A better house to live in, or even just an ice cream cone now and then will help. But generally the direct and continuous pursuit of happiness itself is much less successful in achieving the big result than dedication to worthwhile human values and enterprises, without such overt thought of self-satisfaction. Similarly, while direct and planned development of technology is clearly useful and should not be neglected, efforts confined entirely to this approach will be badly limited. Success can be enormously increased by the stimulation and the discoveries which come from an interested dedication to knowledge and discovery themselves.

Americans are intensely practical, and it is difficult to accept the idea that a result is not best achieved by systematic planning, keeping one's eye on the ball, and good hard work. But we have all too frequently had the experience that in judging the practical value of specific scientific research, and in certain cases even of engineering development, those who would seem to be most knowledgeable and responsible are not able to foresee the most imaginative and important steps. History shows this in many more cases than in quantum electronics. In fact, surprise in the development of technology is our regular fare.

SURPRISE AND NUCLEAR ENERGY

Some of the interesting story of the development of nuclear energy is quite familiar. Einstein's deduction of the equivalence of mass and energy should have given some inkling of the possibilities even early in this century. During the first part of the 1930's, the exciting field of nuclear physics opened up and produced a small flurry of speculation about the possibility of nuclear energy. But the *Herald Tribune* of 1933 carried an assessment of these possibilities under the headline "Lord Rutherford Scoffs at Theory of Harnessing Energy in Laboratories." Rutherford could perhaps fairly be called the greatest experimental physicist of the day and the father of nuclear physics. He had just spoken in Great Britain about the splitting of the atomic nucleus in the same hall where a generation earlier Lord Kelvin, a great physicist of his day, had pronounced the atom indestructible. Rutherford commented, "The energy produced by breaking down of the atom is a very poor kind of thing. Anyone who expects a source of power from the transformation of these atoms is talking moonshine." Professor Rabi of Columbia University, interviewed at the same time, confirmed Rutherford's calculations and hence, apparently, his general conclusions. Professor La Mer, also of Columbia University, was quoted as saying, "I am pleased to see Lord Rutherford call a halt to some of the wild, unbridled speculation in this field." There were indeed some other opinions. Of those interviewed by the *Tribune,* Professors Sheldon of New York University and E. O. Lawrence of the University of California still held out some hope. However, the generation of nuclear energy was not for a few years taken very seriously by the scientific community and was hardly an issue in the support of the study of nuclear physics. In fact, there was considerable concern among physicists, planners, and in industrial circles that too much of physics was swinging toward the nuclear field and that there was too much attention given to this esoteric, relatively useless, aspect of physics. The General Electric Company, deeply involved in power generation, made an overt management decision during this time that the promise of atomic power was not worth its initiating any nuclear research.

Only 5 years after Rutherford's pronouncement, the unlooked-for phenomenon of fission was discovered, and suddenly the whole world of physics saw the possibilities of nuclear energy in a completely different light. Success could not be assured, but there were now straightforward ways of attempting to obtain nuclear energy. The basic knowledge and knowledgeable personnel were fortunately available because of the previous years of intellectual curiosity centered in the universities; this background and help from Europe's intellectuals were crucial to the United States and its allies.

OTHER CASE HISTORIES

The transistor, another outstanding technological triumph, is by contrast quite a different case, and represents one of success in research planning. M. Kelly of the Bell Telephone Laboratories did foresee that solid-state physics was important in a variety of ways to operations of the Bell System, and formed and encouraged a group of physicists interested in basic exploration of this field. At least initially, this was not done with any direct thought of transistor-type amplification. But it was Kelly's plan of basic physical research on solids, in contact with engineering interests and considerably in advance of most other industrial laboratories of the time, which led to the transistor and its many descendents [*sic*].

An interesting example of our difficulty with foresight and imagination in a more engineering domain, and where the basic physical phenomena were rather well known, is the case of the airplane. Lord Rayleigh, one of the greatest physicists of the 19th century and certainly familiar with appropriate fields of physics, commented in 1896, "I have not the smallest molecule of faith in aerial navigation other than ballooning." This was followed by severe congressional criticism over the "waste" of government money on Langley's attempts to build a heavier-than-air machine, and was just 7 years before the Wright brothers successfully "navigated" over the sands of Kitty Hawk. One can trace an interesting and intense argument for some time thereafter over whether or not the airplane would ever amount to much. Eventually, human need for a flying machine and the characteristically surprising power of technology won handsomely again.

WHICH WAY GENUINE REALISM?

The above shows us some of the cases where hard realism wasn't real and dreams were. One might well wonder how we can possibly hope to judge the value of specific basic research for the future of technology, and hence on what we can base our plans. My belief is that knowledgeable and responsible people, in trying to judge carefully and not run too much risk of being wrong, have almost inevitably been too shortsighted. Furthermore, planners, in trying to be realistic and faced with tough budgetary decisions, all too frequently find themselves convinced only about what can be demonstrated, and hence their programs are unhappily limited. Science fiction and human need seem to have frequently been more reliable guides to predicting long-range technological developments than sober scientific statesmen. The progress of technology to a point further than we can see clearly—and this means hardly more than a decade—is always surprising and almost invariably greater than we think.

How can we best foster discovery and useful invention? I certainly would not want to play down the importance of planned research and development toward the shorter-term goals which can be foreseen. For this, organized teams and keeping one's eye on the ball can be very effective, and in some cases are almost essential. On the other hand, an atmosphere where utility is paramount is likely to confine thinking in particular channels, and is too prone to smother and draw attention away from what will produce many of the happy technological surprises and radically new ideas. I can suggest three useful guides.

1. There should be an environment of evident devotion to knowledge and discovery themselves, as well as to practical results.
2. To take best advantage of man's curiosity and his potential for discovery, we must give clear attention to supporting the clever, productive, and dedicated researcher in his own insights above what is interesting or fruitful.
3. If the nation is to ensure itself against missing the most exciting surprises, it must ensure support for those fields, even the nonutilitarian ones, where new understanding (not just new detailed knowledge) is most rapidly developing.

I have purposely concentrated attention on the material results of science, but must at least pause to recognize that this involves the frequent mistake of omitting almost completely other important and perfectly real aspects of science and knowledge—their cultural values. Man's view of his universe and of himself which results from scientific research has a significance considerably beyond what is considered "practical" in the narrow sense. Discovery and understanding give breadth of view and inspiration, the satisfaction of man's innate wonder and intellectual drive, and a sense of creative achievement toward some of his most universal goals and most lasting monuments. As something of a parallel to the limitation of being concerned only with the tangible results of science, consider how far short we would be in explaining the importance of music to mankind in a discussion confined to its practical and economic results. However, basic scientific research does, of course, have a profound effect on man's material productivity and wellbeing, and this can be appropriately discussed as long as we remember that there are also other values at stake.

THE SHORT AND THE LONG RUN

We have done well in basic research and the generation of new ideas during the last two decades. However, I am genuinely concerned about what seems to me a trend in the United States toward emphasis on the shorter-range goals and overconcentration of attention on utility to an extent which may well limit our technological productivity and leadership in the future. Having emphasized man's limitation in predicting the outcome of research, I do not want at this point to try predictions myself, other than to affirm the continuity of history and the constancy of man's nature. However, it is clear that among the many fields where we face decision now are high-energy physics and space exploration. Both are exciting, but expensive. Very little utility can really be predicted for high-energy physics, and little for much of space exploration. Yet we must examine them from both cultural and utilitarian points of view, and with such things on our conscience as the myopic tendencies of the past, our proclivity for taking the lack of foreseeable utility for lack of its real existence, and the ease with which we have disproved the possibility of what only a few years later becomes actuality in this ebullient world of science. And if in these fields or others we are found shortsighted, too lacking in daring, or too indifferent to forward-looking dreams, the pace of science and the impact of technology are now sufficient that our limitations will be obvious not only in the nation's future and the eventual judgment of history, but also to us personally, and in our lifetime.

39. Three Ecologists Suggest Criteria for an Optimum Human Environment. 1970

By the opening of the 1970s, Americans' concern with pollution, technology, and overcrowding had changed into a more general preoccupation with ecological systems. This article shows how three sensitive scientists perceived that "science" as Americans commonly knew it was limited in its ability to provide for man's survival. They suggested a new and more comprehensive use of scientific disciplines to discover the physiological and psychological prerequisites for human survival. Without abandoning the Darwinian ideas of survival and the importance of environment in survival, Iltis, Loucks, and Andrews wrote about a scientific enterprise of a complexity and inclusiveness that was appropriate for a technological society in the last decades of the twentieth century.

HUGH H. ILTIS (born 1925) is a native of Czechoslovakia who was educated at the University of Tennessee and Washington University. He is professor of botany at the University of Wisconsin. ORIE L. LOUCKS was born in Minden, Ontario (in 1931). He took his Ph.D. at the University of Wisconsin and, like Iltis, is professor of botany there. PETER ANDREWS was born in São Paulo, Brazil, in 1940. He holds degrees in forestry from Aberdeen and Toronto and in physical anthropology from Cambridge, where he was a student in St. Johns College when this article was published.

The literature on ecology and the ecological crisis is very extensive. A historical and bibliographical survey is Michael J. Lacey, "Man, Nature, and the Ecological Perspective," *American Studies,* 8 (1970), 13–27.

CRITERIA FOR an Optimum Human Environment*

Hugh H. Iltis, Orie L. Loucks, and Peter Andrews

Almost every current issue of the major science journals contains evidence of an overwhelming interest in one urgent question: Shall a single species of animal, man, be permitted to dominate the earth so that life, as we know it, is threatened? The uniformity of the theme is significant but if there is a consensus, it is only as to the need for concern. Each discipline looks differently at the problem of what to do about man's imminent potential to modify the earth through environmental control. Proposals to study ways of directing present trends in population, space and resource relationships toward an "optimum" for man are so diverse as to bewilder both scientists and the national granting agencies.

ARROGANCE TOWARD NATURE

It is no thirst for argument that compels us to add a further view. Rather it is the sad recognition of major deficiencies in policies guiding support of research on the restoration of the quality of our environment. Many of us find the present situation so desperate that even short-term treatments of the symptoms look attractive. We rapidly lose sight of man's recent origins, probably on the high African plains and the natural environment that shaped him. Part of the scientific community also accepts what Lynn White has called our Judeo-Christian arrogance toward nature, and is gambling that our superior technology will deliver the necessary food, clean water and fresh air. But are these the only necessities? Few research proposals effectively ask whether man has other than these basic needs, or whether there is a limit to the artificiality of the environment that he can tolerate.

In addition, we wish to examine which disciplines have the responsibility to initiate and carry out the research needed to reveal the limits of man's tolerance to environmental modification and control. We are especially concerned that there is, on the one hand, an unfortunate conviction that social

* *Bulletin of the Atomic Scientists,* 26 (1970), 2–5. Slightly abridged. Reprinted by permission of *Science and Public Affairs, the Bulletin of the Atomic Scientists.*

criteria for environmental quality can have no innate biological basis—that they are only conventions. Yet, on the other hand, there is increasing evidence suggesting that mental health and the emotional stability of populations may be profoundly influenced by frustrating aspects of an urban, biologically artificial environment.

There have been numerous proposals for large-scale inter-disciplinary studies of our environment and of the future of man, but such studies must have sufficient breadth to treat conflicting views and to seek to reconcile them. We know of no proposal that would combine the research capabilities of a group studying environmental design with those of a group examining the psychological and mental health responses of man to natural landscapes. The annual mass migration of city man into natural landscapes which provide diversity is a matter of concern to the social scientist, whose research will only be fully satisfactory when joined with studies that quantify the landscape quality, the psychology of individual human response, and the evolutionary basis of man's possible genetic adaptations to nature. The following summary of recent work may provide a basis for scientists in all areas to seek and support even greater breadth in our studies of present and future environments for man.

"WEB OF LIFE"

Two major theses are sufficiently well established to provide the positive foundation of our argument. First we believe the inter-dependency of organisms, popularly known as the "web of life," is essential to maintaining life and a natural environment as we know it. The suffocation of aquatic life in water systems, and the spread of pollutants in the air and on the land, make it clear that the "web of life" for many major ecosystems is seriously threatened. The abrupt extinction of otherwise incidental organisms, or their depletion to the point of no return, threatens permanently to impair our fresh water systems and coastlines, as well as the vegetation of urban regions.

Second, man's recent evolution is now well enough understood for it to play a major part in elucidating the total relation of man to his natural environment. The major selection stresses operating on man's physical evolution have also had some meaning for the development of social structures. These must be considered together with the immense potential of learned adaptations over the entire geologic period of this physical evolution. Unfortunately, scientists, like most of us moderns, are city dwellers dependent on social conventions, and so have become progressively more and more isolated from the landscape where man developed, and where the benchmarks pointing to man's survival may now be found. They, of all men, must recognize that drastic environmental manipulations by modern man must be examined as part of a continuing evolutionary sequence.

The immediacy of problems relating to environmental control is so startling that the threat of a frightening and unwanted future is another point of departure for our views. At the present rate of advance in technology and agriculture, with an unabated expansion of population, it will be only a few years until all of life, even in the atmosphere and the oceans, will be under the conscious dictates of man. While this general result must be accepted by all of us as inevitable, the methods leading to its control offer some flexibility. It is among these that we must weigh and reweigh the cost-benefit ratios, not only for the next 25 or 50 years, but for the next 25,000 years or more. The increasing scope of the threat to man's existence within this controlled environment demands radically new criteria for judging "benefits to man" and "optimum environments."

It would be perverse not to acknowledge the immense debt of modern man to technological development. In mastering his environment, man has been permitted a cultural explosion and attendant intricate civilization made possible by the very inventiveness of modern agriculture, an inventiveness which must not falter if the world is to feed even its present population. Agricultural technology of the nineteenth and twentieth centuries, from Liebig and the gasoline engine to hybrid corn, weed killers and pesticides, has broken an exploitative barrier leading to greatly increased production and prosperity in favored regions of the world. But this very success has imposed upon man an even greater responsibility for managing all of his hpysical and biotic environment to his best and sustained advantage.

The view also has been expressed recently that the "balance of nature," upset by massive use of non-disintegrating detergents and pesticides, will be restored by "new engineering." Such a view is necessarily based on the assumption that it is only an engineering problem to provide "an environment [for man] relatively free from unwanted man-produced stress." But when the engineering is successful, the very success dissipates our abilities to see the human being as part of a complex biological balance. The more successful technology and agriculture become, the more difficult it is to ask pertinent questions and to expect sensible answers on the long-range stability of the system we build.

THE RIGHT QUESTIONS?

Inspired by recent success, some chemical and agricultural authorities still hold firmly that we can feed the world by using suitable means to increase productivity, and there is a conviction that we can and must bend all of nature to our human will. But if open space were known to be as important to man as is food, would we not find ways to assure both? Who among us has such confidence in modern science and technology that he is satisfied we know enough, or that we are even asking the right questions, to ensure our survival beyond the current technological assault upon our environment.

The optimism of post-World War II days that man can solve his problems—the faith in science that we of Western culture learn almost as infants—appears more and more unfounded.

To answer "what does man now need?" we must ask "where has he come from?" and "what evidence is there of continuing genetic ties to surroundings similar to those of his past?"

Theodosius Dobzhansky and others have stressed that man is indeed unique, but we cannot overlook the fact that the uniqueness does not separate him from animals. Man is the product of over a hundred million years of evolution among mammals, over 45 million years among primates, and over 15 million years among apes. While his morphology has been essentially human for about two million years, the most refined neurological and physical attributes are perhaps but a few hundred thousand years old.

SELECTION AND ADAPTATION

G. G. Simpson notes that those among our primate ancestors with faulty senses, who misjudged distances when jumping for a tree branch or who didn't hear the approach of predators, died. Only those with the agility and alertness that permitted survival in ruthless nature lived to contribute to our present-day gene pool. Such selection pressure continued with little modification until the rise of effective medical treatment and social reforms during the last five generations. In the modern artificial environment it is easy to forget the implications of selection and adaptation. George Schaller points out in "The Year of the Gorilla" that the gorilla behaves in the zoo as a dangerous and erratic brute. But in his natural environment in the tropical forests of Africa, he is shy, mild, alert and well-coordinated. Neither gorilla nor man can be fully investigated without considering the environments to which he is adapted.

Unique as we may think we are, it seems likely that we are genetically programmed to a natural habitat of clean air and a varied green landscape, like any other mammal. To be relaxed and feel healthy usually means simply allowing our bodies to react as evolution has equipped them to do for 100 million years. Physically and genetically we appear best adapted to a tropical savanna, but as a civilized animal we adapt culturally to cities and towns. For scores of centuries in the temperate zones we have tried to imitate in our houses not only the climate, but the setting of our evolutionary past: warm humid air, green plants, and even animal companions. Today those of us who can afford it may even build a greenhouse or swimming pool next to our living room, buy a place in the country, or at least take our children vacationing at the seashore. The specific physiological reactions to natural beauty and diversity, to the shapes and color of nature, especially to green, to the motions and sounds of other animals, we do not comprehend and are reluctant to include in studies of environmental quality. Yet it is evident that nature in our daily lives must be thought of, not as a luxury to be made available if possible, but as part of our inherent indispensable biological need. It must be included in studies of resource policies for man.

DEPENDENCE ON NATURE

Studies in anthropology, psychology, ethology and environmental design have obvious implications for our attempts to structure a biologically sound human environment. Unfortunately, these results frequently are masked by the specifics of the studies themselves. Except for some pioneer work by Konrad Lorenz followed up at several symposia in Europe, nothing has been done to systematize these studies or extend their implications to modern social and economic planning. For example, Robert Ardrey's popular work, "The Territorial Imperative," explores territoriality as a basic animal attribute, and tries to extend it to man. But his evidence is somewhat limited, and we have no clear conception of what the thwarting of this instinct does to decrease human happiness. The more extensive studies on the nature of aggression explore the genetic roots of animal conflicts, roots that were slowly developed by natural selection over millions of generations. These studies suggest that the sources of drive, achievement, and even of conflict within the family and war among men are likely to be related to primitive responses as well as to culture.

Evidence exists that man is genetically adapted to a nomadic hunting life, living in small family groups and having only rare contact with larger groups. As such he led a precarious day-to-day existence, with strong selective removal due to competition with other animals, including other groups of humans. Such was the population structure to which man was ecologically restricted and adapted until as recently as 500 generations ago. Unless there has since been a shift in the major causes of human mortality before the breeding age (and except for resistance to specific diseases there is no such evidence), this period is far too short for any significant changes to have occurred in man's genetic makeup.

Studies of neuro-physiological responses to many characteristics of the environment are also an essential part of investigating genetic dependence on natural as opposed to artificial environment. The rapidly expanding work on electroencephalography in relation to stimuli is providing evidence of a need for frequent change in the environment for at least short periods, or, more specifically, for qualities of diversity in it. There is reason to believe that the electrical rhythms in the brain are highly responsive to changes in surroundings when these take the full attention of the subject. The rise of mechanisms for maintaining constant attention to the surroundings can be seen clearly as a product of long-term selection pressures in a "hunter and hunted" environment. Conversely, a monotonous environment produces wave patterns contributing to fatigue. One wonders what the stimuli of brick and asphalt jungles, or the monotony of corn fields, do to the nervous system. Biotic as well as cultural diversity, from the neurological point of view, may well be fundamental to the general health that figures prominently in the discussions of environmental quality.

RESULTS WITH PATIENTS

The interesting results of Maxwell Weismann in taking chronically hospitalized mental patients camping are also worth noting. Hiking through the woods was the most cherished activity. Some 35 of the 90 patients were returned to their communities within three months after the two-week camping experience. Other studies have shown similar results. Many considerations are involved, but it seems possible that in a person whose cultural load has twisted normal functioning into bizarre reactions, his innate genetic drives still continue to function. Responses attuned to natural adaptations would require no conscious effort. An equally plausible interpretation of Weismann's results is that the direct stimuli of the out-of-doors, of nature alone, produces a response toward the more normal. A definitive investigation of the bases for these responses is needed as guidance to urban planners and public health specialists.

These examples are concerned with the negative effects which many see as resulting from the unnatural qualities of man's present, mostly urban, environment. Aldous Huxley ventures a further opinion as he considers the abnormal adaptation of those hopeless victims of mental illness who appear most normal: "These millions of abnormally normal people, living without fuss in a society to which, if they were fully human beings, they ought not to be adjusted, still cherish 'the illusion of individuality,' but in fact they have been to a great extent deindividualized. Their conformity is developing into something like uniformity. But uniformity and freedom are incompatible as well. . . . Man is not made to be an automaton, and if he becomes one, the basis for mental health is lost."

Clearly, a program of research could tell us more about man's subtle genetic dependence on the environment of his evolution. But of one thing we can be sure: only from study of human behavior in its evolutionary context can we investigate the influence of the environment on the life and fate of modern man. Even now we can see the bases by which to judge quality in our environment, if we are to maintain some semblance of one which is biologically optimum for humans.

We do not plead for a return to nature, but for re-examination of how to use science and technology to create environments for human living. While sociological betterment of the environment can do much to relieve poverty and misery, the argument that an expanding economy and increased material wealth alone would produce a Utopia is now substantially discounted. Instead, a natural concern for the quality of life in our affluent society is evident. But few economists or scientists have tried to identify the major elements of the quality we seek, and no one at all has attempted to use evolutionary principles in the search for quality. Solutions to the problems raised by attempts to evaluate quality will not be found before there is tentative agreement on the basis for judging an optimum human environment. A large body of evidence from studies in evolution, medicine, psychology, sociology, and anthropology suggests clearly that *such an environment will be a compromise between one in which humans have maximum contact with the properties of the environment to which they are innately adapted, and a more urban environment in which learned adaptations and social conventions are relied upon to overcome primitive needs.*

Our option to choose a balance between these two extremes runs out very soon. Awareness of the urgency to do something is national, and initial responses may be noted in several well-established but relatively narrow scientific disciplines. There has been the recent revival of eugenics. . . .

AN "IMPOSSIBLE" CHALLENGE

More extreme views have been expressed that man could be changed genetically to fit any future, but the means to do this and the moral justification of the aims sought are still far from being resolved. Many support the so-called evolutionary and technological optimists who, unlike their forefathers of little more than a generation ago, believe man can be changed radically when the time comes. They show a faith that science has proved its ability to draw on an expanding technology to do the impossible. The technologically impossible seems to have been accomplished time and time again during the past two or three generations, and may happen again. But some important scientific objectives have not been achieved, and we are likely to become more aware of the failures of science, of the truly impossible, as the irreversible disruptions of highly complex biological systems become more evident.

We suggest that the alternative to genetic modification of man is to select a course where the objectives only verge on the impossible. Let us regard the study and documentation of criteria for an environmental optimum as the "impossible" challenge for science and technology in the next two decades! Although considerable research in biology, sociology, and environmental design is already directed to this objective, there are several other types of study required that we outline briefly, simply to indicate the scope of the challenge.

First, a thorough examination must be undertaken of the extent to which man's evolutionary heritage dominates his activity both as an individual and in groups. The survival advantage of certain group activities has clearly figured in his evolutionary success and adaptive culture. Although cultural adaptation now dominates the biological in the evolution of man, his basic animal nature has not changed. Research leading to adequate understanding of the need to meet innate genetic demands lies in the field of biology, and more specifically in a combination of genetics, physical anthropology and ethology.

Second, we need to understand more of how cultural adaptations and social conventions of man permit him to succeed in an artificial environment. Cultural adaptation is the basis of his success as a gregarious social animal, and it will continue to be the basis by which he modifies evolutionarily imposed adaptations. Medical studies suggest there may be a limit to the magnitude of cultural adaptations, and that for some people this is nearly reached. Studies in sociology, cultural anthropology and psychology are all necessary to such research, in combination with environmental design and quantitative analysis of diversity in the native landscape.

Third, relationships between the health of individuals, both mental and physical, and the properties of the environment in which they live should be a fundamental area of research. It is easy to forget that we should expect as much genetic variability in the capacity of individuals to adjust to artificial environments as we find in the physical characteristics of man. Some portions of the population should be expected to have a greater inherent commitment to the natural environment, and will react strongly if deprived of it. Others may be much more neutral. Studies of the population as a whole must take into account the variability in reaction, and must therefore consider population genetics as well as psychiatry and environmental design.

Fourth, environmental qualities should be programed so as to optimize for the maximal expression of evolutionary (i.e., human) capabilities at the weakest link in the ontogenic development of human needs. While there are many critical periods during our life, we believe the ties to natural environments to be most vital during youth. We have abundant evidence on our campuses and in our cities that the dislodgement of youth presents one—if not the most—serious obstacle to successful adoption of more complex social structures. . . .

Young men and women accept many of the modern social conventions, but retain the highly questioning mind that once led to new and better ways to hunt and forage. By early middle age, man's physical and mental agility has changed and he becomes a stronger adherent to the social conventions that make his own society possible. During the rise of modern man on the high African plains, and continuing into modern primitive societies, each community was very much dependent on its young men. They contributed to hunting and community protection through their strength and agility, commodities for which there is declining demand in modern society. Survival in the primitive groups was to some degree dependent on the willingness of youth to innovate and take risks, and this has become a fixed adaptation, requiring outlets of expression.

Over 30 years ago, sociologist W. F. Ogburn suggested that society in the future would require "prolonging infancy to, say, thirty or forty years or even longer." Is not our 20-year educational sequence a poorly-veiled attempt to do just that? From an evolutionary point of view will not this dislodgement of youth present the most serious obstacle to successful adoption of more complex social structures? We are compelled to acknowledge that our overall technological environment for youth has not compensated for the loss of the challenges of the hunt and the freedom of the Veldt. The disruptions on our campuses and in the cities indicate the need to plan environmental optima for this weakest link in the human need for expression of evolutionary capabilities.

Finally, systems ecology is developing the capacity for considering all of the relationships and their interactions simultaneously. The notion of fully describing the optimum for any organism may seem presumptuous. It requires measurement of every type of response, particularly behavioral responses, and their statement as a series of component equations. Synthesis in the form of a complex model permits mathematical examination of an optimum for the system as a whole. Until recently it seemed more reasonable to study such optimization for important resources such as fisheries, but the capability is available and relevant to the study of the environmental optimum of man, and its application must now be pursued vigorously.

These five approaches to the study of human environment provide an objective base for investigating the environmental optimum for man. We cannot close this discussion, however, without pointing out that the final decision, both as to the choice of the optimum and its implementation, is an ethical one. There is an optimum for the sick, and another for the well; there is an optimum for the maladjusted, and another for the well-adjusted. But in treating the problems of the poor and minority groups, in our preoccupation with their immediate relief, we may continue to overlook the ways in which cultural demands of the modern, sub-optimum environment go far beyond the capacity of learned adaptations.

A COMPROMISE

Considering our scientific effort to learn the functions and structure of the human body, and of the physical environment around us, the limited knowledge of man's relationships to his environment is appalling. Because of the very success of our scientific establishment we are faced with population densities and environmental contaminants that have left us no alternative but to undertake control of the environment itself. In this undertaking let us understand the need to choose a humane compromise—a balance between the evolutionary demands we cannot deny except with great emotional and physical misery, and the fruits of an unbelievably varied civilization we are loath to give up.

Yet are we even considering such a compromise? With rare exceptions are we not continuing to destroy much that remains of man's natural environment with little thought for the profit of the remote future? In the conflict between preservationists and industrialists (or agriculturalists) the latter have had it their way, standing as they do for "progress" and "modern living." While the balance between these conflicts is slowly changing, preservationists continue to be regarded as sentimentalists rather than realists.

Theodosius Dobzhansky says that "the preponderance of cultural over biological evolution will continue to increase in the foreseeable future." We could not wish this to be otherwise; adaptation to the environment by culture is more rapid and efficient than biological adaptation. But social structures cannot continue indefinitely to become more complex and further removed from evolutionary forces. At some stage a compromise must be reached with man's innate evolutionary adaptability.

NEED FOR CONTINUING STUDY

We believe that the evidence of man's need for nature, particularly its diversity, is sufficient to justify a determined effort by the scientific community to obtain definitive answers to the questions we have posed. The techniques for studying the problems are to be found in separate disciplines, and there is a sufficient measure of willingness among scientists to undertake the new approaches. But the first step will be faltering and financial support will be slow in coming.

Now that buttercups are rare, at least symbolically, and springs often silent, why study them? Have there not already been several generations for whom the fields and woods are nearly a closed book? We could encourage the book to close forever, and we might succeed, but in so doing we might fail disastrously. The desire to see and smell and know has not yet been suppressed and enthusiasm for natural history continues to bring vitality to millions. Let us recognize that we are a product of evolution, without apology for the close affinities with our primate forebears. We need only prepare consciously to make a compromise between our cultural and our genetic heritage by striking a balance of social structures with maintenance of natural environments. Most important, we must discover the mechanisms of environmental influence on man. There is no other satisfactory approach to an optimum environment.